Elementary Statistics

A Brief Version

D1503602

Elementary Statistics

A Brief Version

Allan G. Bluman
Community College of Allegheny County

Boston   Burr Ridge, IL   Dubuque, IA   Madison, WI   New York   San Francisco   St. Louis
Bangkok   Bogotá   Caracas   Lisbon   London   Madrid   Mexico City   Milan
New Dehli   Seoul   Singapore   Sydney   Taipei   Toronto

All examples and exercises in this textbook (unless cited) are hypothetical and are presented to enable students to achieve a basic understanding of the statistical concepts explained. These examples and exercises should not be used in lieu of medical, psychological, or other professional advice. Neither the author nor the publisher shall be held responsible for any misuse of the information presented in this textbook.

McGraw-Hill Higher Education

A Division of The **McGraw-Hill** *Companies*

ELEMENTARY STATISTICS: A BRIEF VERSION

Copyright © 2000 by The McGraw-Hill Companies, Inc. All rights reserved. Printed in the United States of America. Except as permitted under the United States Copyright Act of 1976, no part of this publication may be reproduced or distributed in any form or by any means, or stored in a data base or retrieval system, without the prior written permission of the publisher.

This book is printed on acid-free paper.

1 2 3 4 5 6 7 8 9 0 VNH/VNH 0 9 8 7 6 5 4 3 2 1 0

ISBN 0–07–234993–X
ISBN 0–07–235459–3 (Instructor's Edition)

Vice president and editorial director: *Kevin T. Kane*
Publisher: *JP Lenney*
Sponsoring editor: *William K. Barter*
Senior developmental editor: *David Dietz*
Marketing manager: *Mary K. Kittell*
Project manager: *Marilyn M. Sulzer*
Production supervisor: *Laura Fuller*
Design director: *Francis Owens*
Senior photo research coordinator: *Lori Hancock*
Supplement coordinator: *Sandra M. Schnee*
Compositor: *GAC–Indianapolis*
Typeface: *10/12 Times Roman*
Printer: *Von Hoffmann Press, Inc.*

Designer: *Jeff Storm*
Cover designer: *Carol Barr*
Photo research: *Shirley Lanners*

Portions of MINITAB Statistical Software input and output contained in this book are printed with permission of Minitab, Inc. MINITAB is a trademark of Minitab, Inc. and is used herein with the owner's permission.

The credits section for this book begins on page G and is considered an extension of the copyright page.

Library of Congress Catalog Card Number: 99–63048

Contents

Preface

Approach

Elementary Statistics: A Brief Version is a shorter version of the popular text *Elementary Statistics: A Step-by-Step Approach,* Third Edition. This softcover edition includes all the features of the longer book, but it is designed for a course in which the time available limits the number of topics covered. It offers a more portable, lower-cost alternative to the standard hardcover textbooks.

Elementary Statistics: A Brief Version is written for students in the beginning statistics course whose mathematical background is limited to basic algebra. The book uses a nontheoretical approach in which concepts are explained intuitively and supported by examples. There are no formal proofs in the book. The applications are general in nature, and the exercises include problems from agriculture, biology, business, economics, education, psychology, engineering, medicine, sociology, and computer science.

About This Book

The learning system found in *Elementary Statistics* provides the student with a valuable framework in which to learn and apply concepts.

- Every copy of *Elementary Statistics: A Brief Version* comes with a **data disk** that provides the data sets used in examples and exercises in a variety of formats including:

  - MINITAB
  - TI Graph Link files for TI-83
  - Excel (for Windows and Macintosh)
  - SPSS
  - Comma-Delimited ASCII

This can save the student using a computer or calculator from having to enter data by hand, which takes up valuable time and increases the chances of error.

- Each chapter begins with an outline. **Learning objectives** have been added and are repeated at the beginning of each section to help students focus on the concepts presented within that section.

4–2

Tree Diagrams and the Multiplication Rule for Counting

Objective 1. Determine the number of outcomes to a sequence of events using a tree diagram.

Many times one wishes to list each possibility of a sequence of events. For example, it would be difficult to list all possible outcomes of a seven-game World Series by guessing alone. Rather than do this listing in a haphazard way, one can use a tree diagram.

A **tree diagram** is a device used to list all possibilities of a sequence of events in a systematic way.

- The outline and learning objectives are followed by a feature titled **Statistics Today,** a real life problem that shows students the relevance of the material in the chapter. This problem is subsequently solved near the end of the chapter using the statistical techniques that were presented in the chapter.

Statistics Today

Are We Flying More?

Many executives need to use statistics to win support for their concerns. For example, David Henson, head of the Federal Aviation Administration, in an article entitled, "Airport Expansion: Doing More with Less," stated that increased air travel is causing congestion at most major airports in the United States. He suggested that airports will have to expand their capacities while facing cutbacks in government spending. In the article, he included the following table showing the increase in air traffic.

Because the data are stated in table form, they do not have as much impact on the reader as if they were presented using a statistical graph.

This chapter will show how to organize data and then construct appropriate graphs to represent the data in a concise, easy-to-understand form. An appropriate graph for this table appears near the end of the chapter.

Air Travel

Here's how many millions of passengers used U.S. planes here and abroad; projections out Friday will show even greater increases in the next decade:

Year	Passengers
1980	287.9
1981	274.7
1982	286.1
1983	308.2
1984	334.0
1985	370.1
1986	404.7
1987	441.2
1988	441.2
1989	443.6
1990	456.6
1991	445.7
1992	463.0
1993	468.1
1994	509.0

Source: FAA

Source: *USA Today,* February 28, 1996. Copyright 1996, USA TODAY. Used with permission.

- Over 200 **examples** are provided, followed by the **solutions** to help students learn to solve problems. Examples are solved by using a step-by-step explanation. Illustrations provide a clear display of results for students.

Example 2–5

Using the frequency distribution given in Example 2–4, construct a frequency polygon.

Solution

STEP 1 Find the midpoints of each class. Recall that midpoints are found by adding the upper and lower boundaries and dividing by 2.

$$\frac{99.5 + 104.5}{2} = 102 \qquad \frac{104.5 + 109.5}{2} = 107$$

And so on. The midpoints are listed next.

Class boundaries	Midpoints	Frequency
99.5–104.5	102	2
104.5–109.5	107	8
109.5–114.5	112	18
114.5–119.5	117	13
119.5–124.5	122	7
124.5–129.5	127	1
129.5–134.5	132	1

STEP 2 Draw the x and y axes. Label the x axis with the midpoint of each class, and then use a suitable scale on the y axis for the frequencies.

STEP 3 Using the midpoints for the x values and the frequencies as the y values, plot the points.

- Numerous **Procedure Tables** summarize processes for the student. All use the step-by-step method.

Procedure Table 5

Finding the Sample Variance and Standard Deviation for Grouped Data

STEP 1 Make a table as shown, and find the midpoint of each class.

A	B	C	D	E
Class	Frequency	Midpoint	$f \cdot X_m$	$f \cdot X_m^2$

STEP 2 Multiply the frequency by the midpoint for each class, and place the products in column D.

STEP 3 Multiply the frequency by the square of the midpoint, and place the products in column E.

STEP 4 Find the sums of columns B, D, and E. (The sum of column B is n. The sum of column D is $\sum f \cdot X_m$. The sum of column E is $\sum f \cdot X_m^2$.)

STEP 5 Substitute in the formula and solve to get the variance.

$$s^2 = \frac{\sum f \cdot X_m^2 - [(\sum f \cdot X_m)^2/n]}{n - 1}$$

STEP 6 Take the square root to get the standard deviation.

- The **Speaking of Statistics** sections invite students to think about poll results and other statistics-related news stories.

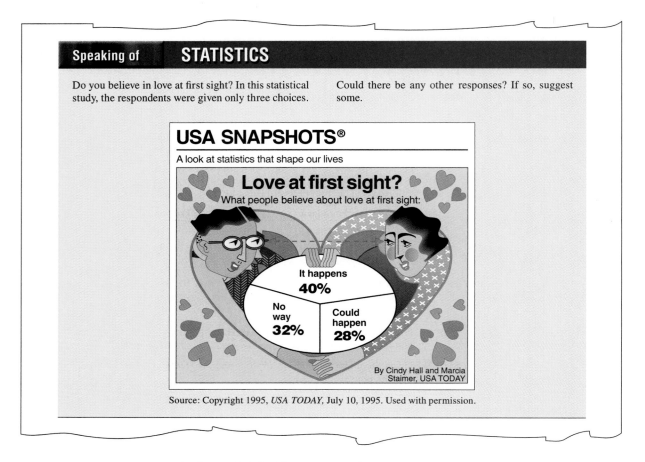

Speaking of **STATISTICS**

Do you believe in love at first sight? In this statistical study, the respondents were given only three choices.

Could there be any other responses? If so, suggest some.

USA SNAPSHOTS®

A look at statistics that shape our lives

Love at first sight?
What people believe about love at first sight:

It happens
40%

No way
32%

Could happen
28%

By Cindy Hall and Marcia Staimer, USA TODAY

Source: Copyright 1995, *USA TODAY,* July 10, 1995. Used with permission.

- **Rules and definitions** are set off for easy referencing by the student.

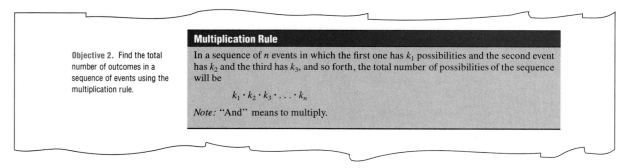

Objective 2. Find the total number of outcomes in a sequence of events using the multiplication rule.

Multiplication Rule

In a sequence of n events in which the first one has k_1 possibilities and the second event has k_2 and the third has k_3, and so forth, the total number of possibilities of the sequence will be

$$k_1 \cdot k_2 \cdot k_3 \cdot \ldots \cdot k_n$$

Note: "And" means to multiply.

- Over 1,200 **exercises** are located at the end of major sections within each chapter.

- **Historical Notes, Unusual Stats,** and **Interesting Facts,** located in the margins, make statistics come alive for the reader.

- **Critical Thinking** sections at the end of each chapter challenge the students to apply what they have learned to new situations.

Critical Thinking Challenges

1. Shake Hands A person decides to shake hands with six different people on a certain day. The next day, each of the six people will shake hands with six different people. The process continues until every person in the United States has shaken someone's hand. How many days will it take until everyone in the United States has shaken hands once? Assume that once a person shakes hands with six different people, he or she does not shake hands again. (*Hint:* The population of the United States is 248,709,873, according to the 1990 census.)

2. How Many Hairs? If it can be assumed that the maximum number of hairs on a human head is about 500,000, explain why at least two people living in Houston (population 1,629,902, according to the 1990 census) have the same number of hairs on their heads.

triangle existed in China in the 1300s. The triangle is formed by adding the two adjacent numbers and writing the sum below in a triangular fashion.

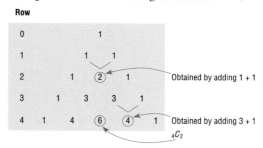

Each number in the triangle represents the number

- Numerous examples and exercises use **real data.**

2–13. The ages of the signers of the Declaration of Independence are shown below. (Age is approximate since only the birth year appeared in the source, and one has been omitted since his birth year is unknown.) Construct a frequency distribution for the data using seven classes. (The data for this exercise will be used for Exercise 3–23.)

41	54	47	40	39	35	50	37	49	42	70	32
44	52	39	50	40	30	34	69	39	45	33	42
44	63	60	27	42	34	50	42	52	38	36	45
35	43	48	46	31	27	55	63	46	33	60	62
35	46	45	34	53	50	50					

Source: John W. Wright, ed., *The Universal Almanac,* Andrews and McMeel, 1994, p. 53. Used with permission.

- In the interest of facilitating student use of the **TI-83 Graphing Calculator,** directions and screen displays are found at the end of appropriate chapters.

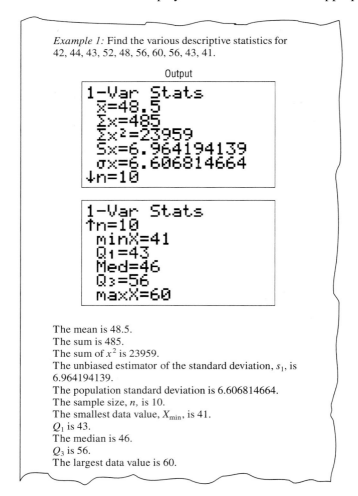

Example 1: Find the various descriptive statistics for 42, 44, 43, 52, 48, 56, 60, 56, 43, 41.

Output

The mean is 48.5.
The sum is 485.
The sum of x^2 is 23959.
The unbiased estimator of the standard deviation, s_1, is 6.964194139.
The population standard deviation is 6.606814664.
The sample size, n, is 10.
The smallest data value, $X_{\min}$, is 41.
Q_1 is 43.
The median is 46.
Q_3 is 56.
The largest data value is 60.

- **MINITAB**™ **computer applications** are found in each chapter to give students hands-on experience in using and interpreting statistical programs. Instructions and sample printouts for **MINITAB Release 12** are provided. These are new to the Brief Version.

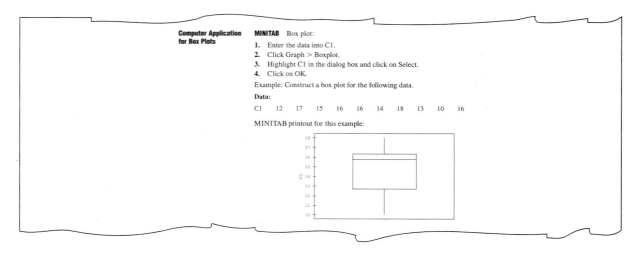

Computer Application for Box Plots

MINITAB   Box plot:

1. Enter the data into C1.
2. Click Graph > Boxplot.
3. Highlight C1 in the dialog box and click on Select.
4. Click on OK.

Example: Construct a box plot for the following data.

Data:

C1   12   17   15   16   16   14   18   13   10   16

MINITAB printout for this example:

- Special sections called **Data Analysis** require students to work with a data set to perform various statistical tests or procedures and then summarize the results.

Data Analysis

1. From the Data Bank located in Appendix D, choose one of the following variables: age, weight, cholesterol level, systolic pressure, IQ, or sodium level. Select at least 30 values. For these values, construct a grouped frequency distribution. Draw a histogram, frequency polygon, and ogive for the distribution. Describe briefly the shape of the distribution.

2. From the Data Bank, choose one of the following variables: educational level, smoking status, or exercise. Select at least 20 values. Construct an ungrouped frequency distribution for the data. For the distribution, draw a Pareto chart and describe briefly the nature of the chart.

3. From the Data Bank, select at least 30 people and construct a categorical distribution for their marital status. Draw a pie chart and describe briefly the findings.

- End-of-chapter **Summaries, Important Terms,** and **Important Formulas** give students a concise summary of the chapter topics and provide a good source for quiz or test preparation.

Hypothesis-Testing Summary 1

1. Comparison of a sample mean with a specific population mean.

Example: $H_0: \mu = 100$

a. Use the z test when σ is known:

$$z = \frac{\overline{X} - \mu}{\frac{\sigma}{\sqrt{n}}}$$

b. Use the t test when σ is unknown:

$$t = \frac{\overline{X} - \mu}{\frac{s}{\sqrt{n}}} \quad \text{with} \quad \text{d.f.} = n - 1$$

2. Comparison of a sample variance or standard deviation with a specific population variance or standard deviation.

Example: $H_0: \sigma^2 = 225$

Use the chi-square test:

$$\chi^2 = \frac{(n - 1)s^2}{\sigma^2} \quad \text{with} \quad \text{d.f.} = n - 1$$

d. Use the t test for dependent samples when the means are related.

Example: $H_0: \mu_D = 0$

$$t = \frac{\overline{D} - \mu_D}{\frac{s_D}{\sqrt{n}}} \quad \text{with} \quad \text{d.f.} = n - 1$$

where n = number of pairs.

4. Comparison of a sample proportion with a specific population proportion.

Example: $H_0: p = 0.32$

Use the z test:

$$z = \frac{X - np}{\sqrt{npq}} \quad \text{or} \quad z = \frac{X - \mu}{\sigma}$$

5. Comparison of two sample proportions.

Example: $H_0: p_1 = p_2$

Use the z test:

$$z = \frac{(\hat{p}_1 - \hat{p}_2) - (p_1 - p_2)}{}$$

- **Review Exercises** are found at the end of each chapter.
- A **reference card** containing the formulas is included with this textbook.

**Content Changes for
the Brief Version**

In response to numerous reviews and other feedback, the following specific changes were made in the coverage of *Elementary Statistics: A Step-by-Step Approach,* Third Edition, to produce this book.

- **Chapter 4:** This chapter has been revised and shortened. It consists of one multiplication rule, one permutation rule, and one combination rule.
- **Chapter 5:** Optional sections on Bayes's Theorem and applying counting techniques to probability were removed.
  Chapter 6: The optional section on the less-used probability distributions was removed.
- **Chapter 7:** The subsection on the finite population correction factor for the Central Limit Theorem was removed.
- **Chapter 9:** A subsection detailing Type-II error and the power of a test was removed.
- **Chapter 11:** The optional section on multiple regression was removed. Prediction intervals using small samples have been included.
- **Chapter 12:** A section on one-way analysis of variance (from Chapter 13 of the third edition) was added.
- **Chapters 13–16:** The Scheffé and Tukey tests and Two-Way ANOVA from Chapter 13, as well as Nonparametric Statistics (Chapter 14), Sampling and Simulation (Chapter 15), and Quality Control (Chapter 16) were eliminated.

Custom Publishing

If you require any topics covered in *Elementary Statistics: A Step-by-Step Approach,* Third Edition, that are not included in *Elementary Statistics: A Brief Version,* you may be able to supplement this book with sections or chapters from the Third Edition. For further information, contact Primis Custom Publishing at 800-228-0634, or visit www.mhhe.com/primis/.

Additional Features

In addition to pedagogical features already described, *Elementary Statistics: A Brief Version* offers the following:

- **Chapter quizzes,** found at the end of each chapter, include multiple choice, true-false, and completion questions along with exercises to test students' knowledge and comprehension of chapter content.
- **Data Projects** at the end of each chapter further challenge students' understanding and application of the material presented in the chapter. Many of these require the student to gather, analyze, and report on real data.
- An explanation of *P*-values has been included in each chapter where appropriate—since they are found in all computer printouts.
- **Hypothesis-Testing Summaries** are found at the end of Chapter 10 (z, t, χ^2 and F tests for testing means, proportions, and variances), and Chapter 12 (chi-square and ANOVA) to show students the different types of hypotheses and the types of tests to use.
- A **Data Bank** listing various attributers (educational level, cholesterol level, gender, etc.) for 100 people is included as Appendix D and referenced in various exercises and projects throughout the book.

Supplements

The text is accompanied by an extensive set of supplements for use by you and your students, all of which are carefully coordinated with the text.

For the Instructor

- *Instructor's Solutions,* by Sally Robinson of South Plains College. This manual includes worked-out solutions to most of the exercises in the text.
- *Critical Thinking Workbook: Instructor's Edition,* contains solutions to the students' version of the *Critical Thinking Workbook* described below.
- *Test Bank,* by Marcel Maupin of Oklahoma State University, contains a variety of questions, including true-false, multiple-choice, short answer and short problems requiring analysis and written answers. The testing material is coded by type of question and level of difficulty.
- *Computest* is a computerized version of the printed test bank. It allows you to efficiently select, add and organize questions, such as by type of question or level of difficulty. Computest also allows for printing tests along with answer keys, as well as editing the original questions. Computest is available for Windows and Mac systems.
- Full-color lecture slides in PowerPoint format highlight chapter concepts, summarize main points, and illustrate examples. These files can be downloaded from the book's website at www.mhhe.com/math/stat/bluman. PowerPoint users can customize the slides to suit the specific needs of their course.
- *Against All Odds* and *Decisions through Data* are video series available to qualified adopters. Please contact your local sales representative for more information about these programs.

For the Student

- *Critical Thinking Workbook,* by James Condor of Manatee Community College, provides a number of additional challenging problems for students to solve that are drawn from real-world applications. Problems are keyed to each chapter and are designed to highlight and emphasize key concepts.
- *Student Study Guide,* by Pat Foard of South Plains College. This guide will assist students in understanding and reviewing key concepts and preparing for exams. It emphasizes all important concepts contained in each chapter, includes explanations, and provides opportunities for students to test their understanding by completing related exercises and problems.
- *Student Solutions Manual,* by Sally Robinson of South Plains College, contains detailed solutions to all odd-numbered text problems.
- *MINITAB—Student Version.* This software and user manual provides the student with how-to information on data and file management, conducting various statistical analyses, and creating presentation-style graphics.

Acknowledgments

I would like to thank the following people and companies for granting permission to reprint their statistical tables and other material:

Addison-Wesley Publishing Company, Inc.
Benjamin/Cummings Publishing Company
CRC Press, Inc.

Institute of Mathematical Statistics

Prentice Hall, Inc.

Texas Instruments

Consumers Union (Copyright 1999 by Consumers Union of U.S., Yonkers, NY 10703-1057. Each data set used by permission from *Consumer Reports*. To subscribe, call 800-234-1645; www.ConsumerReports.org.)

The following people advised me and the publisher on the difficult question of what material from the third edition should be included in this Brief Version, and what could safely be removed. I am grateful for their evaluation of our proposal and their careful advice:

Anne G. Albert, *University of Findlay*

Abraham K. Biggs, *Broward Community College*

Callie Harmon Daniels, *St. Charles County Community College*

Kathleen Fritsch, *University of Tennessee at Martin*

James R. Fryxell, *College of Lake County*

Barney Herron, *Muskegon Community College*

John Jaros, *Molloy College*

Michael J. Keller, *St. Johns River Community College*

Frank Mauz, *Honolulu Community College*

Helen Moshkovich, *University of West Alabama*

Gerry Moultine, *Northwood University*

Sandra Preiss, *Northampton Community College*

Melissa Reeves, *East Texas Baptist University*

Carolyn Showalter, *Ocean County College*

Lynn Smith, *Gloucester County College*

Rena Waller, *Auburn University of Montgomery*

Laurie Sawyer Woodman, *University of New England*

I am also grateful to the many authors and publishers who granted me permission to use their articles and cartoons. I would like to acknowledge the cooperation of Minitab, Inc., in the preparation of this textbook.

I would especially like to thank the more than 60 reviewers of the third edition, whose suggestions and insights have been a positive influence on every page of this Brief Version. They are:

Dan Abbey, *Broward Community College*

Randall Allbritton, *Daytona Beach Community College*

Michael S. Allen, *Glendale Community College*

Mostafa S. Aminzadeh, *Towson State University*

Raymond Badalian, *Los Angeles City College*

Carole Bernett, *William Rainey Harper College*

Rich Campbell, *Butte College*

Mark Carpenter, *Sam Houston State University*

Daniel Cherwien, *Cumberland County College*

James A. Condor, *Manatee Community College*

David T. Cooney, *Polk Community College*
James C. Curl, *Modesto Junior College*
Carol Curtis, *Fresno City College*
Steven Day, *Riverside Community College*
Nirmal Devi, *Embry-Riddle Aeronautical University*
Wayne Ehler, *Anne Arundel College*
Eugene Enneking, *Portland State University*
Michael Eurgubian, *Santa Rosa Junior College*
Ruby Evans, *Santa Fe Community College*
Jeff Gervasi, *Porterville College*
Dawit Getahew, *Chicago State University*
Gary Grimes, *Mt. Hood Community College*
Leslie Grunes, *Mercer County Community College*
Dianne Haber, *Westfield State College*
Ronald Hamill, *Community College of Rhode Island*
Mark Harbison, *Rio Hondo College*
Linda Harper, *Harrisburg Area Community College*
Susan Herring, *Sonoma State University*
Keith A. Hilmer, *Moorpark College*
Shu-ping Hodgson, *Central Michigan University*
Robert L. Horvath, *El Camino College*
K. G. Janardan, *Eastern Michigan University*
Steve Kahn, *Anne Arundel Community College*
David Kozlowski, *Triton College*
Don Krekel, *Southeastern Community College*
Marie Langston, *Palm Beach Community College*
Kaiyang Liang, *Miami-Dade Community College*
Rowan Lindley, *Westchester Community College*
Bill McClure, *Golden West College*
Caren McClure, *Ranch Santiago College*
Rhonda Magel, *North Dakota State University*
Rudy Maglio, *Oakton Community College*
Mary M. Marco, *Bucks County Community College*
Donald K. Mason, *Elmhurst College*
Ed Migliore, *Monterey Peninsula College*
Jeff Mock, *Diablo Valley College*
Charlene Moeckel, *Polk Community College*
Keith Oberlander, *Pasadena City College*
Orlan D. Ohlhausen, *Richland College*
Linda Padilla, *Joliet Junior College*
Marnie Pearson, *Foothill College*
Ronald E. Pierce, *Eastern Kentucky University*
Pervez Rahman, *Truman College*

Mohammed Rajah, *Mira Costa College*

Helen M. Roberts, *Montclair State University*

Martin Sade, *Pima Community College*

Arnold L. Schroeder, *Long Beach City College*

Bruce Sisko, *Belleville Area College*

Aileen Solomon, *Trident Technical College*

Charlotte Stewart, *Southeastern Louisiana University*

Joe Sukta, *Moraine Valley Community College*

James M. Sullivan, *Sierra College*

Mary M. Sullivan, *Curry College*

Arland Thompson, *Community College of Aurora*

Dave Wallach, *University of Findlay*

Sandra A. Weeks, *Johnson & Wales University*

Bob Wendling, *Ashland University*

Two focus groups were held, where additional valuable advice was provided by a panel of instructors and a separate panel of students. Their comments and ideas were extremely helpful and all users of the third edition will benefit from their input. The instructor panel included:

Elizabeth Betzel, *Columbus State Community College*

Graham Chalmers, *Cerritos College*

Pat Deamer, *Skyline College*

Ginny Deus, *Diablo Valley College*

Ronald L. Drayton, *Midlands Technical College*

James Fryxell, *College of Lake County*

Mark Harbison, *Rio Hondo College*

John G. Horner, *Cabrillo College*

Maryann E. Justinger, *Erie Community College–South*

Kathryn Kozak, *Coconino Community College*

Christine Heinecke Lehmann, *Purdue University North Central*

Kathryn T. McClellan, *Tarrant County Junior College–South Campus*

Mike McGaun, *Ventura Community College*

Patty Monroe, *Greenville Technical College*

Suey Quan, *Golden West College*

Patricia A. Rojas, *New Mexico State University*

Natile Woodrow, *Texas State Technical College–Sweetwater*

The student panel included:

Dana Bolton, *Chicago State University*

Frank J. Doyle, *Triton College*

Lowell Lay, *Oakton Community College*

Donna J. Olson, *Joliet Junior College*

Steven Redeker, *William Rainey Harper College*

Nick Trevor, *Moraine Valley Community College*

Sherri Wamstad, *Elmhurst College*
Orville C. Wilson, *Chicago State Univestiy*

I would also like to thank the authors of the supplements that have been developed to accompany the text. Sally Robinson updated the *Instructor's Solutions Manual* and the *Student Solutions Manual.* James Condor developed an all-new *Critical Thinking Workbook*. Pat Foard revised the *Student Study Guide.*

The following people at Laurel Technical Services deserve thanks for their error checking and preparation of the data CD.

Lynda Kay Steele
Jeff Rector
Terri Bittner

Finally, I would like to thank all the people at McGraw-Hill Higher Education for their efforts and support. Thanks go to William K. Barter, Sponsoring Editor; David Dietz, Senior Developmental Editor; and Marilyn Sulzer, Project Manager.

Allan G. Bluman

Elementary Statistics

A Brief Version

chapter

The Nature of Probability and Statistics

Objectives

After completing this chapter, you should be able to

1. Demonstrate knowledge of statistical terms.

2. Differentiate between the two branches of statistics.

3. Identify types of data.

4. Identify the measurement level for each variable.

5. Identify the four basic sampling techniques.

6. Explain the importance of computers and calculators in statistics.

Are We Improving Our Diet?

It has been determined that diets rich in fruits and vegetables are associated with a lower risk of chronic diseases such as cancer. Nutritionists recommend that by the year 2000 Americans should be consuming five or more servings of fruits and vegetables each day. Several researchers from the Division of Nutrition, the National Center for Chronic Disease Control and Prevention, the National Cancer Institute, and the National Institutes of Health decided to use statistical procedures to see how much progress is being made toward this goal.

The procedures they used and the results of the study will be explained in this chapter.

1–1

INTRODUCTION

Most people become familiar with probability and statistics through radio, television, newspapers, and magazines. For example, the following statements were found in newspapers.

A typical one-a-day vitamin and mineral pill boosted certain immune responses in older people by 64 percent, according to researchers at the University of Medicine and Dentistry of New Jersey. (USA Weekend, January 6–8, 1995.)

Of 1,000 households polled nationwide, 40% said they owned at least one cordless phone; 9% had two or more. (Tribune-Review, Greensburg, PA, January 8, 1995, p. B3.)

In 1994 specialty ski shop retailers sold 760,000 parkas at an average price of $153, and 176,200 ski suits at an average price of $280. (Tribune-Review, Greensburg, PA, January 8, 1995, p. G1.)

The average price of real estate sold in Dormont [Pennsylvania] over the past 12 months was $64,304. (Tribune Review, Greensburg, PA, January 8, 1995, p. B3.)

On average, U.S. residents spend $193 a year on sports apparel. (USA Today, January 10, 1995.)

Statistics is used in almost all fields of human endeavor. In sports, for example, a statistician may keep records of the number of yards a running back gains during a football game, or the number of hits a baseball player gets in a season. In other areas, such as public health, an administrator would be concerned with the number of residents who contract a new strain of flu virus during a certain year. In education, the researcher might want to know if new methods of teaching are better than old ones. These are only a few examples of how statistics can be used in various occupations.

Furthermore, statistics is used to analyze the results of surveys and as a tool in scientific research to make decisions based on controlled experiments. Other uses of statistics include operations research, quality control, estimation, and prediction.

Statistics is the science of conducting studies to collect, organize, summarize, analyze, and draw conclusions from data.

Students study statistics for several reasons:

1. Students, like professional people, must be able to read and understand the various statistical studies performed in their field. To have this understanding, they must be knowledgeable about the vocabulary, symbols, concepts, and statistical procedures used in these studies.

2. Students and professional people may be called on to conduct research in their field, since statistical procedures are basic to all research. To accomplish this, they must be able to design experiments; collect, organize, analyze, and summarize data; and possibly make reliable predictions or forecasts for future use. They must also be able to communicate the results of the study in their own words.

3. Students and professional people can also use the knowledge gained from studying statistics to become better consumers and citizens. For example, they can make intelligent decisions about what products to purchase based on consumer studies, government spending based on utilization studies, and so on.

These reasons can be considered the goals for students and professionals who study statistics.

It is the purpose of this chapter to introduce the student to the basic concepts of *probability* and statistics by answering questions such as the following:

What are the branches of statistics?

What are data?

How are samples selected?

This *USA Today* Snapshot shows the favorite top-pings of hamburger lovers. Does it state that 93% of   Americans eat hamburgers? Explain your answer.

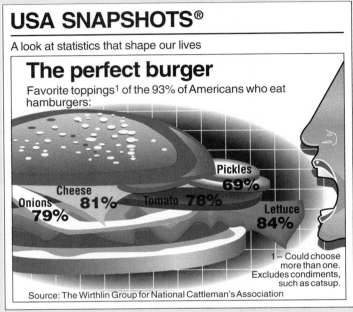

USA SNAPSHOTS®

A look at statistics that shape our lives

The perfect burger

Favorite toppings[1] of the 93% of Americans who eat hamburgers:

Pickles **69%**

Cheese **81%**

Onions **79%**

Tomato **78%**

Lettuce **84%**

1 – Could choose more than one. Excludes condiments, such as catsup.

Source: The Wirthlin Group for National Cattleman's Association

---

1–2

DESCRIPTIVE AND INFERENTIAL STATISTICS

Objective 1. Demonstrate knowledge of statistical terms.

In order to gain information about seemingly haphazard events, statisticians collect data for **variables** used to describe the event. **Data** are the values (measurements or observations) that the variables can assume. Variables whose values are determined by chance are called **random variables.**

Suppose that an insurance company studies its records over the past several years and determines that, on average, 3 out of every 100 automobiles the company insured were involved in an accident during a one-year period. Although there is no way to predict the specific automobiles that will be involved in an accident (random occurrence), the company can adjust its rates accordingly, since the company knows the general pattern over the long run. (That is, on average, 3% of the insured automobiles will be involved in an accident each year.)

A collection of data values forms a **data set.** Each value in the data set is called a **data value** or a **datum.**

Data can be used in different ways. Depending on how they are used, the body of knowledge called statistics is sometimes divided into two main areas.

1. Descriptive statistics
2. Inferential statistics

Speaking of   STATISTICS

Do you believe in love at first sight? In this statistical study, the respondents were given only three choices.

Could there be any other responses? If so, suggest some.

USA SNAPSHOTS®

A look at statistics that shape our lives

Love at first sight?
What people believe about love at first sight:

It happens
40%

No way
32%

Could happen
28%

By Cindy Hall and Marcia Staimer, USA TODAY

Source: Copyright 1995, *USA TODAY*, July 10, 1995. Used with permission.

Objective 2. Differentiate between the two branches of statistics.

Historical Note

The origin of descriptive statistics can be traced to data collection methods used in censuses taken by the Babylonians and Egyptians between 4500 and 3000 B.C. In addition, the Roman Emperor Augustus (27 B.C.–17 A.D.) conducted surveys on births and deaths of the citizens of the empire, as well as the number of livestock each owned and the crops each citizen harvested yearly.

In *descriptive statistics* the statistician tries to describe a situation. Consider the national census conducted by the United States government every 10 years. Results of this census give the average age, income, and other characteristics of the U.S. population. To obtain this information, the Census Bureau must have some means to collect relevant data. Once data are collected, the bureau must organize and summarize them. Finally, the bureau needs a means of presenting the data in some meaningful form, such as charts, graphs, or tables.

Descriptive statistics consists of the collection, organization, summation, and presentation of data.

The second area of statistics is called *inferential statistics*. Here, the statistician tries to make inferences from *samples* to *populations*. Inferential statistics uses **probability**—i.e., the chance of an event occurring. Most people are familiar with the concepts of probability through various forms of gambling. People who play cards, dice, bingo, and lotteries win or lose according to the laws of probability. Probability theory is also used in the insurance industry and other areas.

It is important to distinguish between a sample and a population.

This *USA Today* Snapshot shows the common causes of accidental deaths in the home. Why might insurance companies be interested in this information?

Source: Copyright 1995, *USA TODAY*, March 2, 1995. Used with permission.

A **population** consists of all subjects (human or otherwise) that are being studied.

Sometimes populations can be very large. In order to save time and money, statisticians may study only a part of the population. This part is called a *sample*.

A **sample** is a subgroup of the population.

The area of inferential statistics called **hypothesis testing** is a decision-making process for evaluating claims about a population, based on information obtained from samples. For example, a researcher may wish to know if a new drug will reduce the number of heart attacks in men over 70 years of age. For this study, two groups of men over 70 would be selected. One group would be given the drug, and the other would be given a placebo (a substance with no medical benefits or harm). Later, the number of heart attacks occurring in each group of men would be counted, a statistical test would be run, and a decision would be made about the effectiveness of the drug.

Statisticians also use statistics to determine *relationships* among variables. For example, relationships were the focus of the most noted study in the past few

Historical Note

Inferential statistics originated in the 1600s, when John Graunt published his book on population growth, *Natural and Political Observations Made upon the Bills of Mortality.* About the same time, another mathematician/astronomer, Edmund Halley, published the first complete mortality tables. (Insurance companies use mortality tables to determine life insurance rates.)

decades, "Smoking and Health," published by the surgeon general of the United States in 1964. He stated that after reviewing and evaluating the data, his group found a definite relationship between smoking and lung cancer. He did not say that cigarette smoking actually causes lung cancer but that there is a relationship between smoking and lung cancer. This conclusion was based on a study done in 1958 by Hammond and Horn. In this study, 187,783 men were observed over a period of 45 months. The death rate from lung cancer in this group of volunteers was 10 times as great for smokers as for nonsmokers.

Finally, by studying past and present data and conditions, statisticians try to make predictions based on this information. For example, a car dealer may look at past sales records for a specific month to decide what types of automobiles and how many of each type to order for that month next year.

Inferential statistics consists of generalizing from samples to populations, performing hypothesis testing, determining relationships among variables, and making predictions.

1–3

VARIABLES AND TYPES OF DATA

As stated in the previous section, statisticians gain information about a particular situation by collecting data for random variables. This section will explore in more detail the nature of variables and types of data.

Variables can be classified as qualitative or quantitative. **Qualitative variables** are variables that can be placed into distinct categories, according to some characteristic or attribute. For example, if subjects are classified according to gender (male or female), then the variable "gender" is qualitative. Other examples of qualitative variables are religious preference and geographic locations.

Quantitative variables are numerical in nature and can be ordered or ranked. For example, the variable "age" is numerical, and people can be ranked in order according to the value of their ages. Other examples of quantitative variables are heights, weights, and body temperatures.

Quantitative variables can be further classified into two groups, discrete or continuous. *Discrete variables* can be assigned values such as 0, 1, 2, 3, and are said to be countable. Examples of discrete variables are the number of children in a family, the number of students in a classroom, and the number of calls received by a switchboard operator each day for one month.

Discrete variables assume values that can be counted.

Continuous variables, by comparison, can assume all values between any two specific values. Temperature, for example, is a continuous variable, since the variable can assume all values between any two given temperatures.

Continuous variables can assume all values between any two specific values. They are obtained by measuring.

Since continuous data must be measured, answers must be rounded due to the limits of the measuring device. Usually, answers are rounded to the nearest given unit. For example, heights might be rounded to the nearest inch, weights to the nearest ounce, etc. Hence, a recorded height of 73 inches could mean any measure from 72.5 inches up to but not including 73.5 inches. Thus, the boundary of this measure is given as 72.5–73.5 inches. *Boundaries are written for convenience as*

Speaking of STATISTICS

Many times, researchers overlook factors that influence their statistical studies. In this article, researchers did not take into account the fact that as people age, they lose some height. When the original data were adjusted for height loss due to aging, the results were not significant. Explain how the original study may have been conducted (i.e., what statistical comparisons or tests were used) and how the adjustments for height loss due to aging may have been calculated.

Height Fright Not Right

Turns out a recent study that claimed short men have a greater risk of heart attack may have been a tall tale. According to the scientific journal *Circulation,* the height loss that comes with age accounts for shorter men having an increased risk of coronaries. After the calculations are adjusted for this age-related shrinkage, the tall-guy advantage apparently disappears.

Source: "Height Fright Not Right," *Prime,* Winter 1995, p. 14. Used with permission.

Unusual Stat

According to the *Statistical Abstract of the United States,* 52% of Americans live within 50 miles of a coastal shoreline.

72.5–73.5 *but understood to mean all values up to but not including 73.5.* Actual data values of 73.5 would be rounded to 74 and would be included in a class with boundaries of 73.5 up to but not including 74.5, written as 73.5–74.5. As another example, if a recorded weight is 86 pounds, the exact boundaries are 85.5 up to but not including 86.5, written as 85.5–86.5 pounds. Table 1–1 helps to clarify this concept. The boundaries of a continuous variable are given in one additional decimal place and always end with the digit 5.

Table 1–1	Recorded Values and Boundaries	
Variable	**Recorded value**	**Boundaries**
Length	15 centimeters (cm)	14.5–15.5 cm
Temperature	86 degrees Fahrenheit (°F)	85.5°–86.5°
Time	0.43 second (sec)	0.425–0.435 sec
Weight	1.6 grams (g)	1.55–1.65 g

Objective 3. Identify types of data.

In addition to being classified as qualitative or quantitative, variables can also be classified by how they are categorized, counted, or measured. For example, can the data be organized into specific categories, such as area of residence (rural, suburban, or urban)? Can the data values be ranked, such as first place, second

Objective 4. Identify the measurement level for each variable.

place, etc.? Or are the values obtained from measurement, such as heights, IQs, or temperature? This type of classification—i.e., how variables are categorized, counted, or measured—uses **measurement scales,** and four common types of scales are used: nominal, ordinal, interval, and ratio.

The **nominal level of measurement** classifies data into mutually exclusive (nonoverlapping), exhausting categories in which no order or ranking can be imposed on the data.

A sample of college instructors classified according to subject taught (e.g., English, history, psychology, or mathematics) is an example of nominal-level measurement. Classifying survey subjects as male or female is another example of nominal-level measurement. No ranking or order can be placed on the data. Classifying residents according to zip codes is also an example of the nominal level of measurement. Even though numbers are assigned as zip codes, there is no meaningful order or ranking. Other examples of nominal-level data are political party (Democratic, Republican, Independent), religion (Lutheran, Jewish, Catholic, Methodist, etc.), and marital status (married, divorced, widowed, separated).

The next level of measurement is called the *ordinal level.* Data measured at this level can be placed into categories, and these categories can be ordered, or ranked. For example, from student evaluations, guest speakers might be ranked as superior, average, or poor. Floats in a homecoming parade might be ranked as first place, second place, etc. *Note that precise measurement of differences in the ordinal level of measurement* does not *exist.* For instance, when people are classified according to their build (small, medium, or large), a large variation exists among the individuals in each class.

The **ordinal level of measurement** classifies data into categories that can be ranked; however, precise differences between the ranks do not exist.

Examples of ordinal-measured data are letter grades (A, B, C, D, F), rating scales, and rankings.

The third level of measurement is called the *interval level.* This level differs from the ordinal level in that precise differences do exist between units. For example, many standardized psychological tests yield values measured on an interval scale. IQ is an example of such a variable. There is a meaningful difference of one point between an IQ of 109 and an IQ of 110. Temperature is another example of interval measurement, since there is a meaningful difference of one degree between each unit, such as 72 degrees and 73 degrees. *One property is lacking in the interval scale: There is no true zero.* For example, IQ tests do not measure people who have no intelligence. For temperature, 0 degrees Fahrenheit does not mean no heat.

The **interval level of measurement** ranks data, and precise differences between units of measure do exist; however, there is no meaningful zero.

The highest level of measurement is called the *ratio level.* Examples of ratio scales are those used to measure height, weight, area, and number of phone calls received. Ratio scales have differences between units (1 inch, 1 pound, etc.) and a true zero. In addition, the ratio scale contains a true ratio between values. For example, if one person can lift 200 pounds and another can lift 100 pounds, then the ratio between them is 2 to 1. Put another way, the first person can lift twice as much as the second person.

The **ratio level of measurement** possesses all the characteristics of interval measurement, and there exists a true zero. In addition, true ratios exist when the same variable is measured on two different members of the population.

There is not complete agreement among statisticians about the classification of data into one of the four categories. For example, some researchers classify IQ data as ratio data rather than interval. Also, data can be altered so that they fit into a lower category. For instance, if the incomes of all professors of a college are classified into three categories—low, average, and high—then a ratio variable becomes an ordinal variable. Table 1–2 gives some examples of each type of data.

Table 1–2 Examples of Measurement Scales			
Nominal-level data	**Ordinal-level data**	**Interval-level data**	**Ratio-level data**
Zip code	Grade (A, B, C, D, F)	SAT score	Height
Gender (male, female)	Judging (1st place, 2nd place, etc.)	IQ	Weight
Eye color (blue, brown, green, hazel)	Rating scale (poor, good, excellent)	Temperature	Time
Political affiliation	Ranking of tennis players		Salary
Religious affiliation			Age
Major field (mathematics, computers, etc.)			
Nationality			

DATA COLLECTION AND SAMPLING TECHNIQUES

Objective 5. Identify the four basic sampling techniques.

In research, statisticians use data in many different ways. As stated previously, data can be used to describe situations or events. For example, a manufacturer might want to know something about the consumers who will be purchasing his product so he can plan an effective marketing strategy. In another situation, the management of a company might survey its employees to assess their needs in order to negotiate a new contract with the employees' union. Data can be used to determine whether the educational goals of a school district are being met. Finally, trends in various areas, such as the stock market, can be analyzed, enabling prospective buyers to make more intelligent decisions concerning what stocks to purchase. These examples illustrate a few situations where collecting data will help people make better decisions on courses of action.

Data can be collected in a variety of ways. One of the most common methods is through the use of surveys. Surveys can be done by using a variety of methods. Three of the most common methods are the telephone survey, the mailed questionnaire, and the personal interview.

Telephone surveys have an advantage over personal interview surveys in that they are less costly. Also, people may be more candid in their opinions since there

is no face-to-face contact. A major drawback to the telephone survey is that some people in the population will not have phones or will not be home when the calls are made; hence, not all people have a chance of being surveyed.

Mailed questionnaire surveys can be used to cover a wider geographic area than telephone surveys or personal interviews since they are less expensive to conduct. Also, respondents can remain anonymous if they desire. Disadvantages of mailed questionnaire surveys include a low number of responses and inappropriate answers to questions. Another drawback is that some people may have difficulty reading or understanding the questions.

Personal interview surveys have the advantage of obtaining in-depth responses to questions from the person being interviewed. One disadvantage is that interviewers must be trained in asking questions and recording responses, which makes the personal interview survey more costly than the other two survey methods. Another disadvantage is that the interviewer may be biased in his or her selection of respondents.

Data can also be collected in other ways, such as *surveying records* or *direct observation* of situations.

As stated in Section 1–2, researchers use samples to collect data and information about a particular variable from a large population. Using samples saves time and money and, in some cases, allows the researcher to get more detailed information about a particular subject. Samples cannot be selected in haphazard ways because the information obtained might be biased. For example, interviewing people on a street corner would not include responses from people working in offices at that time or people attending school; hence, all subjects in a particular population would not have a chance of being selected.

In order to obtain samples that are unbiased—i.e., give each subject in the population a chance of being selected—statisticians use four basic methods of sampling: random, systematic, stratified, and cluster sampling.

Random Sampling

Random samples are selected using chance methods or random numbers. One such method is to number each subject in the population. Then place numbered cards in a bowl, mix them thoroughly, and select as many cards as needed. The subjects whose numbers are selected constitute the sample. Since it is difficult to mix the cards thoroughly, there is a chance of obtaining a biased sample. For this reason, statisticians use another method of obtaining numbers. They generate

This *USA Today* Snapshot compares how men and women cope with bad days at work. Explain why the percentages do not add up to 100. Suggest some other ways people can survive a bad day at work. Do you think the survey should include your choices?

USA SNAPSHOTS®

A look at statistics that shape your finances

Surviving bad days at work

Most common ways men and women say they cope with bad workdays.

Talk to friend/family

Men **36%**   **49%** Women

Exercise

Men **21%**   13% {Women

Listen to music

Men 17%   13% {Women

Source: CJ Olson Market Research for Windsor Canadian Supreme

Source: Copyright 1995. *USA TODAY*, August 13, 1995. Used with permission.

random numbers with a computer or calculator. Before the invention of computers, random numbers were obtained from a table of random numbers, similar to the one shown in Appendix C. State lottery numbers are selected at random.

Systematic Sampling   Researchers obtain **systematic samples** by numbering each subject of the population and then selecting every kth number. For example, suppose there are 2000 subjects in the population and a sample of 50 subjects is needed. Since $2000 \div 50 = 40$, then $k = 40$, and every 40th subject would be selected; however, the first subject (numbered between 1 and 40) would be selected at random. Suppose subject 12 was the first subject selected; then the sample would consist of the subjects whose numbers were 12, 52, 92, etc., until 50 subjects were obtained. When using systematic sampling, one must be careful about how the subjects in the population are numbered. If subjects were arranged in a manner such as wife, husband, wife, husband, and every 40th subject was selected, the sample would consist of all husbands.

STATISTICS

This *USA Today* Snapshot shows how often families eat dinner at home together. Can you see anything wrong with the responses? (Hint: How would you define frequently? Occasionally? Always?)

USA SNAPSHOTS®

A look at statistics that shape our lives

How often does your family eat dinner at home together?

Frequently
38%

Always
34%

Don't know
1%

Never
6%

Occasionally
21%

Source:
The
Wirthlin
Group

Source: Copyright 1995. *USA TODAY*, October 23, 1995. Used with permission.

Stratified Sampling

Researchers obtain **stratified samples** by dividing the population into groups (called strata) according to some characteristic that is important to the study, then sampling from each group. For example, suppose the president of a two-year college wants to learn how students feel about a certain issue. Furthermore, the president wishes to see if the opinions of the freshmen differ from those of the sophomores. The president will select students from each group to use in the sample.

Cluster Sampling

Researchers select **cluster samples** by using intact groups called clusters. Suppose a researcher wishes to survey apartment dwellers in a large city. If there are 10 apartment buildings in the city, the researcher can select at random two buildings from the 10 and interview all the residents of these buildings. Cluster sampling is used when the population is large or when it involves subjects residing in a large geographic area. For example, if one wanted to do a study involving the patients in the hospitals in New York City, it would be very costly and time-consuming to try to obtain a random sample of patients since they would be spread over a large area.

Statistical studies often have conflicting conclusions. When this occurs, the researchers attempt to cite reasons for the discrepancies. The studies discussed here suggest several reasons for the different conclusions. Which of the reasons stated do you think was responsible for the discrepancy? Explain your answer.

Shake and Wake

Two major earthquakes, two studies on how traumatic events affect the heart—and two completely different results? The answer may be a testament to what the mind can endure once it adapts to a stressful situation. Or it may hinge on the clock.

At Stanford University's School of Medicine, psychiatrist C. Barr Taylor, M.D., and colleagues looked at the prevalence of heart attacks after the 1989 quake that rocked San Francisco. Medical records at five Bay Area hospitals turned up 16 heart attacks in the week before the disaster—and only 12 in the seven days following.

Surprising, considering the cardiovascular trauma unleashed by the quake: Heart rates skyrocket, blood pressure soars, and breathing becomes labored. "It's as if we gave a stress test to everybody in the Santa Clara Valley," Taylor says of the quake. "Not

to see any demonstrated increase in heart attacks is remarkable."

But the findings were far more grim following the 1994 Northridge quake in Los Angeles. Checking 81 coronary care units around L.A., Good Samaritan Hospital cardiologists Robert A. Kloner, M.D., Ph.D., and Jonathan Leor, M.D., counted a 35 percent jump in heart attacks the week after the quake.

Hospitals within 15 miles of the epicenter were particularly likely to report more cardiac patients. While the physical strain of escaping the devastation was a factor in some cases, in others emotional stress was clearly the trigger, Kloner says.

Why the discrepancy between studies? Are San Franciscans simply more mentally prepared for the Big One—and thus less shocked when a quake hits? Support for this idea comes

courtesy of Israeli researchers who looked at Iraq's missile strikes on Israel during the 1991 Gulf War.

On the day of the first attack, deaths throughout Israel rose by 58 percent, mostly from cardiac-related causes. But in subsequent attacks there was no rise in mortality, suggesting most people adapted to higher levels of stress and fear once they knew more missile strikes were likely.

But Kloner has a different take on the L.A. quake's high coronary toll: It struck at 4 A.M. "Being awakened from a sound sleep," he contends, "is very different from already being up and about." Add to that the physiological reality that heart attacks are already most common around dawn and the killing potential of an early morning quake takes on power that doesn't show up on Mr. Richter's scale.

Source: Reprinted with permission from *PSYCHOLOGY TODAY MAGAZINE* Copyright © 1995 (Sussex Publishers, Inc.).

Historical Note

In 1936, the *Literary Digest,* on the basis of a biased telephone sample of its subscribers, predicted that Alf Landon would defeat Franklin D. Roosevelt in the upcoming presidential election. Roosevelt won by a landslide. The magazine ceased publication the following year.

Instead, a few hospitals could be selected at random and the patients in these hospitals would be interviewed in a cluster.

The four basic sampling methods are summarized in Table 1–3.

Table 1–3	Summary of Sampling Methods
Random	Subjects are selected by random numbers
Systematic	Subjects are selected by using every kth number after the first subject is randomly selected from 1 through k
Stratified	Subjects are selected by dividing up the population into groups (strata) and subjects within groups are randomly selected
Cluster	Subjects are selected by using an intact group that is representative of the population

Other Sampling Methods

In addition to the four basic sampling methods, researchers use other methods to obtain samples. One such method is called a **convenience sample.** Here a researcher uses subjects that are convenient. For example, the researcher may interview subjects entering a local mall to determine the nature of their visit or perhaps what stores they will be patronizing. This sample is probably not representative of the general customers for several reasons. For one thing, it was probably taken at a specific time of day, so not all customers entering the mall have an equal chance of being selected since they were not there when the survey was being conducted. But convenience samples can be representative of the population. If the researcher investigates the characteristics of the population and determines that the sample is representative, then it can be used.

Other sampling techniques, such as *sequential sampling, double sampling,* and *multistage sampling,* can be used. These techniques are beyond the scope of this book.

1–5

Computers and Calculators

Objective 6. Explain the importance of computers and calculators in statistics.

In the past, statistical calculations were done with pencil and paper. However, with the advent of calculators, numerical computations became much easier. Computers do all the numerical calculation. All one does is enter the data into the computer and use the appropriate command; the computer will print the answer or display it on-screen. Now the TI-83 Graphing Calculator accomplishes the same thing.

There are many statistical packages available; this book uses MINITAB™. Instructions for using the TI-83 Graphing Calculator have been placed at the end of each relevant chapter.

Students should realize that the computer and calculator merely give numerical answers and save the time and effort of doing calculations by hand. The student is still responsible for understanding and interpreting each statistical concept. In addition, students should realize that the results come from the data and do not appear magically on the computer. Doing calculations using the procedure tables will help reinforce this idea.

The author has left it up to instructors to choose how much technology they will incorporate into the course.

1–6

Summary

The two major areas of statistics are descriptive and inferential. Descriptive statistics includes the collection, organization, summarization, and presentation of data. Inferential statistics includes making inferences from samples to populations, hypothesis testing, determining relationships, and making predictions. Inferential statistics is based on *probability theory.*

Since in most cases the populations under study are large, statisticians use subgroups called samples to get the necessary data for their studies. There are four basic methods used to obtain samples: random, systematic, stratified, and cluster.

Data can be classified as qualitative or quantitative. Quantitative data can be either discrete or continuous, depending on the values they can assume. Data can also be measured by various scales. The four basic levels of measurement are nominal, ordinal, interval, and ratio.

Finally, the applications of statistics are many and varied. People encounter them in everyday life, such as in reading newspapers or magazines, listening to the radio, or watching television. Since statistics is used in almost every field of endeavor, the educated individual should be knowledgeable about the vocabulary, concepts, and procedures of statistics.

LAFF - A - DAY

"We've polled the entire populace, Your Majesty, and we've come up with exactly the results you ordered!"

Today computers and calculators are used extensively in statistics to facilitate the computations.

Important Terms

Cluster sample 13	Discrete variables 7	Ordinal level of measurement 9	Random variable 4
Convenience sample 15	Hypothesis testing 6	Population 6	Ratio level of measurement 10
Continuous variables 7	Inferential statistics 7	Probability 5	Sample 6
Data 4	Interval level of measurement 9	Qualitative variables 7	Stratified sample 13
Data set 4	Measurement scales 9	Quantitative variables 7	Systematic sample 12
Data value or datum 4	Nominal level of measurement 9	Random sample 11	Variable 4
Descriptive statistics 5			

Review Exercises

1–1. (**W**) Name and define the two areas of statistics.

1–2. What is probability? Name two areas where probability is used.

Note: (**W**) means that some expository writing is necessary to complete this exercise.

Note: All odd-numbered problems and even-numbered problems marked with (**ans**) are included in the answer section at the end of this book.

Note: Problems with an asterisk (*) require more in-depth skills than other problems.

1–3. (**W**) Suggest some ways statistics can be used in everyday life.

1–4. (**W**) Explain the differences between a sample and a population.

1–5. (**W**) Why are samples used in statistics?

1–6. (**ans**) In each statement that follows, tell whether descriptive or inferential statistics were used.
a.   The average price of a home sold in Allegheny County during the week of April 22–28 was $75,328. (*Standard Observer,* Irwin, PA, May 17, 1995.)

b. According to the Census Bureau, 20% of all American workers get to work via carpool.

c. The National Eye Institute has halted a clinical trial on a type of eye surgery, calling it ineffective and possibly harmful to a person's vision. (*USA Today,* February 22, 1995)

d. "Allergy therapy may make bees go away." (*Prevention,* April 1995)

e. The Gallup Poll says 1 out of 10 Americans is a member of a health club or fitness center.

f. Drinking decaffeinated coffee can raise cholesterol levels by 7%. (American Heart Association)

g. According to the Court Administration of Pennsylvania, 14% of trial-ready civil actions and equity cases in Philadelphia during 1993 were decided in less than six months. (*Tribune Review,* Greensburg, PA, May 14, 1995)

h. Cigarettes were associated with 29% of the 4,470 civilian fire deaths in 1989. (Michael D. Shook and Robert L. Shook, *The Book of Odds,* Plume, 1991)

1–7. Classify each as nominal-level, ordinal-level, interval-level, or ratio-level data.
a. Horsepower of motorcycle engines.
b. Ratings of newscasts in Houston (poor, fair, good, excellent).
c. Temperature of automatic popcorn poppers.
d. Time required by drivers to complete a course.
e. Salaries of cashiers of Day-Night grocery stores.
f. Marital status of respondents to a survey on savings accounts.
g. Ages of students enrolled in a martial arts course.
h. Weights of beef cattle fed a special diet.
i. Rankings of weight lifters.
j. Pages in the telephone book for the city of Los Angeles.

1–8. **(ans)** Classify each variable as qualitative or quantitative.
a. Colors of jackets in a men's clothing store.
b. Number of seats in classrooms.
c. Classification of children in a day care center (infant, toddler, preschool).
d. Length of fish caught in a certain stream.
e. Number of students who fail their first statistics test.

1–9. Classify each variable as discrete or continuous.
a. Number of loaves of bread baked each day at a local bakery.
b. Water temperature of the saunas at a given health spa.
c. Incomes of single parents who attend a community college.
d. Lifetimes of batteries in a tape recorder.
e. Weights of newborn infants at a certain hospital.

f. Capacity (in gallons) of water in swimming pools.
g. Number of pizzas sold last year in the United States.

1–10. Give the boundaries for each value.
a. 29.3 miles   *d.* 13 inches
b. 4 tons   *e.* 4.63 ounces
c. 0.12 milliliters   *f.* 67 feet

1–11. **(W)** Name and define the four basic sampling methods.

1–12. **(ans)** Classify each sample as random, systematic, stratified, or cluster.
a. In a large school district, all teachers from two buildings are interviewed to determine whether they believe the students have less homework to do now than in previous years.
b. Every seventh customer entering a shopping mall is asked to select his or her favorite store.
c. Nursing supervisors are selected using random numbers in order to determine annual salaries.
d. Every hundredth hamburger manufactured is checked to determine its fat content.
e. Mail carriers of a large city are divided into four groups according to gender (male or female) and according to whether they walk or ride on their routes. Then 10 are selected from each group and interviewed to determine whether they have been bitten by a dog in the last year.

1–13. **(W)** Give three examples each of nominal, ordinal, interval, and ratio data.

1–14. **(W)** For each of the following statements, define a population and state how a sample might be obtained.
a. The average cost of an airline meal in 1993 was $4.55. (*Everything Has Its Price,* Richard E. Donley, Simon and Schuster, 1995)
b. "More than 1 in 4 United States children have cholesterol levels of 180 milligrams or higher." (The American Health Foundation)
c. "Every 10 minutes, 2 people die in car crashes and 170 are injured." (National Safety Council estimates)
d. "When older people with mild to moderate hypertension were given the mineral salt for six months, the average blood pressure reading dropped by eight points systolic and three points diastolic." (*Prevention,* March 1995, p. 19)
e. "The average amount spent per gift for Mom on Mother's Day is $25.95." (The Gallup Organization)

1–15. **(W)** Select a newspaper or magazine article that involves a statistical study and write a paper answering the following questions.
a. Is this study descriptive or inferential in nature? Explain your answer.

b. What are the variables used in the study? In your opinion, what level of measurement was used to obtain the data from the variables?

c. Does the article define the population? If so, how is it defined? If not, how could it be defined?

d. Does the article state the sample size and how the sample was obtained? If so, explain the size of the sample and how it was selected. If not, suggest a way it could have been obtained.

e. Explain *in your own words* what procedure (survey, comparison of groups, etc.) might have been used to determine the study's conclusions.

f. Do you agree or disagree with the conclusions? State your reasons.

1–16.   (**W**) Information from research studies is sometimes taken out of context. Explain why the claims of these studies might be suspect.

a. The average salary of the graduates of the class of 1980 is $32,500.

b. It is estimated that in Podunk there are 27,256 rats.

c. Only 3% of the men surveyed read *Cosmopolitan* magazine.

d. Based on a recent mail survey, 85% of the respondents favored gun control.

e. A recent study showed that high school dropouts drink more coffee than students who graduated; therefore, coffee dulls the brain.

f. Since most automobile accidents occur within 15 miles of a person's residence, it is safer to make long trips.

*1–17.   Select a newspaper that carries local birth announcements. Keep a record of the ratio of female to male births for one month. Summarize the findings.

Statistics Today

Are We Improving Our Diet?   Revisited

Researchers selected a *sample* of 23,699 adults in the United States, using phone numbers selected at *random,* and conducted a *telephone survey.* All respondents were asked six questions:

1. How often do you drink juices such as orange, grapefruit, or tomato?
2. Not counting juice, how often do you eat fruit?
3. How often do you eat green salad?
4. How often do you eat potatoes (not including french fries, fried potatoes, or potato chips)?
5. How often do you eat carrots?
6. Not counting carrots, potatoes, or salad, how many servings of vegetables do you usually eat?

Researchers found that men consumed fewer servings of fruits and vegetables per day (3.3) than women (3.7). Only 20% of the population consumed the recommended 5 or more daily servings. In addition, they found that youths and less-educated people consume an even lower amount than the average.

Based on this study, they recommend that greater educational efforts are needed to improve fruit and vegetable consumption by Americans and to provide environmental and institutional support to encourage increased consumption.

Source: Mary K. Sendula, M.D., et al., "Fruit and Vegetable Intake among Adults in 16 States: Results of a Brief Telephone Survey," *American Journal of Public Health* 85, no. 2 (February 1995).

Quiz

Determine whether each statement is true or false. If the statement is false, explain why.

1. Probability is used as a basis for inferential statistics.

2. The height of President Lincoln is an example of a variable.

3. The highest level of measurement is the interval level.

4. When the population of college professors is divided into groups according to their rank (instructor, assistant professor, etc.), and then several are selected from each group to make up a sample, the sample is called a cluster sample.

5. The variable of age is an example of a qualitative variable.

6. The weight of pumpkins is considered to be a continuous variable.

7. The boundary of a value such as 6 inches would be 5.9 inches to 6.1 inches.

Select the best answer.

8. The number of absences per year a worker has is an example of what type of data?
a. Nominal
b. Qualitative
c. Discrete
d. Continuous

9. What are the boundaries of 25.6 ounces?
a. 25–26
b. 25.55–25.65
c. 25.5–25.7
d. 20–39

10. A researcher divided subjects into two groups according to gender and then selected members from each group for his sample. What sampling method was the researcher using?
a. Cluster
b. Random
c. Systematic
d. Stratified

11. Data that can be classified according to color are measured on what scale?
a. Nominal
b. Ratio

c. Ordinal
d. Interval

Use the best answer to complete the following statements.

12. Two major branches of statistics are _____ and _____ .

13. Two uses of probability are _____ and _____ .

14. The group of all subjects under a study is called a _____ .

15. A subgroup of subjects selected from the group of all subjects under study is called a _____ .

16. Three reasons why samples are used in statistics are
a. _____   *b.* _____   *c.* _____

17. The four basic sampling methods are
a. _____   *b.* _____   *c.* _____   *d.* _____

18. For each statement, decide whether descriptive or inferential statistics were used.
a. A recent study showed that eating garlic can lower blood pressure.
b. The average number of students in a class at White Oak University is 22.6.
c. It is predicted that the average number of automobiles each household owns will increase next year.
d. Last year's total attendance at Long Run High School's football games was 8,235.
e. The chance that a person will be robbed in a certain city is 15%.

19. Classify each as nominal, ordinal, interval, or ratio level.
a. Number of exams given in a statistics course.
b. Ratings of word-processing programs as user-friendly.
c. Temperatures of a sample of automobile tires tested at 55 miles per hour for six minutes.
d. Weights of suitcases on a selected commercial airline flight.
e. Classification of students according to major field.

20. Classify each variable as discrete or continuous.
a. The time it takes to drive to work.
b. The number of credit cards a person has.
c. The number of employees working in a large department store.
d. The amount of a drug injected into a rat.

e. The amount of sodium contained in a bag of potato chips.
f. The number of cars stolen each week in a large city.

21. Give the boundaries of each.
a. 3.2 quarts   *d.* 0.27 centimeter
b. 18 pounds   *e.* 36 seconds
c. 9 feet

Critical Thinking Challenges

1. High Fat and Your Lower Back Consider the study of heart patients below, reprinted by permission of *Prevention* magazine:

"High fat may hurt your lower back. New research suggests that fatty foods may be the root of your back pain. Studies of the arteries of 86 average-weight men showed that the greater the plaque deposits, the greater the degeneration of the spinal disks." (Copyright 1994 Rodale Press, Inc. All rights reserved.)

a. Do you think a sample of 86 men is large enough to draw the conclusion stated in the first sentence?
b. What other factors might contribute to low back pain?
c. What are the variables used in this study and how can they be defined?
d. Suggest ways low back pain can be measured.
e. Do you think the age or occupation of the individuals could have an effect on back pain?
f. How might the researchers control the effects of the variables of age and occupation?
g. Why do you think only men were used in this study?

2. Adjustment versus Your Heart Read the article below and answer the questions that follow.
The recent study reprinted below reached the following conclusion:

"A heart patient overly dependent on a spouse may have a harder time making necessary lifestyle changes in diet and exercise."

a. Do you agree or disagree with this statement? State your reasons.

Heart Heals the Heart

Can a love sonnet prevent a heart attack? Maybe.

It's not just tissue damage and blocked arteries that determine a heart patient's ability to function. The intangibles of the heart are just as important.

In fact, such nonbiomedical elements as the emotional support of a spouse, feelings of self-efficacy, and absence of depression and anxiety are so crucial that cardiologists are reshaping rehabilitation programs to encompass them.

Perhaps the most stunning news comes from the University of Washington, where an ongoing study finds that the severity of coronary artery disease is a poor predictor of a patient's physical impairment. Psychiatrist Mark Sullivan, M.D., told the American Psychiatric Association that even when arteries were blocked as much as 70% in 231 heart patients aged 45–80, factors like depression, anxiety, a sense of self-efficacy, and the type of spousal support were better predictors of function.

It is surprising that unconditional support from a spouse is not always the best medicine. There's an important distinction between support that enables and that which disables. Putting a spouse to bed and waiting on him hand and foot does not actually help him, says Sullivan. "This is similar to our findings in chronic pain patients."

The best support may be a listening ear, according to cardiologist Martin Sullivan, M.D., who heads the Center for Living at Duke University. "Those patients who have a confidante do much better than those who don't."

Heart to Heart

There is a flurry of new research findings about how heart heals the heart:

• A heart patient overly dependent on a spouse may have a harder time making necessary lifestyle changes in diet and exercise.

• For women heart attack victims, spousal support is critical—but hard to come by. "The family sometimes feels abandoned," explains Martin Sullivan, "and they don't want the woman to take time out of her duties as a wife and mother to make important lifestyle changes. Women are more willing to change for men."

• For men, a heart attack may shatter the sole definition of self (as family provider). The introduction of larger concepts of the self is therapeutic.

• Patients who feel a sense of self-efficacy and control over their disease do better than those who don't.

• Depression and anxiety affect pain perception and the capacity to function in the face of medical symptoms.

• In a study at Stanford University, behavioral counseling after heart attack, especially for hard-driving Type A individuals, lowered the rate of recurrent heart attacks by 45%—the same as the most powerful prescription drugs.

Such findings have led Martin Sullivan to introduce innovative techniques at the Duke Center. These include a program known as PAIRS (Practical Application of Intimacy Relationship Skills), which teaches couples healthy interactive skills, and a meditation program that teaches patients to freeze-frame a moment in time and look at the emotional content of what they are experiencing. Says Sullivan of the Duke Center's work, "We're trying to take the best of everything."

Source: Reprinted with permission from *PSYCHOLOGY TODAY MAGAZINE,* Copyright © 1994 (Sussex Publishers, Inc.).

b. Comment on how this study's conclusion might have been reached.

c. What are the variables used in the study and how might they have been defined?

d. How might the researcher measure the variables?

e. What would the population be for this study?

f. What factors other than dependence on a spouse might have influenced the results of the study?

g. Do you think the gender of the patients would make a difference in the results? Why or why not?

3. Power Lines and Brain Cancer The effect of living near or working on power lines has been a very controversial issue. The article on the next page states several conclusions.

a. Do you think that even if the risk of both cancers (leukemia and brain cancer) is greater, the workers should be more concerned with electrocution or other accidents? Explain your answer.

b. Do you think that "a big, relatively well-done study that doesn't yield persuasive evidence of harm" is a justifiable reason to build power lines near residential areas? Explain your answer.

c. Do you think that since the death rate of industrial workers is lower than in the general population, it is safer to work as an industrial worker than in some other occupation?

d. Do you consider the sample size (138,905 men) large enough to be used to state the conclusion? Explain.

e. What does the statement "But brain cancer risk was 50% higher for men who worked more than five years as a lineman or electrician" mean to you?

f. Do you think the fact that the study was funded by the Electric Power Research Institute could have any effect on its findings?

Power Lines' Link to Brain Cancer but No Tie Found with Leukemia

Long-term exposure to power lines appears to slightly increase risk of brain cancer, but not leukemia, among utility workers, says a large study reported in the *American Journal of Epidemiology.*

Some earlier studies linked electro-magnetic fields to leukemia among utility workers, but no study has been conclusive on the risks.

But even if the risk is greater, both cancers remain rare enough that other dangers to those workers—such as electrocution or other work accidents—should be a greater concern, says the University of North Carolina's David Savitz, one of the study leaders.

Savitz is encouraged by "a big, relatively well-done study that doesn't yield persuasive evidence of harm." But the findings also may help explain why the rate of brain cancer in the general population, though still low, has increased in recent decades, he says.

The study differs from earlier ones because of its size and because it used actual measurements of exposure levels. It looked at 138,905 men who worked for power companies between 1950 and 1986, including 20,733 who died.

Overall death risk among the men was lower than the general population, probably because industrial workers are most likely to be healthy, the researchers report.

But brain cancer risk was 50% higher for men who worked more than five years as a lineman or electrician. Those exposed to the highest levels of magnetic fields had more than double the risk.

There was no association found between amount of magnetic field exposure and leukemia. The research was funded by the Electric Power Research Institute.

Source: Doug Levy, "Power Lines' Link to Brain Cancer. *USA Today,* January 11, 1995. Copyright 1995, USA TODAY. Reprinted with permission.

chapter

Frequency Distributions and Graphs

Objectives

After completing this chapter, you should be able to

1. Organize data using frequency distributions.

2. Represent data in frequency distributions graphically using histograms, frequency polygons, and ogives.

3. Represent data using Pareto charts, time series graphs, and pie graphs.

Are We Flying More?

Many executives need to use statistics to win support for their concerns. For example, David Henson, head of the Federal Aviation Administration, in an article entitled, "Airport Expansion: Doing More with Less," stated that increased air travel is causing congestion at most major airports in the United States. He suggested that airports will have to expand their capacities while facing cutbacks in government spending. In the article, he included the following table showing the increase in air traffic.

Because the data are stated in table form, they do not have as much impact on the reader as if they were presented using a statistical graph.

This chapter will show how to organize data and then construct appropriate graphs to represent the data in a concise, easy-to-understand form. An appropriate graph for this table appears near the end of the chapter.

Air Travel

Here's how many millions of passengers used U.S. planes here and abroad; projections out Friday will show even greater increases in the next decade:

Year	Passengers
1980	287.9
1981	274.7
1982	286.1
1983	308.2
1984	334.0
1985	370.1
1986	404.7
1987	441.2
1988	441.2
1989	443.6
1990	456.6
1991	445.7
1992	463.0
1993	468.1
1994	509.0

Source: FAA

Source: *USA Today,* February 28, 1996. Copyright 1996, USA TODAY. Used with permission.

2–1

Introduction

When conducting a statistical study, the researcher must gather data for the particular variable under study. For example, if a researcher wishes to study the number of people who were bitten by poisonous snakes in a specific geographic area over the past several years, he or she would have to gather the data from various doctors, hospitals, or health departments.

In order to describe situations, draw conclusions, or make inferences about events, the researcher must organize the data in some meaningful way. The most convenient method of organizing data is to construct a *frequency distribution*.

After organizing the data, the researcher must present them so they can be understood by those who will benefit from reading the study. The most useful method of presenting the data is by constructing *statistical charts and graphs*. There are many different types of charts and graphs, and each one has a specific purpose.

This chapter explains how to organize data by constructing frequency distributions and how to present the data by constructing charts and graphs. The charts and graphs illustrated here are histograms, frequency polygons, ogives, pie graphs, Pareto charts, and time series graphs.

2–2

Organizing Data

Objective 1. Organize data using frequency distributions.

Suppose a researcher wished to do a study on the number of miles the employees of a large department store traveled to work each day. The researcher would first have to collect the data by asking each employee the approximate distance the store is from his or her home. When data are collected in original form, they are called **raw data.** In this case, the data are as follows:

1	2	6	7	12	13	2	6	9	5
18	7	3	15	15	4	17	1	14	5
4	16	4	5	8	6	5	18	5	2
9	11	12	1	9	2	10	11	4	10
9	18	8	8	4	14	7	3	2	6

Since little information can be obtained from looking at raw data, the researcher organizes the data by constructing a frequency distribution. The **frequency** is the number of values in a specific class of the distribution. For this data set, a frequency distribution is shown as follows:

Class limits (in miles)	Tally	Frequency
1–3	卅卅 卅卅	10
4–6	卅卅 卅卅 ////	14
7–9	卅卅 卅卅	10
10–12	卅卅 /	6
13–15	卅卅	5
16–18	卅卅	5
		Total 50

Now, some general observations can be obtained from looking at the data in the form of a frequency distribution. For example, the majority of employees live within nine miles of the store.

A **frequency distribution** is the organizing of raw data in table form, using classes and frequencies.

The classes in this distribution are 1–3, 4–6, etc. These values are called class limits; the data values 1, 2, 3 can be tallied in the first class, 4, 5, 6 in the second class, and so on.

There are three basic types of frequency distributions, and there are specific procedures for constructing each type. The three types are *categorical, ungrouped,* and *grouped frequency distributions.* The next three subsections show each type and the procedures for constructing each.

Categorical Frequency Distributions

The categorical frequency distribution is used for data that can be placed in specific categories, such as nominal- or ordinal-level data. For example, data such as political affiliation, religious affiliation, or major field of study would use categorical frequency distributions.

Example 2–1

Twenty-five army inductees were given a blood test to determine their blood type. The data set is as follows:

A	B	B	AB	O
O	O	B	AB	B
B	B	O	A	O
A	O	O	O	AB
AB	A	O	B	A

Construct a frequency distribution for the data.

Solution

Since the data are categorical, discrete classes can be used. There are four blood types: A, B, O, and AB. These types will be used as the classes for the distribution.

The procedure for constructing a frequency distribution for categorical data is given next.

STEP 1   Make a table as shown.

A Class	B Tally	C Frequency	D Percent
A			
B			
O			
AB			

STEP 2   Tally the data and place the results in column B.

STEP 3   Count the tallies and place the results in column C.

STEP 4   Find the percentage of values in each class by using the formula

$$\% = \frac{f}{n} \cdot 100\%$$

where

f = frequency of the class
n = total number of values

For example, in the class of type A blood, the percentage is

$$\% = \frac{5}{25} \cdot 100\% = 20\%$$

Percentages are not normally a part of a frequency distribution, but they can be added since they are used in certain types of graphic presentations, such as pie graphs.

STEP 5 Find the totals for columns C and D (see the completed table that follows).

Class	Tally	Frequency	Percent
A	7╫╫	5	20
B	7╫╫ //	7	28
O	7╫╫ ////	9	36
AB	////	4	16
	Total	25	100

Ungrouped Frequency Distributions

When the data are numerical instead of categorical, the procedure for constructing a frequency distribution is somewhat more complicated, as shown in the next two examples. Example 2–2 describes the procedure for constructing an ungrouped frequency distribution.

Example 2–2

A psychologist administered a test of manual dexterity to 25 third-grade students. The times, in minutes, required to complete the test are given below. Construct a frequency distribution for the data.

4	8	8	9	8
5	9	9	10	11
7	7	8	7	8
4	8	7	5	7
6	5	10	8	9

Solution

STEP 1 Find the range of the data. The **range,** R, is defined as

$$R = \text{highest value} - \text{lowest value}$$

For this data set, the range is $11 - 4$, or 7. Since the range is small, classes consisting of a single data value can be used. They are 4, 5, 6, 7, 8, 9, 10, and 11.

STEP 2 Make a table as shown next.

STEP 3 Tally the data.

STEP 4 Complete the frequency column.

Unusual Stat

According to *100% American*, 59% of Americans think doctors do too many tests because of the fear of malpractice suits.

Class	Tally	Frequency
4	//	2
5	///	3
6	/	1
7	~~////~~	5
8	~~////~~ //	7
9	////	4
10	//	2
11	/	1

The values of the data set are continuous (i.e., are minutes). Recall that in Chapter 1 a value of 4 minutes could theoretically mean any number from 3.5 up to 4.5 minutes. So construct boundaries for each class by subtracting 0.5 from each class value and adding 0.5 to each class value, as shown next.

Class	Class boundaries	Tally	Frequency
4	3.5–4.5	//	2
5	4.5–5.5	///	3
6	5.5–6.5	/	1
7	6.5–7.5	~~////~~	5
8	7.5–8.5	~~////~~ //	7
9	8.5–9.5	////	4
10	9.5–10.5	//	2
11	10.5–11.5	/	1

Add a cumulative frequency (cf) column to the frequency distribution shown above by adding the frequency in each class to the total of the frequencies of the classes preceding that class, as shown next.

Class	Class boundaries	Frequency	Cumulative frequency
4	3.5–4.5	2	$0 + 2 = 2$
5	4.5–5.5	3	$2 + 3 = 5$
6	5.5–6.5	1	$5 + 1 = 6$
7	6.5–7.5	5	$6 + 5 = 11$
8	7.5–8.5	7	$11 + 7 = 18$
9	8.5–9.5	4	$18 + 4 = 22$
10	9.5–10.5	2	$22 + 2 = 24$
11	10.5–11.5	1	$24 + 1 = 25$

Cumulative frequencies are used to show how many values are accumulated up to and including a specific class. For example, 18 students successfully completed the test in 8 minutes or less; 24 students completed the test in 10 minutes or less.

Grouped Frequency Distributions

When the range of the data is large, the data must be grouped into classes that are more than one unit in width. For example, a distribution of the number of hours boat batteries lasted is as follows:

Class limits	Class boundaries	Tally	Frequency	Cumulative frequency
24–30	23.5–30.5	///	3	3
31–37	30.5–37.5	/	1	4
38–44	37.5–44.5	Ⅶⅼ	5	9
45–51	44.5–51.5	Ⅶⅼ ////	9	18
52–58	51.5–58.5	Ⅶⅼ /	6	24
59–65	58.5–65.5	/	1	25
			25	

The procedure for constructing the preceding frequency distribution is given in the next example; however, several things should be noted. In this distribution, the values 24 and 30 of the first class are called *class limits.* The **lower class limit** is 24; it represents the smallest data value that can be included in the class. The **upper class limit** is 30; it represents the largest data value that can be included in the class. The numbers in the second column are called **class boundaries.** These numbers are used to separate the classes so that there are no gaps in the frequency distribution. The gaps are due to the limits; for example, there is a gap between 30 and 31.

Students sometimes have difficulty finding class boundaries when given the class limits. The basic rule of thumb is that *the class limits should have the same decimal place value as the data, but the class boundaries have one additional place value and end in a 5.* For example, if the values in the data set are whole numbers, such as 24, 32, 18, the limits for a class might be 31–37, and the boundaries are 30.5–37.5 Find the boundaries by subtracting 0.5 from 31 (the lower class limit) and adding 0.5 to 37 (the upper class limit)

$$(\text{lower limit}) \quad 31 - 0.5 = 30.5 \quad (\text{lower boundary})$$
$$(\text{upper limit}) \quad 37 + 0.5 = 37.5 \quad (\text{upper boundary})$$

If the data are in tenths, such as 6.2, 7.8, 12.6, the limits for a class hypothetically might be 7.8–8.8, and the boundaries for that class would be 7.75–8.85. Find these values by subtracting 0.05 from 7.8 and adding 0.05 to 8.8.

Finally, the **class width** for a class in a frequency distribution is found by subtracting lower (or upper) class limit of one class minus the lower (or upper) class limit of the previous class. For example, the class width in the preceding distribution is 7, found by subtracting $31 - 24 = 7$. Class width also equals (upper boundary) − (lower boundary).

The researcher must decide how many classes to use and the width of each class. To construct a frequency distribution, follow these rules.

1. *There should be between 5 and 20 classes.* A student would not be in error for having less than 5 classes or more than 20 classes; however, statisticians generally agree on these numbers.
2. *The class width should be an odd number.* This ensures that the midpoint of each class has the same place value as the data. The **class midpoint** X_m is obtained by adding the lower and upper boundaries and dividing by 2, or adding the lower and upper limits and dividing by 2:

$$X_m = \frac{\text{lower boundary} + \text{upper boundary}}{2}$$

or

$$X_m = \frac{\text{lower limit} + \text{upper limit}}{2}$$

For example, the midpoint of the first class is

$$\frac{24 + 30}{2} = 27 \qquad \text{or} \qquad \frac{23.5 + 30.5}{2} = 27$$

The midpoint is the numerical location of the center of the class. Midpoints are necessary for graphing (see Section 2–3) and are used in computing the mean and standard deviation (see Sections 3–2 and 3–3). If the class width is an even number, the midpoint is in tenths. For example, if the class width is 6 and the boundaries are 5.5–11.5, the midpoint is

$$\frac{5.5 + 11.5}{2} = \frac{17}{2} = 8.5$$

Rule 2 is only a suggestion, and it is not rigorously followed, especially when a computer is used to group data.

3.   *The classes must be mutually exclusive.* Mutually exclusive classes have nonoverlapping class limits so that data cannot be placed into two classes. Many times, frequency distributions such as

Age
10–20
20–30
30–40
40–50

are found in the literature or in surveys. If a person is 40 years old, into which class should he or she be placed? A better way to construct a frequency distribution is to use classes such as

Age
10–20
21–31
32–42
43–53

4.   *The classes must be continuous.* Even if there are no values in a class, the class must be included in the frequency distribution. There should be no gaps in a frequency distribution. The only exception occurs when the class with a zero frequency is the first or last class. A class with a zero frequency at either end can be omitted without affecting the distribution.

5.   *The classes must be exhaustive.* There should be enough classes to accommodate all the data.

6.   *The classes must be equal in width.* This avoids a distorted view of the data.

One exception occurs when a distribution is **open-ended**—i.e., it has no specific beginning value or no specific ending value. Following are the class limits for two open-ended distributions.

Age	Minutes
10–20	Below 110
21–31	110–114
32–42	115–119
43–53	120–124
54–64	125–129
65 and above	130–134

The frequency distribution for age is open-ended for the last class, which means that anybody who is 65 years or older will be tallied in the last class. The distribution for minutes is open-ended for the first class, meaning that any minute values below 110 will be tallied in that class.

Example 2–3 shows the procedure for constructing a grouped frequency distribution, i.e., when the classes contain more than one data value.

Example 2–3

The following data represent the record high temperatures for each of the 50 states. Construct a grouped frequency distribution for the data using 7 classes.

112	100	127	120	134	118	105	110	109	112
110	118	117	116	118	122	114	114	105	109
107	112	114	115	118	117	118	122	106	110
116	108	110	121	113	120	119	111	104	111
120	113	120	117	105	110	118	112	114	114

Source: Reprinted with permission from *The World Almanac and Book of Facts 1995*. Copyright © 1994 PRIMEDIA Reference Inc. All rights reserved.

Solution

The procedure for constructing a grouped frequency distribution for numerical data follows.

STEP 1 Find the highest value and lowest value: $H = 134$ and $L = 100$.

STEP 2 Find the range: R = highest value − lowest value.

$$R = 134 - 100 = 34$$

STEP 3 Select the number of classes desired (usually between 5 and 20). In this case, 7 is arbitrarily chosen.

STEP 4 Find the class width by dividing the range by the number of classes.

$$\text{width} = \frac{R}{\text{number of classes}} = \frac{34}{7} = 4.9$$

Round the answer up to the nearest whole number if there is a remainder: $4.9 \approx 5$. (Rounding up is different from rounding off. A number is rounded up if there is any decimal remainder when dividing. For example, $85 \div 6 = 14.167$ and is rounded up to 15. Also, $53 \div 4 = 13.25$ and is rounded up to 14.)

STEP 5 Select a starting point for the lowest class limit. This can be the smallest data value or any convenient number less than the smallest data value. Add the width to the lowest score taken as the starting point to get the lower limit of the next class. Keep adding until there are 7 classes, as shown.

$$100$$
$$105$$
$$110$$
$$115$$
$$120$$
$$125$$
$$130$$

STEP 6 Subtract one unit from the lower limit of the second class to get the upper limit of the first class. Then add the width to each upper limit to get all the upper limits.

$$105 - 1 = 104$$

So the first class is $100 - 104$.

Class limits

100–104
105–109
110–114
115–119
120–124
125–129
130–134

STEP 7 Find the class boundaries by subtracting 0.5 from each lower class limit and adding 0.5 to the upper class limit, as shown.

Class boundaries

99.5–104.5
104.5–109.5
109.5–114.5
114.5–119.5
119.5–124.5
124.5–129.5
129.5–134.5

STEP 8 Tally the data, write the numerical values for the tallies in the frequency column, and find the cumulative frequencies.
The completed frequency distribution follows:

Class limits	Class boundaries	Tally				Frequency	Cumulative frequency
100–104	99.5–104.5	//				2	2
105–109	104.5–109.5	⊬⊬	///			8	10
110–114	109.5–114.5	⊬⊬	⊬⊬	⊬⊬	///	18	28
115–119	114.5–119.5	⊬⊬	⊬⊬	///		13	41
120–124	119.5–124.5	⊬⊬	//			7	48
125–129	124.5–129.5	/				1	49
130–134	129.5–134.5	/				1	50

The frequency distribution shows that the class 109.5–114.5 contains the largest number of temperatures (18) followed by the class

114.5–119.5 with 13 temperatures. Hence, most of the temperatures (31) fall between 109.5° and 119.5°.

The procedure for constructing a grouped frequency distribution is summarized in Procedure Table 1.

Procedure Table 1

Constructing a Grouped Frequency Distribution

1. Find the highest and lowest value.
2. Find the range.
3. Select the number of classes desired.
4. Find the width by dividing the range by the number of classes and rounding up.
5. Select a starting point (usually the lowest value); add the width to get the lower limits.
6. Find the upper class limits.
7. Find the boundaries.
8. Tally the data, find the frequencies, and find the cumulative frequency.

When one is constructing a frequency distribution, the guidelines presented in this section should be followed. However, one can construct several different but correct frequency distributions for the same data by using a different class width, a different number of classes, or a different starting point.

Furthermore, the method shown here for constructing a frequency distribution is not unique, and there are other ways of constructing one. Slight variations exist, especially in computer packages. But regardless of what methods are used, classes should be mutually exclusive, continuous, exhaustive, and of equal width.

In summary, three types of frequency distributions were shown in this section. The first type, shown in Example 2–1, is used when the data are categorical (nominal), such as blood type or political affiliation. This type is called a **categorical distribution.** The second type of distribution is used for numerical data and when the range of data is small, as shown in Example 2–2. Since each class is only one unit, this distribution is called an **ungrouped frequency distribution.** The third type of distribution is used when the range is large and classes several units in width are needed. This type is called a **grouped frequency distribution** and is shown in Example 2–3. All three types of distributions are used frequently in statistics and are helpful when one is organizing and presenting data.

The reasons for constructing a frequency distribution follow:

1. To organize the data in a meaningful, intelligible way.
2. To enable the reader to determine the nature or shape of the distribution.
3. To facilitate computational procedures for measures of average and spread (shown in Sections 3–2 and 3–3).
4. To enable the researcher to draw charts and graphs for the presentation of data (shown in Section 2–3).
5. To enable the reader to make comparisons among different data sets.

Exercises

2–1. (W) List five reasons for organizing data into a frequency distribution.

2–2. (W) Name the three types of frequency distributions and explain when each should be used.

2–3. Find class boundaries, midpoints, and widths for each class.
a. 11–15
b. 17–39
c. 293–353
d. 11.8–14.7
e. 3.13–3.93

2–4. (W) How many classes should frequency distributions have? Why should the class width be an odd number?

2–5. Shown here are four frequency distributions. Each is incorrectly constructed. State the reason why.

a.

Class	Frequency
27–32	1
33–38	0
39–44	6
45–49	4
50–55	2

b.

Class	Frequency
5–9	1
9–13	2
13–17	5
17–20	6
20–24	3

c.

Class	Frequency
123–127	3
128–132	7
138–142	2
143–147	19

d.

Class	Frequency
9–13	1
14–19	6
20–25	2
26–28	5
29–32	9

2–6. (W) What are open-ended frequency distributions? Why are they necessary?

2–7. (W) The following zip codes were obtained from the respondents to a mail survey. Construct a frequency distribution for the data.

15132	15130	15132	15130
15130	15131	15134	15133
15131	15133	15133	15133
15130	15131	15132	15130
15133	15134	15133	15133

2–8. (W) At a college financial aid office, students who applied for a scholarship were classified according to their class rank: Fr = freshman, So = sophomore, Jr = Junior, Se = senior. Construct a frequency distribution for the data.

Fr	Fr	Fr	Fr	Fr
Jr	Fr	Fr	So	Fr
Fr	So	Jr	So	Fr
So	Fr	Fr	Fr	So
Se	Jr	Jr	So	Fr
Fr	Fr	Fr	Fr	So
Se	Se	Jr	Jr	Se
So	So	So	So	So

2–9. (W) A survey taken in a restaurant shows the following number of cups of coffee consumed with each meal. Construct an ungrouped frequency distribution.

0	2	2	1	1	2
3	5	3	2	2	2
1	0	1	2	4	2
0	1	0	1	4	4
2	2	0	1	1	5

2–10. (W) In a survey of 20 patients who smoked, the following data were obtained. Each value represents the number of cigarettes the patient smoked per day. Construct a frequency distribution, using six classes.

10	8	6	14
22	13	17	19
11	9	18	14
13	12	15	15
5	11	16	11

2–11. (W) The numbers of games won by the pitchers who were inducted into the Baseball Hall of Fame through 1992 are shown below. Construct a frequency distribution for the data using 12 classes.

373	254	237	243	308
210	266	253	201	266
239	114	224	373	286
329	236	284	247	273
198	361	416	207	243
326	251	160	360	311
215	189	344	268	363
21	270	165	240	48
150	300	207	314	197
209	210	260	327	

Source: John W. Wright, ed., *The Universal Almanac,* Andrews and McMeel, 1994, pp. 708–709. Used with permission.

2–12. In a study of 32 student grade point averages (GPA), the following data were obtained. Construct a frequency distribution using seven classes. (*Note:* A = 4, B = 3, C = 2, D = 1, F = 0.) (The information in this exercise will be used again for Exercises 2–22 and 3–22.)

3.2	2.0	3.3	2.7	2.1	3.9
1.1	3.5	1.9	1.7	0.8	2.6
0.6	4.0	3.5	2.3	1.6	
2.8	2.6	1.6	1.6	2.4	
2.6	2.3	3.8	2.1	2.9	
3.0	1.7	4.0	1.2	3.1	

2–13. The ages of the signers of the Declaration of Independence are shown below. (Age is approximate since only the birth year appeared in the source, and one has been omitted since his birth year is unknown.) Construct a frequency distribution for the data using seven classes. (The data for this exercise will be used for Exercise 3–23.)

41	54	47	40	39	35	50	37	49	42	70	32
44	52	39	50	40	30	34	69	39	45	33	42
44	63	60	27	42	34	50	42	52	38	36	45
35	43	48	46	31	27	55	63	46	33	60	62
35	46	45	34	53	50	50					

Source: John W. Wright, ed., *The Universal Almanac,* Andrews and McMeel, 1994, p. 53. Used with permission.

2–14. A supervisor wishes to check the miles her commuter buses are driven each day. Her findings are shown below. Construct a frequency distribution using 10 classes. (The data for this exercise will be used for Exercises 2–24, 2–34, and 3–24.)

138	107	136	128	148	118	99
142	129	115	123	133	123	103
121	128	122	144	126	135	107
125	98	117	153	141	126	139
134	115	93	127	118	158	143

2–15. In a study of 40 women, the following data of blood potassium levels, in milliequivalents per liter, were obtained. Construct a frequency distribution using seven classes. (The information in this exercise will be used for Exercises 2–27 and 3–25.)

3.2	4.9	3.8	3.6	3.4
5.8	5.3	4.6	5.1	3.6
6.0	4.7	4.3	4.2	4.4
4.5	5.0	4.4	5.8	3.9
4.2	3.9	3.8	3.7	4.2
4.3	4.9	5.6	3.7	4.1
2.7	4.0	4.2	4.3	4.3
5.1	5.2	4.7	4.5	4.9

2–16. (W) The acreage of the 39 U.S. National Parks under 900,000 acres (in thousands of acres) is shown here. Construct a frequency distribution for the data using eight classes.

41	66	233	775	169
36	338	233	236	64
183	61	13	308	77
520	77	27	217	5
650	462	106	52	52
505	94	75	265	402
196	70	132	28	220
760	143	46	539	

Source: John W. Wright, ed., *The Universal Almanac,* Andrews and McMeel, 1994, p. 45. Used with permission.

2–17. The heights in feet above sea level of the major active volcanoes in Alaska are given here. Construct a frequency distribution for the data using 10 classes.

4265	3545	4025	7050	11,413
3490	5370	4885	5030	6830
4450	5775	3945	7545	8450
3995	10,140	6050	10,265	6965
150	8185	7295	2015	5055
5315	2945	6720	3465	1980
2560	4450	2759	9430	
7985	7540	3540	11,070	
5710	885	8960	7015	

Source: THE UNIVERSAL ALMANAC © by John W. Wright. Reprinted with permission of Andrew McMeel Publishing. All rights reserved.

*2–18. If the data are in units of 10—say 80, 230, 120, 160, etc.—and a hypothetical class is 150–200, what are the boundaries? The following data show the net worth of 20 new-car dealerships in Miami. The amounts are in millions of dollars. Construct a frequency distribution for the data.

130	70	180	160
120	60	150	140
160	80	140	130
210	80	130	140
90	110	120	150

If the data are in units of hundreds—such as 300, 1200, 500, 900—what are the boundaries of a hypothetical class such as 800–1200?

2–3

Histograms, Frequency Polygons, and Ogives

Objective 2. Represent data in frequency distributions using histograms, frequency polygons, and ogives.

After the data have been organized into a frequency distribution, they can be presented in graphic forms. The purpose of graphs in statistics is to convey the data to the viewer in pictorial form. It is easier for most people to comprehend the meaning of data presented graphically than data presented numerically in tables or frequency distributions. This is especially true if they have little or no statistical knowledge.

Statistical graphs can be used to describe the data set or analyze it. Graphs are also useful in getting the audience's attention in a publication or a speaking presentation. They can be used to discuss an issue, reinforce a critical point, or summarize a data set. They can also be used to discover a trend or pattern in a situation over a period of time.

The three most commonly used graphs in research are

1. The histogram.
2. The frequency polygon.
3. The cumulative frequency graph, or ogive (pronounced o-jive).

An example of each type of graph is shown in Figure 2–1. The data for each graph are the distribution of the miles that 20 randomly selected runners ran during a given week.

Figure 2–1
....
Examples of Commonly
Used Graphs

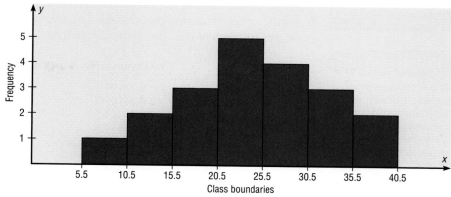

(a) Histogram

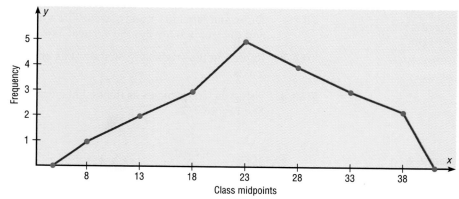

(b) Frequency polygon

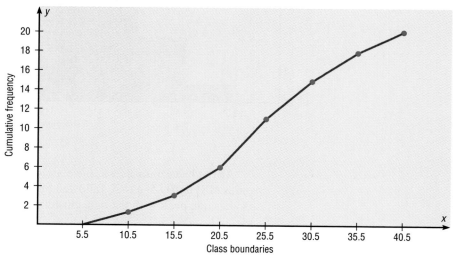

(c) Cumulative frequency graph

The Histogram

The **histogram** is a graph that displays the data by using vertical bars of various heights to represent the frequencies.

Construct a histogram to represent the data shown below for the record high temperatures for each of the 50 states (see Example 2–3).

Class boundaries	Frequency
99.5–104.5	2
104.5–109.5	8
109.5–114.5	18
114.5–119.5	13
119.5–124.5	7
124.5–129.5	1
129.5–134.5	1

Historical Note

Graphs originated when ancient astronomers drew the position of the stars in the heavens. Roman surveyors also used coordinates to locate landmarks on their maps.
    The development of statistical graphs can be traced to William Playfair (1748–1819), an engineer/drafter who used graphs to present economic data pictorially.

Solution

STEP 1   Draw and label the x and y axes. The x axis is always the horizontal axis, and the y axis is always the vertical axis.

STEP 2   Represent the frequency on the y axis and the class boundaries on the x axis.

STEP 3   Using the frequencies as the heights, draw vertical bars for each class. See Figure 2–2.

Figure 2–2

Histogram for Example 2–4

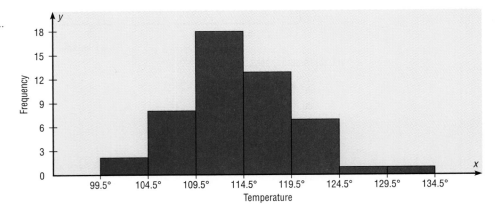

As the histogram shows, the class with the greatest number of data values (18) is 109.5–114.5, followed by 13 for 114.5–119.5. The graph also has one peak with the data clustering around it.

The Frequency Polygon

Another way to represent the same data set is by using a frequency polygon.

The **frequency polygon** is a graph that displays the data by using lines that connect points plotted for the frequencies at the midpoints of the classes. The frequencies represent the heights of the midpoints.

The next example shows the procedure for constructing a frequency polygon.

Example 2–5

Using the frequency distribution given in Example 2–4, construct a frequency polygon.

Solution

STEP 1   Find the midpoints of each class. Recall that midpoints are found by adding the upper and lower boundaries and dividing by 2.

$$\frac{99.5 + 104.5}{2} = 102 \qquad \frac{104.5 + 109.5}{2} = 107$$

And so on. The midpoints are listed next.

Class boundaries	Midpoints	Frequency
99.5–104.5	102	2
104.5–109.5	107	8
109.5–114.5	112	18
114.5–119.5	117	13
119.5–124.5	122	7
124.5–129.5	127	1
129.5–134.5	132	1

STEP 2   Draw the x and y axes. Label the x axis with the midpoint of each class, and then use a suitable scale on the y axis for the frequencies.

STEP 3   Using the midpoints for the x values and the frequencies as the y values, plot the points.

STEP 4   Connect adjacent points with straight lines. Draw a line back to the x axis at the beginning and end of the graph, at the same distance that the previous and next midpoints would be located, as shown in Figure 2–3.

Figure 2–3

Frequency Polygon for Example 2–5

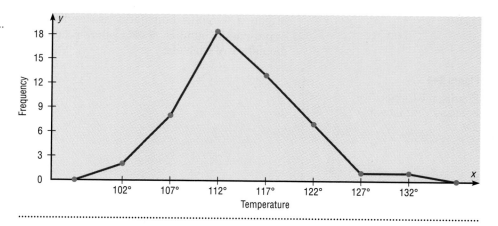

The frequency polygon and the histogram are two different ways to represent the same data set. The choice of which one to use is left to the discretion of the researcher.

The Ogive

The third type of graph that can be used represents the cumulative frequencies for the classes. This type of graph is called the cumulative frequency graph or ogive.

The **cumulative frequency** is the sum of the frequencies accumulated up to the upper boundary of a class in the distribution.

The **ogive** is a graph that represents the cumulative frequencies for the classes in a frequency distribution.

Example 2–6 shows the procedure for constructing an ogive.

Example 2–6

Construct an ogive for the frequency distribution described in Example 2–4.

Solution

STEP 1 Find the cumulative frequency for each class.

Class boundaries	Cumulative frequency
99.5–104.5	2
104.5–109.5	10
109.5–114.5	28
114.5–119.5	41
119.5–124.5	48
124.5–129.5	49
129.5–134.5	50

STEP 2 Draw the x and y axes. Label the x axis with the class boundaries. Use an appropriate scale for the y axis to represent the cumulative frequencies. (Depending on the numbers in the cumulative frequency columns, scales such as 0, 1, 2, 3, . . . , or 5, 10, 15, 20 . . . , or 1000, 2000, 3000, . . . can be used. Do *not* label the y axis with the numbers in the cumulative frequency column.) In this example, a scale of 0, 5, 10, 15, . . . will be used.

STEP 3 Plot the cumulative frequency at each upper class boundary, as shown in Figure 2–4. Upper boundaries are used since the cumulative frequencies represent the number of data values accumulated up to the upper boundary of each class.

Figure 2–4

Plotting the Cumulative Frequency for Example 2–6

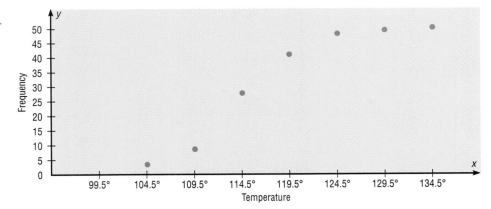

STEP 4 Starting with the first upper class boundary, 104.5, connect adjacent points with straight lines, as shown in Figure 2–5. Then extend the graph to the first lower class boundary, 99.5, on the x axis.

Figure 2–5

Ogive for Example 2–6

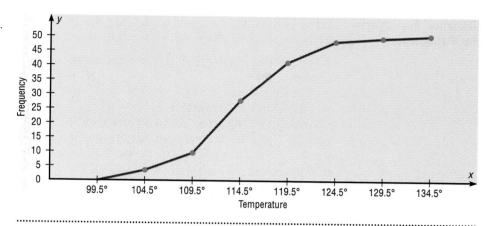

Cumulative frequency graphs are used to visually represent how many values are below a certain upper class boundary. For example, to find out how many record high temperatures are less than 114.5°, locate 114.5° on the x axis, draw a vertical line up until it intersects the graph, and then draw a horizontal line at that point to the y axis. The y axis value is 28, as shown in Figure 2–6.

Figure 2–6

Finding a Specific
Cumulative Frequency

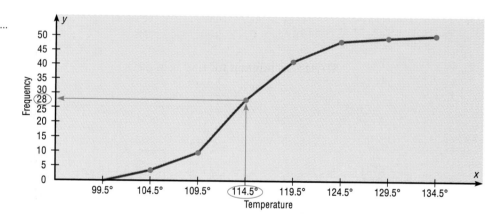

A summary for drawing the three types of graphs is shown in Procedure Table 2.

Procedure Table 2

Constructing Statistical Graphs

1. Draw and label the x and y axes.
2. Choose a suitable scale for the frequencies or cumulative frequencies and label it on the y axis.
3. Represent the class boundaries for the histogram or ogive or the midpoint for the frequency polygon on the x axis.
4. Plot the points and then draw the bars or lines.

Computer Applications for Histograms

MINITAB should be loaded following the procedures given in the MINITAB reference manual. The Windows version is explained here.

When opened, MINITAB will present a split Session window and Worksheet window.

A. To enter data:
1. Click on Worksheet.
2. Click the arrow above the row number so that it points down.
3. Type the data values in the first column and press Enter after each value. Use the arrow keys or mouse to move around on the screen.

B. To construct a histogram:
1. Enter data into C1.
2. Click Graph > Histogram.
3. Highlight C1 in the dialog box and click on Select.
4. Click on OK.

The computer will print a histogram.
Example: Construct a histogram using the following data.

Data:

C1	68	80	69	81	72	100	101	73	102	93
	91	92	93	88	82	83	75	75	89	96
	103	83	84	85	89					

MINITAB printout for this example:

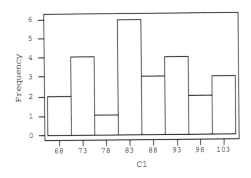

Summary: MINITAB automatically selects class boundaries and represents the frequencies using bars.

Relative Frequency Graphs

The histogram, the frequency polygon, and the ogive shown previously were constructed by using frequencies in terms of the raw data. These distributions can be converted into distributions using *proportions* instead of raw data as frequencies. These types of graphs are called **relative frequency graphs.**

To convert a frequency into a proportion or relative frequency, divide the frequency for each class by the total of the frequencies. The sum of the relative frequencies will always be 1. These graphs are similar to the ones that use raw data as frequencies, but the values on the y axis are in terms of proportions. The next example shows the three types of relative frequency graphs.

Example 2–7

Construct a histogram, frequency polygon, and ogive using relative frequencies for the distribution (shown here) of the miles 20 randomly selected runners ran during a given week.

Class boundaries	Frequency	Cumulative frequency
5.5–10.5	1	1
10.5–15.5	2	3
15.5–20.5	3	6
20.5–25.5	5	11
25.5–30.5	4	15
30.5–35.5	3	18
35.5–40.5	2	20
	20	

Solution

STEP 1   Convert each frequency to a proportion or relative frequency by dividing the frequency for each class by the total number of observations.

For class 5.5–10.5, the relative frequency is $\frac{1}{20} = 0.05$.
For class 10.5–15.5, the relative frequency is $\frac{2}{20} = 0.10$.
For class 15.5–20.5, the relative frequency is $\frac{3}{20} = 0.15$.
And so on.

STEP 2   Using the same procedure, find the relative frequencies for the cumulative frequency column. The relative frequencies are shown here.

Class boundaries	Midpoints	Relative frequency	Cumulative relative frequency
5.5–10.5	8	0.05	0.05
10.5–15.5	13	0.10	0.15
15.5–20.5	18	0.15	0.30
20.5–25.5	23	0.25	0.55
25.5–30.5	28	0.20	0.75
30.5–35.5	33	0.15	0.90
35.5–40.5	38	0.10	1.00
		1.00	

STEP 3   Draw each graph as shown in Figure 2–7. For the histogram and ogive, use the class boundaries along the x axis. For the frequency polygon, use the midpoints on the x axis. The scale on the y axis uses proportions.

Figure 2–7
Graphs for Example 2–7

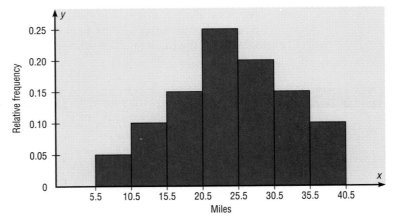

(a) Histogram

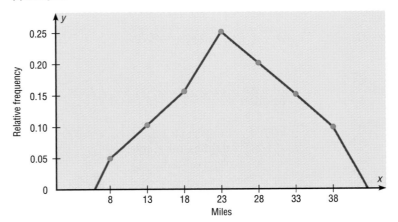

(b) Frequency polygon

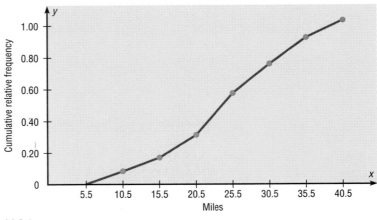

(c) Ogive

When analyzing histograms and frequency polygons, look at the shape of the curve. For example, does it have one peak or two peaks, or is it relatively flat, or is it U-shaped? Are the data values spread out on the graph, or are they clustered around the center? Are there data values in the extreme ends? These may be *outliers*. (See Section 3–4 for an explanation of outliers.) Are there any gaps in the histogram, or does the frequency polygon touch the *x* axis somewhere other than the ends? Finally, are the data clustered at one end or the other, indicating a *skewed distribution*? (See Section 3–2 for an explanation of skewness.)

For example, the histogram for the record high temperatures shown in Figure 2–2 (page 38) shows a single peaked distribution, with the class 109.5–114.5 containing the largest number of temperatures. The distribution has no gaps, and there are fewer temperatures in the highest class than in the lowest class.

Exercises

2–19. For 108 randomly selected college students, the following IQ frequency distribution was obtained. Construct a histogram, frequency polygon, and ogive for the data. (The data for this exercise will be used for Exercise 2–31.)

Class limits	Frequency
90–98	6
99–107	22
108–116	43
117–125	28
126–134	9

2–20. For 75 employees of a large department store, the following distribution for years of service was obtained. Construct a histogram, frequency polygon, and ogive for the data. (The data for this exercise will be used for Exercise 2–32.)

Class limits	Frequency
1–5	21
6–10	25
11–15	15
16–20	0
21–25	8
26–30	6

2–21. (W) Construct a histogram, frequency polygon, and ogive for the data in Exercise 2–11 and analyze the results.

2–22. (W) Construct a histogram, frequency polygon, and ogive for the data in Exercise 2–12 and analyze the results.

2–23. (W) Thirty automobiles were tested for fuel efficiency, in miles per gallon (mpg). The following frequency distribution was obtained. Construct a histogram, frequency polygon, and ogive for the data. (The data for this exercise will be used for Exercise 2–33.)

Class boundaries	Frequency
7.5–12.5	3
12.5–17.5	5
17.5–22.5	15
22.5–27.5	5
27.5–32.5	2

2–24. (W) Construct a histogram, frequency polygon, and ogive for the data in Exercise 2–14 and analyze the results.

2–25. (W) In a class of 35 students, the following grade distribution was found. Construct a histogram, frequency polygon, and ogive for the data. (A = 4, B = 3, C = 2, D = 1, F = 0.) (The data in this exercise will be used for Exercise 2–35.)

Grade	Frequency
0	3
1	6
2	9
3	12
4	5

2–26. (W) In a study of reaction times to a specific stimulus, an animal trainer obtained the following data, given in seconds. Construct a histogram, frequency polygon, and ogive for the data and analyze the results.

Class limits	Frequency
2.3–2.9	10
3.0–3.6	12
3.7–4.3	6
4.4–5.0	8
5.1–5.7	4
5.8–6.4	2

2–27. (**W**) Construct a histogram, frequency polygon, and ogive for the data in Exercise 2–15 and analyze the results.

2–28. (**W**) To determine their lifetimes, 80 randomly selected batteries were tested. The following frequency distribution was obtained. The data values are in hours. Construct a histogram, frequency polygon, and ogive for the data and analyze the results.

Class boundaries	Frequency
63.5–74.5	10
74.5–85.5	15
85.5–96.5	22
96.5–107.5	17
107.5–118.5	11
118.5–129.5	5

2–29. (**W**) Construct a histogram, frequency polygon, and ogive for the data in Exercise 2–16 and analyze the results.

2–30. In a study of reaction times to a specific stimulus, a psychologist obtained the following data, given in seconds. Construct a histogram, frequency polygon, and ogive for the data.

Class limits	Frequency
0.5–0.9	12
1.0–1.4	13
1.5–1.9	7
2.0–2.4	5
2.5–2.9	2
3.0–3.4	0
3.5–3.9	1

2–31. For the data in Exercise 2–19, construct a histogram, frequency polygon, and ogive, using relative frequencies.

2–32. For the data in Exercise 2–20, construct a histogram, frequency polygon, and ogive, using relative frequencies.

2–33. For the data in Exercise 2–23, construct a histogram, frequency polygon, and ogive, using relative frequencies.

2–34. For the data in Exercise 2–14, construct a histogram, frequency polygon, and ogive, using relative frequencies.

2–35. For the data in Exercise 2–25, construct a histogram, frequency polygon, and ogive, using relative frequencies.

*2–36. Using the following histogram:
a. Construct a frequency distribution; include class limits, class frequencies, midpoints, and cumulative frequencies.
b. Construct a frequency polygon.
c. Construct an ogive.

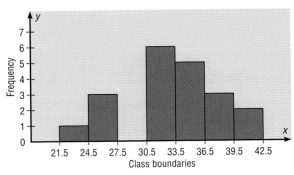

*2–37. Using the results from Exercise 2–36, answer the following questions.
a. How many values are in the class 27.5–30.5?
b. How many values fall between 24.5 and 36.5?
c. How many values are below 33.5?
d. How many values are above 30.5?

*2–38. Using the frequency polygon shown here,
a. Construct a frequency distribution, including class limits, class boundaries, class frequencies, and cumulative frequencies.
b. Construct a histogram.
c. Construct an ogive.

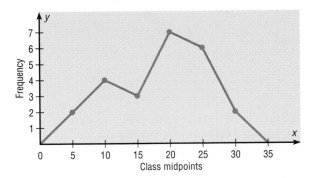

2–4

Other Types of Graphs

In addition to the histogram, the frequency polygon, and the ogive, several other types of graphs are often used in statistics. They are the Pareto chart, the time series graph, and the pie graph. Figure 2–8 shows an example of each type of graph.

Figure 2–8

Other Types of Graphs Used in Statistics

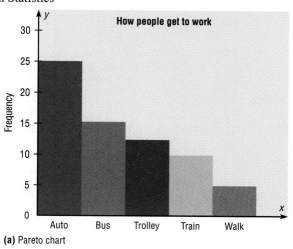

(a) Pareto chart

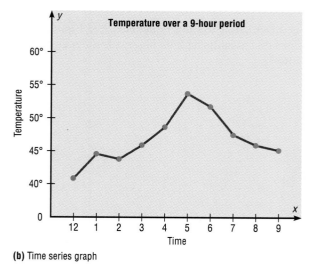

(b) Time series graph

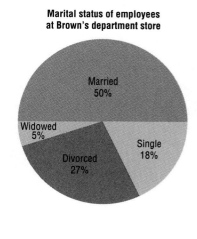

(c) Pie graph

Pareto Charts

Objective 3. Represent data using Pareto charts, time series graphs, and pie graphs.

In the previous section, graphs such as the histogram, frequency polygon, and ogive showed how data can be represented when the variable displayed on the horizontal axis is quantitative, such as heights and weights.

On the other hand, when the variable displayed on the horizontal is qualitative or categorical, a *Pareto chart* can be used.

A **Pareto chart** is used to represent a frequency distribution for a categorical variable, and the frequencies are displayed by the heights of vertical bars.

The following table shows the number of crimes investigated by law enforcement officers in U.S. national parks during 1995. Construct a Pareto chart for the data.

Type	Number
Homicide	13
Rape	34
Robbery	29
Assault	164

Source: "Crime Rate in 1990–95," *USA Today,* June 5, 1995.

Historical Note

Vilfredo Pareto (1848–1923) was an Italian scholar who developed theories in economics, statistics, and social sciences. His contributions to statistics include the development of a mathematical function used in economics. This function has many statistical applications and is called the Pareto Distribution. In addition, he researched income distribution, and his findings became known as Pareto's Law.

Solution

STEP 1 Arrange the data from the largest to smallest according to frequency.

Type	Number
Assault	164
Rape	34
Robbery	29
Homicide	13

STEP 2 Draw and label the *x* and *y* axes.

STEP 3 Draw the bars corresponding to the frequencies. See Figure 2–9. The Pareto chart shows that assaults had the highest frequency and homicides the lowest.

Figure 2–9

Pareto Chart

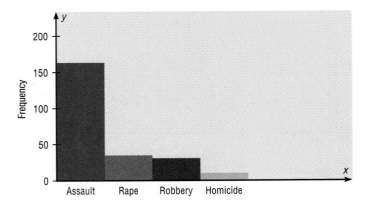

Historical Note

Time series graphs are over 1000 years old. The first ones were used to chart the movements of the planets and the sun.

Suggestions for Drawing Pareto Charts

1. Make the bars the same width.
2. Arrange the data from largest to smallest according to frequencies.
3. Make the units that are used for the frequency equal in size.

When analyzing a Pareto chart, make comparisons by looking at the heights of the bars.

The Time Series Graph   When data are collected over a period of time, they can be represented by a time series graph.

A **time series graph** represents data that occur over a specific period of time.

The next example shows the procedure for constructing a time series graph.

Example 2–9

A transit manager wishes to use the following data for a presentation showing how Port Authority Transit ridership has changed over the years. Draw a time series graph for the data and summarize the findings.

Year	Ridership (in millions)
1990	88.0
1991	85.0
1992	75.7
1993	76.6
1994	75.4

Source: Port Authority Transit, *Tribune-Review* (Greensburg, PA), January 28, 1995.

Solution

STEP 1   Draw and label the x and y axes.

STEP 2   Label the x axis for years and the y axis for the number of riders.

STEP 3   Plot each point according to the table.

STEP 4   Draw straight lines connecting adjacent points. Do not try to fit a smooth curve through the data points. See Figure 2–10.

Figure 2–10

Time Series Graph for Example 2–9

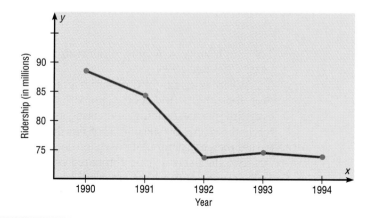

This time series graph compares the coal production of Pennsylvania with that of the Untied States as a whole. Using the information presented in the graph, explain in words how the total coal production of the United States compares with the coal production of Pennsylvania over the years.

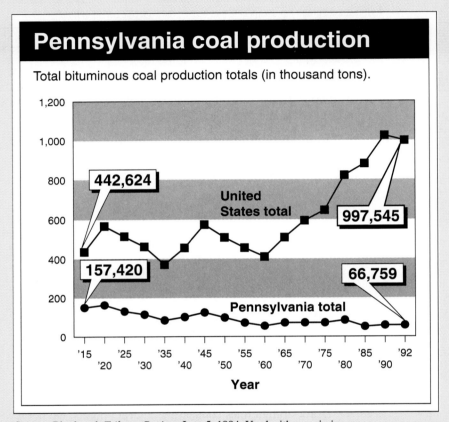

Pennsylvania coal production

Total bituminous coal production totals (in thousand tons).

Source: Pittsburgh *Tribune-Review*, June 5, 1994. Used with permission.

The graph shows a decline in ridership through 1992 and then a leveling off for the years 1993 and 1994.

When analyzing a time series graph, look for a trend or pattern that occurs over the time period. For example, is the line ascending (indicating an increase over time) or descending (indicating a decrease over time)? Another thing to look for is the slope, or steepness, of the line. A line that is steep over a specific time period indicates a rapid increase or decrease over that period.

Two data sets can be compared on the same graph (called a compound time series graph), if two lines are used, as shown in Figure 2–11. This graph shows the number of snow shovels sold at a store for two seasons.

Figure 2–11

Two Time Series Graphs
for Comparisons

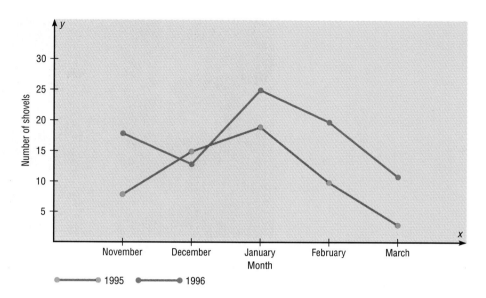

The Pie Graph

Pie graphs are used extensively in statistics. The purpose of the pie graph is to show the relationship of the parts to the whole by visually comparing the sizes of the sectors. Percentages or proportions can be used. The variable is nominal or categorical.

A **pie graph** is a circle that is divided into sections or wedges according to the percentage of frequencies in each category of the distribution.

The next example shows the procedure for constructing a pie graph.

Example 2–10

A car dealer reported information on sales for the month of July. Construct a pie graph to represent the data.

Type	Frequency
Convertibles	3
Station wagons	2
Compacts	5
Coupes	30
Sedans	20
	$n = 60$

Solution

STEP 1 Since there are 360 degrees in a circle, the frequency for each class must be converted into a proportional part of the circle. This conversion is done by using the formula

$$\text{degrees} = \frac{f}{n} \cdot 360°$$

where

  f = frequency for each class
  n = sum of the frequencies

Hence, the following conversions are obtained.

convertibles	$\dfrac{3}{60} \cdot 360°$ =	18°
station wagons	$\dfrac{2}{60} \cdot 360°$ =	12°
compacts	$\dfrac{5}{60} \cdot 360°$ =	30°
coupes	$\dfrac{30}{60} \cdot 360°$ =	180°
sedans	$\dfrac{20}{60} \cdot 360°$ =	<u>120°</u>
		total = 360°

STEP 2   Each frequency must also be converted to a percentage. Recall from Example 2–1 that this conversion is done by using the formula

$$\% = \frac{f}{n} \cdot 100\%$$

Hence, the following percentages are obtained.

convertibles	$\dfrac{3}{60} \cdot 100\%$ =	5%
station wagons	$\dfrac{2}{60} \cdot 100\%$ =	$3\frac{1}{3}\%$
compacts	$\dfrac{5}{60} \cdot 100\%$ =	$8\frac{1}{3}\%$

$$\text{coupes} \qquad \frac{30}{60} \cdot 100\% = 50\%$$

$$\text{sedans} \qquad \frac{20}{60} \cdot 100\% = \underline{33\tfrac{1}{3}\%}$$

$$\text{total} = 100\%$$

The degrees column should add to 360°, and the percentages column should add to 100%.

STEP 3   Next, using a protractor and a compass, draw the graph and label each section with the name and percentages, as shown in Figure 2–12.

Figure 2–12

Pie Graph for Example 2–10

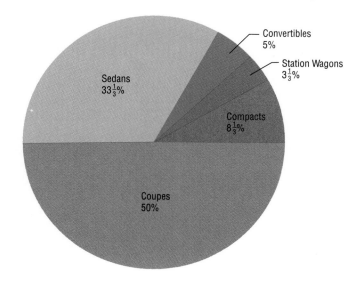

In this case, then, the graph shows that most of the automobiles sold were coupes, followed by sedans.

Example 2–11

Construct a pie graph showing the blood types of the army inductees described in Example 2–1. The frequency distribution is repeated here.

Class	Frequency	Percent
A	5	20%
B	7	28%
O	9	36%
AB	4	16%
	25	100%

Solution

STEP 1   Find the number of degrees for each class, using the formula

$$\text{degrees} = \frac{f}{n} \cdot 360°$$

For each class, then, the following results are obtained.

$$\begin{array}{lll} A & \frac{5}{25} \cdot 360° = & 72° \\ B & \frac{7}{25} \cdot 360° = & 100.8° \\ O & \frac{9}{25} \cdot 360° = & 129.6° \\ AB & \frac{4}{25} \cdot 360° = & 57.6° \end{array}$$

STEP 2   Find the percentages. (This has already been done in Example 2–1.)

STEP 3   Next, using a protractor and a compass, graph each section and write its name and corresponding percentage, as shown in Figure 2–13.

Figure 2–13

Pie Graph for Example 2–11

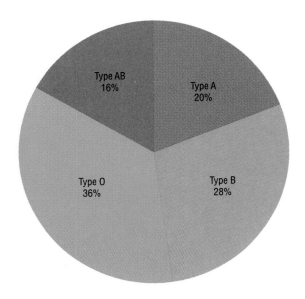

The graph shows that in this case the most common blood type is type O.

To analyze the nature of the data shown in the pie graph, compare the sectors. For example, are any sectors relatively large compared to the rest?

Figure 2–13 shows that among the inductees, type O blood is more prevalent than any other type. People who have type AB blood are in the minority. More than twice as many people have type O blood as type AB.

Misleading Graphs   Graphs give a visual representation that enables readers to analyze and interpret data more easily than they could simply by looking at numbers. However, inappropriately drawn graphs can misrepresent the data and lead the reader to false conclusions. For example, a car manufacturer's ad stated that 98% of the vehicles it had sold in the past 10 years were still on the road. The ad then showed a graph similar to the one in Figure 2–14. The graph shows the percentage of the manufacturer's automobiles still on the road and the percentage of its competitors' automobiles still on the road. Is there a large difference? Not necessarily.

Notice the scale on the vertical axis in Figure 2–14. It has been cut off (or truncated), and it starts at 95%. When the graph is redrawn using a scale that goes from 0% to 100%, as in Figure 2–15, there is hardly a noticeable difference in the

Figure 2–14
....
Graph of Automaker's
Claim, Using a Scale
from 95% to 100%

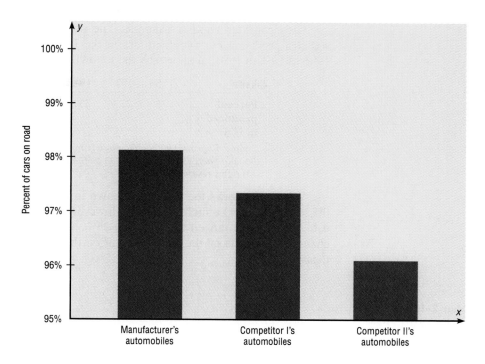

percentages. Thus, changing the units at the starting point on the y axis can convey
a very different visual representation of the data.

It is not wrong to truncate an axis of the graph; many times it is necessary to
do so (see Example 2–9). However, the reader should be aware of this fact and
interpret the graph accordingly. Do not be misled if an inappropriate explanation is
given.

Figure 2–15
....
Graph in Figure 2–14
Redrawn, Using a Scale
from 0% to 100%

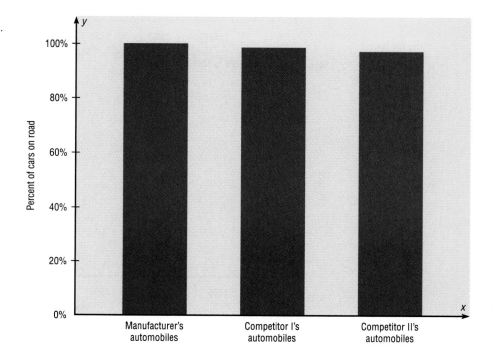

Let's consider another example. The percentage of the world's total motor vehicles produced by manufacturers in the United States declined from 25% in 1986 to 18% in 1991, as shown by the following data.

Year	1986	1987	1988	1989	1990	1991
Percent produced in U.S.	25	23.8	23.3	22.1	20.3	18.0

Source: Reprinted with permission from *The World Almanac and Book of Facts 1993*. Copyright © 1992 PRIMEDIA Reference Inc. All rights reserved.

When one draws the graph, as shown in Figure 2–16a, a scale ranging from 0% to 100% shows a slight decrease. However, this decrease can be emphasized by using a scale that ranged from 15% to 25%, as shown in Figure 2–16b. Again, by changing the units or the starting point on the y axis, one can change the visual message.

Figure 2–16

Percent of World's Motor Vehicles Produced by Manufacturers in the United States

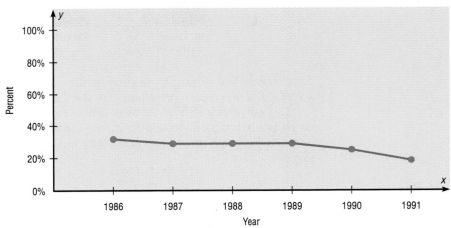

(a) Using a scale from 0% to 100%

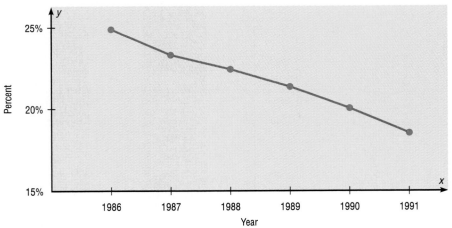

(b) Using a scale from 15% to 25%

Figure 2–17

Comparison of Costs for a 30-Second Super Bowl Commercial

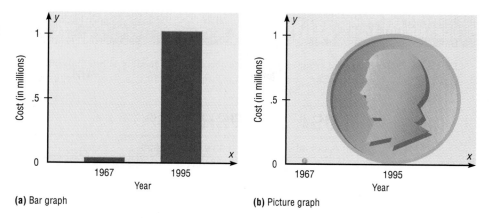

(a) Bar graph

(b) Picture graph

Another misleading graphing technique sometimes used is exaggerating a one-dimensional increase by showing it in two dimensions. For example, the average cost of a 30-second Super Bowl commercial has increased from $40,000 in 1967 to $1 million in 1995. (Source: *USA Today* Snapshot by Cliff Vancura, January 30, 1995; based on information from Nielsen Media Research.)

The increase shown by the bar graph in Figure 2–17a represents the change by a comparison of the heights of the two bars in one dimension. The same data are shown two-dimensionally with circles in Figure 2–17b. Notice that the difference seems much larger because the eye is comparing the areas of the circles rather than the lengths of the diameters.

Note that it is not wrong to use the graphing techniques of truncating the scales or representing data by two-dimensional pictures. But when these techniques are used, the reader should be cautious of the conclusion drawn on the basis of the graphs.

A summary of the types of graphs and their uses is shown in Figure 2–18.

Figure 2–18

Summary of Graphs and Uses of Each

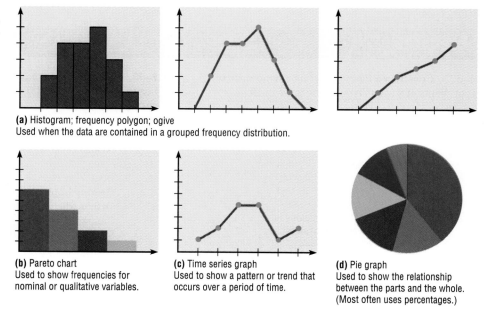

(a) Histogram; frequency polygon; ogive
Used when the data are contained in a grouped frequency distribution.

(b) Pareto chart
Used to show frequencies for nominal or qualitative variables.

(c) Time series graph
Used to show a pattern or trend that occurs over a period of time.

(d) Pie graph
Used to show the relationship between the parts and the whole. (Most often uses percentages.)

This *USA Today* Snapshot compares the costs of driving in several big cities. Can you see anything misleading about the graph? (Note the lengths of the bars for the first two cities in relationship to the numbers.)

Source: *USA Today,* May 3, 1995. Copyright 1995 *USA TODAY.* Used with permission.

Exercises

2–39. Construct a Pareto chart for the number of transplants of various types performed in 1994.

Type	Number
Kidney	11,390
Liver	3653
Pancreas	844
Heart	2340
Lung	737

Source: "Matchmaking at the Heart of Transplant Process," *USA Today.* October 19, 1995.

2–40. Construct a Pareto chart for the number of health conditions per 100 reported by the elderly in a survey.

Condition	Number
Arthritis	48
Hypertension	36
Heart disease	32
Hearing impairments	32
Orthopedic impairments	19
Cataracts	17
Sinusitis	16
Diabetes	11
Visual impairments	9

Source: "The Senior Profile," *USA Today,* July 7, 1995.

2–41. Construct a Pareto chart for the number of registered taxicabs in the selected cities.

City	Number
New York	11,787
Washington, D.C.	8348
Chicago	5300
Philadelphia	1480
Baltimore	1151

Source: "Yellow Cab Service at a Glance," Pittsburgh *Tribune-Review*. Used with permission.

2–42. Construct a Pareto chart for the number of unemployed people in the selected states for June of 1995.

State	Number
Texas	605,000
New York	494,000
Pennsylvania	364,000
Florida	363,000
Ohio	269,000

Source: "Employment in the USA," *USA Today*, July 10, 1995.

2–43. Construct a Pareto chart for the average number of hurricanes reported for the selected months from 1900 to 1995.

Month	Number
June	12
July	16
August	40
September	61
October	23
November	6

Source: *USA Today* Snapshots, September 6, 1995.

2–44. Draw a time series graph to represent the data for the number of worldwide airline fatalities for the given years.

Year	1988	1989	1990	1991	1992	1993	1994
No. of fatalities	699	817	440	510	990	801	732

Source: Reprinted with permission from *The World Almanac and Book of Facts*, 1996. Copyright © 1995 PRIMEDIA Reference Inc. All rights reserved.

2–45. The data here represent the personal consumption expenditures for transportation for the United States (in billions of dollars). Draw a time series graph to represent the data.

Year	1982	1984	1986	1988	1990	1992	1994
Amount	$263	$328	$370	$413	$454	$463	$538

Source: Reprinted with permission from *The World Almanac and Book of Facts*, 1996. Copyright © 1995 PRIMEDIA Reference Inc. All rights reserved.

2–46. (W) The number of operable nuclear power reactors in the United States for the given year is shown below. Draw a time series graph to represent the data and summarize the results.

Year	1984	1986	1988	1990	1992	1994
Number operable	86	100	108	111	109	109

Source: Reprinted with permission from *The World Almanac and Book of Facts*, 1996. Copyright © 1995 PRIMEDIA Reference Inc. All rights reserved.

2–47. (W) A bacteriologist chartered the growth of a certain bacterium over a period of 8 hours. The data are shown here. Construct a time series graph to represent the data and summarize the results.

Hour	1	2	3	4	5	6	7	8
No. of cells	2	5	8	13	17	22	30	38

2–48. (W) In a survey of 100 male smokers concerning the use of tobacco, the following data were obtained. Construct a pie graph for the data and summarize the results.

Type of smoker	Number of users
Cigarettes	63
Cigars	22
Pipe	10
Chewing tobacco	5

2–49. (W) A survey of 500 Philadelphia families were asked the question "Where are you planning to vacation this summer?" It resulted in the following distribution. Construct a pie graph for the data and summarize the results.

Area	Number vacationing
Great Lakes region	37
New England	104
East Coast	206
South	96
West Coast	57

2–50.   The frequency distribution here shows the number of freshmen, sophomores, juniors, and seniors who have part-time jobs after school. Construct a pie graph for the data.

Rank	Frequency
Freshmen	12
Sophomores	25
Juniors	36
Seniors	17

2–51.   In an insurance company study of the causes of 1000 deaths, the following data were obtained. Construct a pie graph to represent the data.

Cause of death	Number of deaths
1. Heart disease	432
2. Cancer	227
3. Stroke	93
4. Accidents	24
5. Other	224
	1000

2–52.   State which graph (Pareto chart, time series graph, or pie graph) would most appropriately represent the given situation.
a.   The number of students enrolled at a local college for each year during the last five years.
b.   The budget for the student activities department at a certain college for each year during the last five years.
c.   The means of transportation the students use to get to school.
d.   The percentage of votes each of the four candidates received in the last election.
e.   The record temperatures of a city for the last 30 years.
f.   The frequency of each type of crime committed in a city during the year.

*2–53.   (**W**) The number of successful space launches by the United States and Japan for the years 1990–94 is shown here. Construct a compound time series graph for the data. What comparison can be made regarding the launches?

Year	1990	1991	1992	1993	1994
U.S.	31	30	27	29	27
Japan	7	2	3	1	4

*2–54.   (**W**) Meat production for veal and lamb for the years 1950–90 is shown here. (Data are in millions of pounds.) Construct a compound time series graph for the data. What comparison can be made regarding meat production?

Year	1950	1960	1970	1980	1990
Veal	1230	1109	588	400	327
Lamb	576	769	551	318	358

Source: Cartoon by Bradford Veley, Marquette, Michigan. Used with permission.

2–5

Summary

When data are collected, they are called raw data. Since very little knowledge can be obtained from raw data, they must be organized in some meaningful way. A frequency distribution using classes is the solution. Once a frequency distribution is constructed, the representation of the data by graphs is a simple task. The most commonly used graphs in research statistics are the histogram, frequency polygon, and ogive. Other graphs, such as the Pareto chart, time series graph, and pie graph, can also be used. Some of these graphs are seen frequently in newspapers, magazines, and various statistical reports.

Speaking of   STATISTICS

This *USA Today* Snapshot compares the budget deficit for fiscal 1995 with that for fiscal 1996. Can you see anything misleading about this graph?

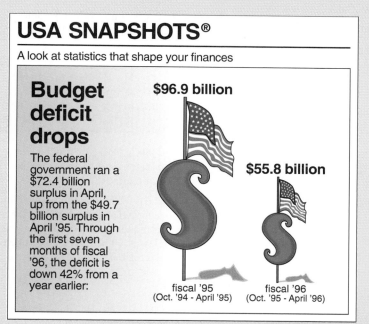

USA SNAPSHOTS®

A look at statistics that shape your finances

Budget deficit drops

The federal government ran a $72.4 billion surplus in April, up from the $49.7 billion surplus in April '95. Through the first seven months of fiscal '96, the deficit is down 42% from a year earlier:

$96.9 billion

$55.8 billion

fiscal '95
(Oct. '94 - April '95)

fiscal '96
(Oct. '95 - April '96)

Source: *USA Today*, May 29, 1996. Copyright 1996 *USA TODAY*. Used with permission.

Important Terms

Categorical frequency distribution 33

Class boundaries 29

Class midpoint 29

Class width 29

Cumulative frequency 40

Frequency 25

Frequency distribution 25

Frequency polygon 38

Grouped frequency distribution 33

Histogram 37

Lower class limit 29

Ogive 40

Open-ended distribution 30

Pareto chart 47

Pie graph 51

Range 27

Raw data 25

Relative frequency graph 42

Time series graph 49

Ungrouped frequency distribution 33

Upper class limit 29

Important Formulas

Formula for the percentage of values in each class:

$$\% = \frac{f}{n} \cdot 100\%$$

where

f = **frequency of the class**
n = **total number of values**

Formula for the range:

$$R = \text{highest value} - \text{lowest value}$$

Formula for the class width:

$$\text{class width} = \text{upper boundary} - \text{lower boundary}$$

Formula for the class midpoint:

$$X_m = \frac{\text{lower boundary} + \text{upper boundary}}{2}$$

or

$$X_m = \frac{\text{lower limit} + \text{upper limit}}{2}$$

Formula for the degrees for each section of a pie graph:

$$\text{degrees} = \frac{f}{n} \cdot 360°$$

Review Exercises

2–55.   A questionnaire about how people get news resulted in the following information from 25 respondents. Construct a frequency distribution for the data (N = newspaper, T = television, R = radio, M = magazine).

N	N	R	T	T
R	N	T	M	R
M	M	N	R	M
T	R	M	N	M
T	R	R	N	N

2–56.   (W) Construct a pie graph for the data in Exercise 2–55 and summarize the results.

2–57.   A sporting-goods store kept a record of sales of five items for one randomly selected hour during a recent sale. Construct a frequency distribution for the data (B = baseballs, G = golf balls, T = tennis balls, S = soccer balls, F = footballs).

F	B	B	B	G	T	F
G	G	F	S	G	T	
F	T	T	T	S	T	
F	S	S	G	S	B	

2–58.   (W) Draw a pie graph for the data in Exercise 2–57 showing the sales of each item and summarize the results.

2–59.   The blood urea nitrogen count (BUN) of 20 randomly selected patients is given here in mg/dl. Construct an ungrouped frequency distribution for the data.

17	18	13	14
12	17	11	20
13	18	19	17
14	16	17	12
16	15	19	22

2–60.   (W) Construct a histogram, frequency polygon, and ogive for the data in Exercise 2–59 and summarize the results.

2–61.   The temperature of Keystone Lake was recorded at noon each day for a month. Construct a frequency distribution for the data. Use five classes. (The data for this exercise will be used for Exercise 2–65.)

52	47	50	47	44
54	46	46	45	46
53	41	50	44	44
49	47	43	48	51
49	40	50	51	43
51	46	44	48	48

2–62.   (W) Construct a histogram, frequency polygon, and ogive for the data in Exercise 2–61 and summarize the results.

2–63.   During July, a private pool recorded the weights of 30 adult males in its water aerobics class. Construct a frequency distribution for the data. Use six classes. (The data for this exercise will be used for Exercise 2–66.)

143	156	156	163	167
142	171	170	169	164
138	158	160	162	164
173	157	158	159	160
138	172	166	166	159
120	125	165	136	168

2–64.   (W) Construct a histogram, frequency polygon, and ogive for the data in Exercise 2–63 and summarize the results.

2–65.   Construct a histogram, frequency polygon, and ogive by using relative frequencies for the data in Exercise 2–61.

2–66. Construct a histogram, frequency polygon, and ogive by using relative frequencies for the data in Exercise 2–63.

2–67. Construct a Pareto chart for the number of homicides reported in 1995 for the following cities.

City	Number
New Orleans	363
Washington, D.C.	352
Chicago	824
Baltimore	323
Atlanta	184

Source: "Homicide, Rape, Robbery: The Numbers Are Fewer," *USA Today,* May 6, 1996.

2–68. Construct a Pareto chart for the number of trial-ready civil action and equity cases decided in less than six months for the selected counties in southwestern Pennsylvania.

County	Number
Westmoreland	427
Washington	298
Green	151
Fayette	106
Somerset	87

Source: "Quick Cases," Pittsburgh *Tribune-Review,* May 14, 1995, p. B1. Used with permission.

2–69. (**W**) The given data represent the federal minimum hourly wage in the years shown. Draw a time series graph to represent the data and summarize the results.

Year	Wage
1960	$1.00
1965	$1.25
1970	$1.60
1975	$2.10
1980	$3.10
1985	$3.35
1990	$3.80
1995	$4.25

Source: Reprinted with permission from *The World Almanac and Book of Facts,* 1996. Copyright © 1995 PRIMEDIA Reference Inc. All rights reserved.

2–70. (**W**) The number of bank failures in the United States during the years 1983–91 is shown below. Draw a time series graph to represent the data and summarize the results.

Year	Number of failures
1983	48
1984	80
1985	120
1986	145
1987	203
1988	221
1989	207
1990	169
1991	127
1992	122
1993	41
1994	13

Source: Reprinted with permission from *The World Almanac and Book of Facts,* 1996. Copyright © 1995 PRIMEDIA Reference Inc. All rights reserved.

2–71. The following data represent the number of strikes (work stoppages) involving 1000 workers or more in the United States for the years shown. Draw a time series graph to represent the data.

Year	Number of strikes
1960	222
1965	268
1970	381
1975	235
1980	187
1985	54
1990	44

Source: Reprinted with permission from *The World Almanac and Book of Facts,* 1993. Copyright © 1992 PRIMEDIA Reference Inc. All rights reserved.

2–72. (**W**) In a study of 100 women, the numbers shown here indicate the major reason why each woman surveyed worked outside the home. Construct a pie graph for the data and summarize the results.

Reason	Number
To support self/family	62
For extra money	18
For something different to do	12
Other	8

2–73. (**W**) A survey of the students in the school of education of a large university obtained the following data for students enrolled in specific fields. Construct a pie graph for the data and summarize the results.

Major field	Number
Preschool	893
Elementary	605
Middle	245
Secondary	1096

Statistics Today

Are We Flying More?   Revisited

The type of graph that can be used to show the increase in the number of passengers on U.S. planes is the time series graph. An example is shown here. The ascending nature of the line over the years shows a major increase in the number of passengers using air travel in the United States. Comparing the height of the beginning point (1980) with the height of the ending point (1994) shows that the number of passengers almost doubled.

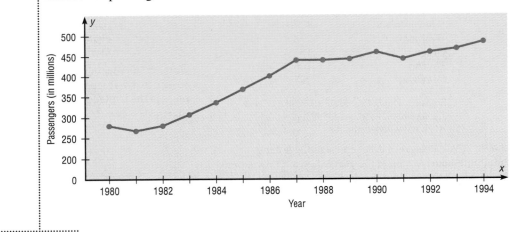

Data Analysis

1. From the Data Bank located in Appendix D, choose one of the following variables: age, weight, cholesterol level, systolic pressure, IQ, or sodium level. Select at least 30 values. For these values, construct a grouped frequency distribution. Draw a histogram, frequency polygon, and ogive for the distribution. Describe briefly the shape of the distribution.

2. From the Data Bank, choose one of the following variables: educational level, smoking status, or exercise. Select at least 20 values. Construct an ungrouped frequency distribution for the data. For the distribution, draw a Pareto chart and describe briefly the nature of the chart.

3. From the Data Bank, select at least 30 people and construct a categorical distribution for their marital status. Draw a pie chart and describe briefly the findings.

Quiz

Determine whether each statement is true or false. If the statement is false, explain why.

1. In the construction of a frequency distribution, it is a good idea to have overlapping class limits, such as 10–20, 20–30, 30–40.

2. Histograms can be drawn by using vertical or horizontal bars.

3. It is not important to keep the width of each class the same in a frequency distribution.

4. Frequency distributions can aid the researcher in drawing charts and graphs.

5. The type of graph used to represent data is determined by the type of data collected and by the researcher's purpose.

6. In construction of a frequency polygon, the class limits are used for the x axis.

7. Data collected over a period of time can be graphed by using a pie graph.

Select the best answer.

8. What is another name for the ogive?
a. histogram
b. frequency polygon
c. cumulative frequency graph
d. Pareto chart

9. What are the boundaries for 8.6–8.8?
a. 8–9
b. 8.5–8.9
c. 8.55–8.85
d. 8.65–9.75

10. What graph should be used to show the relationship between the parts and the whole?
a. histogram
b. pie graph
c. Pareto chart
d. ogive

11. Except for rounding errors, relative frequencies should add up to what sum?
a. 0
b. 1
c. 50
d. 100

Complete the following statements with the best answers.

12. The three types of frequency distributions are _____ , _____ , and _____ .

13. In a frequency distribution, the number of classes should be between _____ and _____ .

14. Data such as blood types (A, B, AB, O) can be organized into a _____ frequency distribution.

15. Data collected over a period of time can be graphed using a _____ graph.

16. When data are first collected, they are called _____ data.

17. On a Pareto chart, the frequencies should be represented on the _____ axis.

18. A questionnaire on housing arrangements showed the following information obtained from 25 respondents. Construct a frequency distribution for the data (H = house, A = apartment, M = mobile home, C = condominium).

H	C	H	M	H	A	C	A	M
C	M	C	A	M	A	C	C	M
C	C	H	A	H	H	M		

19. Construct a pie graph for the data in the previous problem.

20. When 30 randomly selected customers left a convenience store, each was asked the number of items he or she purchased. Construct an ungrouped frequency distribution for the data.

2	9	4	3	6
6	2	8	6	5
7	5	3	8	6
6	2	3	2	4
6	9	9	8	9
4	2	1	7	4

21. Construct a histogram, a frequency polygon, and an ogive for the data in the previous problem.

22. During June, a local theater recorded the following number of patrons per day. Construct a frequency distribution for the data. Use six classes.

102	116	113	132	128	117
156	182	183	171	168	179
170	160	163	187	185	158
163	167	168	186	117	108
171	173	161	163	168	182

23. Construct a histogram, frequency polygon, and ogive for the data in the previous problem.

24. Construct a Pareto chart for the number of taxi licenses issued in 1995 for the cities shown.

City	Number
New York	11,787
Chicago	5,500
Houston	2,049
Los Angeles	1,850
Philadelphia	1,500

Source: "To Be NYC Cabbie, Fare's Not Cheap," *USA Today,* May 20, 1996.

25. These data show the expenses of Chemistry Lab, Inc., for research and development for the years indicated. Each number represents millions of dollars. Draw a time series graph of the data.

Year	Amount
1992	$ 8937
1993	9388
1994	11,271
1995	13,877
1996	19,203

Critical Thinking Challenges

1. Calling Long Distance Using the information given in the table below, draw an appropriate graph or graphs comparing the phone rates. Then write a summary of the data.

2. Inmates and Lawsuits The figure on the next page shows a comparison of the lawsuits filed by inmates in U.S. District Court, Western District of Pennsylvania, and the total number of lawsuits for the years 1982 to 1993.

NEW YORK CALLING L.A.

How typical prepaid phone cards compare with a calling card and a coin phone for three calls.

Length of call	AT&T prepaid card	MCI, Sprint prepaid cards	7-Eleven $20 Phone Card	Regular AT&T Calling Card		Coin Phone (AT&T)	
				Business hrs.	Night/weekend	Business hrs.	Night/weekend
1 min.	$0.45	$0.60	$0.33	$1.08	$0.98	$2.85	$2.45
3 min.	1.35	1.80	1.00	1.64	1.34	2.85	2.45
10 min.	4.50	6.00	3.30	3.60	2.60	5.00	3.85

Source: "New Telephone Calling Cards Let You Pay Before You Dial." Copyright 1995 by Consumers Union of U.S., Inc., Yonkers, NY 10703-1057. Reprinted by permission from CONSUMER REPORTS, January 1995.

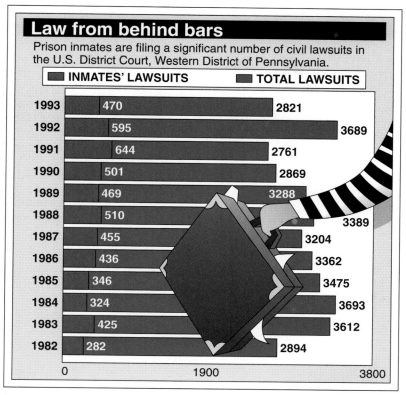

Law from behind bars

Prison inmates are filing a significant number of civil lawsuits in the U.S. District Court, Western District of Pennsylvania.

	INMATES' LAWSUITS	TOTAL LAWSUITS
1993	470	2821
1992	595	3689
1991	644	2761
1990	501	2869
1989	469	3288
1988	510	3389
1987	455	3204
1986	436	3362
1985	346	3475
1984	324	3693
1983	425	3612
1982	282	2894

0   1900   3800

Source: *Tribune-Review,* (Greensburg, PA), December 11, 1994. Used with permission.

a. For *each* year, find the proportion of lawsuits filed by the inmates.

b. In which year was the proportion highest?

c. In which year was the proportion lowest?

d. For the years 1982 through 1993, find the proportion of suits filed by the inmates. (Use the *total* number of lawsuits filed by the inmates divided by the *total* lawsuits filed.)

e. How many years were above the proportion found in Step *d*? How many were below the proportion?

f. Using the proportions instead of the numbers, draw a time series graph for the data.

g. Which graph do you think better represents the data? Why?

3. Great Lakes Statistics Shown below are various statistics about the Great Lakes. Using appropriate graphs (your choice) and summary statements, write a report analyzing the data.

	Superior	Michigan	Huron	Erie	Ontario
Length (miles)	350	307	206	241	193
Breadth (miles)	160	118	183	57	53
Depth (feet)	1330	923	750	210	802
Volume (cubic miles)	2900	1180	850	116	393
Area (square miles)	31,700	22,300	23,000	9910	7550
Shoreline (U.S., miles)	863	1,400	580	431	300

Source: Reprinted with permission from *The World Almanac and Book of Facts,* 1995. Copyright © 1994 PRIMEDIA Reference Inc. All rights reserved.

Data Projects

Where appropriate, use MINITAB, the TI-83, or a computer program of your choice to complete the following exercises.

1. Select a categorical (nominal) variable, such as the colors of cars in the school's parking lot or the major fields of the students in statistics class, and collect data on this variable.
a. State the purpose of the project.
b. Define the population.
c. State how the sample was selected.
d. Show the raw data.
e. Construct a frequency distribution for the variable.
f. Draw an appropriate graph(s) (pie, Pareto, etc.) for the data.
g. Analyze the results.

2. Using an almanac, select a variable that varies over a period of several years (e.g., silver production)

and draw a time series graph for the data. Write a short paragraph interpreting the findings.

3. Select a variable (interval or ratio) and collect at least 30 values. For example, you may ask the students in your class how many hours they study per week, how old they are.
a. State the purpose of the project.
b. Define the population.
c. State how the sample was selected.
d. Show the raw data.
e. Construct a frequency distribution for the data.
f. Draw a histogram, frequency polygon, and ogive for the data.
g. Analyze the results.

TI-83 Calculator

The TI-83 graphing calculator can be used for a variety of statistical graphs and tests. The information is presented at the end of the appropriate chapters and aids the student in graphing and calculations.

General Information

A. To turn calculator on:
Press **ON** key.

B. To turn calculator off:
Press **2nd [OFF]**.

C. To reset defaults only:
1. Press **2nd** then **[MEM]**.
2. Select **5** then **2** then **2**.

(Optional). To reset settings on calculator and clear memory: (*Note:* This will clear all settings and programs in the calculator's memory.)
Press **2nd**, then **[MEM]**. Then press **5**, then **1** then **2**.
(*Note:* The contrast may need to be adjusted after this.)

D. To adjust contrast (if necessary):
Press **2nd**. Then press and hold ▲ to darken or ▼ to lighten contrast.

E. To clear screen:
Press **CLEAR**.
(*Note:* This will return you to the screen you were using.)

F. To display a menu:
Press appropriate menu key. Example: **STAT**.

G. To return to home screen:
Press **2nd**, then **[QUIT]**.

H. To move around on the screens:
Use the arrow keys.

I. To select items on the menu:
Press the corresponding number or move the cursor to the item using the arrow keys. Then press **ENTER**.
Note: In some cases, you do not have to press **ENTER**, and in other cases you may need to press **ENTER** twice.

Entering Data

A. To enter single variable data (if necessary, clear the old list):
1. Press **STAT** to display the Edit menu.
2. Press **ENTER** to select 1:Edit.

3. Enter the data in L_1 and press **ENTER** after each value.

4. After all data values are entered, press **STAT** to get back to the `Edit` menu or **2nd [QUIT]** to end.

Example 1   Enter the following data values in L_1: 213, 208, 203, 215, 222

Output

```
L1      L2      L3       1
213     ------  ------
208
203
215
222
------
L1(6)=
```

B.   To enter multiple variable data:

The TI-83 will take up to six lists designated L_1, L_2, L_3, L_4, L_5, and L_6.

1. To enter more than one set of data values, complete the steps in part A. Then move the cursor to L_2 by pressing the ▶ key

2. Repeat the steps in part A.

Editing Data

A.   To correct a data value before pressing **ENTER**, use ◀ and retype the value and press **ENTER**.

B.   To correct a data value in a list after pressing **ENTER**, move cursor to incorrect value in list and type in the correct value. Then press **ENTER**.

C.   To delete a data value in a list:

Move cursor to value and press **DEL**.

D.   To insert a data value in a list:

1. Move cursor to position where data value is to be inserted, then press **2nd [INS]**.

2. Type data value; then press **ENTER**.

E.   To clear a list:

1. Press **STAT** then **4**.

2. Enter list to be cleared. Example: to clear L_1, press **2nd [L_1]**. Then press **ENTER**.

Note: To clear several lists, follow step 1, but enter each list to be cleared, separating them with a comma.

Example 2   **2nd [L_1], 2nd [L_2] ENTER.**

Sorting Data

To sort the data in a list:

1. Enter the data in L_1.

2. Press **STAT 2** to get `Sort A` to sort list in ascending order.

3. Then press **2nd [L_1]) ENTER**.

The calculator will display `Done`.

4. Press **STAT ENTER** to display sorted list.

(*Note:* The `Sort D` or **3** sorts the list in descending order.)

Example 3   Sort in ascending order the data values entered in Example 1.

Output

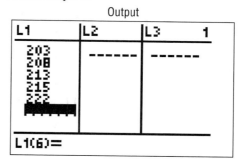

```
L1      L2      L3       1
203     ------  ------
208
213
215
222
L1(6)=
```

Histogram

In order to display the graphs on the screen, enter the appropriate values in the calculator using the Window menu. The default values are $X_{min} = -10$, $X_{max} = +10$, $Y_{min} = -10$, and $Y_{max} = +10$.

The X_{scl} changes the distance between the tick marks on the x-axis and can be used to change the class width for the histogram.

A.   To change the values in the `Window`:

1. Press **WINDOW**.

2. Move the cursor to the value that needs to be changed. Then type in the desired value and press **ENTER**.

3. Continue until all values are appropriate.

4. Press **2nd [QUIT]** to leave the `Window` menu.

B.   To plot the histogram:

1. Enter the data in L_1.

2. Make sure Window values are appropriate for the histogram.

3. Press **2nd [STAT PLOT] ENTER**.

4. Press **ENTER** to turn the plot on, if necessary.

5. Move cursor to the `Histogram` symbol and press **ENTER**, if necessary.

6. Move cursor to L_1 on the `Xlist`. Press **ENTER**, if necessary.

7. Move cursor to 1 on `Freq`. Press **ENTER**, if necessary.

8. Press **GRAPH** to display the histogram.

9. To obtain various coordinates, press the **TRACE** key, followed by ◀ or ▶ keys.

Example 4   Plot a histogram for the following data.

68	80	69	81	72	100	101	73	102	93	91
92	93	88	82	83	75	75	89	96	103	83
84	85	89								

Set the Window values as follows.

$X_{min} = 60$
$X_{max} = 105$
$Y_{min} = 0$
$Y_{max} = 5$

Input

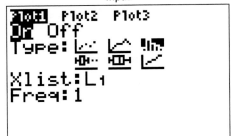

Input

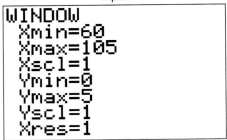

Output

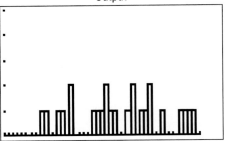

chapter

3

Data Description

Objectives

After completing this chapter, you should be able to

1. Summarize data using the measures of central tendency, such as the mean, median, mode, and midrange.

2. Describe data using the measures of variation, such as the range, variance, and standard deviation.

3. Identify the position of a data value in a data set using various measures of position, such as percentiles, deciles, and quartiles.

4. Use the techniques of exploratory data analysis, including stem and leaf plots, box plots, and five-number summaries to discover various aspects of data.

Who Is a Typical First-Time Home Buyer?

Real estate agents, bankers, and insurance company executives can use statistics to obtain a profile of the typical first-time home buyer. With this knowledge, they can tailor their advertising to target certain groups and provide their best services to their customers. This *USA Today* Snapshot shows some statistics on first-time home buyers.

   This chapter will show you how to obtain these descriptive statistics and explain how each can be used to create a profile of a typical first-time home buyer.

Source: *USA Today*, February 22, 1995. Copyright 1995, USA TODAY. Reprinted with permission.

3–1

Introduction

The previous chapter showed how one can gain useful information from raw data by organizing it into a frequency distribution, then presenting the data by using various graphs. This chapter shows the statistical methods that can be used to summarize data. The most familiar of these methods is finding averages.

For example, one may read that the average speed of a car crossing midtown Manhattan during the day is 5.3 miles per hour or that the average number of minutes an American father of a four-year-old spends alone with his child each day is 42.[1]

In the book *American Averages* by Mike Feinsilber and William B. Meed, the authors state:

"Average" when you stop to think of it is a funny concept. Although it describes all of us it describes none of us . . . While none of us wants to be the average American, we all want to know about him or her.

The authors go on to give examples of averages:

The average American man is five feet, nine inches tall; the average woman is five feet, 3.6 inches.

The average American is sick in bed seven days a year missing five days of work.

On the average day, 24 million people receive animal bites.

By his or her 70th birthday, the average American will have eaten 14 steers, 1050 chickens, 3.5 lambs, and 25.2 hogs.[2]

In these examples, the word *average* is ambiguous, since several different methods can be used to obtain an average. Loosely stated, the average means the center of the distribution or the most typical case. Measures of average are also called *measures of central tendency* and include the *mean, median, mode,* and *midrange.*

Knowing the average of a data set is not enough to describe the data set entirely. Even though a shoe-store owner knows that the average size of a man's shoe is size 10, she would not be in business very long if she ordered only size 10 shoes.

[1]"Harper's Index," *Harper's* magazine, June 1995, p. 11. All rights reserved.

[2]Mike Feinsilber and William B. Meed, *American Averages* (New York: Bantam Doubleday Dell, 1980). Used with permission.

As this example shows, in addition to knowing the average, one must know how the data values are dispersed. That is, do the data values cluster around the mean, or are they spread more evenly throughout the distribution? The measures that determine the spread of the data values are called *measures of variation* or *measures of dispersion*. These measures include the *range, variance,* and *standard deviation.*

Finally, another set of measures is necessary to describe data. These measures are called *measures of position.* They tell where a specific data value falls within the data set or its relative position in comparison with other data values. The most common position measures are *percentiles, deciles,* and *quartiles.* These measures are used extensively in psychology and education. Sometimes they are referred to as *norms.*

The measures of central tendency, variation, and position explained in this chapter are part of what is called *traditional statistics.*

The last section of this chapter shows the techniques of what is called *exploratory data analysis.* These techniques include the *stem and leaf plot,* the *box plot,* and the *five-number summary.* They can be used to explore data to see what they show (as opposed to the traditional techniques, which are used to confirm conjectures about the data).

3–2

Measures of Central Tendency

Objective 1. Summarize data using measures of central tendency.

Chapter 1 stated that statisticians use samples taken from populations; however, when populations are small, it is not necessary to use samples since the entire population can be used to gain information. For example, suppose an insurance manager wanted to know the average weekly sales of all the company's representatives. If the company employed a large number of salespeople, say nationwide, he would have to use a sample and make an inference to the entire sales force. But if the company had only a few salespeople, say only 87 agents, he would be able to use all representatives' sales for a randomly chosen week and thus use the entire population.

Measures taken by using all the data values in the populations are called *parameters.* Measures obtained by using the data values of samples are called *statistics.* Hence, the average of the sales from a sample of representatives is called a *statistic,* and the average of sales obtained from the entire population is called a *parameter.*

A **statistic** is a characteristic or measure obtained by using the data values from a sample.

A **parameter** is a characteristic or measure obtained by using all the data values for a specific population.

These concepts as well as the symbols used to represent them will be explained in detail in this chapter.

Rounding Rule

In statistics the basic rounding rule is that when computations are done in the calculation, rounding should not be done until the final answer is calculated. When rounding is done in the intermediate steps, it tends to increase the difference between that answer and the exact one. But in the textbook and solutions manual, it is not practical to show long decimals in the intermediate calculations;

hence, the values in the examples are carried out enough places (usually three or four) to obtain the same answer a calculator would give after rounding on the last step.

The Mean

Historical Note

In 1796, Adolphe Quetelet investigated the character-istics (heights, weights, etc.) of French conscripts to determine the "average man." Florence Nightin-gale was so influenced by Quetelet's work that she began collecting and analyzing medical records in the military hospitals during the Crimean War. Based on her work, hospi-tals began keeping accurate records on their patients.

The *mean,* also known as the arithmetic average, is found by adding the values of the data and dividing by the total number of values. For example, the mean of 3, 2, 6, 5, and 4 is found by adding $3 + 2 + 6 + 5 + 4 = 20$ and dividing by 5; hence, the mean of the data is $20 \div 5 = 4$. The values of the data are represented by X's. In this data set, $X_1 = 3$, $X_2 = 2$, $X_3 = 6$, $X_4 = 5$, and $X_5 = 4$. To show a sum of the total X values, the symbol Σ (the capital letter sigma) is used, and ΣX means to find the sum of the X values in the data set. The summation notation is explained in Appendix A.

The **mean** is the sum of the values divided by the total number of values. The symbol $\overline{X}$ represents the sample mean.

$$\overline{X} = \frac{X_1 + X_2 + X_3 + \cdots + X_n}{n} = \frac{\Sigma X}{n}$$

where n represents the total number of values in the sample.

For a population, the Greek letter μ (mu) is used for the mean.

$$\mu = \frac{X_1 + X_2 + X_3 + \cdots + X_N}{N} = \frac{\Sigma X}{N}$$

where N represents the total number of values in the population.

In statistics, Greek letters are used to denote parameters and Roman letters are used to denote statistics. Assume that the data are obtained from samples unless otherwise specified.

Example 3–1

The ages in weeks of six kittens at an animal shelter are 3, 8, 5, 12, 14, and 12. Find the mean.

Solution

$$\overline{X} = \frac{\Sigma X}{n} = \frac{3 + 8 + 5 + 12 + 14 + 12}{6} = \frac{54}{6} = 9 \text{ weeks}$$

Example 3–2

The fat contents in grams for one serving of 11 brands of packaged foods, as determined by the U.S. Department of Agriculture, are given as follows. Find the mean.

6.5, 6.5, 9.5, 8.0, 14.0, 8.5, 3.0, 7.5, 16.5, 7.0, 8.0

Source: *Consumer Reports,* June 1995, p. 391.

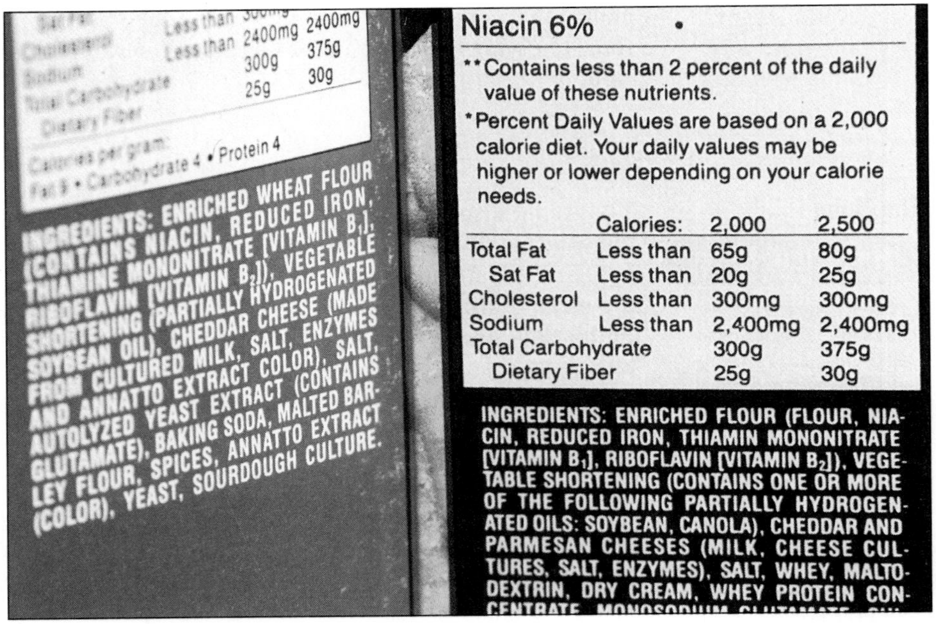

Solution

$$\overline{X} = \frac{\sum X}{n} = \frac{6.5 + 6.5 + 9.5 + 8.0 + 14.0 + 8.5 + 3.0 + 7.5 + 16.5 + 7.0 + 8.0}{11}$$

$$= \frac{95}{11} = 8.64 \text{ grams}$$

Hence, the mean fat content is 8.64 grams.

The mean, in most cases, is not an actual data value.

Rounding Rule for the Mean The mean should be rounded to one more decimal place than occurs in the raw data. For example, if the raw data are given in whole numbers, the mean should be rounded to the nearest tenth. If the data are given in tenths, the mean should be rounded to the nearest hundredth, and so on.

For data in an ungrouped frequency distribution, the mean can be found as shown in the next example.

Example 3–3

The scores for 25 students on a 5-point quiz are shown below. Find the mean.

Score	Frequency
0	1
1	2
2	6
3	12
4	3
5	1
	$n = 25$

Statistics are used quite often in weather forecasting. Shown in this Snapshot are the five windiest cities in the United States. Using an almanac, find the average wind speed in a city near you and compare it to those of the top five cities.

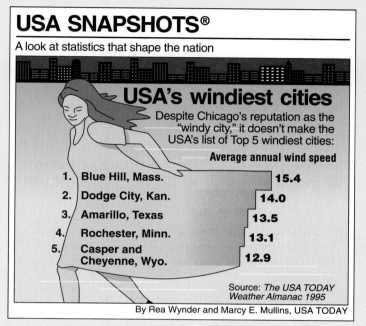

USA SNAPSHOTS®
A look at statistics that shape the nation

USA's windiest cities
Despite Chicago's reputation as the "windy city," it doesn't make the USA's list of Top 5 windiest cities:

Average annual wind speed

1. Blue Hill, Mass. — 15.4
2. Dodge City, Kan. — 14.0
3. Amarillo, Texas — 13.5
4. Rochester, Minn. — 13.1
5. Casper and Cheyenne, Wyo. — 12.9

Source: *The USA TODAY Weather Almanac 1995*
By Rea Wynder and Marcy E. Mullins, USA TODAY

Source: *USA Today*, July 26, 1995. Copyright 1995, USA TODAY. Reprinted with permission.

Solution

STEP 1   Multiply the score by the frequency for each class, and place the result in column C as shown here.

A Score (X)	B Frequency (f)	C $f \cdot X$
0	1	$0 \cdot 1 = 0$
1	2	$1 \cdot 2 = 2$
2	6	$2 \cdot 6 = 12$
3	12	$3 \cdot 12 = 36$
4	3	$4 \cdot 3 = 12$
5	1	$5 \cdot 1 = 5$
	$n = 25$	$\sum f \cdot X = 67$

Note: The notation $\sum f \cdot X$ means to find the sum of the products of f and X. Technically, it should be written $\sum (f \cdot X)$, but to keep the formulas as uncluttered as possible, it is written as $\sum f \cdot X$.

In this case, X represents the class value rather than the individual data value.

STEP 2   Find the sum of column C, as shown above.

STEP 3   Divide this sum of the products by n, as shown:

$$\overline{X} = \frac{67}{25} = 2.7$$

Hence, the mean of the scores is 2.7. The formula is

$$\overline{X} = \frac{\Sigma f \cdot X}{n}$$

Note: The denominator in this formula is obtained by summing the frequencies ($\Sigma f = n$). It is not the number of classes.

··

The procedure for finding the mean for grouped data is similar to that for ungrouped data, except that the midpoints of the classes are used for the X values.

··

| Example 3–4 | Using the frequency distribution for Example 2–7 in Chapter 2, find the mean. The data represent the number of miles run during one week for a sample of 20 runners. |

Solution

The procedure for finding the mean for grouped data is given here.

STEP 1   Make a table as shown.

A Class	B Frequency (f)	C Midpoint (X_m)	D $f \cdot X_m$
5.5–10.5	1		
10.5–15.5	2		
15.5–20.5	3		
20.5–25.5	5		
25.5–30.5	4		
30.5–35.5	3		
35.5–40.5	2		
	$n = 20$		

STEP 2   Find the midpoints of each class and enter them in column C.

$$X_m = \frac{5.5 + 10.5}{2} = 8, \qquad \frac{10.5 + 15.5}{2} = 13, \qquad \text{etc.}$$

STEP 3   For each class, multiply the frequency by the midpoint, as shown below, and place the product in column D.

$$1 \cdot 8 = 8, \qquad 2 \cdot 13 = 26, \qquad \text{etc.}$$

The completed table is shown here.

A Class	B Frequency (f)	C Midpoint (X_m)	D $f \cdot X_m$
5.5–10.5	1	8	8
10.5–15.5	2	13	26
15.5–20.5	3	18	54
20.5–25.5	5	23	115
25.5–30.5	4	28	112
30.5–35.5	3	33	99
35.5–40.5	2	38	76
	$n = 20$		$\Sigma f \cdot X_m = 490$

STEP 4   Find the sum of column D, as shown above.

STEP 5   Divide the sum by n to get the mean.

$$\overline{X} = \frac{\Sigma f \cdot X_m}{n} = \frac{490}{20} = 24.5 \text{ miles}$$

Example 3–5

A random sample of the life expectancy of residents for 25 countries in Asia was selected, and the following frequency distribution was obtained. Find the mean.

Class	Frequency
42.5–47.5	2
47.5–52.5	2
52.5–57.5	3
57.5–62.5	3
62.5–67.5	4
67.5–72.5	6
72.5–77.5	3
77.5–82.5	2

Source: THE UNIVERSAL ALAMANC © by John W. Wright. Reprinted with permission of Andrews McMeel Publishing. All rights reserved.

Solution

STEP 1   Make a table, as shown in Step 1 of Example 3–4.

STEP 2   Find the midpoints of each class, and place the values in column C.

$$X_m = \frac{42.5 + 47.5}{2} = 45, \qquad \text{etc.}$$

STEP 3   Multiply the midpoint by the frequency for each class, and place the products in column D.

$$2 \cdot 45 = 90, \qquad 2 \cdot 50 = 100, \qquad \text{etc.}$$

STEP 4   Find the sum of column D.

$$\Sigma f \cdot X_m = 1600$$

STEP 5 Divide the sum by n.

$$\bar{X} = \frac{\Sigma f \cdot X_m}{n} = \frac{1600}{25} = 64 \text{ years}$$

Hence, the mean life expectancy of the residents in the random sample is 64 years. These steps are summarized in the following table.

A Class	B Frequency (f)	C Midpoint (X_m)	D $f \cdot X_m$
42.5–47.5	2	45	90
47.5–52.5	2	50	100
52.5–57.5	3	55	165
57.5–62.5	3	60	180
62.5–67.5	4	65	260
67.5–72.5	6	70	420
72.5–77.5	3	75	225
77.5–82.5	2	80	160
	$n = 25$		$\Sigma f \cdot X_m = 1600$

The procedure for finding the mean for grouped data assumes that the midpoints of all of the raw data values in each class are equal to the midpoint of the class. In reality, this is not true, since the average of the raw data values in each class will not be exactly equal to the midpoint. However, using this procedure will give an acceptable approximation of the mean, since some values fall above the midpoint and some values fall below the midpoint for each class, and the midpoint represents an estimate of all values in the class.

The procedure for finding the mean for grouped data is summarized in Procedure Table 3.

Procedure Table 3

Finding the Mean for Grouped Data

1. Make a table as shown.

A Class	B Frequency (f)	C Midpoint (X_m)	D $f \cdot X_m$

2. Find the midpoints of each class and place them in column C.
3. Multiply the frequency by the midpoint for each class and place the product in column D.
4. Find the sum of column D.
5. Divide the sum obtained in column D by the sum of the frequencies obtained in column B.

The formula for the mean is

$$\bar{X} = \frac{\Sigma f \cdot X_m}{n}$$

The Median

An article recently reported that the median income for college professors was $43,250. This measure of average means that half of all the professors surveyed earned more than $43,250, and half earned less than $43,250.

The *median* is the halfway point in a data set. Before one can find this point, the data must be arranged in order. When the data set is ordered, it is called a **data array.** The median either will be a specific value in the data set or will fall between two values, as shown in the following examples.

The **median** is the midpoint of the data array. The symbol for the median is MD.

Steps in Computing the Median of a Data Array

STEP 1 Arrange the data in order.

STEP 2 Select the middle point.

Example 3–6

The weights (in pounds) of seven army recruits are 180, 201, 220, 191, 219, 209, and 186. Find the median.

Solution

STEP 1 Arrange the data in order.

180, 186, 191, 201, 209, 219, 220

STEP 2 Select the middle value.

180, 186, 191, (201,) 209, 219, 220
↑
Median

Example 3–7

Find the median for the ages of seven preschool children. The ages are 1, 3, 4, 2, 3, 5, and 1.

Solution

1, 1, 2, (3,) 3, 4, 5
↑
Median

Hence, the median age is 3 years.

Each of these examples had an odd number of values in the data set; hence, the median was an actual data value. When there is an even number of values in the data set, the median will fall between two given values, as illustrated in the following examples.

Example 3–8

The number of tornadoes that have occurred in the United States in the last 8 years follows. Find the median.

$$684, 764, 656, 702, 856, 1133, 1132, 1303$$

Source: THE UNIVERSAL ALMANAC © by John W. Wright. Reprinted with permission of Andrews McMeel Publishing. All rights reserved.

Solution

$$656, 684, 702, 764, 856, 1132, 1133, 1303$$
$$\uparrow$$
$$\text{Median}$$

Since the middle point falls halfway between 764 and 856, find the median by adding the two values and dividing by 2.

$$MD = \frac{764 + 856}{2} = \frac{1620}{2} = 810$$

The median number of tornadoes is 810.

Example 3–9

The ages of 10 college students are given below. Find the median.

$$18, 24, 20, 35, 19, 23, 26, 23, 19, 20$$

Solution

$$18, 19, 19, 20, 20, 23, 23, 24, 26, 35$$
$$\uparrow$$
$$\text{Median}$$

$$MD = \frac{20 + 23}{2} = 21.5$$

Hence, the median age is 21.5 years.

Example 3–10

Six customers purchased the following number of magazines: 1, 7, 3, 2, 3, 4. Find the median.

Solution

$$1, 2, 3, 3, 4, 7 \qquad MD = \frac{3 + 3}{2} = 3$$
$$\uparrow$$
$$\text{Median}$$

Hence, the median number of magazines purchased is 3.

For an ungrouped frequency distribution, find the median by examining the cumulative frequencies to locate the middle value, as shown in the next example.

Example 3–11

Holmes Appliance recorded the number of videocassette recorders (VCRs) sold per month over a two-year period. Find the median.

Number of sets sold	Frequency (months)
1	3
2	8
3	5
4	4
5	2
6	1
7	1
	$n = 24$

Solution

To locate the middle point, divide n by 2; $24 \div 2 = 12$. Then locate the point where 12 values would fall below and 12 values would fall above. To do this, look at the cumulative frequency.

Class	Frequency	Cumulative frequency
1	3	3
2	8	11
3	5	16 ← (This class contains the
4	4	20 12th through the 16th
5	2	22 values.)
6	1	23
7	1	24

The twelfth and thirteenth values fall in class 3. Hence, the median is 3.

The procedure for finding the median for grouped data is shown in the next example.

Example 3–12

Using the frequency distribution for Example 2–7 in Chapter 2, find the median.

Solution

The procedure for finding the median for grouped data is shown here.

STEP 1 Make a table as shown next.

A Class boundaries	B Frequency	C Cumulative frequency
5.5–10.5	1	1
10.5–15.5	2	3
15.5–20.5	3	6
20.5–25.5	5	11
25.5–30.5	4	15
30.5–35.5	3	18
35.5–40.5	2	20
	$n = 20$	

STEP 2   Divide n (the sum of column B) by 2 to find the halfway point.

$$\frac{20}{2} = 10$$

STEP 3   Find the class that contains the tenth value by using the cumulative frequency distribution. This class is called the *median class;* it contains the median.

Class	Frequency	Cumulative frequency
5.5–10.5	1	1
⋮	⋮	⋮
15.5–20.5	3	6 ← cf
$L_m \rightarrow$ 20.5–25.5	5 ← f	11 ← Median class
⋮	⋮	⋮
	$n = \overline{20}$	

STEP 4   Substitute in the formula

$$\text{MD} = \frac{(n/2) - \text{cf}}{f}(w) + L_m$$

where

$$n = \text{sum of the frequencies} = 20$$
$$\text{cf} = \text{cumulative frequency of the class immediately}$$
$$\text{preceding the median class} = 6$$
$$f = \text{frequency of the median class} = 5$$
$$w = \text{width of the median class} = 25.5 - 20.5 = 5$$
$$L_m = \text{lower boundary of the median class} = 20.5$$

STEP 5   Solve for the median.

$$\text{MD} = \frac{(20/2) - 6}{5}(5) + 20.5 = 24.5$$

The median is 24.5.

The reasoning behind this procedure is as follows: Since the halfway point falls in the class 20.5–25.5, and there are five values in this class, it is assumed that these five values are evenly distributed within the class, as shown in Figure 3–1.

Figure 3–1

Median Class

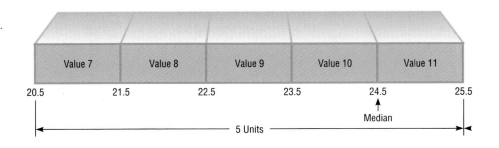

The middle point is between the tenth and eleventh value, which is 24.5. The formula

$$\frac{(n/2) - cf}{f}$$

gives the fractional part of the class where the median lies. To get the middle point, one must go $\frac{4}{5}$ of the way into the class. Since the class width is 5, $\frac{4}{5}$ of 5 is 4. This value (4) is added to the lower boundary (20.5) to get the median, 24.5.

One can compute the median for open-ended frequency distributions as long as the middle score does not occur in the open-ended class. If the median does fall in the open-ended class, the data probably have not been grouped correctly.

The procedure for finding the median for grouped data is summarized in Procedure Table 4.

Procedure Table 4

Finding the Median for Grouped Data

1. Make a table as shown below.

A	B	C
Class boundaries	Frequency	Cumulative frequency

2. Divide *n*, the sum of the frequencies (column B), by 2 to get the halfway point.
3. Locate the median class in column C.
4. Substitute in the formula

$$MD = \frac{(n/2) - cf}{f}(w) + L_m$$

where

n = sum of the frequencies
cf = cumulative frequency of the class immediately preceding the median class
f = frequency of the median class
w = class width
L_m = lower boundary of the median class

5. Solve for the median.

The Mode

The third measure of average is called the *mode*. The mode is the value that occurs most often in the data set. It is sometimes said to be the most typical case.

The value that occurs most often in a data set is called the **mode**.

A data set can have more than one mode or no mode at all. These situations will be shown in some of the examples that follow.

Medians are often used in real estate statistics. In this survey on rents, the median rents were used. Suggest several reasons for using medians instead of the mean for rents.

Study: Pittsburgh Has Lowest Median Rents

WASHINGTON (AP) — Miami-area residents spend the greatest share of their incomes on rental housing, while those in Houston spend the least, according to a big-city survey released Monday by the Census Bureau.

Among the 46 U.S. metropolitan areas with a million or more residents, Miami-Hialeah ranked first in share of income at 31.3 percent. Median monthly rent there was $493, meaning half paid more and half less.

Houston's $406 median rent was just 23.2 percent of income there.

The highest median rental costs were in Anaheim-Santa Ana, Calif., at $790, but that took 29 percent of median income for residents of that area.

The lowest median rents were in Pittsburgh at $365 a month, taking 26.6 percent of residents' income.

Here is a rundown of the 46 metropolitan areas ranked by percentage of median income required for rentals.

Metro Area	Rent	Percent	Metro Area	Rent	Percent
Miami	$493	31.3	Cleveland	$406	26.3
San Diego	611	29.8	Newark, N.J.	583	26.2
Riverside, Calif.	562	29.7	Bergen-Passaic, N.J.	646	26.2
Los Angeles	626	29.5	Atlanta	529	26.0
Sacramento, Calif.	531	29.3	Milwaukee	447	26.0
Anaheim, Calif.	790	29.0	Washington	667	25.9
Nassau-Suffolk, N.Y.	778	29.0	Seattle	516	25.8
Ft. Lauderdale, Fla.	575	29.0	New York	503	25.7
Oakland, Calif.	642	28.6	St. Louis	415	25.7
New Orleans	397	28.5	Denver	531	25.5
San Fransisco	709	28.0	Middlesex, N.J.	680	25.3
Rochester, N.Y.	466	27.9	Baltimore	490	25.3
Phoenix	465	27.5	San Antonio	380	25.2
San Jose, Calif.	773	27.4	Portland, Ore.	437	24.9
Philadelphia	516	27.4	Columbus, Ohio	421	24.6
Detroit	455	27.3	Cincinnati	367	24.6
Orlando, Fla.	524	27.2	Kansas City	425	24.5
Norfolk-Va. Beach, Va.	480	27.2	Dallas	456	24.3
Tampa-St. Pete, Fla.	448	27.2	Fort Worth, Texas	428	24.0
Minneapolis-St. Paul	479	26.9	Indianapolis	413	24.0
Boston	656	26.7	Salt Lake City	379	23.7
Pittsburgh	365	26.6	Charlotte, N.C.	426	23.6
Chicago	491	26.3	Houston	406	23.2

Source: *The Daily News,* McKeesport, PA, November 8, 1994. Used with permission.

Example 3–13

The following data represent the duration (in days) of U.S. space shuttle voyages for the years 1992–94. Find the mode.

8, 9, 9, 14, 8, 8, 10, 7, 6, 9, 7, 8, 10, 14, 11, 8, 14, 11

Source: THE UNIVERSAL ALMANAC © by John W. Wright. Reprinted with permission of Andrews McMeel Publishing. All rights reserved.

Solution

It is helpful to arrange the data in order, although it is not necessary.

6, 7, 7, 8, 8, 8, 8, 8, 9, 9, 9, 10, 10, 11, 11, 14, 14, 14

Since 8-day voyages occurred five times—a frequency larger than any other number—the mode for the data set is 8.

Example 3–14

Six strains of bacteria were tested to see how long they could remain alive outside their normal environment. The time, in minutes, is recorded below. Find the mode.

2, 3, 5, 7, 8, 10

Solution

Since each value occurs only once, there is no mode.

Note: Do not say that the mode is zero. That would be incorrect, because in some data, such as temperature, zero can be an actual value.

Example 3–15

Eleven different automobiles were tested at a speed of 15 miles per hour for stopping distances. The data, in feet, are shown below. Find the mode.

15, 18, 18, 18, 20, 22, 24, 24, 24, 26, 26

Solution

Since 18 and 24 both occur three times, the modes are 18 and 24 feet. This data set is said to be *bimodal*.

The mode for grouped data is the modal class. The **modal class** is the class with the largest frequency.

Example 3–16

Find the modal class for the frequency distribution of miles 20 runners ran in one week, used in Example 2–7 in Chapter 2.

Class	Frequency
5.5–10.5	1
10.5–15.5	2
15.5–20.5	3
20.5–25.5	5 ← Modal class
25.5–30.5	4
30.5–35.5	3
35.5–40.5	2

Solution

The modal class is 20.5–25.5, since it has the largest frequency. Sometimes the midpoint of the class is used rather than the boundaries; hence, the mode could also be given as 23 miles per week.

The mode is the only measure of central tendency that can be used in finding the most typical case when the data are nominal or categorical.

Example 3–17

A survey showed the following distribution for the number of students enrolled in each field. Find the mode.

Business	1425
Liberal arts	878
Computer science	632
Education	471
General studies	95

Solution

Since the category with the highest frequency is business, the most typical case is a business major.

The next example shows how to find the mode for an ungrouped frequency distribution.

Example 3–18

The following data were collected on the number of blood tests a local hospital conducted for a random sample of 50 days. Find the mode.

Number of tests per day	Frequency (days)
26	5
27	9
28	12
29	18
30	5
31	0
32	1

Solution

Since 29 tests were given on 18 days (the number of tests that occurred most often), the mode is 29.

For a data set, the mean, median, and mode can be quite different. Consider the following example.

Example 3–19

A small company consists of the owner, the manager, the salesperson, and two technicians, all of whose annual salaries are listed here. (Assume that this is the entire population.)

Staff	Salary
Owner	$50,000
Manager	20,000
Salesperson	12,000
Technician	9,000
Technician	9,000

Many statistical studies involve typical individuals. Here is a profile of a "typical" bus tour traveler. What descriptive statistics are used here to estimate the parameters of the population and hence describe those travelers?

USA SNAPSHOTS®

A look at statistics that shape our lives

Leave the driving to . . .
Profile of typical bus tour traveler:

Female
Married
Age 35-54
College educated
Income $25,000-plus

Source: United Bus Owners of America

By Cindy Hall and Cliff Vancura, USA TODAY

Source: *USA Today,* April 28, 1995, p. 1D. Copyright 1995, USA TODAY. Reprinted with permission.

Find the mean, median, and mode.

Solution

$$\mu = \frac{\sum X}{N} = \frac{50{,}000 + 20{,}000 + 12{,}000 + 9{,}000 + 9{,}000}{5} = \$20{,}000$$

Hence, the mean is $20,000, the median is $12,000, and the mode is $9,000.

In this example, the mean is much higher than the median or the mode. This is because the extremely high salary of the owner tends to raise the value of the mean. In this and similar situations, the median should be used as the measure of central tendency.

The Midrange

The *midrange* is a rough estimate of the middle. It is found by adding the lowest and highest values in the data set and dividing by 2. It is a very rough estimate of the average and can be affected by one extremely high or low value.

The **midrange** is defined as the sum of the lowest and highest values in the data set divided by 2. The symbol MR is used for the midrange.

$$MR = \frac{\text{lowest value} + \text{highest value}}{2}$$

Example 3–20

Last winter, the city of Brownsville, Minnesota, reported the following number of water-line breaks per month. Find the midrange.

2, 3, 6, 8, 4, 1

Solution

$$MR = \frac{1 + 8}{2} = \frac{9}{2} = 4.5$$

Hence, the midrange is 4.5.

If the data set contains one extremely large value or one extremely small value, a higher or lower midrange value will result and may not be a typical description of the middle.

Example 3–21

Suppose the number of water-line breaks was as follows: 2, 3, 6, 16, 4, and 1. Find the midrange.

Solution

$$MR = \frac{1 + 16}{2} = \frac{17}{2} = 8.5$$

Hence, the midrange is 8.5. The value 8.5 is not typical of the average monthly number of breaks, since an excessively high number of breaks, 16, occurred in one month.

In statistics, several measures can be used for an average. The most common measures are the mean, the median, the mode, and the midrange. Each has its own specific purpose and use. Exercises 3–39 through 3–41 show examples of other averages—such as the harmonic mean, the geometric mean, and the quadratic mean. Their applications are limited to specific areas, as shown in the exercises.

The Weighted Mean

Sometimes, one must find the mean of a data set in which not all values are equally represented. Consider the case of finding the average cost of a gallon of gasoline for three taxis. Suppose the drivers buy gasoline at three different service stations at a cost of $1.19, $1.27, and $1.32 per gallon. One might try to find the average by using the formula

$$\overline{X} = \frac{\sum X}{n}$$

$$= \frac{1.19 + 1.27 + 1.32}{3} = \$1.26$$

But not all drivers purchased the same number of gallons. Hence, to find the true average cost per gallon, one must take into consideration the number of gallons each driver purchased.

The type of mean that considers an additional factor is called the *weighted mean,* and it is used when the values are not all equally represented.

Find the **weighted mean** of a variable X by multiplying each value by its corresponding weight and dividing the sum of the products by the sum of the weights.

$$\overline{X} = \frac{w_1 X_1 + w_2 X_2 + \cdots + w_n X_n}{w_1 + w_2 + \cdots + w_n} = \frac{\sum wX}{\sum w}$$

where $w_1, w_2, \ldots, w_n$ are the weights and $X_1, X_2, \ldots, X_n$ are the values.

The next example shows how the weighted mean is used to compute a grade point average. Since courses vary in their credit value, the number of credits must be used as weights.

Example 3–22

A student received an A in English Composition I (3 credits), a C in Introduction to Psychology (3 credits), a B in Biology I (4 credits), and a D in Physical Education (2 credits). Assuming A = 4 grade points, B = 3 grade points, C = 2 grade points, D = 1 grade point, and F = 0 grade points, find the student's grade point average.

Solution

Course	Credits (w)	Grade (X)
Eng Comp I	3	A (4 points)
Intro to Psych	3	C (2 points)
Biology I	4	B (3 points)
Phys Ed	2	D (1 point)

$$\overline{X} = \frac{\sum wX}{\sum w} = \frac{3 \cdot 4 + 3 \cdot 2 + 4 \cdot 3 + 2 \cdot 1}{3 + 3 + 4 + 2} = \frac{32}{12} = 2.7$$

The grade point average is 2.7.

Table 3–1 summarizes the measures of central tendency.

Table 3–1	Summary of Measures of Central Tendency	
Measure	**Definition**	**Symbol(s)**
Mean	Sum of values divided by total number of values	$\mu, \overline{X}$
Median	Middle point in data set that has been ordered	MD
Mode	Most frequent data value	none
Midrange	(Lowest value plus highest value) divided by 2	MR

Researchers and statisticians must know which measure of central tendency is being used and when to use each measure of central tendency. The properties and uses of the three measures of central tendency are summarized here.

Properties and Uses of Central Tendency

The Mean

1. One computes the mean by using all the values of the data.
2. The mean varies less than the other two measures when samples are taken from the same population and all three measures are computed for these samples.
3. The mean is used in computing other statistics, such as the variance.
4. The mean for the data set is unique, and not necessarily one of the data values.
5. The mean cannot be computed for an open-ended frequency distribution.
6. The mean is affected by extremely high or low values and may not be the appropriate average to use in these situations.

The Median

1. The median is used when one must find the center or middle value of a data set.
2. The median is used when one must determine whether the data values fall into the upper half or lower half of the distribution.
3. The median is used to find the average of an open-ended distribution.
4. The median is affected less than the mean by extremely high or extremely low values.

The Mode

1. The mode is used when the most typical case is desired.
2. The mode is the easiest average to compute.
3. The mode can be used when the data are nominal, such as religious preference, gender, or political affiliation.
4. The mode is not always unique. A data set can have more than one mode, or the mode may not exist for a data set.

The Midrange

1. The midrange is easy to compute.
2. The midrange gives the midpoint.
3. The midrange is affected by extremely high or low values in a data set.

Distribution Shapes

Frequency distributions can assume many shapes. The three most important shapes are positively skewed, symmetrical, and negatively skewed. Figure 3–2 shows histograms of each.

Figure 3–2

Figure 3–2

Types of Distributions

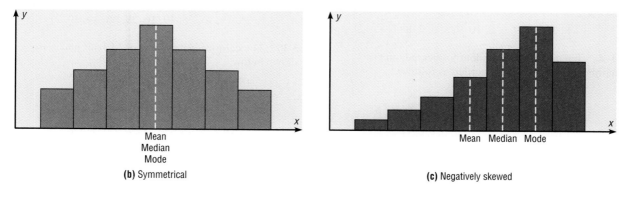

(a) Positively skewed

(b) Symmetrical

(c) Negatively skewed

In a **positively skewed distribution,** the majority of the data values fall to the left of the mean and cluster at the lower end of the distribution; the "tail" is to the right. Also, the mean is to the right of the median, and the mode is to the left of the median.

For example, if an instructor gave an examination and most of the students did poorly, their scores would tend to cluster on the left side of the distribution. A few high scores would constitute the tail of the distribution, which would be on the right side. Another example of a positively skewed distribution is the incomes of the population of the United States. Most of the incomes cluster about the low end of the distribution; those with high incomes are in the minority and are in the tail at the right of the distribution.

In a **symmetrical distribution,** the data values are evenly distributed on both sides of the mean. In addition, when the distribution is unimodal, the mean, median, and mode are the same and are at the center of the distribution. Examples of symmetrical distributions are IQ scores and heights of adult males.

When the majority of the data values fall to the right of the mean and cluster at the upper end of the distribution, with the tail to the left, the distribution is said to be **negatively skewed.** Also, the mean is to the left of the median, and the mode is to the right of the median. As an example, a negatively skewed distribution results if the majority of students score very high on an instructor's examination. These scores will tend to cluster to the right of the distribution.

When a distribution is extremely skewed, the value of the mean will be pulled toward the tail, but the majority of the data values will be greater than the mean; hence, the median rather than the mean is a more appropriate measure of central tendency. An extremely skewed distribution can also affect other statistics.

A measure of skewness for a distribution is discussed in Exercise 3–88 in Section 3–3.

Exercises

For Exercises 3–1 through 3–11, find (a) the mean, (b) the median, (c) the mode, and (d) the midrange.

3–1. Twelve secretaries were given a typing test, and the times (in minutes) to complete it were as follows:

    8, 12, 15, 9, 6, 8, 10, 9, 8, 6, 7, 8

3–2. Ten novels were randomly selected, and the numbers of pages were recorded as follows:

    415, 398, 402, 399, 400,
    405, 395, 401, 412, 407

3–3. Twelve members of the high school cross-country team were asked how many minutes each ran during practice sessions. Their answers are recorded here.

    32, 28, 35, 37, 43, 51, 61, 39, 48, 51, 53, 49

3–4. Nine physicians were selected and given a blood pressure check. Their systolic pressures are recorded below.

    135, 120, 116, 119, 121, 125, 132, 136, 124

3–5. The manager of a sports shop recorded the number of baseball caps he sold during the week. The data are shown here.

    132, 121, 119, 116, 130, 121, 131

3–6. The heights (in inches) of six female police officer candidates are shown below.

    62, 64, 63, 61, 62, 66

3–7. The grade point averages of 10 students who applied for financial aid are shown below.

    3.15, 3.62, 2.54, 2.81, 3.97,
    1.85, 1.93, 2.63, 2.50, 2.80

3–8. The number of calls received by the Internal Revenue Service office each business day in July are shown below.

    18, 12, 25, 16, 27, 32, 25, 15, 23, 22, 37, 16,
    25, 19, 16, 25, 19, 16, 29, 38, 29, 30, 21

3–9. The calories per serving of 11 fruit juices are shown below.

    150, 110, 100, 35, 60, 130,
    40, 140, 120, 160, 110

Source: *Consumer Reports,* February 1995, p. 79.

3–10. The exam scores of 18 English composition students were recorded as follows:

    78, 62, 98, 90, 88, 73, 79, 86, 81,
    84, 93, 97, 63, 59, 78, 82, 87, 93

3–11. During 1993, the major earthquakes had Richter magnitudes as shown below.

    7.0, 6.2, 7.7, 8.0, 6.4, 6.2,
    7.2, 5.4, 6.4, 6.5, 7.2, 5.4

Source: THE UNIVERSAL ALMANAC © by John W. Wright. Reprinted with permission of Andrews McMeel Publishing. All rights reserved.

For Exercises 3–12 through 3–21, find (a) the mean, (b) the median, and (c) the modal class.

3–12. For 108 randomly selected college students, the following IQ frequency distribution was obtained.

Class limits	Frequency
90–98	6
99–107	22
108–116	43
117–125	28
126–134	9

3–13. For 50 antique car owners, the following distribution of the cars' ages was obtained.

Class limits	Frequency
16–18	20
19–21	18
22–24	8
25–27	4

3–14. Thirty automobiles were tested for fuel efficiency (in miles per gallon). The following frequency distribution was obtained.

Class boundaries	Frequency
7.5–12.5	3
12.5–17.5	5
17.5–22.5	15
22.5–27.5	5
27.5–32.5	2

3–15. The following numbers of books were read by each of the 28 students in a literature class.

Number of books	Frequency
0	2
1	6
2	12
3	5
4	3

3–16. In a study of the time it takes an untrained mouse to run a maze, a researcher recorded the following data (in seconds).

Class limits	Frequency
2.1–2.7	5
2.8–3.4	7
3.5–4.1	12
4.2–4.8	14
4.9–5.5	16
5.6–6.2	8

3–17. Eighty randomly selected lightbulbs were tested to determine their lifetimes (in hours). The following frequency distribution was obtained.

Class boundaries	Frequency
52.5–63.5	6
63.5–74.5	12
74.5–85.5	25
85.5–96.5	18
96.5–107.5	14
107.5–118.5	5

3–18. The following data represent the net worth (in millions of dollars) of 45 national corporations.

Class limits	Frequency
10–20	2
21–31	8
32–42	15
43–53	7
54–64	10
65–75	3

3–19. The cost per load (in cents) of 35 laundry detergents tested by a consumer organization is shown below. (The data in this exercise will be used for Exercise 3–61.)

Class limits	Frequency
13–19	2
20–26	7
27–33	12
34–40	5
41–47	6
48–54	1
55–61	0
62–68	2

Source: *Consumer Reports,* February 1995.

3–20. The following frequency distribution represents the commission earned (in dollars) by 100 salespeople employed at several branches of a large chain store.

Class limits	Frequency
150–158	5
159–167	16
168–176	20
177–185	21
186–194	20
195–203	15
204–212	3

3–21. This frequency distribution represents the data obtained from a sample of 75 copying machine service technicians. The values represent the days between service calls for various copying machines.

Class boundaries	Frequency
15.5–18.5	14
18.5–21.5	12
21.5–24.5	18
24.5–27.5	10
27.5–30.5	15
30.5–33.5	6

3–22. Find the mean, median, and modal class for the data in Exercise 2–12, Chapter 2.

3–23. Find the mean, median, and modal class for the data in Exercise 2–13, Chapter 2.

3–24. Find the mean, median, and modal class for the data in Exercise 2–14, Chapter 2.

3–25. Find the mean, median, and modal class for the data in Exercise 2–15, Chapter 2.

3–26. Find the weighted mean price of three models of automobiles sold. The number and price of each model sold are shown in the following list.

Model	Number	Price
A	8	$10,000
B	10	$12,000
C	12	$ 8,000

3–27. Using the weighted mean, find the average number of grams of fat in meat a person would consume over a five-day period if he ate the following:

Meat	Fat (grams/oz.)
3 oz. fried shrimp	3.33
3 oz. veal cutlet (broiled)	3.00
2 oz. roast beef (lean)	2.50
2.5 oz. fried chicken drumstick	4.40
4 oz. tuna (canned in oil)	1.75

Source: Reprinted with permission from *The World Almanac and Book of Facts,* 1995. Copyright © 1994 PRIMEDIA Reference Inc. All rights reserved.

3–28. A recent survey of a new diet cola reported the following percentages of people who liked the taste. Find the weighted mean of the percentages.

Area	% Favored	Number surveyed
1	40	1000
2	30	3000
3	50	800

3–29. The costs of three models of helicopters are shown below. Find the weighted mean of the costs of the models.

Model	Number sold	Cost
Sunscraper	9	$427,000
Skycoaster	6	365,000
Highflyer	12	725,000

3–30. An instructor grades as follows: exams, 20%; term paper, 30%; final exam, 50%. A student had grades of 83, 72, and 90, respectively, for exams, term paper, and final exam. Find the student's final average. Use the weighted mean.

3–31. Another instructor gives four 1-hour exams and one final exam, which counts as two 1-hour exams. Find a student's grade if she received 62, 83, 97, and 90 on the hour exams and 82 on the final exam.

3–32. For the following situations, state which measure of central tendency—mean, median, or mode—should be used.
a. The most typical case is desired.
b. The distribution is open-ended.
c. There is an extreme value in the data set.
d. The data are categorical.
e. Further statistical computations will be needed.
f. The values are to be divided into two approximately equal groups, one group containing the larger values and one containing the smaller values.

3–33. Describe which measure of central tendency—mean, median, or mode—was probably used in each situation.
a. Half of the factory workers make more than $5.37 per hour and half make less than $5.37 per hour.
b. The average number of children per family in the Plaza Heights Complex is 1.8.
c. Most people prefer red convertibles over any other color.
d. The average person cuts the lawn once a week.
e. The most common fear today is fear of speaking in public.
f. The average age of college professors is 42.3 years.

3–34. What types of symbols are used to represent sample statistics? Give an example.

3–35. What types of symbols are used to represent population parameters? Give an example.

*3–36. If the mean of five values is 64, find the sum of the values.

*3–37. If the mean of five values is 8.2, and four of the values are 6, 10, 7, and 12, find the fifth value.

*3–38. (**W**) Find the mean of 10, 20, 30, 40, and 50.
a. Add 10 to each value, and find the mean.
b. Subtract 10 from each value, and find the mean.
c. Multiply each value by 10, and find the mean.
d. Divide each value by 10, and find the mean.
e. Make a general statement about each situation.

*3–39. The *harmonic mean* (HM) is defined as the number of values divided by the sum of the reciprocals of each value. The formula is

$$\text{HM} = \frac{n}{\sum 1/X}$$

For example, the harmonic mean of 1, 4, 5, and 2 is

$$HM = \frac{4}{\frac{1}{1} + \frac{1}{4} + \frac{1}{5} + \frac{1}{2}} = 2.05$$

This mean is useful for finding the average speed. Suppose a person drove 100 miles at 40 miles per hour and returned driving 50 miles per hour. The average miles per hour is *not* 45 miles per hour, which is found by adding 40 and 50 and dividing by 2. The average is found as follows.

Since

$$time = distance \div rate$$

then

$$time\ 1 = \frac{100}{40} = 2.5 \text{ hours to make the trip}$$

$$time\ 2 = \frac{100}{50} = 2 \text{ hours to return}$$

Hence, the total time is 4.5 hours and the total miles driven are 200. Now, the average speed is

$$rate = \frac{distance}{time} = \frac{200}{4.5} = 44.44 \text{ miles per hour}$$

This value can also be found by using the harmonic mean formula:

$$HM = \frac{2}{\frac{1}{40} + \frac{1}{50}} = 44.44$$

Using the harmonic mean, find each of the following.
a. A salesperson drives 300 miles round trip at 30 miles per hour going to Chicago and 45 miles per hour returning home. Find the average miles per hour.
b. A bus driver drives the 50 miles to West Chester at 40 miles per hour and returns driving 25 miles per hour. Find the average miles per hour.
c. A carpenter buys $500 worth of nails at $50 per pound and $500 worth of nails at $10 per pound. Find the average cost of a pound of nails.

*3–40. The *geometric mean* (GM) is defined as the *n*th root of the product of *n* values. The formula is

$$GM = \sqrt[n]{(X_1)(X_2)(X_3) \cdots (X_n)}$$

The geometric mean of 4 and 16 is

$$GM = \sqrt{(4)(16)} = \sqrt{64} = 8$$

The geometric mean of 1, 3, and 9 is

$$GM = \sqrt[3]{(1)(3)(9)} = \sqrt[3]{27} = 3$$

The geometric mean is useful in finding the average of percentages, ratios, indexes, or growth rates. For example, if a person receives a 20% raise after one year of service and a 10% raise after the second year of service, the average percentage raise per year is not 15% but 14.89%, as shown.

$$GM = \sqrt{(1.2)(1.1)} = 1.1489$$

or $$GM = \sqrt{(120)(110)} = 114.89\%$$

His salary is 120% at the end of the first year and 110% at the end of the second year. This is equivalent to an average of 14.89%, since 114.89% − 100% = 14.89%.

This answer can also be shown by assuming that the person makes $10,000 to start and receives two raises of 20% and 10%.

$$raise\ 1 = 10,000 \cdot 20\% = \$2000$$
$$raise\ 2 = 12,000 \cdot 10\% = \$1200$$

His total salary raise is $3200. This total is equivalent to

$$\$10,000 \cdot 14.89\% = \$1489.00$$
$$\underline{\$11,489 \cdot 14.89\% = \$1710.71}$$
$$\$3199.71 \approx \$3200$$

Find the geometric mean of each of the following.
a. The growth rates of the Living Life Insurance Corporation for the past three years were 35%, 24%, and 18%.
b. A person received the following percentage raises in salary over a four-year period: 8%, 6%, 4%, and 5%.
c. A stock increased each year for five years at the following percentages: 10%, 8%, 12%, 9%, and 3%.
d. The price increases, in percentages, for the cost of food in a specific geographic region for the past three years were 1%, 3%, and 5.5%.

*3–41. A useful mean in the physical sciences is the *quadratic mean* (QM), which is found by taking the square root of the sum of the average of the squares of each value. The formula is

$$QM = \sqrt{\frac{\sum X^2}{n}}$$

The quadratic mean of 3, 5, 6, and 10 is

$$QM = \sqrt{\frac{3^2 + 5^2 + 6^2 + 10^2}{4}}$$
$$= \sqrt{42.5} = 6.52$$

Find the quadratic mean of 8, 6, 3, 5, and 4.

3–3

Measures of Variation

In statistics, in order to describe the data set accurately, statisticians must know more than the measures of central tendency. Consider the following example.

Example 3–23

Objective 2. Describe data using measures of variation.

A testing lab wishes to test two experimental brands of outdoor paint to see how long each would last before fading. The testing lab makes six gallons of each paint to test. Since different chemical agents are added to each group and only six cans are involved, these two groups constitute two small populations. The results (in months) follow. Find the mean of each group.

Brand A	Brand B
10	35
60	45
50	30
30	35
40	40
20	25

Solution

The mean for brand A is

$$\mu = \frac{\sum X}{N} = \frac{210}{6} = 35 \text{ months}$$

The mean for brand B is

$$\mu = \frac{\sum X}{N} = \frac{210}{6} = 35 \text{ months}$$

Since the means are equal in Example 3–23, one might conclude that both brands of paint last equally well. However, when the data sets are examined graphically, a somewhat different conclusion might be drawn. See Figure 3–3.

As Figure 3–3 shows, even though the means are the same for both brands, the spread, or variation, is quite different. The figure shows that brand B performs more consistently; it is less variable. For the spread or variability of a data set, three measures are commonly used: *range, variance,* and *standard deviation.* Each measure will be discussed in this section.

Range

The range is the simplest of the three measures and is defined next.

The **range** is the highest value minus the lowest value. The symbol R is used for the range.

$$R = \text{highest value} - \text{lowest value}$$

Figure 3–3
....
Examining Data Sets
Graphically

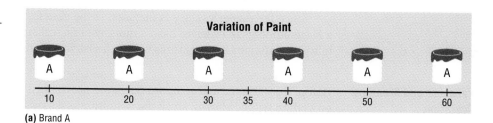

(a) Brand A

(b) Brand B

| Example 3–24 |

Find the ranges for the paints in Example 3–23.

Solution

For brand A, the range is

$$R = 60 - 10 = 50 \text{ months}$$

For brand B, the range is

$$R = 45 - 25 = 20 \text{ months}$$

Make sure the range is given as a single number ($45-25$ is incorrect).

The range for brand A shows that 50 months separate the largest data value from the smallest data value. For brand B, 20 months separate the largest data value from the smallest data value, which is less than half of brand A's range.

One extremely high or one extremely low data value can affect the range markedly, as shown in the next example.

| Example 3–25 |

The salaries for the staff of the XYZ Manufacturing Co. follow. Find the range.

Staff	Salary
Owner	$100,000
Manager	40,000
Sales representative	30,000
Workers	25,000
	15,000
	18,000

Solution

The range is $R = \$100{,}000 - \$15{,}000 = \$85{,}000$.

Since the owner's salary is included in the data for Example 3–25, the range is a large number. In order to have a more meaningful statistic to measure the variability, statisticians use measures called the *variance* and *standard deviation.*

Variance and Standard Deviation

Before the variance and standard deviation are defined formally, the computational procedure will be shown, since the definition is derived from the procedure.

Rounding Rule for the Standard Deviation   The rounding rule for the standard deviation is the same as for the mean. The final answer should be rounded to one more decimal place than the original data.

Example 3–26

Find the variance and standard deviation for the data set for brand A paint in Example 3–23.

$$10, 60, 50, 30, 40, 20$$

Solution

STEP 1   Find the mean for the data.

$$\mu = \frac{\Sigma X}{N} = \frac{10 + 60 + 50 + 30 + 40 + 20}{6} = \frac{210}{6} = 35$$

STEP 2   Subtract the mean from each data value.

$$10 - 35 = -25 \qquad 50 - 35 = +15 \qquad 40 - 35 = +5$$
$$60 - 35 = +25 \qquad 30 - 35 = -5 \qquad 20 - 35 = -15$$

STEP 3   Square each result.

$$(-25)^2 = 625 \qquad (+15)^2 = 225 \qquad (+5)^2 = 25$$
$$(+25)^2 = 625 \qquad (-5)^2 = 25 \qquad (-15)^2 = 225$$

STEP 4   Find the sum of the squares.

$$625 + 625 + 225 + 25 + 25 + 225 = 1750$$

STEP 5   Divide the sum by N to get the variance.

$$\text{variance} = 1750 \div 6 = 291.7$$

STEP 6   Take the square root of the variance to get the standard deviation.

Hence, the standard deviation equals $\sqrt{291.7}$, or 17.1. It is helpful to make a table, as follows:

A Values (X)	B $X - \mu$	C $(X - \mu)^2$
10	−25	625
60	+25	625
50	+15	225
30	−5	25
40	+5	25
20	−15	225
		1750

Column A contains the raw data, X. Column B contains the differences obtained in Step 2, $X - \mu$. Column C contains the squares of the differences obtained in Step 3.

....

The preceding computational procedure reveals several things. First, the square root of the variance gives the standard deviation; vice versa, squaring the standard deviation gives the variance. Second, the variance is actually the average of the square of the distance that each value is from the mean. Therefore, if the values are near the mean, the variance will be small. In contrast, if the values are farther away from the mean, the variance will be large.

One might wonder why the squared distances are used instead of the actual distances. The reason is that the sum of the distances will always be zero. To verify this result for a specific case, add the values in column B of the table in Example 3–26. When each value is squared, the negative signs are eliminated.

Finally, why is it necessary to take the square root? The reason is that since the distances were squared, the units of the resultant numbers are the squares of the units of the original raw data. Taking the square root of the variance puts the standard deviation in the same units as the raw data.

When taking the square root, always use its positive or principal value, since the variance and standard deviation of a data set can never be negative.

The **variance** is the average of the squares of the distance each value is from the mean. The symbol for the population variance is σ^2 (σ is the Greek lowercase letter sigma).

The formula for the population variance is

$$\sigma^2 = \frac{\Sigma (X - \mu)^2}{N}$$

where

  X = individual value

  μ = population mean

  N = population size

The **standard deviation** is the square root of the variance. The symbol for the population standard deviation is σ.

The corresponding formula for the standard deviation is

$$\sigma = \sqrt{\sigma^2} = \sqrt{\frac{\Sigma (X - \mu)^2}{N}}$$

....

Example 3–27

Find the variance and standard deviation for brand B paint data in Example 3–23.

  35, 45, 30, 35, 40, 25

Solution

STEP 1   Find the mean.

$$\mu = \frac{\Sigma X}{N} = \frac{35 + 45 + 30 + 35 + 40 + 25}{6} = \frac{210}{6} = 35$$

STEP 2   Subtract the mean from each value and place the result in column B of the table.

STEP 3   Square each result and place the squares in column C of the table.

A	B	C
X	$X - \mu$	$(X - \mu)^2$
35	0	0
45	10	100
30	-5	25
35	0	0
40	5	25
25	-10	100

STEP 4   Find the sum of the squares in column C.

$$\Sigma (X - \mu)^2 = 0 + 100 + 25 + 0 + 25 + 100 = 250$$

STEP 5   Divide the sum by N to get the variance.

$$\sigma^2 = \frac{\Sigma (X - \mu)^2}{N} = \frac{250}{6} = 41.7$$

STEP 6   Take the square root to get the standard deviation.

$$\sigma = \sqrt{\frac{\Sigma (X - \mu)^2}{N}} = \sqrt{41.7} = 6.5$$

Hence, the standard deviation is 6.5.

Since the standard deviation of brand A is 17.1 (see Example 3–26), and the standard deviation of brand B is 6.5, the data are more variable for brand A. In summary, when the means are equal, the larger the variance or standard deviation is, the more variable the data are.

....

The Unbiased Estimator

When computing the variance for a sample, one might expect the following formula would be used:

$$\frac{\Sigma (X - \overline{X})^2}{n}$$

where $\overline{X}$ is the sample mean and n is the sample size. *This formula is not usually used, however, since in most cases the purpose of calculating the statistic is to estimate the corresponding parameter.* For example, the sample mean $\overline{X}$ is used to estimate the population mean μ. The formula

$$\frac{\Sigma (X - \overline{X})^2}{n}$$

produces what is called a *biased estimate* of the population variance—an estimate that is different from the expected value of a population parameter. When the population is large and the sample is small (usually less than 30), the variance computed by this formula would underestimate the population variance. Therefore, instead of dividing by n, find the variance of the sample by dividing by $n - 1$, giving a slightly larger value and an *unbiased* estimate of the population variance.

Interesting Fact

Each person receives on average 598 pieces of mail per year. (*The Harper's Index Book*, p. 22)

The **unbiased estimator** of the population variance is a statistic whose value approximates the expected value of a population variance. It is denoted by s^2, and its formula is

$$s^2 = \frac{\Sigma (X - \bar{X})^2}{n - 1}$$

where

$\bar{X}$ = sample mean

n = sample size

To find the standard deviation of a sample, one must take the square root of the sample variance, which was found by using the preceding formula.

Formula for the Sample Standard Deviation

The standard deviation of a sample (denoted by s) is

$$s = \sqrt{s^2} = \sqrt{\frac{\Sigma (X - \bar{X})^2}{n - 1}}$$

where

X = individual value

$\bar{X}$ = sample mean

n = sample size

Shortcut formulas for computing the variance and standard deviation are presented next and will be used in the remainder of the chapter and in the exercises. These formulas are mathematically equivalent to the preceding formulas and do not involve using the mean. They save time when repeated subtracting and squaring occur in the original formulas.

Shortcut Formulas for s^2 and s

The shortcut formulas for computing the variance and standard deviation for data obtained from samples are as follows.

Variance	**Standard deviation**
$s^2 = \dfrac{\Sigma X^2 - (\Sigma X)^2/n}{n - 1}$	$s = \sqrt{\dfrac{\Sigma X^2 - (\Sigma X)^2/n}{n - 1}}$

The next two examples explain how to use the shortcut formulas.

Example 3–28

For a newly created position, the manager interviewed the following numbers of applicants each day over a five-day period: 16, 19, 15, 15, and 14. Find the variance and standard deviation.

Solution

STEP 1   Find the sum of the values.

$$\Sigma X = 16 + 19 + 15 + 15 + 14 = 79$$

STEP 2   Square each value and find the sum.

$$\Sigma X^2 = 16^2 + 19^2 + 15^2 + 15^2 + 14^2 = 1263$$

STEP 3   Substitute in the formulas and solve.

$$s^2 = \frac{\Sigma X^2 - [(\Sigma X)^2/n]}{n-1} = \frac{1263 - [(79)^2/5]}{4} = 3.7$$
$$s = \sqrt{3.7} = 1.9$$

Hence, the sample standard deviation is 1.9.

....

Note that ΣX^2 is not the same as $(\Sigma X)^2$. The notation ΣX^2 means to square the values first, then sum; $(\Sigma X)^2$ means to sum the values first, then square.

Finding the Variance and Standard Deviation for Grouped Data

The procedure for finding the variance and standard deviation for grouped data is similar to that for finding the mean for grouped data, and it uses the midpoints of each class.

....

Example 3–29

Find the variance and the standard deviation for the frequency distribution of the data in Example 2–7 in Chapter 2. The data represent the number of miles 20 runners ran during one week.

Class	Frequency	Midpoint
5.5–10.5	1	8
10.5–15.5	2	13
15.5–20.5	3	18
20.5–25.5	5	23
25.5–30.5	4	28
30.5–35.5	3	33
35.5–40.5	2	38

Solution

STEP 1   Make a table as shown, and find the midpoint of each class.

A	B	C	D	E
Class	Frequency (f)	Midpoint (X_m)	$f \cdot X_m$	$f \cdot X_m^2$
5.5–10.5	1	8		
10.5–15.5	2	13		
15.5–20.5	3	18		
20.5–25.5	5	23		
25.5–30.5	4	28		
30.5–35.5	3	33		
35.5–40.5	2	38		

STEP 2   Multiply the frequency by the midpoint for each class and place the products in column D.

$$1 \cdot 8 = 8 \qquad 2 \cdot 13 = 26 \qquad \ldots \qquad 2 \cdot 38 = 76$$

STEP 3   Multiply the frequency by the square of the midpoint and place the products in column E.

$$1 \cdot 8^2 = 64 \qquad 2 \cdot 13^2 = 338 \qquad \ldots \qquad 2 \cdot 38^2 = 2888$$

STEP 4   Find the sum of columns B, D, and E. The sum of column B is n, the sum of column D is $\sum f \cdot X_m$, and the sum of column E is $\sum f \cdot X_m^2$. The completed table follows.

A Class	B Frequency	C Midpoint	D $f \cdot X_m$	E $f \cdot X_m^2$
5.5–10.5	1	8	8	64
10.5–15.5	2	13	26	338
15.5–20.5	3	18	54	972
20.5–25.5	5	23	115	2645
25.5–30.5	4	28	112	3136
30.5–35.5	3	33	99	3267
35.5–40.5	2	38	76	2888
	$n = 20$		$\sum f \cdot X_m = 490$	$\sum f \cdot X_m^2 = 13{,}310$

STEP 5   Substitute in the formula and solve for s^2 to get the variance.

$$s^2 = \frac{\sum f \cdot X_m^2 - [(\sum f \cdot X_m)^2/n]}{n-1}$$
$$= \frac{13{,}310 - [(490)^2/20]}{20-1} = 68.7$$

STEP 6   Take the square root to get the standard deviation.

$$s = \sqrt{68.7} = 8.3$$

Be sure to use the number found in the sum of column B (i.e., the sum of the frequencies) for n. Do not use the number of classes.

When the data are ungrouped, the same procedure is used, except that the class value, X, is used as the midpoint. The next example illustrates this procedure.

Example 3–30

Holmes Appliance reported the number of television sets sold per month over a two-year period. Find the variance and standard deviation for the data.

Number of sets sold	Frequency (months)
5	2
6	3
7	8
8	1
9	6
10	4

Solution

STEP 1   Make a table as shown.

A Sets	B Frequency	C $f \cdot X$	D $f \cdot X^2$
5	2		
6	3		
7	8		
8	1		
9	6		
10	4		

STEP 2 Multiply the frequency by the values in column A, and place the products in column C.

$$2 \cdot 5 = 10 \qquad 3 \cdot 6 = 18 \qquad \text{etc.}$$

STEP 3 Multiply the frequency by the values squared, and place the products in column D.

$$2 \cdot 5^2 = 50 \qquad 3 \cdot 6^2 = 108 \qquad \text{etc.}$$

STEP 4 Find the sums of columns B, C, and D. The completed table follows.

A Sets	B Frequency	C $f \cdot X$	D $f \cdot X^2$
5	2	10	50
6	3	18	108
7	8	56	392
8	1	8	64
9	6	54	486
10	4	40	400
	$n = 24$	$\Sigma f \cdot X = 186$	$\Sigma f \cdot X^2 = 1500$

STEP 5 Substitute in the formula to get the variance.

$$s^2 = \frac{\Sigma f \cdot X^2 - [(\Sigma f \cdot X)^2/n]}{n-1} = \frac{1500 - [(186)^2/24]}{23} = 2.5$$

STEP 6 Take the square root to get the standard deviation.

$$s = \sqrt{2.5} = 1.6$$

The procedure for finding the variance and standard deviation for grouped data is summarized in Procedure Table 5.

Procedure Table 5

Finding the Sample Variance and Standard Deviation for Grouped Data

STEP 1 Make a table as shown, and find the midpoint of each class.

A Class	B Frequency	C Midpoint	D $f \cdot X_m$	E $f \cdot X_m^2$

STEP 2 Multiply the frequency by the midpoint for each class, and place the products in column D.

STEP 3 Multiply the frequency by the square of the midpoint, and place the products in column E.

STEP 4 Find the sums of columns B, D, and E. (The sum of column B is n. The sum of column D is $\Sigma f \cdot X_m$. The sum of column E is $\Sigma f \cdot X_m^2$.)

STEP 5 Substitute in the formula and solve to get the variance.

$$s^2 = \frac{\Sigma f \cdot X_m^2 - [(\Sigma f \cdot X_m)^2/n]}{n-1}$$

STEP 6 Take the square root to get the standard deviation.

Table 3–2 summarizes the measures of variation.

Table 3–2 Summary of Measures of Variation		
Measure	**Definition**	**Symbol(s)**
Range	Distance between highest value and lowest value	R
Variance	Average of the squares of the distance each value is from the mean	σ^2, s^2
Standard deviation	Square root of the variance	σ, s

Uses of the Variance and Standard Deviation

1. As previously stated, variances and standard deviations can be used to determine the spread of the data. If the variance or standard deviation is large, the data are more dispersed. This information is useful in comparing two or more data sets to determine which is more (most) variable.
2. The measures of variance and standard deviation are used to determine the consistency of a variable. For example, in the manufacture of fittings, like nuts and bolts, the variation in the diameters must be small or the parts will not fit together.
3. The variance and standard deviation are used to determine the number of data values that fall within a specified interval in a distribution. For example, Chebyshev's theorem (explained later) shows that for any distribution, at least 75% of the data values will fall within two standard deviations of the mean.
4. Finally, the variance and standard deviation are used quite often in inferential statistics. These uses will be shown in later chapters of this textbook.

Coefficient of Variation   Whenever two samples have the same units of measure, the variance and standard deviation for each can be compared directly. For example, suppose an automobile dealer wanted to compare the standard deviation of miles driven for the cars he received as trade-ins on new cars. He found that for a specific year, the standard deviation for Buicks was 422 miles and the standard deviation for Cadillacs was 350 miles. He could say that the variation in mileage was greater in the Buicks. But what if a manager wanted to compare the standard deviations of two different variables, like the number of sales per salesperson over a three-month period and the commissions made by these salespeople?

A statistic that allows one to compare standard deviations when the units are different, as in this example, is called the *coefficient of variation*.

The **coefficient of variation** is the standard deviation divided by the mean. The result is expressed as a percentage.

For samples,   For populations,

$$\text{CVar} = \frac{s}{\bar{X}} \cdot 100\% \qquad \text{CVar} = \frac{\sigma}{\mu} \cdot 100\%$$

Example 3–31

The mean of the number of sales of cars over a three-month period is 87 and the standard deviation is 5. The mean of the commissions is $5225 and the standard deviation is $773. Compare the variations of the two.

Solution

The coefficients of variation are

$$\text{CVar} = \frac{s}{\bar{X}} = \frac{5}{87} \cdot 100\% = 5.7\% \qquad \text{sales}$$

$$\text{CVar} = \frac{773}{5225} \cdot 100\% = 14.8\% \qquad \text{commissions}$$

Since the coefficient of variation is larger for commissions, the commissions are more variable than the sales.

Example 3–32

The mean for the number of pages of a sample of women's fitness magazines is 132, with a variance of 23; the mean for a sample of men's fitness magazines is 182, with a variance of 62. Compare the variations of the two magazines.

Solution

The cofficients of variation are

$$\text{CVar} = \frac{\sqrt{23}}{132} \cdot 100\% = 3.6\% \qquad \text{women's}$$

$$\text{CVar} = \frac{\sqrt{62}}{182} \cdot 100\% = 4.3\% \qquad \text{men's}$$

The number of pages in the men's magazines is more variable than in the women's, since the coefficient of variation is larger for the men's magazines.

Chebyshev's Theorem

As stated previously, the variance and standard deviation of a variable can be used to determine the spread or dispersion of a variable. That is, the larger the variance or standard deviation, the more the data values are dispersed. For example, if two variables measured in the same units have the same mean, say 70, and variable one has a standard deviation of 1.5 while variable two has a standard deviation of 10, then the data for variable two will be more spread out than the data for variable one. **Chebyshev's theorem,** developed by the Russian mathematician Chebyshev, specifies the proportions of the spread in terms of the standard deviation.

Chebyshev's theorem   The proportion of values from a data set that will fall within k standard deviations of the mean will be at least $1 - 1/k^2$, where k is a number greater than 1 (k is not necessarily an integer).

This theorem states that at least $\frac{3}{4}$, or 75%, of the data values will fall within two standard deviations of the mean of the data set. This result is found by substituting $k = 2$ in the formula.

$$1 - \frac{1}{k^2} \qquad \text{or} \qquad 1 - \frac{1}{2^2} = 1 - \frac{1}{4} = \frac{3}{4} = 75\%$$

In this example, since variable one has a mean of 70 and a standard deviation of 1.5, at least $\frac{3}{4}$, or 75%, of the data values fall between 67 and 73. These values are found by adding two standard deviations to the mean and subtracting two standard deviations from the mean, as shown.

$$70 + 2(1.5) = 70 + 3 = 73$$

and

$$70 - 2(1.5) = 70 - 3 = 67$$

For variable two, at least $\frac{3}{4}$, or 75%, of the data values fall between 50 and 90. Again, these values are found by adding and subtracting, respectively, two standard deviations to and from the mean.

$$70 + 2(10) = 70 + 20 = 90$$

and

$$70 - 2(10) = 70 - 20 = 50$$

Furthermore, the theorem states that at least $\frac{8}{9}$, or 88.89%, of the data values will fall within three standard deviations of the mean. This result is found by letting $k = 3$ and substituting in the formula.

$$1 - \frac{1}{k^2} \quad \text{or} \quad 1 - \frac{1}{3^2} = 1 - \frac{1}{9} = \frac{8}{9} = 88.89\%$$

For variable one, at least $\frac{8}{9}$, or 88.89%, of the data values fall between 65.5 and 74.5, since

$$70 + 3(1.5) = 70 + 4.5 = 74.5$$

and

$$70 - 3(1.5) = 70 - 4.5 = 65.5$$

For variable two, at least $\frac{8}{9}$, or 88.89%, of the data values fall between 40 and 100.

This theorem can be applied to any distribution regardless of its shape (see Figure 3–4).

Figure 3–4
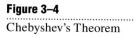
Chebyshev's Theorem

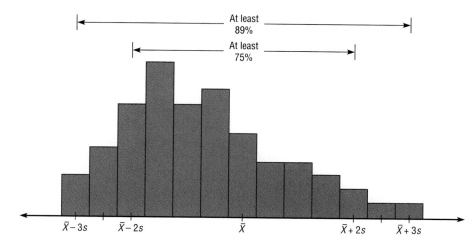

The next two examples illustrate the application of Chebyshev's theorem.

Example 3–33

The mean price of houses in a certain neighborhood is $50,000, and the standard deviation is $10,000. Find the price range for which at least 75% of the houses will sell.

Solution

Chebyshev's theorem states that $\frac{3}{4}$, or 75%, of the data values will fall within two standard deviations of the mean. Thus,

$$\$50,000 + 2(\$10,000) = \$50,000 + \$20,000 = \$70,000$$

and

$$\$50,000 - 2(\$10,000) = \$50,000 - \$20,000 = \$30,000$$

Hence, at least 75% of all homes sold in the area will have a price range from $30,000 to $70,000.

Chebyshev's theorem can be used to find the minimum percentage of data values that will fall between any two given values. The procedure is shown in the next example.

Example 3–34

A survey of local companies found that the mean amount of travel allowance for executives was $0.25 per mile. The standard deviation was $0.02. Using Chebyshev's theorem, find the minimum percentage of the data values that will fall between $0.20 and $0.30.

Solution

STEP 1 Subtract the mean from the larger value.

$$\$0.30 - \$0.25 = \$0.05$$

STEP 2 Divide the difference by the standard deviation to get k.

$$k = \frac{0.05}{0.02} = 2.5$$

STEP 3 Use Chebyshev's theorem to find the percentage.

$$1 - \frac{1}{k^2} = 1 - \frac{1}{2.5^2} = 1 - \frac{1}{6.25} = 1 - 0.16 = 0.84 \quad \text{or} \quad 84\%$$

Hence, at least 84% of the data values will fall between $0.20 and $0.30

The Empirical (Normal) Rule

Chebyshev's theorem applies to any distribution regardless of its shape. However, when a distribution is *bell-shaped* (or what is called *normal*), the following statements, which make up the **empirical rule,** are true.

Approximately 68% of the data values will fall within one standard deviation of the mean.

Approximately 95% of the data values will fall within two standard deviations of the mean.

Approximately 99.7% of the data values will fall within three standard deviations of the mean.

For example, suppose that the scores on a national achievement exam have a mean of 480 and a standard deviation of 90. If these scores are normally distributed, then approximately 68% will fall between 390 and 570 (480 + 90 = 570 and 480 − 90 = 390). Approximately 95% of the scores will fall between 300 and 660 (480 + 2 · 90 = 660 and 480 − 2 · 90 = 300). Approximately 99.7% will fall between 210 and 750 (480 + 3 · 90 = 750 and 480 − 3 · 90 = 210). See Figure 3–5.

Figure 3–5

The Empirical Rule

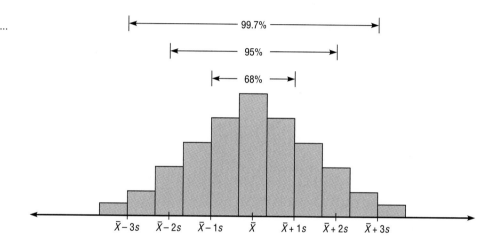

Exercises

3–42. Name three measures of variation.

3–43. (**W**) What is the relationship between the variance and the standard deviation?

3–44. (**W**) Why might the range *not* be the best estimate of variability?

3–45. What are the symbols used to represent the population variance and standard deviation?

3–46. What are the symbols used to represent the sample variance and standard deviation?

3–47. (**W**) Why is the unbiased estimator of variance used?

3–48. The three data sets below have the same mean and range, but is the variation the same? Prove your answer by computing the standard deviation. Assume the data were obtained from samples.

a. 5, 7, 9, 11, 13, 15, 17
b. 5, 6, 7, 11, 15, 16, 17
c. 5, 5, 5, 11, 17, 17, 17

For Exercises 3–49 through 3–59, find the range, variance, and standard deviation. Assume the data represent samples, and use the shortcut formula for the unbiased estimator to compute the variance and standard deviation.

3–49. Twelve students were given an arithmetic test, and the times (in minutes) to complete it were as follows.

10, 9, 12, 11, 8, 15, 9, 7, 8, 6, 12, 10

3–50. Ten used trail bikes are randomly selected, and the odometer reading of each is recorded as follows.

1902, 103, 653, 1901, 788,
361, 216, 363, 223, 656

3–51. Fifteen students were selected and asked how many hours each studied for the final exam in statistics. Their answers are recorded here.

8, 6, 3, 0, 0, 5, 9, 2, 1, 3, 7, 10, 0, 3, 6

3–52. The weights of nine football players are recorded as follows.

206, 215, 305, 297, 265, 282, 301, 255, 261

3–53. Shown below are the numbers of stories in the 11 tallest buildings in St. Paul, Minnesota.

32, 36, 46, 20, 32, 18, 16, 34, 26, 27, 26

Source: Reprinted with permission from *The World Almanac and Book of Facts*, 1995. Copyright © 1994 PRIMEDIA Reference Inc. All rights reserved.

3–54. The heights (in inches) of nine male army recruits are shown here.

78, 72, 68, 73, 75, 69, 74, 73, 72

3–55. The manufacturer's ratings (in amps) of 22 upright vacuum cleaners are shown here.

7.8, 7, 12, 10, 12, 7, 7.2, 7.3, 11, 7, 12,
11, 9.5, 9, 9, 10, 7.2, 9.5, 11, 10, 7.5, 10

Source: *Consumer Reports*, April 1995, p. 274.

3–56. The number of calls received by the Internal Revenue Service office each business day in July appears here.

18, 12, 25, 16, 27, 32, 25, 15, 23, 22, 37, 16,
25, 19, 16, 25, 19, 16, 29, 38, 29, 30, 21

3–57. The number of calories in 12 randomly selected microwave dinners is shown here.

560, 832, 780, 650, 470, 920,
1090, 970, 495, 550, 605, 735

3–58. The rated tread wear of 11 automobile tires tested by a consumer group is shown here.

280, 260, 300, 360, 160, 320,
300, 240, 340, 300, 300

Source: *Consumer Reports*, January 1995, pp. 40–41.

3–59. The number of words printed in each of 12 randomly selected third-grade storybooks is listed below.

502, 213, 335, 197, 414, 469,
497, 367, 409, 297, 309, 414

For Exercises 3–60 through 3–69, find the variance and standard deviation.

3–60. For 108 randomly selected college students, the following IQ frequency distribution was obtained.

Class limits	Frequency
90–98	6
99–107	22
108–116	43
117–125	28
126–134	9

3–61. The costs per load (in cents) of 35 laundry detergents tested by a consumer organization are shown below.

Class limits	Frequency
13–19	2
20–26	7
27–33	12
34–40	5
41–47	6
48–54	1
55–61	0
62–68	2

Source: *Consumer Reports*, February 1995.

3–62. Thirty automobiles were tested for fuel efficiency (in miles per gallon). The following frequency distribution was obtained.

Class boundaries	Frequency
7.5–12.5	3
12.5–17.5	5
17.5–22.5	15
22.5–27.5	5
27.5–32.5	2

3–63. In a class of 29 students, the following grade distribution was recorded (A = 4, B = 3, C = 2, D = 1, F = 0).

Grade	Frequency
0	1
1	3
2	5
3	14
4	6

3–64. In a study of reaction times to a specific stimulus, a psychologist recorded the following data (in seconds).

Class limits	Frequency
2.1–2.7	12
2.8–3.4	13
3.5–4.1	7
4.2–4.8	5
4.9–5.5	2
5.6–6.2	1

3–65. Eighty randomly selected lightbulbs were tested to determine their lifetimes (in hours). The following frequency distribution was obtained.

Class boundaries	Frequency
52.5–63.5	6
63.5–74.5	12
74.5–85.5	25
85.5–96.5	18
96.5–107.5	14
107.5–118.5	5

3–66. The following data represent the net worth (in millions of dollars) of 50 businesses in a large city.

Class limits	Frequency
10–20	5
21–31	10
32–42	3
43–53	7
54–64	18
65–75	7

3–67. The following data represent the scores (in words per minute) of 25 typists on a speed test.

Class limits	Frequency
54–58	2
59–63	5
64–68	8
69–73	0
74–78	4
79–83	5
84–88	1

3–68. The following frequency distribution represents the commission earned (in dollars) by 100 salespeople employed at several branches of a large chain store.

Class limits	Frequency
150–158	5
159–167	16
168–176	20
177–185	21
186–194	20
195–203	15
204–212	3

3–69. This frequency distribution represents the data obtained from a sample of word-processor repairers. The values are the days between service calls on 80 machines.

Class boundaries	Frequency
25.5–28.5	5
28.5–31.5	9
31.5–34.5	32
34.5–37.5	20
37.5–40.5	12
40.5–43.5	2

3–70. The average score of the students in one calculus class is 110, with a standard deviation of 5; the average score of students in a statistics class is 106, with a standard deviation of 4. Which class is more variable in terms of scores?

3–71. The average price of the Panther convertible is $40,000, with a standard deviation of $4000. The average price of the Suburban station wagon is $20,000, with a standard deviation of $2000. Compare the variances of both models.

3–72. The average score on an English final examination was 85, with a standard deviation of 5; the average score on a history final exam was 110, with a standard deviation of 8. Which class was more variable?

3–73. The average age of the accountants at Three Rivers Corp. is 26, with a standard deviation of 6; the average salary of the accountants is $31,000, with a standard deviation of $4000. Compare the variations of age and income.

3–74. Using Chebyshev's theorem, solve the following problems for a distribution with a mean of 80 and a standard deviation of 10.
a. At least what percentage of values will fall between 60 and 100?
b. At least what percentage of values will fall between 65 and 95?

3–75. The mean of a distribution is 20 and the standard deviation is 2. Answer each. Use Chebyshev's theorem.
a. At least what percentage of the values will fall between 10 and 30?
b. At least what percentage of the values will fall between 12 and 28?

3–76. In a distribution of 200 values, the mean is 50 and the standard deviation is 5. Answer each. Use Chebyshev's theorem.

a. At least how many values will fall between 30 and 70?

b. At most how many values will be less than 40 or more than 60?

3–77. A sample of the hourly wages of employees who work in restaurants in a large city has a mean of $5.02 and a standard deviation of $0.09. Using Chebyshev's theorem, find the range in which at least 75% of the data values will fall.

3–78. A sample of the labor costs per hour to assemble a certain product has a mean of $2.60 and a standard deviation of $0.15. Using Chebyshev's theorem, find the values in which at least 88.89% of the data will lie.

3–79. A survey of a number of the leading brands of cereal shows that the mean content of potassium per serving is 95 milligrams, and the standard deviation is 2 milligrams. Find the values in which at least 88.89% of the data will fall. Use Chebyshev's theorem.

3–80. The average score on a special test of knowledge of wood refinishing has a mean of 53 and a standard deviation of 6. Using Chebyshev's theorem, find the range of values in which at least 75% of the scores will lie.

3–81. The average of the number of trials it took a sample of mice to learn to traverse a maze was 12. The standard deviation was 3. Using Chebyshev's theorem, find the minimum percentage of data values that will fall in the range of 4 to 20 trials.

3–82. The average cost of a certain type of grass seed is $4.00 per box. The standard deviation is $0.10. Using Chebyshev's theorem, find the minimum percentage of data values that will fall in the range of $3.82 to $4.18.

*3–83. (**W**) For the following data set, find the mean and standard deviation of the variable. The data represent the serum cholesterol level of 30 individuals. Count the number of data values that fall within two standard deviations of the mean. Compare this with the number obtained from Chebyshev's theorem. Comment on the answer.

211	240	255	219	204
200	212	193	187	205
256	203	210	221	249
231	212	236	204	187
201	247	206	187	200
237	227	221	192	196

*3–84. (**W**) For the following data set, find the mean and standard deviation of the variable. The data represent the ages of 30 customers who ordered a product advertised on television. Count the number of data values that fall within two standard deviations of the mean. Compare this with the number obtained from Chebyshev's theorem. Comment on the answer.

42	44	62	35	20
30	56	20	23	41
55	22	31	27	66
21	18	24	42	25
32	50	31	26	36
39	40	18	36	22

*3–85. Using Chebyshev's theorem, complete the table to find the minimum percentage of the data values that fall within k standard deviations of the mean.

k	1.5	2	2.5	3	3.5
Percent					

*3–86. (**W**) Use the following data set: 10, 20, 30, 40, 50.

a. Find the standard deviation.

b. Add 5 to each value, and then find the standard deviation.

c. Subtract 5 from each value, and find the standard deviation.

d. Multiply each value by 5, and find the standard deviation.

e. Divide each value by 5, and find the standard deviation.

f. Generalize the results of parts b through e.

g. Compare these results with those in Exercise 3–38.

*3–87. The **mean deviation** is found by using the following formula:

$$\text{mean deviation} = \frac{\Sigma |X - \overline{X}|}{n}$$

where

X = value

$\overline{X}$ = mean

n = number of values

$|\ \ |$ = absolute value

Find the mean deviation for the following data.

5, 9, 10, 11, 11, 12, 15, 18, 20, 22

*3–88. A measure to determine the skewness of a distribution is called the *Pearson coefficient of skewness.* The formula is

$$\text{skewness} = \frac{3(\overline{X} - MD)}{s}$$

The values of the coefficient usually range from −3 to +3. When the distribution is symmetrical, the coefficient is zero; when the distribution is positively skewed, it is positive; and when the distribution is negatively skewed, it is negative.

Using the formula, find the coefficient of skewness for each distribution, and describe the shape of the distribution.

a. Mean = 10, median = 8, standard deviation = 3.
b. Mean = 42, median = 45, standard deviation = 4.
c. Mean = 18.6, median = 18.6, standard deviation = 1.5.
d. Mean = 98, median = 97.6, standard deviation = 4.

3–4

Measures of Position

Objective 3. Identify the position of data in a data set.

In addition to measures of central tendency and measures of variation, there are also measures of position or location. These measures include standard scores, percentiles, deciles, and quartiles. They are used to locate the relative position of a data value in the data set. For example, if a value is located at the 80th percentile, it means that 80% of the values fall below it in the distribution and 20% of the values fall above it. The *median* is the value that corresponds to the 50th percentile, since half of the values fall below it and half of the values fall above it. This section discusses these measures of position.

Standard Scores

There is an old saying that states, "You can't compare apples and oranges." But with the use of statistics, it can be done to some extent. Suppose that a student scored 90 on a music test and 45 on an English exam. Direct comparison of raw scores is impossible, since the exams are not equivalent in terms of number of questions, value of each question, and so on. However, a comparison of a relative standard similar to both can be made. This comparison uses the mean and standard deviation and is called a standard score or *z* score.

A **standard score** or ***z* score** for a value is obtained by subtracting the mean from the value and dividing the result by the standard deviation. The symbol for a standard score is *z*. The formula is

$$z = \frac{\text{value} - \text{mean}}{\text{standard deviation}}$$

For samples, the formula is

$$z = \frac{X - \bar{X}}{s}$$

For populations, the formula is

$$z = \frac{X - \mu}{\sigma}$$

The *z* score represents the number of standard deviations a data value falls above or below the mean.

Example 3–35

A student scored 65 on a calculus test that had a mean of 50 and a standard deviation of 10; she scored 30 on a history test with a mean of 25 and a standard deviation of 5. Compare her relative positions on the two tests.

Solution

First, find the z scores. For calculus the z score is

$$z = \frac{X - \overline{X}}{s} = \frac{65 - 50}{10} = 1.5$$

For history the z score is

$$z = \frac{30 - 25}{5} = 1.0$$

Since the z score for calculus is larger, her relative position in the calculus class is higher than her relative position in the history class.

Note that if the z score is positive, the score is above the mean. If the z score is 0, the score is the same as the mean. And if the z score is negative, the score is below the mean.

Example 3–36

Find the z score for each test and state which is higher.

Test A	$X = 38$	$\overline{X} = 40$	$s = 5$
Test B	$X = 94$	$\overline{X} = 100$	$s = 10$

Solution

For test A,

$$z = \frac{X - \overline{X}}{s} = \frac{38 - 40}{5} = -0.4$$

For test B,

$$z = \frac{94 - 100}{10} = -0.6$$

The score for test A is relatively higher than the score for test B.

When all data for a variable are transformed into z scores, the resulting distribution will have a mean of 0 and a standard deviation of 1. A z score, then, is actually the number of standard deviations each variable is from the mean for a specific distribution. In Example 3–35, the calculus score of 65 was actually 1.5 standard deviations above the mean of 50. This will be explained in more detail in Chapter 7.

Percentiles

Percentiles are position measures used in educational and health-related fields to indicate the position of an individual in a group.

In many situations, the graphs and tables showing the percentiles for various measures such as test scores, heights, or weights have already been completed.

Table 3–3 shows the percentile ranks for scaled scores on the Test of English as a Foreign Language. If a student had a scaled score of 58 for Section 1 (listening and comprehension), that student would have a percentile rank of 81. Hence, that student did better than 81% of the students who took Section 1 of the exam.

Table 3–3	Percentile Ranks and Scaled Scores on the Test of English as a Foreign Language*				
Scaled score	Section 1: Listening comprehension	Section 2: Structure and written expression	Section 3: Vocabulary and reading comprehension	Total scaled score	Percentile rank
68	99	98			
66	98	96	98	660	99
64	96	94	96	640	97
62	92	90	93	620	94
60	87	84	88	600	89
→58	81	76	81	580	82
56	73	68	72	560	73
54	64	58	61	540	62
52	54	48	50	520	50
50	42	38	40	500	39
48	32	29	30	480	29
46	22	21	23	460	20
44	14	15	16	440	13
42	9	10	11	420	9
40	5	7	8	400	5
38	3	4	5	380	3
36	2	3	3	360	1
34	1	2	2	340	1
32		1	1	320	
30		1	1	300	
Mean	51.5	52.2	51.4	Mean	517
S.D.	7.1	7.9	7.5	S.D.	68

*Based on the total group of 1,178,193 examinees tested from July 1989 through June 1991.
Source: Reprinted by permission of Educational Testing Service, the copyright owner.

Figure 3–6 shows percentiles in graphic form of weights of girls from ages 2 to 18. To find the percentile rank of an 11-year-old who weighs 82 pounds, start at the 82-pound weight on the left axis and move horizontally to the right. Find the 11 on the horizontal axis and move up vertically. The two lines meet at the 50th percentile curved line; hence, an 11-year-old girl who weighs 82 pounds is in the 50th percentile for her age group. If the lines do not meet exactly on one of the curved percentile lines, then the percentile rank must be approximated.

Percentiles are also used to compare an individual's test score with the national norm. For example, tests such as the National Educational Development Test

Figure 3–6

Weights of Girls by Age
and Percentile Rankings

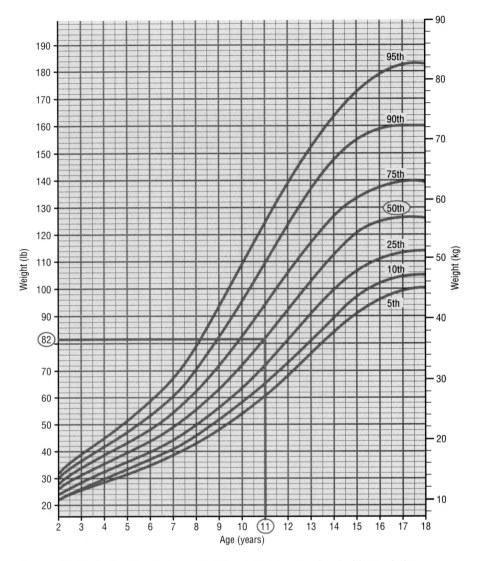

Source: Distributed by Mead Johnson Nutritional Division. Reprinted with permission.

(NEDT) are taken by students in ninth or tenth grade. A student's scores are compared with those of other students locally and nationally by using percentile ranks. A similar test for elementary school students is called the California Achievement Test.

Percentiles are not the same as percentages. That is, if a student gets 72 correct answers out of a possible 100, she obtains a percentage score of 72. There is no indication of her position with respect to the rest of the class. She could have scored the highest, the lowest, or somewhere in between. On the other hand, if a raw score of 72 corresponds to the 64th percentile, then she did better than 64% of the students in her class.

Percentiles are symbolized by

$$P_1, P_2, P_3, \ldots, P_{99}$$

and divide the distribution into 100 groups. Percentile graphs can be constructed as shown in the next example. Percentile graphs use the same values as the cumulative relative frequency graphs described in Section 2–3, except that the proportions have been converted to percents.

Example 3–37

The frequency distribution for the systolic blood pressure readings (in millimeters of mercury, mmHg) of 200 randomly selected college students follows. Construct a percentile graph.

A Class boundaries	B Frequency	C Cumulative frequency	D Cumulative percent
89.5–104.5	24		
104.5–119.5	62		
119.5–134.5	72		
134.5–149.5	26		
149.5–164.5	12		
164.5–179.5	4		
	200		

Solution

STEP 1 Find the cumulative frequencies and place them in column C.

STEP 2 Find the cumulative percentages and place them in column D. To do this step, use the formula

$$\text{cumulative \%} = \frac{\text{cumulative frequency}}{n} \cdot 100\%$$

For the first class,

$$\text{cumulative \%} = \frac{24}{200} \cdot 100\% = 12\%$$

The completed table is shown next.

A Class boundaries	B Frequency	C Cumulative frequency	D Cumulative percent
89.5–104.5	24	24	12
104.5–119.5	62	86	43
119.5–134.5	72	158	79
134.5–149.5	26	184	92
149.5–164.5	12	196	98
164.5–179.5	4	200	100
	200		

STEP 3 Graph the data, using class boundaries for the x axis and the percentages for the y axis, as shown in Figure 3–7.

Figure 3–7

Percentile Graph for
Example 3–37

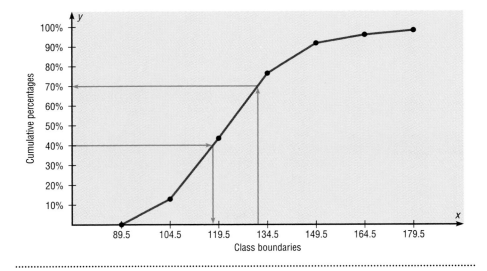

Once a cumulative percentage graph has been constructed, one can find the approximate corresponding percentile ranks for given blood pressure values and find approximate blood pressure values for given percentile ranks.

For example, to find the percentile rank of a blood pressure reading of 130, find 130 on the *x* axis of Figure 3–7, and draw a vertical line to the graph. Then move horizontally to the value on the *y* axis. Note that a blood pressure of 130 corresponds to approximately the 70th percentile.

If the value that corresponds to the 40th percentile is desired, start on the *y* axis at 40 and draw a horizontal line to the graph. Then draw a vertical line to the *x* axis, and read the value. In Figure 3–7, the 40th percentile corresponds to a value of approximately 118. Thus, if a person has a blood pressure of 118, he or she is at the 40th percentile.

Finding values and the corresponding percentile ranks by using a graph yields only approximate answers. Several mathematical methods exist for computing percentiles for data. They can be used to find the approximate percentile rank of a data value or to find a data value corresponding to a given percentile. When the data set is large (100 or more), these methods yield better results. The next several examples show these methods.

Percentile Formula

The percentile corresponding to a given value (X) is computed by using the following formula:

$$\text{percentile} = \frac{\text{number of values below } X + 0.5}{\text{total number of values}} \cdot 100\%$$

Example 3–38

A teacher gives a 20-point test to 10 students. The scores are shown below. Find the percentile rank of a score of 12.

18, 15, 12, 6, 8, 2, 3, 5, 20, 10

Solution

Arrange the data in order from lowest to highest.

$$2, 3, 5, 6, 8, 10, 12, 15, 18, 20$$

Then substitute in the formula.

$$\text{percentile} = \frac{\text{number of values below } X + 0.5}{\text{total number of values}} \cdot 100\%$$

Since there are six values below a score of 12, the solution is

$$\text{percentile} = \frac{6 + 0.5}{10} \cdot 100\% = 65\text{th percentile}$$

Thus, a student whose score was 12 did better than 65% of the class.

....

Note: One assumes that a score of 12 in Example 3–38, for instance, means theoretically any value between 11.5 and 12.5.

Example 3–39

Using the data in Example 3–38, find the percentile rank for a score of 6.

Solution

There are three values below 6. Thus

$$\text{percentile} = \frac{3 + 0.5}{10} \cdot 100\% = 35\text{th percentile}$$

A student who scored 6 did better than 35% of the class.

....

The next two examples show a procedure for finding a value corresponding to a given percentile.

Example 3–40

Using the scores in Example 3–38, find the value corresponding to the 25th percentile.

Solution

STEP 1   Arrange the data in order from lowest to highest.

$$2, 3, 5, 6, 8, 10, 12, 15, 18, 20$$

STEP 2   Compute

$$c = \frac{n \cdot p}{100}$$

where

$$n = \text{total number of values}$$
$$p = \text{percentile}$$

Thus,

$$c = \frac{10 \cdot 25}{100} = 2.5$$

STEP 3 If c is not a whole number, round it up to the next whole number; in this case, $c = 3$. (If c is a whole number, see the next example.) Start at the lowest value and count over to the third value, which is 5. Hence, the value 5 corresponds to the 25th percentile.

Example 3–41

Using the data set in Example 3–38, find the value that corresponds to the 60th percentile.

Solution

STEP 1 Arrange the data in order from smallest to largest.

2, 3, 5, 6, 8, 10, 12, 15, 18, 20

STEP 2 Substitute in the formula.

$$c = \frac{n \cdot p}{100} = \frac{10 \cdot 60}{100} = 6$$

STEP 3 If c is a whole number, use the value halfway between the c and $c + 1$ values when counting up from the lowest value—in this case, the 6th and 7th values.

2, 3, 5, 6, 8, 10, 12, 15, 18, 20
            ↗ ↘
    6th value   7th value

The value halfway between 10 and 12 is 11. Find it by adding the two values and dividing by 2.

$$\frac{10 + 12}{2} = 11$$

Hence, 11 corresponds to the 60th percentile. Anyone scoring 11 would have done better than 60% of the class.

The procedure for finding a value corresponding to a given percentile is summarized on the next page.

Finding a Data Value Corresponding to a Given Percentile

STEP 1   Arrange the data in order from lowest to highest.

STEP 2   Substitute in the formula

$$c = \frac{n \cdot p}{100}$$

where

n = total number of values
p = percentile

STEP 3A   If c is not a whole number, round up to the next whole number. Starting at the lowest value, count over to the number that corresponds to the rounded-up value.

STEP 3B   If c is a whole number, use the value halfway between c and $c + 1$ when counting up from the lowest value.

Deciles and Quartiles

Deciles divide the distribution into 10 groups. **Quartiles** divide the distribution into four groups.

Deciles are denoted by $D_1, D_2, D_3, \ldots, D_9$

and they correspond to $P_{10}, P_{20}, P_{30}, \ldots, P_{90}$

Quartiles are denoted by Q_1, Q_2, Q_3

and they correspond to P_{25}, P_{50}, P_{75}

The median is the same as P_{50} or Q_2

The position measures are summarized in Table 3–4.

Table 3–4   Summary of Position Measures

Measure	Definition	Symbol
Standard score or z score	Number of standard deviations a data value is above or below the mean	z
Percentile	Position in hundredths a data value is in the distribution	P_n
Decile	Position in tenths a data value is in the distribution	D_n
Quartile	Position in fourths a data value is in the distribution	Q_n

Outliers

A data set should be checked for extremely high or extremely low values. These values are called *outliers*.

An **outlier** is an extremely high or an extremely low data value when compared with the rest of the data values.

There are several ways to check for outliers. One method is shown in the next example.

Example 3–42

Check the following data set for outliers.

$$5, 6, 12, 13, 15, 18, 22, 50$$

Solution

The data value 50 is extremely suspect. The steps in checking for an outlier follow.

STEP 1 Find Q_1 and Q_3. This is done by using the procedure shown in Examples 3–40 and 3–41. Q_1 responds to the 25th percentile. Hence,

$$c = \frac{n \cdot p}{100} = \frac{8 \cdot 25}{100} = 2$$

Since c is a whole number, find the value halfway between the second and third values, counting from the left in the data set. This corresponds to 9 $\left(\text{i.e., } \dfrac{6 + 12}{2} \right)$, as shown below.

$$5, 6, 12, 13, 15, 18, 22, 50$$
$$\uparrow$$
$$Q_1 = 9$$

Q_3 corresponds to the 75th percentile. Hence,

$$c = \frac{n \cdot p}{100} = \frac{8 \cdot 75}{100} = 6$$

Find the value halfway between the 6th and 7th values. In this case, it is 20 $\left(\text{i.e., } \dfrac{18 + 22}{2} \right)$, as shown.

$$5, 6, 12, 13, 15, 18, 22, 50$$
$$\uparrow$$
$$Q_3 = 20$$

STEP 2 Find the **interquartile range (IQR),** which is $Q_3 - Q_1$.

$$\text{IQR} = Q_3 - Q_1 = 20 - 9 = 11$$

STEP 3 Multiply this value by 1.5.

$$1.5(11) = 16.5$$

STEP 4 Subtract the value obtained in Step 3 from Q_1 and add the value obtained in Step 3 to Q_3.

$$9 - 16.5 = -7.5 \qquad \text{and} \qquad 20 + 16.5 = 36.5$$

STEP 5 Check the data set for any data values that fall outside the range of -7.5 to 36.5 range. The value 50 is outside this range; hence, it can be considered an outlier.

There are several reasons to check a data set for outliers. First, the data value may have resulted from a measurement or observational error. Perhaps the researcher measured the variable incorrectly. Second, the data value may have resulted from a recording error. That is, it may have been written or typed incorrectly.

Third, the data value may have been obtained from a subject that is not in the defined population. For example, suppose test scores were obtained from a seventh-grade class, but a student in that class was actually in the sixth grade and had special permission to attend the class. This student might have scored extremely low on that particular exam on that day. Fourth, the data value might be a legitimate value that occurred by chance (although the probability is extremely small).

There are no hard-and-fast rules on what to do with outliers, nor is there complete agreement among statisticians on ways to identify them. Obviously, if they occurred as a result of an error, an attempt should be made to correct the error or else the data value should be omitted entirely. When they occur naturally by chance, the statistician must make a decision about whether to include them in the data set.

When a distribution is normal or bell-shaped, data values that are beyond three standard deviations of the mean can be considered suspected outliers.

Computer Application for Descriptive Statistics

Data Description

1. Enter the data into C1.
2. Click Stat > Basic Statistics > Display Descriptive Statistics
3. Highlight C1 in the dialog box and click on Select.
4. Click OK.

The computer will print various descriptive statistics.

Example: Perform a data description for the following data.

Data:

C1   12   17   15   16   16   14   18   13   10   16

MINITAB printout for this example:

Descriptive Statistics

Variable	N	Mean	Median	TrMean	StDev	SE Mean
C1	10	14.700	15.500	14.875	2.452	0.775

Variable	Minimum	Maximum	Q1	Q3
C1	10.000	18.000	12.750	16.250

Summary: The descriptive statistics are shown above. The sample size is 10. The mean is 14.7. The median is 15.5. The standard deviation is 2.452. The largest data value is 18, and the smallest data value is 10. Q_1 is 12.750, and Q_3 is 16.250. The value 14.875 is the trimmed mean and is obtained by omitting the smallest and largest 5% of the data values. SE Mean is the standard error of the mean. This concept is explained in Chapter 7.

Exercises

3–89. What is a z score?

3–90. Define *percentile rank*.

3–91. (W) What is the difference between a percentage and a percentile?

3–92. (**W**) Define *quartile*.

3–93. (**W**) What is the relationship between quartiles and percentiles?

3–94. (**W**) What is a decile?

3–95. (**W**) How are deciles related to percentiles?

3–96. To which percentile, quartile, and decile does the median correspond?

3–97. If an IQ test has a mean of 100 and a standard deviation of 15, find the corresponding z score for each IQ.

a. 115
b. 122
c. 93
d. 100
e. 85

3–98. The reaction time to a stimulus for a certain test has a mean of 2.5 seconds and a standard deviation of 0.3 second. Find the corresponding z score for each reaction time.

a. 2.7
b. 3.9
c. 2.8
d. 3.1
e. 2.2

3–99. A final examination for a psychology course has a mean of 84 and a standard deviation of 4. Find the corresponding z score for each raw score.

a. 87
b. 79
c. 93
d. 76
e. 82

3–100. An aptitude test has a mean of 220 and a standard deviation of 10. Find the corresponding z score for each exam score.

a. 200
b. 232
c. 218
d. 212
e. 225

3–101. Which of the following exam grades has a better relative position?

a. A grade of 43 on a test with $\overline{X} = 40$ and $s = 3$.
b. A grade of 75 on a test with $\overline{X} = 72$ and $s = 5$.

3–102. A student scores 60 on a mathematics test that has a mean of 54 and a standard deviation of 3, and she scores 80 on a history test with a mean of 75 and a standard deviation of 2. On which test did she do better than the rest of the class?

3–103. Which score indicates the highest relative position?

a. A score of 3.2 on a test with $\overline{X} = 4.6$ and $s = 1.5$.
b. A score of 630 on a test with $\overline{X} = 800$ and $s = 200$.
c. A score of 43 on a test with $\overline{X} = 50$ and $s = 5$.

3–104. The following distribution represents the data for weights of fifth-grade boys. Find the approximate weights corresponding to each percentile given by constructing a percentile graph.

Weight (pounds)	Frequency
52.5–55.5	9
55.5–58.5	12
58.5–61.5	17
61.5–64.5	22
64.5–67.5	15

a. 25th
b. 60th
c. 80th
d. 95th

3–105. For the data in Exercise 3–104, find the approximate percentile ranks of the following weights.

a. 57 pounds
b. 62 pounds
c. 64 pounds
d. 59 pounds

3–106. (**ans**) The data below represent the scores on a national achievement test for a group of tenth-grade students. Find the approximate percentile ranks of the following scores by constructing a percentile graph.

a. 220
b. 245
c. 276
d. 280
e. 300

Score	Frequency
196.5–217.5	5
217.5–238.5	17
238.5–259.5	22
259.5–280.5	48
280.5–301.5	22
301.5–322.5	6

3–107. For the data in Exercise 3–106, find the approximate scores that correspond to the following percentiles.

a. 15th
b. 29th
c. 43rd
d. 65th
e. 80th

3–108. (**ans**) The airborne speeds in miles per hour of 21 planes are shown next. Find the approximate values that correspond to the given percentiles by constructing a percentile graph.

Class	Frequency
366–386	4
387–407	2
408–428	3
429–449	2
450–470	1
471–491	2
492–512	3
513–533	4
	21

a. 9th
b. 20th
c. 45th

d. 60th
e. 75th

3–109. Using the data in Exercise 3–108, find the approximate percentile ranks of the following miles per hour.

a. 380 mph
b. 425 mph
c. 455 mph

d. 505 mph
e. 525 mph

3–110. Find the percentile ranks of each weight in the data set. The weights are in pounds.

78, 82, 86, 88, 92, 97

3–111. In Exercise 3–110, what value corresponds to the 30th percentile?

3–112. Find the percentile rank for each test score in the data set.

12, 28, 35, 42, 47, 49, 50

3–113. In Exercise 3–112, what value corresponds to the 60th percentile?

3–114. Find the percentile rank for each test score in the data set.

5, 12, 15, 16, 20, 21

3–115. What test score in Exercise 3–114 corresponds to the 33rd percentile?

3–116. Using the procedure shown in Example 3–42, check each data set for outliers.
a. 16, 18, 22, 19, 3, 21, 17, 20
b. 24, 32, 54, 31, 16, 18, 19, 14, 17, 20
c. 321, 343, 350, 327, 200
d. 88, 72, 97, 84, 86, 85, 100
e. 145, 119, 122, 118, 125, 116
f. 14, 16, 27, 18, 13, 19, 36, 15, 20

*3–117. Another measure of average is called the *midquartile*; it is the numerical value halfway between Q_1 and Q_3, and the formula is

$$\text{midquartile} = \frac{Q_1 + Q_3}{2}$$

Using this formula and other formulas, find Q_1, Q_2, Q_3, the midquartile, and the interquartile range for each data set.
a. 5, 12, 16, 25, 32, 38
b. 53, 62, 78, 94, 96, 99, 103

3–5

Exploratory Data Analysis

Objective 4. Use the techniques of exploratory data analysis to discover various aspects of data.

In traditional statistics, data are organized by using a frequency distribution. From this distribution various graphs such as the histogram, frequency polygon, and ogive can be constructed to determine the shape or nature of the distribution. In addition, various statistics such as the mean and standard deviation can be computed to summarize the data.

The purpose of traditional analysis is to confirm various conjectures about the nature of the data. For example, from a carefully designed study, a researcher might want to know if the proportion of Americans who are exercising today has increased from 10 years ago. This study would contain various assumptions about the population, various definitions such as exercise, and so on.

In **exploratory data analysis (EDA),** data are organized using a *stem and leaf plot*. The summary statistics used are the *median* and *interquartile range*. Finally, a *box plot* can be constructed to determine visually the nature of the distribution. The purpose of exploratory data analysis is to examine data in order to find out what information can be discovered. For example, are there any gaps in the data? Can any patterns be discerned? Here the researcher starts out with few or no assumptions.

Exploratory data analysis was developed by John Tukey and presented in his book *Exploratory Data Analysis* (Addison-Wesley, 1977).

Stem and Leaf Plots   The stem and leaf plot is a method of organizing data and is a combination of sorting and graphing. It has the advantage over grouped frequency distribution of retaining the actual data while showing them in graphic form.

A **stem and leaf plot** is a data plot that uses part of a data value as the stem and part of the data value as the leaf to form groups or classes.

Example 3–43 shows the procedure for constructing a stem and leaf plot.

Example 3–43

At an outpatient testing center, a sample of 20 days showed the following number of cardiograms done each day. Construct a stem and leaf plot for the data.

25	31	20	32	13
14	43	02	57	23
36	32	33	32	44
32	52	44	51	45

Solution

STEP 1   Arrange the data in order:

02, 13, 14, 20, 23, 25, 31, 32, 32, 32,
32, 33, 36, 43, 44, 44, 45, 51, 52, 57

Note: Arranging the data in order is not essential, but it is helpful in construction of the plot.

STEP 2   Separate the data according to the first digit, as shown.

02   13, 14   20, 23, 25   31, 32, 32, 32, 32, 33, 36
43, 44, 44, 45   51, 52, 57

STEP 3   Since the values range from 2 to 57, a display can be made by using the leading digit as a *stem* and the trailing digit as the *leaf.* For example, for the value 32, the leading digit, 3, is the stem and the trailing digit, 2, is the leaf. For the value 14, the 1 is the stem and the 4 is the leaf. Now a plot can be constructed as shown.

Figure 3–8

Stem and Leaf Plot for Example 3–43

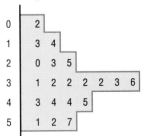

Leading digit (stem)	Trailing digit (leaf)
0	2
1	3 4
2	0 3 5
3	1 2 2 2 2 3 6
4	3 4 4 5
5	1 2 7

A graph can be drawn over the plot as shown in Figure 3–8. This graph is optional and the plot can be drawn without it.

Figure 3–8 shows that the distribution peaks in the center and that there are no gaps in the data. For 7 of the 20 days, the number of patients receiving cardiograms was between 31 and 36. The plot also shows that the testing center treated from a minimum of 2 patients to a maximum of 57 patients in any one day.

Example 3–44

An insurance company researcher conducted a survey on the number of car thefts in a large city for a period of 30 days last summer. The raw data are shown below. Construct a stem and leaf plot by using classes 50–54, 55–59, 60–64, 65–69, 70–74, and 75–79.

52	62	51	50	69
58	77	66	53	57
75	56	55	67	73
79	59	68	65	72
57	51	63	69	75
65	53	78	66	55

Solution

STEP 1   Arrange the data in order.

50, 51, 51, 52, 53, 53, 55, 55, 56, 57, 57, 58, 59, 62, 63, 65, 65, 66, 66, 67, 68, 69, 69, 72, 73, 75, 75, 77, 78, 79

STEP 2   Separate the data according to the classes.

50, 51, 51, 52, 53, 53   55, 55, 56, 57, 57, 58, 59
62, 63   65, 65, 66, 66, 67, 68, 69, 69   72, 73
75, 75, 77, 78, 79

Figure 3–9

Stem and Leaf Plot for Example 3–44

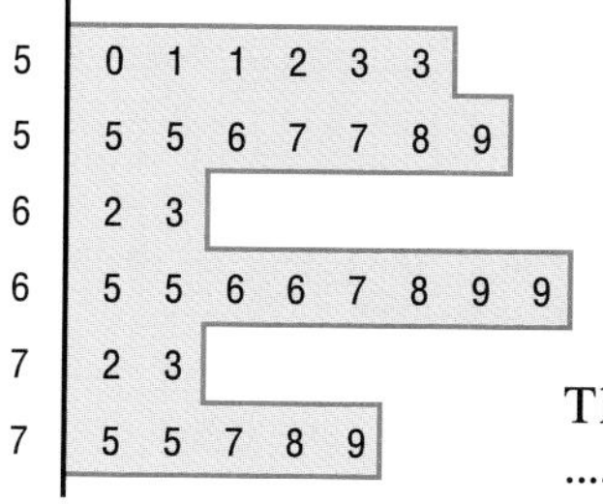

Stem	Leaf
5	0 1 1 2 3 3
5	5 5 6 7 7 8 9
6	2 3
6	5 5 6 6 7 8 9 9
7	2 3
7	5 5 7 8 9

STEP 3   Plot the data as shown here.

Leading digit (Stem)	Trailing digit (Leaf)
5	0 1 1 2 3 3
5	5 5 6 7 7 8 9
6	2 3
6	5 5 6 6 7 8 9 9
7	2 3
7	5 5 7 8 9

The graph for this plot is shown in Figure 3–9.

When the values are in the hundreds, such as 325, the stem is 32 and the leaf is 5. When data are grouped into classes, such as 20–24, 25–29, and so on, a stem and leaf plot might look like this:

Class	Leading digit (Stem)	Trailing digit (Leaf)	Value
20–24	2	0 1 3 4	20, 21, 23, 24
25–29	2	5 6	25, 26

Of course, the class and value columns would be left out of the final plot.

Analyzing the stem and leaf plot is similar to analyzing the histogram. Look for the nature of the leaves. For example, does the distribution have a single horizontal peak? Are the data clustered about certain peaks?

Computer Application for Stem and Leaf Plot

MINITAB   Stem and leaf plot:

1. Enter the data into C1.
2. Click Graph > Stem-and-Leaf.
3. Highlight C1 in the dialog box and click on Select.
4. Click on OK.

The computer will print a stem and leaf display.

Example: Construct a stem and leaf display for the following data.

Data:

C1	68	80	69	81	72	100	101	73	102	93
	91	92	93	88	82	83	75	75	89	96
	103	83	84	85	88					

MINITAB printout for this example:

```
Stem and leaf of C1   N = 25
Leaf Unit = 1.0

   2     6 89
   4     7 23
   6     7 55
  12     8 012334
  (4)    8 5889
   9     9 1233
   5     9 6
   4    10 0123
```

Summary: The stems are 6, 7, 8, 9, and 10. The leaves are to the right of the stems. The first column shows the cumulative depth up to the median class in parentheses. The next technique used in exploratory data analysis is called a *box plot*.

Box Plots

When the data set contains a small number of values, a **box plot** is used to graphically represent the data set. These plots involve five specific values:

1. The lowest value of the data set (i.e., minimum)
2. The lower hinge
3. The median
4. The upper hinge
5. The highest value of the data set (i.e., maximum)

These values are called a **five-number summary** of the data set.

The **lower hinge** is defined as the median of all values less than or equal to the median when the data set has an odd number of values, or as the median of all values less than the median when the data set has an even number of values. The symbol for lower hinge is LH.

The **upper hinge** is defined as the median of all values greater than or equal to the median when the data set has an odd number of values, or as the median of all values greater than the median when the data set has an even number of values. The symbol for upper hinge is UH.

Example 3–45

Find the lower hinge and the upper hinge for the following data set.

3, 8, 10, 12, 15, 18, 20

Solution

Since the median is 12, the data set is divided into two subsets:

3, 8, 10, 12   and   12, 15, 18, 20

The median 12 is included in both subsets, since there is an odd number of data values. Next, the median of the lower subset is found as shown.

3, 8, 10, 12
↑
9

And the median of the upper subset is found as shown.

12, 15, 18, 20
↑
16.5

Hence, the lower hinge is 9 and the upper hinge is 16.5

In this example, there was an odd number of values in the data set. When there is an even number of values in the data set, the lower hinge is the same as Q_1 and the upper hinge is the same as Q_3, as shown in the next example.

Example 3–46

Find the lower and upper hinges for the data set shown here.

18, 20, 25, 31, 33, 38, 42, 50

Solution

The median is 32, the lower hinge is 22.5, and the upper hinge is 40, as shown.

18	20	25	31	33	38	42	50

22.5   32   40
LH   Median   UH

$$\text{MD} = \frac{31 + 33}{2} = 32$$

$$\text{LH} = \frac{20 + 25}{2} = 22.5$$

$$\text{UH} = \frac{38 + 42}{2} = 40$$

The next two examples show the procedure for constructing a box plot.

Example 3–47

A stockbroker recorded the number of clients she saw each day over an 11-day period. The data are shown below. Construct a box plot for the data.

$$33, 38, 43, 30, 29, 40, 51, 27, 42, 23, 31$$

Solution

STEP 1 Arrange the data in order.

$$23, 27, 29, 30, 31, 33, 38, 40, 42, 43, 51$$

STEP 2 Find the median.

$$23, 27, 29, 30, 31, 33, 38, 40, 42, 43, 51$$
$$\uparrow$$
Median

STEP 3 Find the lower hinge.

$$23, 27, 29, 30, 31, 33$$
$$\uparrow$$
29.5

STEP 4 Find the upper hinge.

$$33, 38, 40, 42, 43, 51$$
$$\uparrow$$
41

STEP 5 Draw a scale for the data on the x axis.

STEP 6 Locate the lowest value, the lower hinge, the median, the upper hinge, and the highest value on the scale.

STEP 7 Draw a box around the hinges, draw a vertical line through the median, and connect the upper and lower values, as shown in Figure 3–10.

Figure 3–10

Box Plot for
Example 3–47

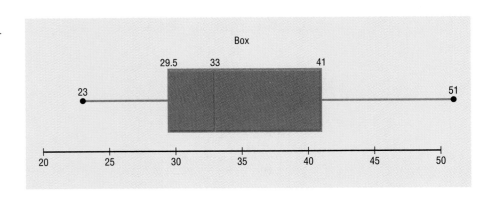

The box in Figure 3–10 represents the middle 50% of the data, and the lines represent the lower and upper ends of the data.

Information Obtained from a Box Plot

1. *a.* If the median is near the center of the box, the distribution is approximately symmetric.
   b. If the median falls to the left of the center of the box, the distribution is positively skewed.
   c. If the median falls to the right of the center, the distribution is negatively skewed.
2. *a.* If the lines are about the same length, the distribution is approximately symmetric.
   b. If the right line is larger than the left line, the distribution is positively skewed.
   c. If the left line is larger than the right line, the distribution is negatively skewed.

The box plot in Figure 3–10 indicates that the distribution is slightly positively skewed.

Example 3–48

Another stockbroker recorded the number of clients he saw each day for 10 days. The data follow. Construct a box plot for the data set.

11, 14, 18, 5, 16, 8, 19, 10, 17, 20

Solution

STEP 1   Arrange the data in order.

5, 8, 10, 11, 14, 16, 17, 18, 19, 20

STEP 2   Find the median and the hinges.

$$5, 8, \underset{\substack{\uparrow \\ \text{LH}}}{10,} 11, 14, \underset{\substack{\uparrow \\ 15 = \text{MD}}}{16,} 17, \underset{\substack{\uparrow \\ \text{UH}}}{18,} 19, 20$$

Note that in this case, the lower hinge and the upper hinge correspond to Q_1 and Q_3.

STEP 3   Draw the graph. See Figure 3–11.

Figure 3–11

Box Plot for
Example 3–48

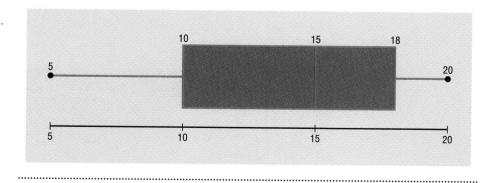

The plot in Figure 3–11 shows that the distribution is somewhat negatively skewed.

Box plots can be drawn vertically and can be used to compare several small data sets. Also, many books and computer programs refer to Q_1 and Q_3 instead of LH and UH.

Computer Application for Box Plots

MINITAB   Box plot:

1. Enter the data into C1.
2. Click Graph > Boxplot.
3. Highlight C1 in the dialog box and click on Select.
4. Click on OK.

Example: Construct a box plot for the following data.

Data:

C1   12   17   15   16   16   14   18   13   10   16

MINITAB printout for this example:

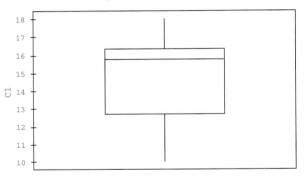

Summary: The box plot is shown above. The computer does not give the exact values for the data. Obtain them by using the commands shown in Data Description.

Another important point to remember is that the summary statistics (median and interquartile range) used in exploratory data analysis are said to be resistant statistics. A *resistant statistic* is relatively less affected by outliers than a *nonresistant statistic*. The mean and standard deviation are nonresistant statistics. Sometimes when a distribution is skewed or contains outliers, the median and interquartile range may more accurately summarize the data than the mean and standard deviation, since the mean and standard deviation are more affected in this case. Table 3–5 compares the traditional versus the exploratory data analysis approach.

Table 3–5 Traditional versus EDA Techniques	
Traditional	**Exploratory data analysis**
Frequency distribution	Stem and leaf plot
Histogram	Box plot
Mean	Median
Standard deviation	Interquartile range

Exercises

3–118. What is the purpose of exploratory data analysis?

3–119. What four statistical techniques are used in exploratory data analysis?

3–120. What five statistics are used in the five-number summaries?

3–121. What is meant by resistant statistics?

3–122. Twenty-nine executives reported the following number of telephone calls made during a randomly selected week. Construct a stem and leaf plot for the data and analyze the results.

22	14	12	9	54	12
16	12	14	49	10	14
8	21	37	28	36	22
9	33	58	31	41	19
3	18	25	28	52	

3–123. Twenty salespeople reported the following number of sales calls completed last month. Construct a stem and leaf plot for the data.

72	102	88	103
93	107	100	93
82	97	91	73
81	119	83	89
82	86	102	106

3–124. The growth (in centimeters) of a certain variety of plant after 20 days is shown below. Construct a stem and leaf plot for the data.

20	12	39	38
41	43	51	52
59	55	53	59
50	58	35	38
23	32	43	53

3–125. (W) Construct a stem and leaf plot for the following results of blood calcium level (in milligrams per deciliter) in 24 nurses, and analyze the results.

7.2	10.2	11.2	10.3
9.0	10.7	7.3	9.1
8.1	9.6	9.6	9.2
9.7	8.2	9.8	9.9
7.3	9.3	9.4	8.5
10.0	11.1	8.3	10.3

3–126. Shown next are the number of farms in six counties in western Pennsylvania. Construct a box plot for the data and comment on the shape of the distribution.

338, 767, 633, 747, 1369, 1139

Source: *Pittsburgh Tribune-Review,* August 28, 1994.

3–127. Shown next are the sizes of the police forces in the 10 largest cities in the United States in 1993 (the numbers represent hundreds). Construct a box plot for the data and comment on the shape of the distribution.

29.3, 7.6, 12.1, 4.7, 6.2, 1.9, 3.9, 2.8, 2.0, 1.7

Source: *USA Today,* February 17, 1995.

3–128. Construct a box plot for the number of calculators sold during a randomly selected week. Comment on the shape of the distribution.

8, 12, 23, 5, 9, 15, 3

3–129. Construct a box plot for the data in exercise 3–123.

3–130. Construct a box plot for the data in exercise 3–124.

3–6 Summary

This chapter explains the basic ways to summarize data. These include measures of central tendency, measures of variation or dispersion, and measures of position. The three most commonly used measures of central tendency are the mean, median, and mode. The midrange is also used occasionally to represent an average. The three most commonly used measurements of variation are the range, variance, and standard deviation.

The most common measures of position are percentiles, quartiles, and deciles. This chapter explains how data values are distributed according to Chebyshev's theorem and the empirical rule. The coefficient of variation is used to describe the standard deviation in relationship to the mean. These methods are commonly called traditional statistical methods and are primarily used to confirm various conjectures about the nature of the data.

Other methods, such as the stem and leaf plot, the box plot, and five-number summaries, are part of exploratory data analysis; they are used to examine data to see what they reveal.

After learning the techniques presented in Chapter 2 and this chapter, students will have a substantial knowledge of descriptive statistics. That is, they will be able to collect, organize, summarize, and present data.

Important Terms

Box plot 130
Chebyshev's theorem 108
Coefficient of variation 107
Data array 81
Decile 123
Empirical rule 110
Exploratory data analysis 127

Five-number summary 130
Interquartile range 124
Lower hinge 130
Mean 75
Median 81
Midrange 90
Modal class 87
Mode 85

Negatively skewed distribution 93
Outlier 123
Parameter 74
Percentile 119
Positively skewed distribution 93
Quartile 123
Range 98
Standard deviation 101

Statistic 74
Stem and leaf plot 128
Symmetrical distribution 93
Unbiased estimator 103
Upper hinge 130
Variance 101
Weighted mean 91
z score or standard score 115

Important Formulas

Formula for the mean for individual data:

$$\bar{X} = \frac{\sum X}{n} \qquad \mu = \frac{\sum X}{N}$$

Formula for the mean for grouped data:

$$\bar{X} = \frac{\sum f \cdot X_m}{n}$$

Formula for the median for grouped data:

$$MD = \frac{(n/2) - cf}{f} \cdot w + L_m$$

Formula for the weighted mean:

$$\bar{X} = \frac{\sum w \cdot X}{\sum w}$$

Formula for the midrange:

$$MR = \frac{\text{lowest value} + \text{highest value}}{2}$$

Formula for the range:

$$R = \text{highest value} - \text{lowest value}$$

Formula for the variance for population data:

$$\sigma^2 = \frac{\sum (X - \mu)^2}{N}$$

Formula for the variance for sample data (shortcut formula for the unbiased estimator):

$$s^2 = \frac{\sum X^2 - [(\sum X)^2/n]}{n - 1}$$

Formula for the variance for grouped data:

$$s^2 = \frac{\sum f \cdot X_m^2 - [(\sum f \cdot X_m)^2/n]}{n - 1}$$

Formula for the standard deviation for population data:

$$\sigma = \sqrt{\frac{\sum (X - \mu)^2}{N}}$$

Formula for the standard deviation for sample data (shortcut formula):

$$s = \sqrt{\frac{\sum X^2 - [(\sum X)^2/n]}{n - 1}}$$

Formula for the standard deviation for grouped data:

$$s = \sqrt{\frac{\sum f \cdot X_m^2 - [(\sum f \cdot X_m)^2/n]}{n-1}}$$

Formula for the coefficient of variation:

$$\text{CVar} = \frac{s}{\overline{X}} \cdot 100\% \quad \text{or} \quad \text{CVar} = \frac{\sigma}{\mu} \cdot 100\%$$

Formula for Chebyshev's theorem: The proportion of values from a data set that will fall within k standard deviations of the mean will be at least

$$1 - \frac{1}{k^2} \text{ where } k \text{ is a number greater than 1}$$

Formula for the z score (standard score):

$$z = \frac{X - \mu}{\sigma} \quad \text{or} \quad z = \frac{X - \overline{X}}{s}$$

Formula for the cumulative percentage:

$$\text{cumulative } \% = \frac{\text{cumulative frequency}}{n} \cdot 100\%$$

Formula for the percentile rank of a value X:

$$\text{percentile} = \frac{\text{number of values below } X + 0.5}{\text{total number of values}} \cdot 100\%$$

Formula for finding a value corresponding to a given percentile:

$$c = \frac{n \cdot p}{100}$$

Formula for interquartile range:

$$\text{IQR} = Q_3 - Q_1$$

Review Exercises

3–131. The following data represent the number of deer killed by motor vehicles during 1994 for eight counties in southwestern Pennsylvania.

2343, 1240, 1088, 600,
497, 1925, 1480, 458

Source: *Pittsburgh Post-Gazette*, June 6, 1995, p. 1.

Find each of the following.
a. Mean
b. Median
c. Mode
d. Midrange
e. Range
f. Variance
g. Standard deviation

3–132. Ten motors were tested and the following data were obtained for the number of revolutions per minute the flywheel attached to the motors turned.

215, 308, 225, 236, 272,
300, 254, 260, 232, 216

Find each of the following.
a. Mean
b. Median
c. Mode
d. Midrange
e. Range
f. Variance
g. Standard deviation

3–133. Twelve batteries were tested after being used for one hour. The output (in volts) is shown below.

Volts	Frequency
2	1
3	4
4	5
5	1
6	1

Find each of the following.
a. Mean
b. Median
c. Mode
d. Range
e. Variance
f. Standard deviation

3–134. A survey of 40 clothing stores reported the following number of sales held in the month of November.

Number of sales	Frequency
5	3
6	5
7	12
8	9
9	8
10	3

Find each of the following.
a. Mean
b. Median
c. Mode
d. Range
e. Variance
f. Standard deviation

3–135. Shown below is a frequency distribution for the rise in tides at 30 selected locations in the United States.

Rise in tides (inches)	Frequency
12.5–27.5	6
27.5–42.5	3
42.5–57.5	5
57.5–72.5	8
72.5–87.5	6
87.5–102.5	2

Find each of the following.
a. Mean
b. Median
c. Modal class
d. Variance
e. Standard deviation

3–136. The fuel capacity in gallons of 50 randomly selected 1995 cars is shown below.

Class	Frequency
10–12	6
13–15	4
16–18	14
19–21	15
22–24	8
25–27	2
28–30	1
	50

Source: *Consumer Reports,* April 1995, pp. 274–79.

Find each of the following.
a. Mean
b. Median
c. Modal class
d. Variance
e. Standard deviation

3–137. In a dental survey of third-grade students, the following distribution was obtained for the number of cavities found. Find the average number of cavities for the class. Use the weighted mean.

Number of students	Number of cavities
12	0
8	1
5	2
5	3

3–138. An investor calculated the following percentages of each of three stock investments with payoffs as shown. Find the average payoff. Use the weighted mean.

Stock	Percent	Payoff
A	30%	$10,000
B	50%	$ 3,000
C	20%	$ 1,000

3–139. In an advertisement, a transmission service center stated that the average years of service of its employees was 13. The distribution is shown below. Using the weighted mean, calculate the correct average.

Number of employees	Years of service
8	3
1	6
1	30

3–140. The average number of textbooks in professors' offices is 16, and the standard deviation is 5. The average age of the professors is 43, with a standard deviation of 8. Which data set is more variable?

3–141. A survey of bookstores showed that the average number of magazines carried is 56, with a standard deviation of 12. The same survey showed that the average length of time each store had been in business was 6 years, with a standard deviation of 2.5 years. Which is more variable, the number of magazines or the number of years?

3–142. The number of previous jobs held by each of six applicants is shown here.

2, 4, 5, 6, 8, 9

a. Find the percentile for each value.
b. What value corresponds to the 30th percentile?
c. Construct a box plot and comment on the nature of the distribution.

3–143. The weights in pounds of 30 one-room air conditioners are shown below.

Class	Frequency
46–54	3
55–63	12
64–72	10
73–81	3
82–90	1
91–99	0
100–108	1
	30

Source: *Consumer Reports,* June 1995, p. 407.

a. Construct a percentile graph.
b. Find the values that correspond to the 35th, 65th, and 85th percentiles.
c. Find the percentile of values 48, 54, and 62.

3–144. Check each data set for outliers.
a. 506, 511, 517, 514, 400, 521
b. 3, 7, 9, 6, 8, 10, 14, 16, 20, 12
c. 14, 18, 27, 26, 19, 13, 5, 25
d. 112, 157, 192, 116, 153, 129, 131

3–145. A survey of car rental agencies shows that the average cost of a car rental is $0.32 per mile. The standard deviation is $0.03. Using Chebyshev's theorem, find the range in which at least 75% of the data values will fall.

3–146. The average cost of a certain type of seed per acre is $42. The standard deviation is $3. Using Chebyshev's theorem, find the range in which at least 88.89% of the data values will fall.

3–147. The average labor charge for automobile mechanics if $54 per hour. The standard deviation is $4. Find the minimum percentage of data values that will fall within the range of $48 to $60. Use Chebyshev's theorem.

3–148. For a certain type of job, it costs a company an average of $231 to train an employee to perform the task. The standard deviation is $5. Find the minimum percentage of data values that will fall in the range of $219 to $243. Use Chebyshev's theorem.

3–149. The average delivery charge for a refrigerator is $32. The standard deviation is $4. Find the minimum percentage of data values that will fall in the range of $20 to $44. Use Chebyshev's theorem.

3–150. Which of the following exam grades has a better relative position?
a. A grade of 37 on a test with $\overline{X} = 42$ and $s = 5$.
b. A grade of 72 on a test with $\overline{X} = 80$ and $s = 6$.

3–151. The number of visitors to the Railroad Museum during 24 randomly selected hours is shown here. Construct a stem and leaf plot for the data.

67	62	38	73	34	43	72	35
53	55	58	63	47	42	51	62
32	29	47	62	29	38	36	41

3–152. (**W**) The data set shown below represents the number of hours 25 part-time employees worked at the Sea Side Amusement Park during a randomly selected week in June. Construct a stem and leaf plot for the data and summarize the results.

16	25	18	39	25	17	29	14	37
22	18	12	23	32	35	24	26	
20	19	25	26	38	38	33	29	

3–153. (**W**) A special aptitude test is given to job applicants. The data shown below represent the scores of 30 applicants. Construct a stem and leaf plot for the data and summarize the results.

204	210	227	218	254
256	238	242	253	227
251	243	233	251	241
237	247	211	222	231
218	212	217	227	209
260	230	228	242	200

3–154. The data set shown here represents the number of hours 25 part-time employees worked at the Superior Super Market during the week before Christmas. Construct a box plot for the data.

19	27	16	23	38	33
20	16	23	36	38	27
22	23	24	16	32	38
31	39	35	34	39	26
30					

3–155. The data here represent the scores of 30 students on a special aptitude test. Construct a box plot for the data.

206	217	222	249	212	209
247	253	209	231	241	218
242	210	238	242	248	210
204	240	217	219	211	212
250	251	226	218	227	222

Statistics Today

Who Is a Typical First-Time Home Buyer?   Revisited

The *USA Today* Snapshot shows several measures of average and two percentages (63% and 82%). Although it is not possible to know which measures were used, one could make an educated guess. For example, the mean was probably used for family size (2.7 people) since it is in decimal form. The age of the buyer (32) is probably a median. The time (5 months) it took to find a home is probably a mode.

The information presented in this Snapshot is very vague. After mastering the concepts in this chapter, you should be able to present a more accurate profile if given the raw data.

Data Analysis

The Data Bank is Appendix D in this book.

1. From the Data Bank, choose one of the following variables: age, weight, cholesterol level, systolic pressure, IQ, or sodium level. Select at least 30 values and find the mean, median, mode, and midrange. State which measurement of central tendency best describes the average and why.

2. Find the range, variance, and standard deviation for the data selected in Exercise 1.

3. From the Data Bank, choose 10 values from any variable, construct a box plot, and interpret the results.

Quiz

Determine whether each statement is true or false. If the statement is false, explain why.

1. When the mean is computed for individual data, all values in the data set are used.

2. The median cannot be found for group data when there is an open class.

3. A single extremely large value can affect the median more than the mean.

4. Half of all the data values will fall above the mode and half will fall below the mode.

5. In a data set, the mode will always be unique.

6. The range and midrange are both measures of variation.

7. One disadvantage of the median is that it is not unique.

8. The mode and midrange are both measures of variation.

9. If a person's score on an exam corresponds to the 75th percentile, then that person obtained 75 correct answers out of 100 questions.

Select the best answer.

10. What is the value of the mode when all values in the data set are different?
a. Zero
b. One
c. There is no mode.
d. It cannot be determined unless the data values are given.

11. When data are categorized as, for example, places of residence (rural, suburban, urban), the most appropriate measure of central tendency is
a. Mean   *c.* Mode
b. Median   *d.* Midrange

12. P_{50} corresponds to
a. Q_2 *c.* IQR
b. D_5 *d.* Midrange

13. Which is not part of the five-number summary?
a. The upper and lower hinges
b. The mean
c. The median
d. The smallest and the largest data values

14. A statistic that tells the number of standard deviations a data value is above or below the mean is called
a. A quartile
b. A percentile
c. A coefficient of variation
d. A z score

15. When a distribution is bell-shaped, approximately what percentage of data values will fall within one standard deviation of the mean?
a. 50%   c. 95%
b. 68%   d. 99.7%

Complete the following statements with the best answer.

16. A measure obtained from sample data is called a _____.

17. Generally, Greek letters are used to represent _____, and Roman letters are used to represent _____.

18. The positive square root of the variance is called the _____.

19. The symbol for the population standard deviation is _____.

20. When the sum of the lowest data value plus the highest data value is divided by 2, the measure is called _____.

21. If the mode is to the left of the median and the mean is to the right of the median, then the distribution is _____ skewed.

22. An extremely high or extremely low data value is called an _____.

23. The following temperatures were recorded in Pasadena for a week in April.

87, 85, 80, 78, 83, 86, 90

Find each of the following.
a. Mean   e. Range
b. Median   f. Variance
c. Mode   g. Standard deviation
d. Midrange

24. The outputs in volts of ten 9-volt batteries after 6 hours of use are shown here.

Volts	Frequency
3	1
4	3
5	4
6	1
7	1

Find each of the following.
a. Mean   d. Range
b. Median   e. Variance
c. Mode   f. Standard deviation

25. Shown here is a frequency distribution for the number of inches of rain received in one year in 25 selected cities in the United States.

Number of inches	Frequency
5.5–20.5	2
20.5–35.5	3
35.5–50.5	8
50.5–65.5	6
65.5–80.5	3
80.5–95.5	3

Find each of the following.
a. Mean   d. Variance
b. Median   e. Standard deviation
c. Modal class

26. A survey of 36 selected recording companies showed the following number of days it took to receive a shipment from the day it was ordered.

Days	Frequency
1–3	6
4–6	8
7–9	10
10–12	7
13–15	0
16–18	5

Find each of the following.
a. Mean   d. Variance
b. Median   e. Standard deviation
c. Modal class

27. In a survey of third-grade students, the following distribution was obtained for the number of "best friends" each had.

Number of students	Number of best friends
8	1
6	2
5	3
3	0

Find the average number of best friends for the class. Use the weighted mean.

28. In an advertisement, a retail store stated that its employees averaged nine years of service. The distribution is shown here.

Number of employees	Years of service
8	2
2	6
3	10

Using the weighted mean, calculate the correct average.

29. The average number of newspapers for sale in an airport newsstand is 12, and the standard deviation is 4. The average age of the pilots is 37, with a standard deviation of 6. Which data set is more variable?

30. A survey of grocery stores showed that the average number of brands of toothpaste carried was 16, with a standard deviation of 5. The same survey showed the average length of time each store was in business was 7 years, with a standard deviation of 1.6 years. Which is more variable, the number of brands or the number of years?

31. A student scored 76 on a general science test where the class mean and standard deviation were 82 and 8, respectively; he also scored 53 on a psychology test where the class mean and standard deviation were 58 and 3, respectively. In which class was his relative position higher?

32. Which score has the highest relative position?
a. $X = 12$ $\bar{X} = 10$ $s = 4$
b. $X = 170$ $\bar{X} = 120$ $s = 32$
c. $X = 180$ $\bar{X} = 60$ $s = 8$

33. (W) The number of credits in business courses eight job applicants had is shown here.

9, 12, 15, 27, 33, 45, 63, 72

a. Find the percentile for each value.
b. What value corresponds to the 40th percentile?
c. Construct a box plot and comment on the nature of the distribution.

34. On a philosophy comprehensive exam, the following distribution was obtained from 25 students.

Score	Frequency
40.5–45.5	3
45.5–50.5	8
50.5–55.5	10
55.5–60.5	3
60.5–65.5	1

a. Construct a percentile graph.
b. Find the values that correspond to the 22nd, 78th, and 99th percentiles.
c. Find the percentile of the values 52, 43, and 64.

35. The number of visitors to the Historic Museum for 25 randomly selected hours is shown below. Construct a stem and leaf plot for the data.

15	53	48	19	38
86	63	98	79	38
62	89	67	39	26
28	35	54	88	76
31	47	53	41	68

Critical Thinking Challenges

1. Averaging Averages Consider the following problem: Mary and Bill play basketball. In the first game, Mary had a foul shot average of 0.60 (she made 12 of 20 shots), and Bill had a foul shot average of 0.50 (he made 5 of 10 shots). In the second game, Mary had a foul shot average of 0.80 (she made 8 out of 10 shots), and Bill had a foul shot average of 0.75 (he made 15 out of 20 shots). Now, who do you think has the best overall average? The answer may surprise you. Hint: Compute the averages for both games based on 30 shots for each player.

2. Average American A *USA Today* Snapshot, shown on next page, states that, "The average American spent $3,299 on health care in 1993."
a. Explain what is meant by an "average American."

b. Why would it be better to state "Americans spent an average of $3,299 on health care in 1993"?
c. How might this average have been derived?
d. Is it representative of what you spent on health care? Explain.

3. American Students Study Less The graph shown on the next page indicates the number of hours students spent studying during four years of high school. Use it to answer the following questions.
a. Which average do you think was used (mean, median, or mode)? Why do you think this average was used?
b. If a mean was used, how would it be calculated?
c. Can you estimate the average number of hours per day the students studied? (*Note:* You would not use

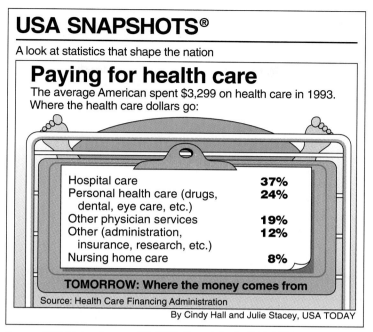

USA SNAPSHOTS®

A look at statistics that shape the nation

Paying for health care

The average American spent $3,299 on health care in 1993. Where the health care dollars go:

Hospital care	**37%**
Personal health care (drugs, dental, eye care, etc.)	**24%**
Other physician services	**19%**
Other (administration, insurance, research, etc.)	**12%**
Nursing home care	**8%**

TOMORROW: Where the money comes from

Source: Health Care Financing Administration

By Cindy Hall and Julie Stacey, USA TODAY

Source: *USA Today,* January 12, 1995. Copyright © 1995, USA TODAY. Reprinted with permission.

1 year = 365 days, since school is not in session 365 days per year.)

d. Do you think the length of the school year in each country is the same? How could you find out?

e. Do you think the length of the school year would affect the results of the study? Explain your answer.

f. In your own words, write a brief summary of the study results.

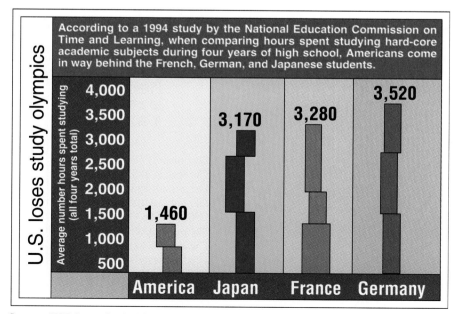

U.S. loses study olympics

According to a 1994 study by the National Education Commission on Time and Learning, when comparing hours spent studying hard-core academic subjects during four years of high school, Americans come in way behind the French, German, and Japanese students.

Average number hours spent studying (all four years total)

America	Japan	France	Germany
1,460	3,170	3,280	3,520

Source: "U.S. Loses Study Olympics," *tell* 4, no. 3, (Fall 1994), p. 23. Used with permission.

Data Projects

Where appropriate, use MINITAB, the TI-83, or a computer program of your choice to complete the following exercises.

1. Select a variable and collect about 10 values for two groups. (For example, you may want to ask 10 men how many cups of coffee they drink per day and 10 women the same question.)
a. Define the variable.
b. Define the populations.
c. Describe how the samples were selected.

d. Write a paragraph describing the similarities and differences between the two groups, using appropriate descriptive statistics such as means, standard deviations, and so on.

2. Collect data consisting of at least 30 values.
a. State the purpose of the project.
b. Define the population.
c. State how the sample was selected.
d. Using appropriate descriptive statistics, write a paragraph summarizing the data.

TI-83 Calculator

Descriptive Statistics

To calculate various descriptive statistics:

1. Enter data into L_1. (See Chapter 2.)

2. Press **STAT** to get the menu.

3. Press ▶ to move cursor to CALC; then press **ENTER**.

4. Press **2nd [L_1]** then **ENTER**.

The calculator will display

$\bar{X}$	sample mean
ΣX	sum of the data values
ΣX^2	sum of the squares of the data values
S_x	sample standard deviation
σ_x	population standard deviation
n	number of data values
minX	smallest data value
Q_1	lower quartile
Med	median
Q_3	upper quartile
maxX	largest data value

Example 5 Find the various descriptive statistics for 42, 44, 43, 52, 48, 56, 60, 56, 43, 41.

Output

```
1-Var Stats
 x̄=48.5
 Σx=485
 Σx²=23959
 Sx=6.964194139
 σx=6.606814664
↓n=10
```

```
1-Var Stats
↑n=10
 minX=41
 Q₁=43
 Med=46
 Q₃=56
 maxX=60
```

The mean is 48.5.
The sum is 485.
The sum of x^2 is 23959.
The unbiased estimator of the standard deviation, s_1, is 6.964194139.
The population standard deviation is 6.606814664.
The sample size, n, is 10.
The smallest data value, $X_{\min}$, is 41.
Q_1 is 43.
The median is 46.
Q_3 is 56.
The largest data value is 60.

Box Plot

To plot a box graph:

1. Enter data into L_1. (See Chapter 2.)

2. Change values in WINDOW menu, if necessary. (*Note:* Make X_{min} somewhat smaller than the smallest data value and X_{max} somewhat larger than the largest data value.) Change Y_{min} to 0 and Y_{max} to 10.

3. Press **2nd [STAT PLOT]** then either **ENTER** or **1**.

4. Press **ENTER** to turn plot On.

5. Move cursor to Box Plot symbol on the Type: line; then press **ENTER**.

6. Move cursor to L_1 on X list: line, and press **ENTER**.

7. Move cursor to 1 for Freq: line and press **ENTER**.

8. Press **GRAPH** to display the box plot.

9. Press **TRACE** followed by ◀ or ▶ to obtain the values or the box plot.

To display two box plots on the same display, follow the above steps and use the 2:Plot 2 and L_2 symbols.

Example 6 Construct a box plot for the data values in Example 5. Change the settings as follows:

$X_{min} = 40$
$X_{max} = 70$
$Y_{min} = 0$
$Y_{max} = 10$

Input

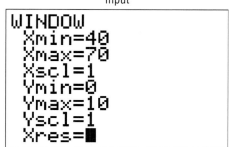

Input

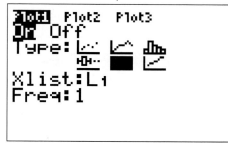

Output

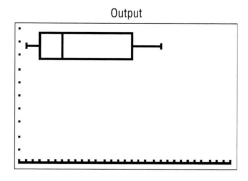

The minimum value is 41.
Q_1 is 43.
The median is 46.
Q_3 is 56.
The maximum value is 60.

chapter

4

Counting Techniques

Objectives

After completing this chapter, you
should be able to

1. Determine the number of
   outcomes to a sequence of
   events using a tree diagram.

2. Find the total number
   of outcomes in a sequence
   of events using the
   multiplication rule.

3. Find the number of ways
   r objects can be selected
   from n objects using the
   permutation rule.

4. Find the number of ways
   r objects can be selected
   from n objects without
   regard to order using the
   combination rule.

Statistics Today

Why Are We Running Out of 800 Numbers?

Phone companies and other agencies that deal in numbers need to know how many phone numbers, ID tags, or license plates they can issue using certain combinations of various letters and numbers. The article shown below explains that the phone companies are running out of toll-free 800 numbers. The question is: How many phone numbers with the 800 prefix can be issued in the United States?

Toll-free call? Get ready to dial 888

By Becky Beyers
USA TODAY

Get ready to keep your finger on the 8 when you make a toll-free call.

Phone companies will run out of 800 numbers early next year and start issuing toll-free numbers beginning with 888.

Use of 800 numbers has grown so fast, "we're a victim of our own success," says Dennis Byrne of the U.S. Telephone Association.

Only about 1.7 million of the 7.6 million possible 800-prefix combinations are still available. Why so few are left:

► Demand has taken off since May 1993, when the government allowed users to keep 800 numbers if they changed long-distance carriers.

► 800 numbers aren't just for big companies anymore. Small businesses use them, as do residential customers so family members can call home more cheaply than collect.

Such customers may pay 25 cents a minute for each call plus a monthly fee of $5.

► Some toll-free numbers are hoarded for promotional value or occasional use.

The industry's numbering council — phone companies and associations that set phone-number policies — is asking that little-used numbers be returned so they can be reissued.

Setting up a new toll-free access code involves the entire phone industry, Byrne says.

All internal systems must upgrade switching equipment so they can handle 888 calls.

What happens when the 888s are used up? It's on to 877, 866, and all the way down to 822.

Source: *USA Today,* February 13, 1995. Copyright 1995 *USA TODAY.* Used with permission.

In this chapter, you will learn the rule for counting, the differences between permutations and combinations, and how to figure out how many different combinations for specific situations exist.

Introduction

Many problems in probability and statistics require a careful analysis of the outcomes of a sequence of events. A sequence of events occurs when one or more events follow one another. For example, a sales representative may wish to select the most efficient way to visit several different stores in four cities. Sometimes, in order to determine costs or rates, a manager must know all possible outcomes of a classification scheme. Or an insurance company may wish to classify its drivers according to the following classes:

1. Gender (male, female).
2. Age (under 25, between 25 and 60, over 60).
3. Area of residence (rural, suburban, urban).
4. Distance driven to work (under 4 miles, between 4 and 10 miles, over 10 miles).
5. Value of the vehicle (under $5000, between $5000 and $10,000, over $10,000).

In a psychological study, a researcher might attempt to train a rat to run a maze. To determine the rat's success, the researcher must know the number of possible choices the rat can make to traverse the maze. Then the researcher can differentiate between actual learning and chance successes.

On a game show, a contestant might be required to arrange four digits correctly to guess the exact price of a new car. To determine the probability of his guessing the correct answer, one must know the number of possible ways the four digits can be arranged.

The vice president of a company might wish to know the number of different possible ways four employees can be selected from a group of 10 in order to be transferred to a new location.

Sometimes the total number of possible outcomes is enough; other times a list of all outcomes is needed. One can use several rules of counting here: the multiplication rule, the permutation rule, and the combination rule.

Tree Diagrams and the Multiplication Rule for Counting

Objective 1. Determine the number of outcomes to a sequence of events using a tree diagram.

Tree Diagrams

Many times one wishes to list each possibility of a sequence of events. For example, it would be difficult to list all possible outcomes of a seven-game World Series by guessing alone. Rather than do this listing in a haphazard way, one can use a tree diagram.

A **tree diagram** is a device used to list all possibilities of a sequence of events in a systematic way.

Tree diagrams are also useful in determining the probabilities of events, as will be shown in the next chapter.

Example 4–1

Suppose a sales rep can travel from New York to Pittsburgh by plane, train, or bus, and from Pittsburgh to Cincinnati by bus, boat, or automobile. List all possible ways he can travel from New York to Cincinnati.

Solution

A tree diagram can be drawn to show the possible ways. First, the salesman can travel from New York to Pittsburgh by three methods. The tree diagram for this situation is shown in Figure 4–1.

Figure 4–1

Tree Diagram for
New York–Pittsburgh
Trips in Example 4–1

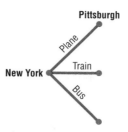

Then the salesman can travel from Pittsburgh to Cincinnati by bus, boat, or automobile. This tree diagram is shown in Figure 4–2.

Figure 4–2

Tree Diagram for
Pittsburgh–Cincinnati
Trips in Example 4–1

Next, the second branch is paired up with the first branch in three ways, as shown in Figure 4–3.

Figure 4–3

Complete Tree Diagram
for Example 4–1

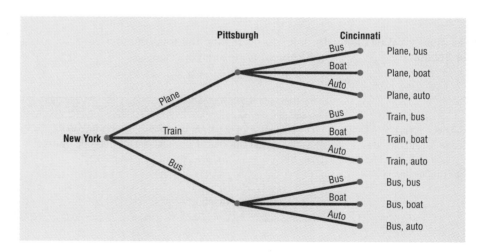

Finally, all outcomes can be listed by starting at New York and following the branches to Cincinnati, as shown at the right end of the tree in Figure 4–3. There are nine different ways.

Example 4–2

A coin is tossed and a die is rolled. Find all possible outcomes of this sequence of events.

Solution

Since the coin can land either heads up or tails up, and since the die can land with any one of six numbers shown face up, the outcomes can be represented as shown in Figure 4–4.

Figure 4–4

Complete Tree Diagram
for Example 4–2

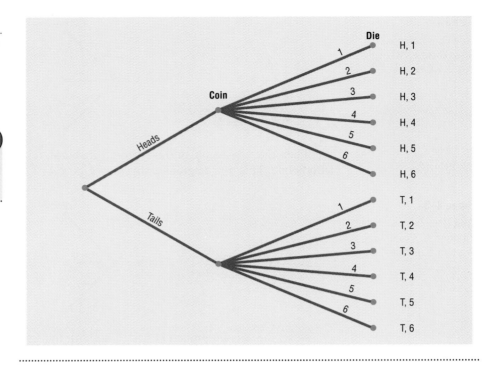

The Multiplication Rule for Counting

In order to determine the total number of outcomes in a sequence of events, the *multiplication* rule can be used.

Objective 2. Find the total number of outcomes in a sequence of events using the multiplication rule.

> ## Multiplication Rule
>
> In a sequence of n events in which the first one has k_1 possibilities and the second event has k_2 and the third has k_3, and so forth, the total number of possibilities of the sequence will be
>
> $$k_1 \cdot k_2 \cdot k_3 \cdot \ldots \cdot k_n$$
>
> *Note:* "And" means to multiply.

The next examples illustrate the multiplication rule.

Example 4–3

A paint manufacturer wishes to manufacture several different paints. The categories include

Color	Red, blue, white, black, green, brown, yellow
Type	Latex, oil
Texture	Flat, semigloss, high gloss
Use	Outdoor, indoor

How many different kinds of paint can be made if a person can select one color, one type, one texture, and one use?

Solution

A person can choose one color and one type and one texture and one use. Since there are seven color choices, two type choices, three texture choices, and two use choices, the total number of possible different paints is

Color		Type		Texture		Use	
7	·	2	·	3	·	2	= 84

Example 4–4

There are four blood types, A, B, AB, and O. Blood can also be Rh+ and Rh−. Finally, a blood donor can be classified as either male or female. How many different ways can a donor have his or her blood labeled?

Solution

Since there are four possibilities for blood type, two possibilities for the Rh factor, and two possibilities for the gender of the donor, there are $4 \cdot 2 \cdot 2$, or 16, different classification categories.

When determining the number of different possibilities of a sequence of events, one must know whether repetitions are permissible.

Example 4–5

The digits 0, 1, 2, 3, and 4 are to be used in a four-digit ID card. How many different cards are possible if repetitions are permitted?

Solution

Since there are four spaces to fill and five choices for each space, the solution is

$$5 \cdot 5 \cdot 5 \cdot 5 = 5^4 = 625$$

Now, what if repetitions are not permitted? For Example 4–5, the first digit can be chosen in five ways. But the second digit can be chosen in only four ways, since there are only four digits left; etc. Thus, the solution is

$$5 \cdot 4 \cdot 3 \cdot 2 = 120$$

The same situation occurs when one is drawing balls from an urn or cards from a deck. If the ball or card is replaced before the next one is selected, then repetitions are permitted, since the same one can be selected again. But if the selected ball or card is not replaced, then repetitions are not permitted, since the same ball or card cannot be selected the second time.

Exercises

4–1. By means of a tree diagram, find all possible outcomes for the genders of the children in a family that has three children.

4–2. Bill's Burger Joint sells hot dogs and hamburgers plain or with a choice of any *one* of the following items: tomato, onion, or mustard. Draw a tree diagram to represent all possible selections Bill can serve. (*Hint:* The customer can either select an item or not. If she does not select the item, use "plain.")

4–3. A quiz consists of four true–false questions. How many possible answer keys are there? Use a tree diagram.

4–4. Students are classified according to eye color (blue, brown, green), gender (male, female), and major (chemistry, mathematics, physics, business). How many possible different classifications are there? Use a tree diagram.

4–5. A box contains a $1 bill, a $5 bill, and a $10 bill. Two bills are selected in succession, without the first bill being replaced. Draw a tree diagram and represent all possible amounts of money that can be selected.

4–6. The Eagles and the Hawks play a hockey tournament. The first team to win two out of three games wins the tournament. Draw a tree diagram to represent the outcomes of the tournament.

4–7. An inspector selects three batteries from a lot, then tests each to see whether each is overcharged, normal, or undercharged. Draw a tree diagram to represent all possible outcomes.

4–8. List all possible outcomes of a five-game series played between the Pirates and the Yankees. (*Note:* The winner must win three games.)

4–9. A coin is tossed. If it comes up heads, it is tossed again. If it lands tails, a die is rolled. Find all possible outcomes of this sequence of events.

4–10. A person has a chance of obtaining a degree from each category listed below. Draw a tree diagram showing all possible ways a person could obtain these degrees.

Bachelor's	Master's	Doctor's
B.S.	M.S.	Ph.D.
B.A.	M.Ed.	D.Ed.
	M.A.	

4–11. If blood types can be A, B, AB, and O, and Rh+ and Rh−, draw a tree diagram for the possibilities.

4–12. A woman has three skirts, five blouses, and four scarves. How many different outfits can she wear, assuming that they are color-coordinated?

4–13. How many five-digit zip codes are possible if digits can be repeated? If there cannot be repetitions?

4–14. How many ways can a baseball manager arrange a batting order of nine players?

4–15. How many different ways can seven floral arrangements be arranged in a row on a single display shelf?

4–16. A store manager wishes to display six different kinds of laundry soap in a semicircle. How many ways can this be done?

4–17. A store manager wishes to display eight different brands of shampoo in a row. How many ways can this be done?

4–18. There are eight different statistics books, six different geometry books, and three different trigonometry books. A student must select one book of each type. How many different ways can this be done?

4–19. At a local cheerleaders' camp, five routines must be practiced. How many different ways can these five routines be presented?

4–20. The call letters of a radio station must have four letters. The first letter must be a *K* or a *W*. How many different station call letters can be made if repetitions are not allowed? If repetitions are allowed?

4–21. How many different three-digit identification tags can be made if the digits can be used more than once? If the first digit must be a 5 and repetitions are not permitted?

4–22. How many different ways can nine trophies be arranged on a shelf?

4–23. If a baseball manager has five pitchers and two catchers, how many different possible pitcher–catcher combinations can he field?

4–24. There are three roads from city X to city Y and four roads from city Y to city Z. How many different trips can be made from city X to city Z passing through city Y?

*4–25. Pine Pizza Palace sells pizza plain or with one or more of the following toppings: pepperoni, sausage, mushrooms, olives, onions, or anchovies. How many different pizzas can be made? (*Hint:* A person can select or not select each item.)

*4–26. Generalize Exercise 4–25 for *n* different toppings. (*Hint:* For example, there are two ways to select pepperoni: either take it or not take it. For two toppings, a person can select none, both, or one of the two. Continue this reasoning for three toppings, etc.)

*4–27. How many different ways can a person select one or more coins if he has two nickels, one dime, and one half-dollar?

*4–28. A photographer has five photographs that she can mount on a page in her portfolio. How many different ways can she mount her photographs?

*4–29. In a barnyard there is an assortment of chickens and cows. Counting heads, one gets 15; counting legs, one gets 46. How many of each are there?

*4–30. How many committees of two or more people can be formed from four people? (*Hint:* Make a list using the letters A, B, C, and D to represent the people.)

4–3

Permutations and Combinations

Factorial Notation

Two other rules that can be used to determine the total number of possibilities of a sequence of events are the permutation rule and the combination rule.

These rules use the *factorial notation*. The factorial notation uses the exclamation point.

$$5! \text{ means } 5 \cdot 4 \cdot 3 \cdot 2 \cdot 1$$
$$9! = 9 \cdot 8 \cdot 7 \cdot 6 \cdot 5 \cdot 4 \cdot 3 \cdot 2 \cdot 1$$

In order to use the formulas in the permutation and combination rules, a special definition of 0! is needed. It is equal to one.

Factorial Formulas

For any counting n

$$n! = n(n-1)(n-2)\ldots 1$$
$$0! = 1$$

A list of factorials is found in Table A of Appendix C.

Permutations

A **permutation** is an arrangement of n objects in a specific order.

The next two examples illustrate permutations.

Example 4–6

Suppose a business owner has a choice of five locations in which to establish her business. She decides to rank each location according to certain criteria, such as price of the store and parking facilities. How many different ways can she rank the five locations?

Solution

There are

$$5! = 5 \cdot 4 \cdot 3 \cdot 2 \cdot 1 = 120$$

different possible rankings.

In the previous example, all objects were used up. But what happens when all objects are not used up? The answer to this question is given in Example 4–7.

Example 4–7

Suppose the business owner in Example 4–6 wishes to rank only the top three locations. How many different ways can she rank them?

Solution

Using the multiplication rule, she can select any one of the five for first choice, then any one of the remaining four locations for her second choice, and finally, any one of the remaining three locations for her third choice, as shown.

First choice	Second choice	Third choice	
5	· 4	· 3	= 60

The solutions in Examples 4–6 and 4–7 are permutations.

Permutation Rule

Objective 3. Find the number of ways r objects can be selected from n objects using the permutation rule.

The arrangement of n objects in a specific order using r objects at a time is called a *permutation of n objects taking r objects at a time.* It is written as $_nP_r$, and the formula is

$$_nP_r = \frac{n!}{(n-r)!}$$

The notation $_nP_r$ is used for permutations.

$$_6P_4 \text{ means } \frac{6!}{(6-4)!} \text{ or } \frac{6!}{2!} \text{ and is equal to } 360.$$

A list of permutations can be found in Table C in Appendix C.

Although Examples 4–6 and 4–7 were solved by the multiplication rule, they can now be solved by the permutation rule.

In Example 4–6, five locations were taken and then arranged in order; hence,

$$_5P_5 = \frac{5!}{(5-5)!} = \frac{5!}{0!} = 120$$

(Recall $0! = 1$)

In Example 4–7, three locations were selected from five locations, so $n = 5$ and $r = 3$; hence

$$_5P_3 = \frac{5!}{(5-3)!} = \frac{5!}{2!} = 60$$

The next two examples illustrate the permutation rule.

Example 4–8

How many different arrangements of three boxcars can be selected from eight boxcars for a train? The order is important since each boxcar is to be delivered to a different location.

Interesting Fact

Solution

$$_8P_3 = \frac{8!}{(8-3)!} = \frac{8!}{5!} = 336$$

Example 4–9

How many different ways can a chairperson and an assistant chairperson be selected for a research project if there are seven scientists available?

Solution

$$_7P_2 = \frac{7!}{(7-2)!} = \frac{7!}{5!} = 42$$

Combinations

Objective 4. Find the number of ways *r* objects can be selected from *n* objects without regard to order using the combination rule.

Suppose a dress designer wishes to select two colors of material to design a new dress, and she has on hand four colors. How many different possibilities can there be in this situation?

This type of problem differs from previous ones in that the order of selection is not important. That is, if the designer selects yellow and red, this selection is the same as the selection red and yellow. This type of selection is called a *combination*. The difference between a permutation and a combination is that in a combination, the order or arrangement of the objects is not important; by contrast, order *is* important in a permutation. The next example illustrates this difference.

A selection of distinct objects without regard to order is called a **combination.**

Example 4–10

Given the letters A, B, C, and D, list the permutations and combinations for selecting two letters.

Solution

The listings follow.

Permutations				Combinations	
AB	BA	CA	DA	AB	BC
AC	BC	CB	DB	AC	BD
AD	BD	CD	DC	AD	CD

Note that in permutations, AB is different from BA. But in combinations, AB is the same as BA.

Combinations are used when the order or arrangement is not important, as in the selecting process. Suppose a committee of 5 students is to be selected from 25 students. The five students represent a combination, since it does not matter who is selected first, second, etc.

Combination Rule

The number of combinations of r objects selected from n objects is denoted by $_nC_r$ and is given by the formula

$$_nC_r = \frac{n!}{(n-r)!r!}$$

Example 4–11

How many combinations of four objects are there taken two at a time?

Solution

Since this is a combination problem, the answer is

$$_4C_2 = \frac{4!}{(4-2)!2!} = \frac{4!}{2!2!} = 6$$

This is the same result shown in Example 4–10.

Notice that the formula for $_nC_r$ is

$$\frac{n!}{(n-r)!r!}$$

which is the formula for permutations,

$$\frac{n!}{(n-r)!}$$

with an $r!$ in the denominator. This $r!$ divides out the duplicates from the number of permutations, as shown in Example 4–10. For each two letters, there are two permutations but only one combination. Hence, dividing the number of permutations by $r!$ eliminates the duplicates. This result can be verified for other values of n and r. *Note:* $_nC_n = 1$.

A list of combinations can be found in Table C in Appendix C.

Example 4–12

In order to survey the opinions of customers at local malls, a researcher decides to select 5 malls from a total of 12 malls in a specific geographic area. How many different ways can the selection be made?

Solution

$$_{12}C_5 = \frac{12!}{(12-5)!5!} = \frac{12!}{7!5!} = 792$$

Example 4–13

In a club there are 7 women and 5 men. A committee of 3 women and 2 men is to be chosen. How many different possibilities are there?

Solution

Here, one must select 3 women from 7 women, which can be done in $_7C_3$, or 35, ways. Next, 2 men must be selected from 5 men, which can be done in $_5C_2$, or 10, ways. Finally, by the multiplication rule, the total number of different ways is $35 \cdot 10 = 350$, since one is choosing both men and women. Using the formula

$$_7C_3 \cdot {_5C_2} = \frac{7!}{(7-3)!3!} \cdot \frac{5!}{(5-2)!2!} = 350$$

Table 4–1 summarizes the counting rules.

Table 4–1   Summary of Counting Rules

Rule	Definition	Formula
Multiplication rule	The number of ways a sequence of n events can occur if the first event can occur in k_1 ways, the second event can occur in k_2 ways, etc.	$k_1 \cdot k_2 \cdot k_3 \cdot \ldots \cdot k_n$
Permutation rule	The number of permutations of n objects taking r objects at a time	$_nP_r = \dfrac{n!}{(n-r)!}$
Combination rule	The number of combinations (order is not important) of r objects taken from n objects	$_nC_r = \dfrac{n!}{(n-r)!r!}$

Exercises

4–31. Evaluate each:
a. 8!
b. 10!
c. 4!
d. 1!
e. 0!

4–32. (Ans.) Evaluate each expression.
a. $_8P_2$
b. $_7P_5$
c. $_{12}P_4$
d. $_5P_3$
e. $_6P_0$
f. $_6P_6$
g. $_8P_0$
h. $_8P_8$
i. $_{11}P_3$
j. $_6P_2$

4–33. How many different four-letter permutations can be formed from the letters in the word *decagon?*

4–34. In a board of directors composed of eight people, how many ways can a chief executive officer, a director, and a treasurer be selected?

4–35. How many different ID cards can be made if there are six digits on a card and no digit can be used more than once?

4–36. How many ways can seven different types of soaps be displayed on a shelf in a grocery store?

4–37. How many different four-color code stripes can be made on a sports car if each code consists of the colors green, red, blue, and white? All colors are used only once.

4–38. An inspector must select three tests to perform in a certain order on a manufactured part. He has a choice of seven tests. How many ways can he perform three different tests?

4–39. The Anderson Research Co. decides to test-market a product in six areas. How many different ways can three areas be selected in a certain order for the first test?

4–40. How many different ways can a visiting nurse see six patients if she sees them all in one day?

4–41. How many different ways can five radio commercials be run during an hour of time?

4–42. How many different ways can four tickets be selected from 50 tickets if each ticket wins a different prize?

4–43. How many different ways can a researcher select five rats from 20 rats and assign each to a different test?

4–44. How many different signals can be made by using at least three distinct flags if there are five different flags from which to select?

4–45. An investigative agency has seven cases and five agents. How many different ways can the cases be assigned if only one case is assigned to each agent?

4–46. (Ans.) Evaluate each expression.

a. $_5C_2$	*d.* $_6C_2$	*g.* $_3C_3$	*j.* $_4C_3$
b. $_8C_3$	*e.* $_6C_4$	*h.* $_9C_7$	
c. $_7C_4$	*f.* $_3C_0$	*i.* $_{12}C_2$	

4–47. How many ways can 3 cards be selected from a standard deck of 52 cards disregarding the order of selection?

4–48. How many ways are there to select 3 bracelets from a box of 10 bracelets disregarding the order of selection?

4–49. How many ways can 4 baseball players and 3 basketball players be selected from 12 baseball players and 9 basketball players?

4–50. How many ways can a committee of 4 people be selected from a group of 10 people?

4–51. If a person can select 3 presents from 10 presents under a Christmas tree, how many different combinations are there?

4–52. How many different tests can be made from a test bank of 20 questions if the test consists of 5 questions?

4–53. The general manager of a fast-food restaurant chain must select 6 restaurants from 11 for a promotional program. How many different possible ways can this selection be done?

4–54. How many ways can 3 cars and 4 trucks be selected from 8 cars and 11 trucks to be tested for a safety inspection?

4–55. In a train yard there are 4 tank cars, 12 boxcars, and 7 flatcars. How many ways can a train be made up consisting of 2 tank cars, 5 boxcars, and 3 flatcars? (In this case order is not important.)

4–56. There are seven women and five men in a department. How many ways can a committee of four people be selected? How many ways can this committee be selected if there must be two men and two women on the committee? How many ways can this committee be selected if there must be at least two women on the committee?

4–57. Wake Up cereal comes in two types, crispy and crunchy. If a researcher has 10 boxes of each, how many ways can she select 3 boxes of each for a quality control test?

4–58. How many ways can a dinner patron select three appetizers and two vegetables if there are six appetizers and five vegetables on the menu?

4–59. How many ways can a jury of 6 men and 6 women be selected from 12 men and 10 women?

4–60. How many ways can a foursome of 2 men and 2 women be selected from 10 men and 12 women in a golf club?

4–61. The state narcotics bureau must form a 5-member investigative team. If it has 25 agents from which to choose, how many different possible teams can be formed?

4–62. How many different ways can an instructor select 2 textbooks from a possible 17?

4–63. The Environmental Protection Agency must investigate nine mills for complaints of air pollution. How many different ways can a representative select five of these to investigate this week?

4–64. How many ways can a person select 7 television commercials from 11 television commercials?

4–65. How many ways can a person select 8 videotapes from 10 tapes?

4–66. A buyer decides to stock 8 different posters. How many ways can she select these 8 if there are 20 from which to choose?

4–67. An advertising manager decides to have an ad campaign in which 8 special calculators will be hidden at various locations in a shopping mall. If he has 17 locations from which to pick, how many different possible combinations can he choose?

*4–68. How many different ways can five people—A, B, C, D, and E—sit in a row at a movie theater if (a) A and B must sit together; (b) C must sit to the right of, but not necessarily next to, B; (c) D and E will not sit next to each other?

*4–69. Using combinations, calculate the number of each poker hand in a deck of cards.
a. Royal flush   *c.* Four of a kind
b. Straight flush   *d.* Full house

4–4

Summary

This chapter illustrates how one can find the outcomes of a sequence of events. A tree diagram can be used when a list of all possible outcomes is necessary. When only the total number of outcomes is needed, the multiplication rule, the permutation rule, and the combination rule can be used.

Using these rules, statisticians can find the solutions to a variety of problems in which they must know the number of possibilities that can occur. These rules can also be used to determine the probabilities of events.

Important Terms

Combination 155   Permutation 153   Tree diagram 148

Important Formulas

Multiplication rule: In a sequence of n events in which the first one has k_1 possibilities, the second event has k_2 possibilities, the third has k_3 possibilities, etc., the total number of possibilities of the sequence will be

$$k_1 \cdot k_2 \cdot k_3 \cdot \ldots \cdot k_n$$

Permutation rule: The number of permutations of n objects taken r objects at a time is

$$_nP_r = \frac{n!}{(n-r)!}$$

Combination rule: The number of combinations of r objects selected from n objects is

$$_nC_r = \frac{n!}{(n-r)!r!}$$

Review Exercises

4–70. An automobile license plate consists of three letters followed by four digits. How many different plates can be made if repetitions are allowed? If repetitions are not allowed? If repetitions are allowed in the letters but not in the digits?

4–71. How many different arrangements of the letters in the word *bread* can be made?

4–72. How many different three-digit combinations can be made by using the numbers 1, 3, 5, 7, and 9 without repetitions if the "right" combination can be

used to open a safe? (*Hint:* Does a combination lock really use combinations?)

4–73. How many two-card pairs are there in a poker deck?

4–74. A quiz consists of six multiple-choice questions. Each question has three possible answer choices. How many different answer keys can be made?

4–75. How many ways can 5 different television programs be selected from 12 programs?

4–76. How many different ways can a buyer select four television models from a possible choice of six models?

4–77. Draw a tree diagram to show all possible outcomes when a coin is flipped four times.

4–78. How many ways can three outfielders and four infielders be chosen from five outfielders and seven infielders?

4–79. How many different ways can eight computer operators be seated in a row?

4–80. How many ways can a student select 2 electives from a possible choice of 10 electives?

4–81. There are six Republican, five Democrat, and four Independent candidates. How many different ways can a committee of three Republicans, two Democrats, and one Independent be selected?

4–82. Using Exercise 4–81, how many ways can a committee of four people be selected if they are all from the same party?

4–83. Employees can be classified according to gender (male, female), income (low, medium, high), and rank (assistant, instructor, dean). Draw a tree diagram and show all possible outcomes.

4–84. A disc jockey can select 4 records from 10 to play in one segment. How many ways can this selection be done? (*Note:* Order is important.)

4–85. A judge is to rank six brands of cookies according to their flavor. How many different ways can this ranking be done?

4–86. A vending machine servicer must restock and collect money from 20 machines, each one at a different location. How many different ways can he select 5 machines to service in one day?

4–87. How many different ways can four paintings be arranged on a wall?

4–88. How many different ways can 3 fraternity members be selected from 10 members if one must be president, one must be vice president, and one must be secretary/treasurer?

4–89. How many different ways can two balls be drawn from a bag containing five balls? Each ball is a different color, and the first ball is replaced before the second one is selected. How many different ways are there if the first ball is not replaced before the second one is selected?

4–90. A restaurant offers three choices of meat, two choices of potatoes, four choices of vegetables, and five choices of dessert. How many different possible meals can be made if a customer must select one item from each category?

4–91. If someone wears a blouse or a sweater and a pair of slacks or a skirt, how many different outfits can she wear?

4–92. How many different computer passwords are possible if each consists of four symbols and if the first one must be a letter and the other three must be digits?

4–93. If a student has a choice of five computers, three printers, and two monitors, how many ways can she select a computer system?

4–94. A combination lock consists of the numbers 0 to 39. If no number can be used twice, how many different combinations are possible using three numbers? Remember, a combination lock is really a permutation lock.

4–95. There are 12 applicants for financial aid. The Seabest Scholarship allows four students to receive the money. How many different ways can these scholarships be awarded?

4–96. A candy store allows customers to select three different candies to be packaged and mailed. If there are 13 varieties available, how many possible selections can be made?

4–97. If a student can select 5 novels from a reading list of 20 for a course in literature, how many different possible ways can this selection be done?

4–98. If a student can select one of three language courses, one of five mathematics courses, and one of four history courses, how many different schedules can be made?

Statistics Today

Why Are We Running Out of 800 Numbers?   Revisited

Since each 800 number is followed by a seven-digit number, multiplication rule 1 can be used to determine the total number of 800 phone numbers that are available.

Since there are 10 digits (0 through 9) that can be used for each digit of the seven-digit number, the answer is 10^7, or 10,000,000, starting with 800-0000000 and ending with 800-9999999. (Since the telephone industry forbids usage of certain numbers, the article that opens this chapter refers to a smaller number of "possible" 800 numbers.)

Quiz

Determine whether each statement is true or false. If the statement is false, explain why.

1. If there are 5 contestants in a race, the number of different ways the first- and second-place winners can be selected is 25.

2. If a true–false exam contains 10 questions, there are 20 different ways to answer all the questions.

3. Some permutation problems can be solved by the multiplication rule.

4. The arrangement ABC is the same as BAC for combinations.

5. When objects are arranged in a specific order, the arrangement is called a combination.

Select the best answer.

6. What is $_nP_0$?
a. 0
b. 1
c. n
d. It cannot be determined.

7. What is the number of permutations of six different objects taken all together?
a. 0
b. 1
c. 36
d. 720

8. How many permutations of the letters in the word *tide* are there?
a. 1
b. 6
c. 12
d. 24

9. What is 0!?
a. 0
b. 1
c. undefined
d. 10

10. What is $_nC_n$?
a. 0
b. 1
c. n
d. It cannot be determined.

Complete the following statements with the best answer.

11. A device that is helpful in listing the outcomes of a sequence of events is called a _____.

12. When a coin is tossed and a die is rolled, there are _____ outcomes.

13. If in a sequence of k events each event can occur the same number of ways (n), then the total number of outcomes is _____.

14. The number of permutations of n objects taken all together is _____.

15. Telephone numbers are examples of _____.

16. One company's ID cards consist of five letters followed by two digits. How many cards can be made if repetitions are allowed? If repetitions are not allowed?

17. How many different arrangements of the letters in the word *number* can be made?

18. A physics test consists of 25 true–false questions. How many different possible answer keys can be made?

19. How many different ways can four radios be selected from a total of seven radios?

20. The National Bridge Association can select one of four cities for its playoff tournament next year. The cities are Pasadena, Wilmington, Chicago, and

Charleston. The following year, it can hold the tournament in Hyattsville or Green Springs. How many different possibilities are there for the next two years? Draw a tree diagram and show all possibilities.

21.   How many ways can five sopranos and four altos be selected from seven sopranos and nine altos?

22.   How many different ways can eight kindergarten children be seated in a row?

23.   Employees can be classified according to gender (male, female), income (low, medium, high), and rank (staff nurse, charge nurse, head nurse). Draw a tree diagram and show all possible outcomes.

24.   A soda machine servicer must restock and collect money from 15 machines, each one at a different location. How many ways can she select four machines to service in one day?

25.   How many different ways can three cubes be drawn from a bag containing four cubes? Each cube is a different color, and the first cube is replaced before

the second one is selected. How many different ways are there if the first cube is not replaced before the second one is selected?

26.   If a man can wear a shirt or a sweater and a pair of dress slacks or a pair of jeans, how many different outfits can he wear?

27.   A beachfront candy store allows customers to select three different candies to be packaged and mailed. If there are 10 varieties available, how many possible selections can be made?

28.   How many different ways can an avid reader select 6 novels from a shelf containing 15?

29.   If a woman can select one of three nail-polish colors, one of five lipstick shades, and one of four blush shades, how many different combinations can she make?

30.   Find the number of different snacks, consisting of one or more of these crackers—whole wheat, onion, buttercrisp, or sesame—that can be served to guests.

Critical Thinking Challenges

1.   Shake Hands   A person decides to shake hands with six different people on a certain day. The next day, each of the six people will shake hands with six different people. The process continues until every person in the United States has shaken someone's hand. How many days will it take until everyone in the United States has shaken hands once? Assume that once a person shakes hands with six different people, he or she does not shake hands again. (*Hint:* The population of the United States is 248,709,873, according to the 1990 census.)

2.   How Many Hairs?   If it can be assumed that the maximum number of hairs on a human head is about 500,000, explain why at least two people living in Houston (population 1,629,902, according to the 1990 census) have the same number of hairs on their heads.

3.   Combinations and Pascal's Triangle   A mathematician named Pascal wrote a treatise showing how combinations can be derived from a triangular array of numbers. This triangle became known as Pascal's triangle, although there is evidence that the

triangle existed in China in the 1300s. The triangle is formed by adding the two adjacent numbers and writing the sum below in a triangular fashion.

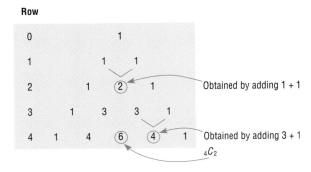

Each number in the triangle represents the number of combinations of n objects taken r at a time. For example, $_4C_2 = 6$, which is the third value found in row 4.

Complete Pascal's triangle for nine rows and verify the answers using combinations.

TI-83 Calculator

Permutations, Combinations, and Factorials

A. To find the value of a permutation:
Example $_5P_3$
1. Enter **5** on the home screen.
2. Press **MATH** and move the cursor to PRB.
3. Press **2**, then **3**, then **ENTER**.
The calculator will display the answer, 60.

B. To find the value of a combination:
Example $_8C_5$
1. Enter **8** on the home screen.
2. Press **MATH**, move the cursor to PRB, then press **3**.

3. Then press **5** and **ENTER**.
The calculator will display the answer, 56.

C. To find a factorial of a number:
Example 5!
1. Enter **5** on the home screen.
2. Press **MATH** and move the cursor to PRB.
3. Press **4**, then **ENTER**.
The calculator will display the answer, 120.

chapter

5 Probability

Objectives

After completing this chapter, you should be able to

1. Determine sample spaces and find the probability of an event using classical probability or empirical probability.

2. Find the probability of compound events using the addition rules.

3. Find the probability of compound events using the multiplication rules.

4. Find the conditional probability of an event.

Would You Bet Your Life?

Humans not only bet money when they gamble, but they also bet their lives by engaging in unhealthy activities such as smoking, drinking, using drugs, and exceeding the speed limit when driving. Many people don't care about the risks involved in these activities since they do not understand the concepts of probability. On the other hand, people may fear activities that involve little risk to health or life because these activities have been sensationalized by the press and media.

In his book *Probabilities in Everyday Life* (Ivy Books, 1986, p. 191), John D. McGervey states,

> *When people have been asked to estimate the frequency of death from various causes, the most overestimated categories are those involving pregnancy, tornadoes, floods, fire, and homicide. The most underestimated categories include deaths from diseases such as diabetes, strokes, tuberculosis, asthma, and stomach cancer (although cancer in general is overestimated).*

The question then is: Would you feel safer if you flew across the United States on a commercial airline or if you drove? How much greater is the risk of one way to travel over the other?

In this chapter, you will learn about probability—its meaning, how it is computed, and how to evaluate it in terms of the likelihood of an event actually happening.

Introduction

A cynical person once said, "The only two sure things are death and taxes." This philosophy no doubt arose because so much in people's lives is affected by chance. From the time a person awakes until he or she goes to bed, that person makes decisions regarding the possible events that are governed at least in part by chance. For example, should I carry an umbrella to work today? Will my car battery last until spring? Should I accept that new job?

Probability as a general concept can be defined as the chance of an event occurring. Most people are familiar with probability from observing or playing games of chance, such as card games, slot machines, or lotteries. In addition to being used in games of chance, probability theory is used in the fields of insurance, investments, and weather forecasting, and in various other areas. Finally, as stated in Chapter 1, probability is the basis of inferential statistics. For example, predictions are based on probability, and hypotheses are tested by using probability.

The basic concepts of probability are explained in this chapter. These concepts include *probability experiments, sample spaces,* the *addition and multiplication rules,* and the *probabilities of complementary events.*

Sample Spaces and Probability

Objective 1. Determine sample spaces and find the probability of an event using classical probability or empirical probability.

The theory of probability grew out of the study of various games of chance using coins, dice, and cards. Since these devices lend themselves well to the application of concepts of probability, they will be used in this chapter as examples. This section begins by explaining some basic concepts of probability. Then the types of probability and probability rules are discussed.

Basic Concepts

Processes such as flipping a coin, rolling a die, or drawing a card from a deck are called *probability experiments.*

A **probability experiment** is a process that leads to well-defined results called outcomes.

An **outcome** is the result of a single trial of a probability experiment.

A trial means flipping a coin once, rolling one die once, or the like. When a coin is tossed, there are two possible outcomes: head or tail. (*Note:* We exclude the possibility of a coin landing on its edge.) In the roll of a single die, there are six possible outcomes: 1, 2, 3, 4, 5, or 6. In any experiment, the set of all possible outcomes is called the *sample space.*

A **sample space** is the set of all possible outcomes of a probability experiment.

Some sample spaces for various probability experiments are shown here.

Experiment	Sample space
Toss one coin	Head, tail
Roll a die	1, 2, 3, 4, 5, 6
Answer a true–false question	True, false
Toss two coins	Head-head, tail-tail, head-tail, tail-head

It is important to realize that when two coins are tossed, there are *four* possible outcomes, as shown in the fourth experiment above. Both coins could fall heads up. Both coins could fall tails up. Coin 1 could fall heads up and coin 2 tails up. Or coin 1 could fall tails up and coin 2 heads up. Heads and tails will be abbreviated as H and T throughout this chapter.

Example 5–1

Find the sample space for rolling two dice.

Solution

Since each die can land in six different ways, and two dice are rolled, the sample space can be presented by a rectangular array, as shown in Figure 5–1. The sample space is the list of pairs of numbers in the chart.

Figure 5–1

Sample Space for
Rolling Two Dice
(Example 5–1)

Die 1	Die 2					
	1	2	3	4	5	6
1	(1, 1)	(1, 2)	(1, 3)	(1, 4)	(1, 5)	(1, 6)
2	(2, 1)	(2, 2)	(2, 3)	(2, 4)	(2, 5)	(2, 6)
3	(3, 1)	(3, 2)	(3, 3)	(3, 4)	(3, 5)	(3, 6)
4	(4, 1)	(4, 2)	(4, 3)	(4, 4)	(4, 5)	(4, 6)
5	(5, 1)	(5, 2)	(5, 3)	(5, 4)	(5, 5)	(5, 6)
6	(6, 1)	(6, 2)	(6, 3)	(6, 4)	(6, 5)	(6, 6)

Example 5–2

Find the sample space for drawing one card from an ordinary deck of cards.

Solution

Since there are four suits (hearts, clubs, diamonds, and spades) and 13 cards for each suit (ace through king), there are 52 outcomes in the sample space. See Figure 5–2.

Figure 5–2

Sample Space for
Drawing a Card
(Example 5–2)

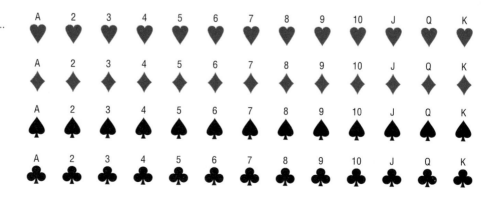

Example 5–3

Find the sample space for the gender of the children if a family has three children. Use B for boy and G for girl.

Historical Note

The famous Italian astronomer Galileo (1564–1642) found that a sum of nine occurs more often than any other sum when three dice are tossed. Previously, it was thought that a sum of 10 occurred more often than any other sum.

Solution

There are two genders, male and female, and each child could be either gender. Hence, there are eight possibilities, as shown here.

BBB   BBG   BGB   GBB   GGG   GGB   GBG   BGG

In the previous examples, the sample spaces were found by observation and reasoning; however, a tree diagram can also be used. In Chapter 4, the tree diagram was used to show all possible outcomes in a sequence of events. *The tree diagram can also be used as a systematic way to find all possible outcomes of a probability experiment.*

Example 5–4

Use a tree diagram to find the sample space for the gender of three children in a family, as in Example 5–3.

Solution

There are two possibilities for the first child, two for the second, and two for the third. Hence, the tree diagram can be drawn as shown in Figure 5–3.

Figure 5–3

Tree Diagram for Example 5–4

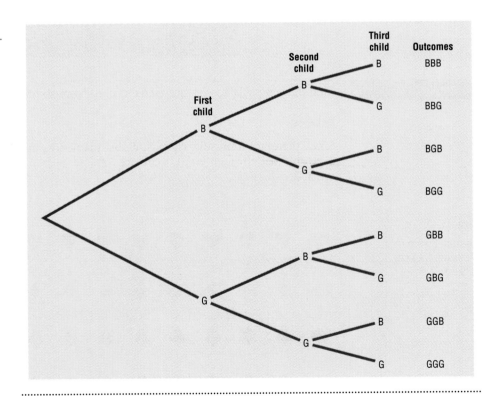

An outcome was defined previously as the result of a single trial of a probability experiment. In many problems, one must find the probability of two or more outcomes. For this reason, it is necessary to distinguish between an outcome and an event.

An **event** consists of one or more outcomes of a probability experiment.

An event can be one outcome or more than one outcome. For example, if a die is rolled and a 6 shows, this result is called an *outcome,* since it is a result of a single trial. An event with one outcome is called a **simple event.** The event of getting an odd number is called a **compound event,** since it consists of three outcomes or three simple events. In general, a compound event consists of two or more outcomes or simple events.

There are three basic types of probability:

1. Classical probability
2. Empirical or relative frequency probability
3. Subjective probability

Classical Probability

Classical probability uses sample spaces to determine the numerical probability that an event will happen. One does not actually have to perform the experiment to determine that probability. Classical probability is so named because it was the first type of probability studied formally by mathematicians in the 17th and 18th centuries.

Classical probability assumes that all outcomes in the sample space are equally likely to occur. For example, when a single die is rolled, each outcome has the same probability of occurring. Since there are six outcomes, each outcome has a probability of $\frac{1}{6}$. When a card is selected from an ordinary deck of 52 cards, one assumes that the deck has been shuffled, and each card has the same probability of being selected. In this case, it is $\frac{1}{52}$.

Equally likely events are events that have the same probability of occurring.

Formula for Classical Probability

The probability of any event E is

$$\frac{\text{number of outcomes in } E}{\text{total number of outcomes in the sample space}}$$

This probability is denoted by

$$P(E) = \frac{n(E)}{n(S)}$$

This probability is called *classical probability,* and it uses the sample space S.

Probabilities can be expressed as fractions, decimals, or—where appropriate—percentages. If one asks, "What is the probability of getting a head when a coin is tossed?" typical responses can be any of the following three.

"One-half."
"Point five."
"Fifty percent."

These answers are all equivalent. In most cases, the answers to examples and exercises given in this chapter are expressed as fractions or decimals, but percentages are used where appropriate.

Rounding Rule for Probabilities

Probabilities should be expressed as reduced fractions or rounded to two or three decimal places. However, when obtaining probabilities from one of the tables in Appendix C, use the number of decimal places given in the table. If decimals are converted to percentages to express probabilities, move the point two places to the right and add a percent sign.

Example 5–5

For a card drawn from an ordinary deck, find the probability of getting a queen.

Solution

Since there are 4 queens and 52 cards, P (queen) $= \frac{4}{52} = \frac{1}{13}$.

Example 5–6

If a family has three children, find the probability that all the children are girls.

Solution

The sample space for the gender of children for a family that has three children is BBB, BBG, BGB, GBB, GGG, GGB, GBG, and BGG (see Examples 5–3 and 5–4). Since there is one way in eight possibilities for all three children to be girls,

$$P(\text{GGG}) = \tfrac{1}{8}$$

Example 5–7

A card is drawn from an ordinary deck. Find these probabilities.

a. Of getting a jack.
b. Of getting the 6 of clubs.
c. Of getting a 3 or a diamond.

Historical Note

Ancient Greeks and Romans made crude dice from animal bones, various stones, minerals, and ivory. When they were tested mathematically, some were found to be quite good.

Solution

a. Refer to the sample space in Figure 5–2. There are 4 jacks and 52 possible outcomes. Hence,

$$P(\text{jack}) = \tfrac{4}{52} = \tfrac{1}{13}$$

b. Since there is only one 6 of clubs, the probability of getting a 6 of clubs is

$$P(\text{6 of clubs}) = \tfrac{1}{52}$$

c. There are four 3s and 13 diamonds, but the 3 of diamonds is counted twice in this listing. Hence, there are 16 possibilities of drawing a 3 or a diamond, so

$$P(\text{3 or diamond}) = \tfrac{16}{52} = \tfrac{4}{13}$$

See the sample space shown in Figure 5–2.

The probability of any event will always be a number (*either a fraction or a decimal*) between and including 0 and one. This is denoted mathematically as $0 \le P(E) \le 1$. Probabilities cannot be negative or greater than one.

When an event cannot occur (i.e., the event contains no members in the sample space), the probability is zero. This is shown in the next sample.

Example 5–8

When a single die is rolled, find the probability of getting a 9.

Solution

Since the sample space is 1, 2, 3, 4, 5, and 6, it is impossible to get a 9. Hence, the probability is $P(9) = \frac{0}{6} = 0$.

When the event is certain to occur, the probability is one, as shown in the next example.

Example 5–9

When a single die is rolled, what is the probability of getting a number less than 7?

Solution

Since all outcomes, 1, 2, 3, 4, 5, and 6, are less than 7, the probability is

$$P(\text{number less than 7}) = \frac{6}{6} = 1$$

The event of getting a number less than 7 is certain.

Finally, the sum of the probabilities of all outcomes in a sample space is one. For example, in the roll of a fair die, each outcome in the sample space has a probability of $\frac{1}{6}$. Hence, the sum of the probabilities of the outcomes is as shown.

Outcome	1	2	3	4	5	6
Probability	$\frac{1}{6}$	$\frac{1}{6}$	$\frac{1}{6}$	$\frac{1}{6}$	$\frac{1}{6}$	$\frac{1}{6}$
Sum	$\frac{1}{6}$ +	$\frac{1}{6}$ +	$\frac{1}{6}$ +	$\frac{1}{6}$ +	$\frac{1}{6}$ +	$\frac{1}{6} = \frac{6}{6} = 1$

Historical Note

Paintings in tombs excavated in Egypt show that the Egyptians played games of chance. One game called **Hounds and Jackals** played in 1800 B.C. is similar to the present-day game of **Snakes and Ladders**.

Complementary Events

Another important concept in probability theory is that of *complementary events*. When a die is rolled, for instance, the sample space consists of the outcomes 1, 2, 3, 4, 5, and 6. The event E of getting odd numbers consists of the outcomes 1, 3, and 5. The event of not getting an odd number is called the *complement* of event E, and it consists of the outcomes 2, 4, and 6.

The **complement of an event** E is the set of outcomes in the sample space that are not included in the outcomes of event E. The complement of E is denoted by $\overline{E}$ (read "E bar").

The next example further illustrates the concept of complementary events.

Example 5–10

Find the complement of each event.

a. Rolling a die and getting a 4.

b. Selecting a letter of the alphabet and getting a vowel.

c. Selecting a month and getting a month that begins with a *J*.

d. Selecting a day of the week and getting a weekday.

Solution

a. Getting a 1, 2, 3, 5, or 6.

b. Getting a consonant (assume y is a consonant).

c. Getting February, March, April, May, August, September, October, November, or December.

d. Getting Saturday or Sunday.

....

The outcomes of an event and the outcomes of the complement make up the entire sample space. For example, if two coins are tossed, the sample space is HH, HT, TH, and TT. The complement of "getting all heads" is not "getting all tails," since the event "all heads" is HH, and the complement of HH is HT, TH, and TT. Hence, the complement of the event "all heads" is the event "getting at least one tail."

Since the event and its complement make up the entire sample space, it follows that the sum of the probability of the event and the probability of its complement will equal 1. That is, $P(E) + P(\overline{E}) = 1$. In the previous example, let E = all heads, or HH, and let $\overline{E}$ = at least one tail, or HT, TH, TT. Then $P(E) = \frac{1}{4}$ and $P(\overline{E}) = \frac{3}{4}$; hence, $P(E) + P(\overline{E}) = \frac{1}{4} + \frac{3}{4} = 1$.

The rule for complementary events can be stated algebraically in three ways.

Rule for Complementary Events

$$P(\overline{E}) = 1 - P(E) \text{ or } P(E) = 1 - P(\overline{E}) \text{ or } P(E) + P(\overline{E}) = 1$$

Stated in words, the rule is: *If the probability of an event or the probability of its complement is known, then the other can be found by subtracting the probability from 1.* This rule is important in probability theory because at times the best solution to a problem is to find the probability of the complement of an event and then subtract from 1 to get the probability of the event itself.

....

Example 5–11

If the probability that a person lives in an industrialized country of the world is $\frac{1}{5}$, find the probability that a person does not live in an industrialized country.

Source: *Harper's Index* 289, no. 1737 (February 1995), p. 11.

Solution

P (not living in an industrialized country) = $1 - P$ (living in an industrialized country) = $1 - \frac{1}{5} = \frac{4}{5}$.

....

Probabilities can be represented pictorially by **Venn diagrams.** Figure 5–4(a) shows the probability of a simple event E. The area inside the circle represents the probability of event E—that is, $P(E)$. The area inside the rectangle represents the probability of all the events in the sample space, $P(S)$.

The Venn diagram that represents the probability of the complement of an event $P(\overline{E})$ is shown in Figure 5–4(b). In this case, $P(\overline{E}) = 1 - P(E)$, which is the area inside the rectangle but outside the circle representing $P(E)$. Recall that $P(S) = 1$ and $P(E) = 1 - P(\bar{E})$. The reasoning is that $P(E)$ is represented by the area of the circle and $P(\bar{E})$ is the probability of the events that are outside the circle.

Figure 5–4

Venn Diagram for the Probability and Complement

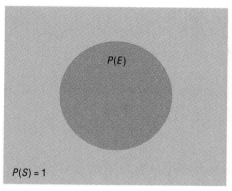

$P(E)$

$P(S) = 1$

(a) Simple probability

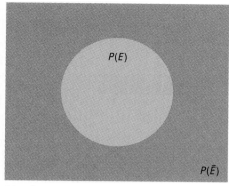

$P(E)$

$P(\bar{E})$

(b) $P(\bar{E}) = 1 - P(E)$

Empirical Probability

The difference between classical and **empirical probability** is that classical probability assumes that certain outcomes are equally likely (such as the outcomes when a die is rolled) while empirical probability relies on actual experience to determine the likelihood of outcomes. In empirical probability, one might actually roll a given die 6000 times and observe the relative frequencies and use these frequencies to determine the probability of an outcome. Suppose, for example, that a researcher asked 25 people if they liked the taste of a new soft drink. The responses were classified as "yes," "no," or "undecided." The results were categorized in a frequency distribution, as shown.

Response	Frequency
Yes	15
No	8
Undecided	2
Total	25

Probabilities now can be compared for various categories. For example, the probability of selecting a person who liked the taste is $\frac{15}{25}$, or $\frac{3}{5}$, since 15 out of 25 people in the survey answered "yes."

Formula for Empirical Probability

Given a frequency distribution, the probability of an event being in a given class is

$$P(E) = \frac{\text{frequency for the class}}{\text{total frequencies in the distribution}} = \frac{f}{n}$$

This probability is called *empirical probability* and is based on observation.

Example 5–12

In the soft-drink survey just described, find the probability that a person responded "no."

Solution

$$P(E) = \frac{f}{n} = \frac{8}{25}$$

Example 5–13

In a sample of 50 people, 21 had type O blood, 22 had type A blood, 5 had type B blood, and 2 had type AB blood. Set up a frequency distribution and find the following probabilities:

a. A person has type O blood.
b. A person has type A or type B blood.
c. A person has neither type A nor type O blood.
d. A person does not have type AB blood.

Source: Based on American Red Cross figures presented in *The Book of Odds* by Michael D. Shook and Robert L. Shook (Plume, a division of Penguin Books, 1991).

Solution

Type	Frequency
A	22
B	5
AB	2
O	21
Total	50

a. $P(\text{O}) = \dfrac{f}{n} = \dfrac{21}{50}$

b. $P(\text{A or B}) = \dfrac{22}{50} + \dfrac{5}{50} = \dfrac{27}{50}$
    (Add the frequencies of the two classes.)

c. $P(\text{neither A nor O}) = \dfrac{5}{50} + \dfrac{2}{50} = \dfrac{7}{50}$
    (Neither A nor O means that a person has either type B or type AB blood.)

d. $P(\text{not AB}) = 1 - P(\text{AB}) = 1 - \dfrac{2}{50} = \dfrac{48}{50} = \dfrac{24}{25}$
    (Find the probability of not AB by subtracting the probability of type AB from 1.)

Example 5–14

Hospital records indicated that maternity patients stayed in the hospital for the number of days shown in the distribution.

Number of days stayed	Frequency
3	15
4	32
5	56
6	19
7	5
	127

Find these probabilities.

a. A patient stayed exactly 5 days.
b. A patient stayed less than 6 days.
c. A patient stayed at most 4 days.
d. A patient stayed at least 5 days.

Solution

a. $P(5) = \dfrac{56}{127}$

b. $P(\text{less than 6 days}) = \dfrac{15}{127} + \dfrac{32}{127} + \dfrac{56}{127} = \dfrac{103}{127}$
(Less than 6 days means either 3, or 4, or 5 days.)

c. $P(\text{at most 4 days}) = \dfrac{15}{127} + \dfrac{32}{127} = \dfrac{47}{127}$
(At most 4 days means 3 or 4 days.)

d. $P(\text{at least 5 days}) = \dfrac{56}{127} + \dfrac{19}{127} + \dfrac{5}{127} = \dfrac{80}{127}$
(At least 5 days means either 5, or 6, or 7 days.)

Empirical probabilities can also be found using a relative frequency distribution, as shown in Section 2–3.

For example, the relative frequency distribution for Example 5–12 is

Response	Frequency	Relative frequency
Yes	15	0.60
No	8	0.32
Undecided	2	0.08
	25	1.00

Hence, the probability that a person responded "no" is 0.32, which is equal to $\frac{8}{25}$.

Law of Large Numbers

When a coin is tossed one time, it is common knowledge that the probability of getting a head is $\frac{1}{2}$. But what happens when the coin is tossed 50 times? Will it come up heads 25 times? Not all of the time. One should expect about 25 heads if the coin is fair. But due to the chance variation, 25 heads will not occur most of the time.

If the empirical probability of getting a head is computed using a small number of trials, it is usually not exactly $\frac{1}{2}$. However, as the number of trials increases, the empirical probability of getting a head will approach the theoretical probability of

$\frac{1}{2}$, if in fact the coin is fair (i.e., balanced). This phenomenon is known as the **law of large numbers.** In other words, if one tosses a coin enough times, the number of heads and tails will tend to "even out." This law holds for any type of gambling game—tossing dice, playing roulette, and so on.

It should be pointed out that the probabilities which the proportions steadily approach may or may not agree with those theorized in the classical model. If not, it can have important implications, such as "the die is not fair." Pit bosses in Las Vegas watch for empirical trends that do not agree with classical theories, and they will sometimes take a set of dice out of play if observed frequencies are too far out of line with classical expected frequencies.

Subjective Probability

The third type of probability is called *subjective probability*. **Subjective probability** uses a probability value based on an educated guess or estimate, employing opinions and inexact information.

In subjective probability, a person or group makes an educated guess at the chance that an event will occur. This guess is based on the person's experience and evaluation of a solution. For example, a sportswriter may say that there is a 70% probability that the Pirates will win the pennant next year. A physician might say that on the basis of her diagnosis, there is a 30% chance the patient will need an operation. A seismologist might say there is an 80% probability that an earthquake will occur in a certain area. These are only a few examples of how subjective probability is used in everyday life.

All three types of probability (classical, empirical, and subjective) are used to solve a variety of problems in business, engineering, and other fields.

Exercises

5–1. (W) What is a probability experiment?

5–2. (W) Define *sample space*.

5–3. (W) What is the difference between an outcome and an event?

5–4. (W) What are equally likely events?

5–5. What is the range of the values of a probability event?

5–6. When an event is certain to occur, what is its probability?

5–7. If an event cannot happen, what value is assigned to its probability?

5–8. What is the sum of the probabilities of all of the outcomes in a sample space?

5–9. If the probability that it will rain in your town tomorrow is 0.45, what is the probability that it will not rain?

5–10. A probability experiment is conducted. Which of the following cannot be considered a probability of an outcome?

a. $\frac{1}{3}$ *d.* -0.59 *g.* 1
b. $-\frac{1}{5}$ *e.* 0   *h.* 33%
c. 0.80   *f.* 1.45   *i.* 112%

5–11. Classify each statement as an example of classical probability, empirical probability, or subjective probability.
a. The probability that a person will watch the 6:00 evening news is 0.15.
b. The probability of winning at a chuck-a-luck game is $\frac{5}{36}$.
c. The probability that a bus will be in an accident on a specific run is about 6%.
d. The probability of getting a royal flush when five cards are selected at random is $\frac{1}{649,740}$.
e. The probability that a student will get a C or better in a statistics course is about 70%.
f. The probability that a new fast-food restaurant will be a success in Chicago is 35%.
g. The probability that interest rates will rise in the next six months is 0.50.

5–12. (ans) If a die is rolled one time, find these probabilities.

a. Of getting a 4.

b. Of getting an even number.

c. Of getting a number greater than 4.

d. Of getting a number less than 7.

e. Of getting a number greater than 0.

f. Of getting a number greater than 3 or an odd number.

g. Of getting a number greater than 3 and an odd number.

5–13. If two dice are rolled one time, find the probability of getting these results.

a. A sum of 6.

b. Doubles.

c. A sum of 7 or 11.

d. A sum greater than 9.

e. A sum less than or equal to 4.

5–14. **(ans)** If one card is drawn from a deck, find the probability of getting these results.

a. An ace.

b. A diamond.

c. An ace of diamonds.

d. A 4 or a 6.

e. A 4 or a club.

f. A 6 or a spade.

g. A heart or a club.

h. A red queen.

i. A red card or a 7.

j. A black card and a 10.

5–15. A box contains five red, two white, and three green marbles. If a marble is selected at random, find these probabilities.

a. That it is red.

b. That it is green.

c. That it is red or white.

d. That it is not green.

e. That it is not red.

5–16. In an office there are five women and four men. If one person is selected, find the probability that the person is a woman.

5–17. If there are 50 tickets sold for a raffle and one person buys 7 tickets, what is the probability of that person winning the prize?

5–18. In an office there are seven women and nine men. If one person is promoted, find the probability that the person is a man.

5–19. A survey found that 53% of Americans think U.S. military forces should be used to "protect the interest of U.S. corporations" in other countries. If an American is selected at random, find the probability that he or she will disagree or have no opinion on the issue.

5–20. A certain brand of grass seed has an 86% probability of germination. If a lawn care specialist plants 9000 seeds, find the number that should germinate.

5–21. A couple plans to have three children. Find each probability.

a. Of all boys.

b. Of all girls or all boys.

c. Of exactly two boys or two girls.

d. Of at least one child of each gender.

5–22. In the game craps using two dice, a person wins on the first roll if a 7 or an 11 is rolled. Find the probability of winning on the first roll.

5–23. In a game of craps, a player loses on the roll if a 2, 3, or 12 is tossed on the first roll. Find the probability of losing on the first roll.

5–24. When 350 students were interviewed, 186 stated that they preferred the lecture presentation to a group discussion. Find the probability that if a student were randomly selected, he or she would prefer the lecture presentation method.

5–25. A roulette wheel has 38 spaces numbered 1 through 36, 0, and 00. Find the probability of getting these results.

a. An odd number.

b. A number greater than 25.

c. A number less than 15 not counting 0 and 00.

5–26. Thirty-nine of 50 states are currently under court order to alleviate overcrowding and poor conditions in one or all of their prisons. If a state is selected at random, find the probability that it is currently under such a court order.

Source: *Harper's Index* 289, no. 1736 (January 1995), p. 11.

5–27. A baseball player's batting average is 0.331. If she is at bat 53 times during the season, find the approximate number of times she gets to first base safely. Walks do not count.

5–28. In a survey, 16 percent of American children said they use flattery to get their parents to buy them things. If a child is selected at random, find the probability that the child said he or she does not use parental flattery.

Source: *Harper's Index* 289, no. 1735 (December 1994), p. 13.

5–29. If three dice are rolled, find the probability of getting triples—e.g., 1, 1, 1; 2, 2, 2.

5–30. Among 100 students at a small school, 50 are mathematics majors, 30 are English majors, and 20 are history majors. If a student is selected at random, find the probability that she is neither a math major nor an English major.

5–31.   A record store receives orders from the following age groups.

Age	Percentage
Under 20	30
20–29	13
30–39	20
40–49	27
50 and over	10

If a customer is selected at random, find the probability that he or she is in the following age groups. Use percentages.
a.   Between 30 and 39 years of age.
b.   Under 30 years of age.
c.   Over 29 and under 50 years of age.
d.   Under 20 or over 49 years of age.

*5–32.   A person flipped a coin 100 times and obtained 73 heads. Can the person conclude that the coin was unbalanced?

*5–33.   A medical doctor stated that with a certain treatment, a patient has a 50% chance of recovering without surgery. That is, "Either he will get well or he won't get well." Comment on his statement.

*5–34.   The wheel spinner shown below is spun twice. Find the sample space, and then determine the probability of the following events.

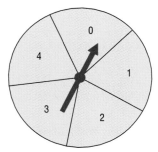

a.   An odd number on the first spin and an even number on the second spin. (*Note:* 0 is considered even.)
b.   A sum greater than 4.
c.   Even numbers on both spins.
d.   A sum that is odd.
e.   The same number on both spins.

*5–35.   Roll a die 180 times and record the number of 1s, 2s, 3s, 4s, 5s, and 6s. Compute the probabilities of each, and compare these probabilities with the theoretical results.

*5–36.   Toss two coins 100 times and record the number of heads (0, 1, 2). Compute the probabilities of each outcome, and compare these probabilities with the theoretical results.

*5–37.   Odds are used in gambling games to make them fair. For example, if a person rolled a die and won every time he or she rolled a 6, then the person would win on the average of once every 6 times. So that the game is fair, the odds of 5 to 1 are given. This means that if the person bet $1 and won, he or she could win $5. On the average, the player would win $5 once in 6 rolls and lose $1 on the other 5 rolls—hence the term *fair game.*

In most gambling games, the odds given are not fair. For example, if the odds of winning are really 20 to 1, the house might offer 15 to 1 in order to make a profit.

Odds can be expressed as a fraction or as a ratio, such as $\frac{5}{1}$, 5:1, or 5 to 1. Odds are computed in favor of the event or against the event. The formulas for odds are

$$\text{odds in favor} = \frac{P(E)}{1 - P(E)}$$

$$\text{odds against} = \frac{P(\overline{E})}{1 - P(\overline{E})}$$

In the die example,

$$\text{odds in favor of a } 6 = \frac{\frac{1}{6}}{\frac{5}{6}} = \frac{1}{5} \text{ or } 1{:}5$$

$$\text{odds against a } 6 = \frac{\frac{5}{6}}{\frac{1}{6}} = \frac{5}{1} \text{ or } 5{:}1.$$

Find the odds in favor of and against each event.
a.   Rolling a die and getting a 2.
b.   Rolling a die and getting an even number.
c.   Drawing a card from a deck and getting a spade.
d.   Drawing a card and getting a red card.
e.   Drawing a card and getting a queen.
f.   Tossing two coins and getting two tails.
g.   Tossing two coins and getting one tail.

5–3

The Addition Rules for Probability

Objective 2. Find the probability of compound events using the addition rules.

Many problems involve finding the probability of two or more events. For example, at a large political gathering, one might wish to know, for a person selected at random, the probability that the person is a female or is a Republican. In this case, there are three possibilities to consider:

1. The person is a female.
2. The person is a Republican.
3. The person is both a female and a Republican.

Consider another example. At the same gathering there are Republicans, Democrats, and Independents. If a person is selected at random, what is the probability that the person is a Democrat or an Independent? In this case, there are only two possibilities:

1. The person is a Democrat.
2. The person is an Independent.

The difference between the two examples is that in the first case, the person selected can be a female and a Republican at the same time. In the second case, the person selected cannot be both a Democrat and an Independent at the same time. In the second case, the two events are said to be mutually exclusive; in the first case, they are not mutually exclusive.

Two events are **mutually exclusive** if they cannot occur at the same time (i.e., they have no outcomes in common).

In another situation, the events of getting a 4 and getting a 6 when a single card is drawn from a deck are mutually exclusive events, since a single card cannot be both a 4 and a 6. On the other hand, the events of getting a 4 and getting a heart on a single draw are not mutually exclusive, since one can select the 4 of hearts when drawing a single card from an ordinary deck.

Example 5–15

Determine which events are mutually exclusive and which are not when a single die is rolled.

a. Getting an odd number and getting an even number.
b. Getting a 3 and getting an odd number.
c. Getting an odd number and getting a number less than 4.
d. Getting a number greater than 4 and getting a number less than 4.

Solution

a. The events are mutually exclusive, since the first event can be 1, 3, or 5, and the second event can be 2, 4, or 6.

b. The events are not mutually exclusive, since the first event is a 3 and the second can be 1, 3, or 5. Hence, 3 is contained in both outcomes.

c. The events are not mutually exclusive, since the first event can be 1, 3, or 5, and the second can be 1, 2, or 3. Hence, 1 and 3 are contained in both outcomes.

d. The events are mutually exclusive, since the first event can be 5 or 6, and the second event can be 1, 2, or 3.

Example 5–16

Determine which events are mutually exclusive and which are not when a single card is drawn from a deck.

a. Getting a 7 and getting a jack.

b. Getting a club and getting a king.

c. Getting a face card and getting an ace.

d. Getting a face card and getting a spade.

Solution

Only the events in parts *a* and *c* are mutually exclusive.

The probability of two or more events can be determined by the *addition rules*. The first addition rule is used when the events are mutually exclusive.

Addition Rule 1

When two events A and B are mutually exclusive, the probability that A or B will occur is

$$P(A \text{ or } B) = P(A) + P(B)$$

Example 5–17

A drawer contains three pairs of red socks, two pairs of black socks, and four pairs of brown socks. If a person in a dark room selects a pair of socks, find the probability that the pair will be either black or brown. (*Note:* The socks are folded together in matching pairs.)

Solution

Since there are nine pairs of socks,

$$P(\text{black or brown}) = P(\text{black}) + P(\text{brown}) = \tfrac{2}{9} + \tfrac{4}{9} = \tfrac{6}{9} = \tfrac{2}{3}$$

The events are assumed to be mutually exclusive.

Example 5–18

At a political rally, there are 20 Republicans, 13 Democrats, and 6 Independents. If a person is selected, find the probability that he or she is either a Democrat or an Independent.

$$P(\text{Democrat or Independent}) = P(\text{Democrat}) + P(\text{Independent})$$
$$= \tfrac{13}{39} + \tfrac{6}{39} = \tfrac{19}{39}$$

Example 5–19

A day of the week is selected at random. Find the probability that it is a weekend day.

$$P(\text{Saturday or Sunday}) = P(\text{Saturday}) + P(\text{Sunday}) = \tfrac{1}{7} + \tfrac{1}{7} = \tfrac{2}{7}$$

When two events are not mutually exclusive, one must subtract the probabilities of the outcomes that are common to both events, since they have been counted twice. This technique is illustrated in the next example.

Example 5–20

A single card is drawn from a deck. Find the probability that it is a king or a club. Since the king of clubs is counted twice, the probability must be subtracted, as shown.

$$P(\text{king or club}) = P(\text{king}) + P(\text{club}) - P(\text{king of clubs})$$
$$= \tfrac{4}{52} + \tfrac{13}{52} - \tfrac{1}{52} = \tfrac{16}{52} = \tfrac{4}{13}$$

When events are not mutually exclusive, addition rule 2 can be used to find the probability of the events.

Addition Rule 2

If A and B are *not* mutually exclusive, then

$$P(A \text{ or } B) = P(A) + P(B) - P(A \text{ and } B)$$

Note: This rule can also be used when the events are mutually exclusive, since $P(A \text{ and } B)$ will always equal 0. However, it is important to make a distinction between the two situations.

Example 5–21

In a hospital unit there are eight nurses and five physicians. Seven nurses and three physicians are females. If a staff person is selected, find the probability that the subject is a nurse or a male.

Solution

The sample space is shown below.

Staff	Females	Males	Total
Nurses	7	1	8
Physicians	3	2	5
Total	10	3	13

The probability is

$$P(\text{nurse or male}) = P(\text{nurse}) + P(\text{male}) - P(\text{male nurse})$$

$$= \tfrac{8}{13} + \tfrac{3}{13} - \tfrac{1}{13} = \tfrac{10}{13}$$

Example 5–22

On New Year's Eve, the probability of a person driving while intoxicated is 0.32, the probability of a person having a driving accident is 0.09, and the probability of a person having a driving accident while intoxicated is 0.06. What is the probability of a person driving while intoxicated or having a driving accident?

Solution

$$P(\text{intoxicated or accident}) = P(\text{intoxicated}) + P(\text{accident})$$
$$- P(\text{intoxicated and accident})$$
$$= 0.32 + 0.09 - 0.06 = 0.35$$

The probability rules can be extended to three or more events. For three mutually exclusive events, A, B, and C,

$$P(A \text{ or } B \text{ or } C) = P(A) + P(B) + P(C)$$

For three events that are *not* mutually exclusive,

$$P(A \text{ or } B \text{ or } C) = P(A) + P(B) + P(C) - P(A \text{ and } B) - P(A \text{ and } C)$$
$$- P(B \text{ and } C) + P(A \text{ and } B \text{ and } C)$$

See Exercises 5–61, 5–62, and 5–64.

Figure 5–5(a) shows a Venn diagram that represents two mutually exclusive events, A and B. In this case, $P(A \text{ or } B) = P(A) + P(B)$, since these events are mutually exclusive and do not overlap. In other words, the probability of event A or event B occurring is the sum of the areas of the two circles.

Figure 5–5

Venn Diagrams for the
Addition Rules

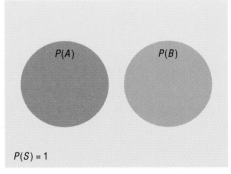

$P(S) = 1$

(a) Mutually exclusive events
  $P(A \text{ or } B) = P(A) + P(B)$

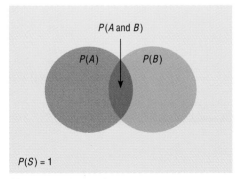

$P(A \text{ and } B)$

$P(S) = 1$

(b) Non-mutually exclusive events
  $P(A \text{ or } B) = P(A) + P(B) - P(A \text{ and } B)$

Figure 5–5(b) represents the probability of two events that are *not* mutually exclusive. In this case, $P(A \text{ or } B) = P(A) + P(B) - P(A \text{ and } B)$. The area in the intersection or overlapping part of both circles corresponds to $P(A \text{ and } B)$, and when the area of circle A is added to the area of circle B, the overlapping part is counted twice. It must therefore be subtracted once to get the correct area or probability.

Note: Venn diagrams were developed by mathematician John Venn (1834–1923) and are used in set theory and symbolic logic. They have been adapted to probability theory also. In set theory, the symbol $\cup$ represents the union of two sets and $A \cup B$ corresponds to A or B. The symbol $\cap$ represents the intersection of two sets, and $A \cap B$ corresponds to A and B. Venn diagrams show only a general picture of the probability rules and do not portray all situations, such as $P(A) = 0$, accurately.

Exercises

5–38. (W) Define *mutually exclusive events.*

5–39. (W) Give an example of two events that are mutually exclusive and two events that are not mutually exclusive.

5–40. Determine whether the following events are mutually exclusive.
a. Roll a die: Get an even number, and get a number less than 3.
b. Roll a die: Get a prime number (2, 3, 5), and get an odd number.
c. Roll a die: Get a number greater than 3, and get a number less than 3.
d. Select a student in your class: The student has blond hair, and the student has blue eyes.
e. Select a student in your college: The student is a sophomore, and the student is a business major.
f. Select any course: It is a calculus course, and it is an English course.
g. Select a registered voter: The voter is a Republican, and the voter is a Democrat.

5–41. A ski shop decides to select a month for its annual sale. Find the probability that it will be November or December. Assume that all months have an equal probability of being selected.

5–42. In a large department store, there are two managers, four department heads, 16 clerks, and four stockers. If a person is selected at random, find the probability that the person is either a clerk or a manager.

5–43. At a convention there are seven mathematics instructors, five computer science instructors, three statistics instructors, and four science instructors. If an instructor is selected, find the probability of getting a science instructor or a math instructor.

5–44. In a pet store, there are 12 puppies, eight kittens, and nine gerbils. If a pet is selected at random, find the probability of getting a puppy or a kitten.

5–45. On a small college campus, there are five English professors, four mathematics professors, two science professors, three psychology professors, and three history professors. If a professor is selected at random, find the probability that the professor is the following:
a. An English or psychology professor.
b. A mathematics or science professor.
c. A history, science, or mathematics professor.
d. An English, mathematics, or history professor.

5–46. The probability of a California teenager owning a surfboard is 0.43, of owning a skateboard is 0.38, and of owning both is 0.28. If a California teenager is selected at random, find the probability that he or she owns a surfboard or a skateboard.

5–47. The probability of a tourist visiting Indian Caverns is 0.80 and of visiting the Safari Zoo is 0.55. The probability of visiting both places on the same day is 0.42. Find the probability that a tourist visits Indian Caverns or visits Safari Zoo.

5–48. A single card is drawn from a deck. Find the probability of selecting the following:
a. A 4 or a diamond
b. A club or a diamond
c. A jack or a black card

5–49. In a statistics class there are 18 juniors and 10 seniors; 6 of the seniors are females, and 12 of the juniors are males. If a student is selected at random, find the probability of selecting the following:
a. A junior or a female
b. A senior or a female
c. A junior or a senior

5–50. A large department store has 500 employees. There are 350 females and 200 of them are under the age of 25. There are 75 males under 25. If an employee is selected for promotion, find the probability that the employee will be the following:
a. Under 25 or a female

b. Over 24 or a female
c. Male or over 24

5–51. A woman's clothing store owner buys from three companies: A, B, and C. The most recent purchases are shown here.

Product	Company A	Company B	Company C
Dresses	24	18	12
Blouses	13	36	15

If one item is selected at random, find the following probabilities.
a. It was purchased from company A or is a dress.
b. It was purchased from company B or company C.
c. It is a blouse or was purchased from company A.

5–52. In a recent study, the following data were obtained in response to the question, "Do you favor the school combining the elementary and middle school students in one building?"

	Yes	No	No opinion
Males	72	81	5
Females	103	68	7

If a person is selected at random, find these probabilities.
a. The person has no opinion.
b. The person is a male or is against the issue.
c. The person is a female or favors the issue.

5–53. A grocery store employs cashiers, stock clerks, and deli personnel. The distribution of employees according to marital status is shown next.

Marital status	Cashiers	Stock clerks	Deli personnel
Married	8	12	3
Not married	5	15	2

If an employee is selected at random, find these probabilities:
a. The employee is a stock clerk or married.
b. The employee is not married.
c. The employee is a cashier or is not married.

5–54. In a certain geographic region, newspapers are classified as being published daily morning, daily evening, and weekly. Some have a comics section and some do not. The distribution is shown next.

Have comics section	Morning	Evening	Weekly
Yes	2	3	1
No	3	4	2

If a newspaper is selected at random, find these probabilities.
a. The newspaper is a weekly publication.
b. The newspaper is a daily morning publication or has comics.
c. The newspaper is published weekly or does not have comics.

5–55. Three cable channels (6, 8, and 10) have quiz shows, comedies, and dramas. The number of each is shown here.

Type of show	Channel 6	Channel 8	Channel 10
Quiz show	5	2	1
Comedy	3	2	8
Drama	4	4	2

If a show is selected at random, find these probabilities.
a. The show is a quiz show or it is shown on channel 8.
b. The show is a drama or a comedy.
c. The show is shown on channel 10 or it is a drama.

5–56. A local postal carrier distributes first-class letters, advertisements, or magazines. For a certain day, she distributed the following number of each type of item.

Delivered to	First-class letters	Ads	Magazines
Home	325	406	203
Business	732	1021	97

If an item of mail is selected at random, find these probabilities.
a. The item went to a home.
b. The item was an ad or it went to a business.
c. The item was a first-class letter or it went to a home.

5–57. The frequency distribution shown here illustrates the number of medical tests conducted on 30 randomly selected emergency patients.

Number of tests performed	Number of patients
0	12
1	8
2	2
3	3
4 or more	5

If a patient is selected at random, find these probabilities.
a. The patient has had exactly two tests done.

b. The patient has had at least two tests done.

c. The patient has had at most three tests done.

d. The patient has had three or fewer tests done.

e. The patient has had one or two tests done.

5–58. The following distribution represents the length of time a patient spends in a hospital.

Days	Frequency
0–3	2
4–7	15
8–11	8
12–15	6
16+	9

If a patient is selected, find these probabilities.

a. The patient spends 3 days or less in the hospital.

b. The patient spends less than 8 days in the hospital.

c. The patient spends 16 or more days in the hospital.

d. The patient spends a maximum of 11 days in the hospital.

5–59. A sales representative who visits customers at home finds she sells 0, 1, 2, 3, or 4 items according to the following frequency distribution.

Items sold	Frequency
0	8
1	10
2	3
3	2
4	1

Find the probability that she sells the following.

a. Exactly one item

b. More than two items

c. At least one item

d. At most three items

5–60. A recent study of 300 patients found that of 100 alcoholic patients, 87 had elevated cholesterol levels, and of 200 nonalcoholic patients, 43 had elevated cholesterol levels. If a patient is selected at random, find the probability that the patient is the following.

a. An alcoholic with elevated cholesterol level.

b. A nonalcoholic.

c. A nonalcoholic with normal cholesterol level.

5–61. If one card is drawn from an ordinary deck of cards, find the probability of getting the following.

a. A king or a queen or a jack.

b. A club or a heart or a spade.

c. A king or a queen or a diamond.

d. An ace or a diamond or a heart.

e. A 9 or a 10 or a spade or a club.

5–62. Two dice are rolled. Find the probability of getting the following.

a. A sum of 6 or 7 or 8.

b. Doubles or a sum of 4 or 6.

c. A sum greater than 9 or less than 4 or a 7.

5–63. An urn contains six red balls, two green balls, one blue ball, and one white ball. If a ball is drawn, find the probability of getting a red or a white ball.

5–64. Three dice are rolled. Find the probability of getting the following.

a. Triples   *b.* A sum of 5

*5–65. The probability that a customer selects a pizza with mushrooms or pepperoni is 0.55, and the probability that the customer selects mushrooms only is 0.32. If the probability that he or she selects pepperoni only is 0.17, find the probability of the customer selecting both items.

*5–66. In building new homes, a contractor finds that the probability of a home buyer selecting a two-car garage is 0.70 and of selecting a one-car garage is 0.20. Find the probability that the buyer will select no garage. The builder does not build houses with three-car garages.

*5–67. In Exercise 5–66, find the probability that the buyer will not want a two-car garage.

"I know you haven't had an accident in thirteen years. We're raising your rates because you're about due one."

The previous section showed that the addition rules are used to compute probabilities for mutually exclusive and not mutually exclusive events. This section introduces two more rules, the multiplication rules.

The Multiplication Rules

Objective 3. Find the probability of compound events using the multiplication rules.

The multiplication rules can be used to find the probability of two or more events that occur in sequence. For example, if a coin is tossed and then a die is rolled, one can find the probability of getting a head on the coin *and* a 4 on the die. These two events are said to be *independent* since the outcome of the first event (tossing a coin) does not affect the probability outcome of the second event (rolling a die).

Two events A and B are **independent** if the fact that A occurs does not affect the probability of B occurring.

Here are other examples of independent events:

Rolling a die and getting a 6, and then rolling a second die and getting a 3.
    Drawing a card from a deck and getting a queen, replacing it, and drawing a second card and getting a queen.

In order to find the probability of two independent events that occur in sequence, one must find the probability of each event occurring separately and then multiply the answers. For example, if a coin is tossed twice, the probability of getting two heads is $\frac{1}{2} \cdot \frac{1}{2} = \frac{1}{4}$. This result can be verified by looking at the sample space, HH, HT, TH, TT. Then $P(\text{HH}) = \frac{1}{4}$.

Multiplication Rule 1

When two events are independent, the probability of both occurring is

$$P(A \text{ and } B) = P(A) \cdot P(B)$$

Example 5–23

A coin is flipped and a die is rolled. Find the probability of getting a head on the coin and a 4 on the die.

Solution

$$P(\text{head and } 4) = P(\text{head}) \cdot P(4) = \frac{1}{2} \cdot \frac{1}{6} = \frac{1}{12}$$

Note that the sample space for the coin is H, T; and for the die it is 1, 2, 3, 4, 5, 6.

The problem in Example 5–23 can also be solved by using the sample space:

H1   H2   H3   H4   H5   H6   T1   T2   T3   T4   T5   T6

The solution is $\frac{1}{12}$, since there is only one way to get the head-4 outcome.

Example 5–24

A card is drawn from a deck and replaced; then a second card is drawn. Find the probability of getting a queen and then an ace.

Solution

Since the card is replaced, the probability of getting a queen is $\frac{4}{52}$, and the probability of getting an ace is $\frac{4}{52}$. Hence, the probability of getting a queen and an ace is

$$P(\text{queen and ace}) = P(\text{queen}) \cdot P(\text{ace}) = \frac{4}{52} \cdot \frac{4}{52} = \frac{16}{2704} = \frac{1}{169}$$

Example 5–25

An urn contains three red balls, two blue balls, and five white balls. A ball is selected and its color noted. Then it is replaced. A second ball is selected and its color noted. Find the probability of each of the following.

a. Selecting two blue balls.
b. Selecting a blue ball and then a white ball.
c. Selecting a red ball and then a blue ball.

Solution

a. $P(\text{blue and blue}) = P(\text{blue}) \cdot P(\text{blue}) = \dfrac{2}{10} \cdot \dfrac{2}{10} = \dfrac{4}{100} = \dfrac{1}{25}$

b. $P(\text{blue and white}) = P(\text{blue}) \cdot P(\text{white}) = \dfrac{2}{10} \cdot \dfrac{5}{10} = \dfrac{10}{100} = \dfrac{1}{10}$

c. $P(\text{red and blue}) = P(\text{red}) \cdot P(\text{blue}) = \dfrac{3}{10} \cdot \dfrac{2}{10} = \dfrac{6}{100} = \dfrac{3}{50}$

Multiplication rule 1 can be extended to three or more independent events by using the formula

$$P(A \text{ and } B \text{ and } C \text{ and } \ldots \text{ and } K) = P(A) \cdot P(B) \cdot P(C) \cdot \ldots \cdot P(K)$$

When a small sample is selected from a large population and the subjects are not replaced, the probability of the event occurring changes so slightly that for the most part, it is considered to remain the same. The next two examples illustrate this concept.

Example 5–26

A Harris poll found that 46% of Americans say they suffer great stress at least once a week. If three people are selected at random, find the probability that all three will say that they suffer stress at least once a week.

Source: *100% American* by Daniel Evan Weiss (Poseidon Press, 1988).

Solution

Let S denote stress. Then

$$P(S \text{ and } S \text{ and } S) = P(S) \cdot P(S) \cdot P(S)$$
$$= (0.46)(0.46)(0.46) = 0.097$$

Example 5–27

The probability that a specific medical test will show positive is 0.32. If four people are tested, find the probability that all four will show positive.

Solution

Let T be the symbol for a positive test result. Then

$$P(T \text{ and } T \text{ and } T \text{ and } T) = P(T) \cdot P(T) \cdot P(T) \cdot P(T)$$
$$= (0.32)(0.32)(0.32)(0.32) = 0.010$$

In the previous examples, the events were independent of each other, since the occurrence of the first event in no way affected the outcome of the second event. On the other hand, when the occurrence of the first event changes the probability of the occurrence of the second event, the two events are said to be *dependent*. For example, suppose a card is drawn from a deck and *not* replaced, and then a second card is drawn. What is the probability of selecting an ace on the first card and a king on the second card?

Before an answer to the question can be given, one must realize that the events are dependent. The probability of selecting an ace on the first draw is $\frac{4}{52}$. If that card is *not* replaced, the probability of selecting a king on the second card is $\frac{4}{51}$, since there are 4 kings and 51 cards remaining. The outcome of the first draw has affected the outcome of the second draw.

Dependent events are formally defined next.

When the outcome or occurrence of the first event affects the outcome or occurrence of the second event in such a way that the probability is changed, the events are said to be **dependent.**

Here are some examples of dependent events:

Drawing a card from a deck, not replacing it, and then drawing a second card.
Selecting a ball from an urn, not replacing it, and then selecting a second ball.
Being a lifeguard and getting a suntan.
Having high grades and getting a scholarship.
Parking in a no-parking zone and getting a parking ticket.

In order to find probabilities when events are dependent, use the multiplication rule with a modification in notation. For the problem just discussed, the probability of getting an ace on the first draw is $\frac{4}{52}$, and the probability of getting a king on the second draw is $\frac{4}{51}$. By the multiplication rule, the probability of both events occurring is

$$\frac{4}{52} \cdot \frac{4}{51} = \frac{16}{2652} = \frac{4}{663}$$

The event of getting a king on the second draw *given* that an ace was drawn the first time is called a *conditional probability*.

The **conditional probability** of an event B in relationship to an event A is the probability that event B occurs after event A has already occurred. The notation for conditional probability is $P(B|A)$. This notation does not mean that B is divided by A; rather, it means the probability that event B occurs given that event A has already occurred. In the card example, $P(B|A)$ is the probability that the second card is a king given that the first card is an ace, and it is equal to $\frac{4}{51}$ since the first card was *not* replaced.

> **Multiplication Rule 2**
>
> When two events are dependent, the probability of both occurring is
>
> $$P(A \text{ and } B) = P(A) \cdot P(B|A)$$

Example 5–28

In a shipment of 25 microwave ovens, 2 are defective. If two ovens are randomly selected and tested, find the probability that both are defective if the first one is not replaced after it has been tested.

Solution

Since the events are dependent,

$$P(D_1 \text{ and } D_2) = P(D_1) \cdot P(D_2|D_1) = \frac{2}{25} \cdot \frac{1}{24} = \frac{2}{600} = \frac{1}{300}$$

Example 5–29

The World Wide Insurance Company found that 53% of the residents of a city had homeowner's insurance with its company. Of these clients, 27% also had automobile insurance with the company. If a resident is selected at random, find the probability that the resident has both homeowner's and automobile insurance with the World Wide Insurance Company.

Solution

$$P(H \text{ and } A) = P(H) \cdot P(A|H) = (0.53)(0.27) = 0.1431$$

This multiplication rule can be extended to three or more events, as shown in the next example.

Example 5–30

Three cards are drawn from an ordinary deck and not replaced. Find the probability of the following.

a. Getting three jacks.
b. Getting an ace, a king, and a queen in order.
c. Getting a club, a spade, and a heart in order.
d. Getting three clubs.

Solution

a. $P(3 \text{ jacks}) = \dfrac{4}{52} \cdot \dfrac{3}{51} \cdot \dfrac{2}{50} = \dfrac{24}{132,600} = \dfrac{1}{5525}$

b. $P(\text{ace and king and queen}) = \dfrac{4}{52} \cdot \dfrac{4}{51} \cdot \dfrac{4}{50} = \dfrac{64}{132,600} = \dfrac{8}{16,575}$

c. $P(\text{club and spade and heart}) = \dfrac{13}{52} \cdot \dfrac{13}{51} \cdot \dfrac{13}{50} = \dfrac{2197}{132,600} = \dfrac{169}{10,200}$

d. $P(3 \text{ clubs}) = \dfrac{13}{52} \cdot \dfrac{12}{51} \cdot \dfrac{11}{50} = \dfrac{1716}{132,600} = \dfrac{11}{850}$

Tree diagrams can be used as an aid to finding the solution to probability problems when the events are sequential. The next example illustrates the use of tree diagrams.

Example 5–31

Box 1 contains two red balls and one blue ball. Box 2 contains three blue balls and one red ball. A coin is tossed. If it falls heads up, box 1 is selected and a ball is drawn. If it falls tails up, box 2 is selected and a ball is drawn. Find the probability of selecting a red ball.

Solution

With the use of a tree diagram, the sample space can be determined as shown in Figure 5–6. First, assign probabilities to each branch. Next, using the multiplication rule, multiply the probabilities for each branch.

Figure 5–6

Tree Diagram for
Example 5–31

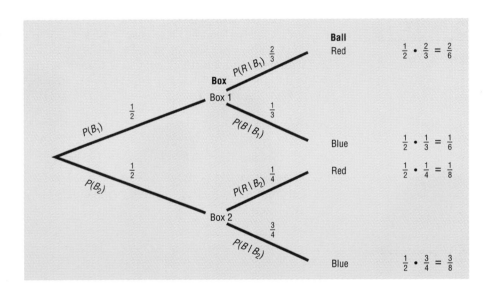

Finally, use the addition rule, since a red ball can be obtained from box 1 or box 2.

$$P(\text{red}) = \tfrac{2}{6} + \tfrac{1}{8} = \tfrac{8}{24} + \tfrac{3}{24} = \tfrac{11}{24}$$

(*Note:* The sum of all final probabilities will always be equal to 1.)

Tree diagrams can be used when the events are independent or dependent, and they can also be used for sequences of three or more events.

Conditional Probability

Objective 4. Find the conditional probability of an event.

The conditional probability of an event B in relationship to an event A was defined as the probability that event B occurs after event A has already occurred.

The conditional probability of an event can be found by dividing both sides of the equation for multiplication rule 2 by $P(A)$, as shown:

$$P(A \text{ and } B) = P(A) \cdot P(B|A)$$

$$\frac{P(A \text{ and } B)}{P(A)} = \frac{\cancel{P(A)} \cdot P(B|A)}{\cancel{P(A)}}$$

$$\frac{P(A \text{ and } B)}{P(A)} = P(B|A)$$

Formula for Conditional Probability

The probability that the second event B occurs given that the first event A has occurred can be found by dividing the probability that both events occurred by the probability that the first event has occurred. The formula is

$$P(B|A) = \frac{P(A \text{ and } B)}{P(A)}$$

The next three examples illustrate the use of this rule.

Example 5–32

A box contains black chips and white chips. A person selects two chips without replacement. If the probability of selecting a black chip *and* a white chip is $\frac{15}{56}$, and the probability of selecting a black chip on the first draw is $\frac{3}{8}$, find the probability of selecting the white chip on the second draw, *given* that the first chip selected was a black chip.

Solution

Let

$$B = \text{selecting a black chip} \qquad W = \text{selecting a white chip}$$

Then

$$P(W|B) = \frac{P(B \text{ and } W)}{P(B)} = \frac{\frac{15}{56}}{\frac{3}{8}}$$

$$= \frac{15}{56} \div \frac{3}{8} = \frac{15}{56} \cdot \frac{8}{3} = \frac{\overset{5}{\cancel{15}}}{\underset{7}{\cancel{56}}} \cdot \frac{\overset{1}{\cancel{8}}}{\underset{1}{\cancel{3}}} = \frac{5}{7}$$

Hence, the probability of selecting a white chip on the second draw given that the first chip selected was black is $\frac{5}{7}$.

Example 5–33

The probability that Sam parks in a no-parking zone *and* gets a parking ticket is 0.06, and the probability that Sam cannot find a legal parking space and has to park in the no-parking zone is 0.20. On Tuesday, Sam arrives at school and has to park in a no-parking zone. Find the probability that he will get a parking ticket.

Solution

Let

$$N = \text{parking in a no-parking zone} \qquad T = \text{getting a ticket}$$

Then

$$P(T|N) = \frac{P(N \text{ and } T)}{P(N)} = \frac{0.06}{0.20} = 0.30$$

Hence, Sam has a 0.30 probability of getting a parking ticket, given that he parked in a no-parking zone.

....

The conditional probability of events occurring can also be computed when the data is given in table form, as shown in the next example.

....

Example 5–34

A recent survey asked 100 people if they thought women in the armed forces should be permitted to participate in combat. The results of the survey are shown in the table.

Gender	Yes	No	Total
Male	32	18	50
Female	8	42	50
Total	40	60	100

Find these probabilities.

a.   The respondent answered "yes," given that the respondent was a female.
b.   The respondent was a male, given that the respondent answered "no."

Solution

Let M = respondent was a male Y = respondent answered "yes"
          F = respondent was a female N = respondent answered "no"

a.   The problem is to find $P(Y|F)$. The rule states

$$P(Y|F) = \frac{P(F \text{ and } Y)}{P(F)}$$

The probability $P(F \text{ and } Y)$ is the number of females who responded "yes" divided by the total number of respondents:

$$P(F \text{ and } Y) = \frac{8}{100}$$

The probability $P(F)$ is the probability of selecting a female:

$$P(F) = \frac{50}{100}$$

Then $P(Y|F) = \frac{P(F \text{ and } Y)}{P(F)} = \frac{8/100}{50/100}$

$$= \frac{8}{100} \div \frac{50}{100} = \frac{\overset{4}{\cancel{8}}}{\underset{1}{\cancel{100}}} \cdot \frac{\overset{1}{\cancel{100}}}{\underset{25}{\cancel{50}}} = \frac{4}{25}$$

b. The problem is to find $P(M|N)$.

$$P(M|N) = \frac{P(N \text{ and } M)}{P(N)} = \frac{18/100}{60/100}$$

$$= \frac{18}{100} \div \frac{60}{100} = \frac{\overset{3}{\cancel{18}}}{\underset{1}{\cancel{100}}} \cdot \frac{\overset{1}{\cancel{100}}}{\underset{10}{\cancel{60}}} = \frac{3}{10}$$

The Venn diagram for conditional probability is shown in Figure 5–7. In this case,

$$P(B|A) = \frac{P(A \text{ and } B)}{P(A)}$$

which is represented by the area in the intersection or overlapping part of the circles A and B divided by the area of circle A. The reasoning here is that if one assumes A has occurred, then A becomes the sample space for the next calculation and is the denominator of the probability fraction $\dfrac{P(A \text{ and } B)}{P(A)}$. The numerator $P(A \text{ and } B)$ represents the probability of the part of B that is contained in A. Hence, $P(A \cap B)$ becomes the numerator of the probability fraction $\dfrac{P(A \text{ and } B)}{P(A)}$. Imposing a condition reduces the sample space.

Figure 5–7

Venn Diagram for
Conditional Probability

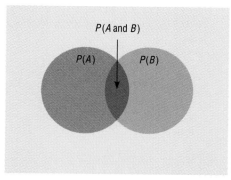

$$P(B|A) = \frac{P(A \text{ and } B)}{P(A)}$$

Probabilities for "At Least"

The multiplication rules can be used with the complementary event rule to simplify solving probability problems involving "at least." The next three examples illustrate how this is done.

Example 5–35

A game is played by drawing four cards from an ordinary deck and replacing each card after it is drawn. Find the probability of winning if at least one ace is drawn.

Solution

It is much easier to find the probability that no aces are drawn (i.e., losing) and then subtract from 1 than to find the solution directly, because that would involve finding the probability of getting one ace, two aces, three aces, and four aces and then adding the results.

Let E = at least one ace is drawn and $\overline{E}$ = no aces drawn. Then

$$P(\overline{E}) = \frac{48}{52} \cdot \frac{48}{52} \cdot \frac{48}{52} \cdot \frac{48}{52}$$

$$= \frac{12}{13} \cdot \frac{12}{13} \cdot \frac{12}{13} \cdot \frac{12}{13} = \frac{20{,}736}{28{,}561}$$

Hence,

$$P(E) = 1 - P(\overline{E})$$

$$P(\text{winning}) = 1 - P(\text{losing}) = 1 - \frac{20{,}736}{28{,}561} = \frac{7825}{28{,}561} \approx 0.27$$

or a hand with at least one ace will win about 27% of the time.

Example 5–36

A coin is tossed five times. Find the probability of getting at least one tail.

Solution

It is easier to find the probability of the complement of the event, which is "all heads," and then subtract the probability from 1 to get the probability of at least one tail.

$$P(E) = 1 - P(\overline{E})$$

$$P(\text{at least 1 tail}) = 1 - P(\text{all heads})$$

$$P(\text{all heads}) = \left(\tfrac{1}{2}\right)^5 = \tfrac{1}{32}$$

Hence,

$$P(\text{at least 1 tail}) = 1 - \tfrac{1}{32} = \tfrac{31}{32}$$

Example 5–37

It has been found that 40% of all people over the age of 85 suffer from Alzheimer's disease. If three people over 85 are selected at random, find the probability that at least one person does not suffer from Alzheimer's disease.

Solution

The complement of "at least one person" is "no persons," and the probability that a person over 85 does *not* suffer from Alzheimer's disease is $1 - 0.40 = 0.60$, or

60%. Hence the probability that at least one person does not suffer from the disease is

$$1 - P(\text{no persons have}) = 1 - (0.60)^3 = 1 - 0.216 = 0.784, \text{ or } 78.4\%.$$

Similar methods can be used for problems involving "at most."

Exercises

5–68. (**W**) What is the difference between independent and dependent events? Give an example of each.

5–69. State which events are independent and which are dependent.
a. Tossing a coin and drawing a card from a deck.
b. Drawing a ball from an urn, not replacing it, and then drawing a second ball.
c. Getting a raise in salary and purchasing a new car.
d. Driving on ice and having an accident.
e. Having a large shoe size and having a high IQ.
f. A father being left-handed and a daughter being left-handed.
g. Smoking excessively and having lung cancer.
h. Eating an excessive amount of ice cream and smoking an excessive amount of cigarettes.

5–70. If 18% of all Americans are underweight, find the probability that if three Americans are selected at random, all will be underweight.
Source: *100% American* by Daniel Evan Weiss (New York: Poseidon Press, 1988).

5–71. A national study of patients who were overweight found that 56% also had elevated blood pressure. If two overweight patients are selected, find the probability that both have elevated blood pressure.

5–72. The Gallup Poll reported that 52% of Americans used a seat belt the last time they got into a car. If four people are selected at random, find the probability that they all used a seat belt the last time they got into a car.
Source: *100% American* by Daniel Evan Weiss (New York: Poseidon Press, 1988).

5–73. An automobile saleswoman finds that the probability of making a sale is 0.23. If she talks to four customers today, find the probability that she will sell four cars.

5–74. If 25% of U.S. federal prison inmates are not U.S. citizens, find the probability that two randomly selected federal prison inmates will not be U.S. citizens.
Source: *Harper's Index* 290, no. 1740 (May 1995), p. 11.

5–75. Find the probability of selecting two people at random who were born in the same month.

5–76. If two people are selected at random, find the probability that they have the same birthday (both month and day).

5–77. If three people are selected, find the probability that all three were born in March.

5–78. If half of Americans believe that the federal government should take "primary responsibility" for eliminating poverty, find the probability that three randomly selected Americans will agree that it is the federal government's responsibility to eliminate poverty.
Source: *Harper's Index* 289, no. 1735 (December 1994), p. 13.

5–79. What is the probability that a husband, wife, and daughter have the same birthday?

5–80. A flashlight has six batteries, two of which are defective. If two are selected at random without replacement, find the probability that both are defective.

5–81. In Exercise 5–80, find the probability that the first battery tests good and the second one is defective.

5–82. The U.S. Department of Justice reported that 6% of all American murders are committed without a weapon. If three murder cases are selected at random, find the probability that a weapon was not used in any one of them.
Source: *100% American* by Daniel Evan Weiss (New York: Poseidon Press, 1988).

5–83. In a department store there are 120 customers, 90 of whom will buy at least one item. If five customers are selected at random, one by one, find the probability that all will buy at least one item.

5–84.   Three cards are drawn from a deck *without* replacement. Find these probabilities.
a.   All are jacks.
b.   All are clubs.
c.   All are red cards.

5–85.   In a scientific study there are eight guinea pigs, five of which are pregnant. If three are selected at random without replacement, find the probability that all are pregnant.

5–86.   In Exercise 5–85, find the probability that none are pregnant.

5–87.   In a class consisting of 15 men and 12 women, two homework papers were selected at random. Find the probability that both papers belonged to women.

5–88.   In Exercise 5–87, find the probability that both papers belonged to men.

5–89.   A manufacturer makes two models of an item: model I, which accounts for 80% of unit sales, and model II, which accounts for 20% of unit sales. Because of defects, the manufacturer has to replace (or exchange) 10% of its model I and 18% of its model II. If a model is selected at random, find the probability that it will be defective.

5–90.   An automobile manufacturer has three factories, A, B, and C. They produce 50%, 30%, and 20%, respectively, of a specific model of car. Thirty percent of the cars produced in factory A are white, 40% of those produced in factory B are white, and 25% produced in factory C are white. If an automobile produced by the company is selected at random, find the probability that it is white.

5–91.   An insurance company classifies drivers as low-risk, medium-risk, and high-risk. Of those insured, 60% are low-risk, 30% are medium-risk, and 10% are high-risk. After a study, the company finds that during a one-year period, 1% of the low-risk drivers have an accident, 5% of the medium-risk drivers have an accident, and 9% of the high-risk drivers have an accident. If a driver is selected at random, find the probability that the driver will have an accident during the year.

5–92.   In a certain geographic location, 25% of the wage earners have a college degree and 75% do not. Of those who have a college degree, 5% earn more than

$100,000 a year. Of those who do not have a college degree, 2% earn more than $100,000 a year. If a wage earner is selected at random, find the probability that he or she earns more than $100,000 a year.

5–93.   Urn 1 contains five red balls and three black balls. Urn 2 contains three red balls and one black ball. Urn 3 contains four red balls and two black balls. If an urn is selected at random and a ball is drawn, find the probability it will be red.

5–94.   At a small college, the probability that a student takes physics and sociology is 0.092. The probability that a student takes sociology is 0.73. Find the probability that the student is taking physics, given that he or she is taking sociology.

5–95.   In a certain city, the probability that an automobile will be stolen and found within one week is 0.0009. The probability that an automobile will be stolen is 0.0015. Find the probability that a stolen automobile will be found within one week.

5–96.   A circuit to run a model railroad has eight switches. Two are defective. If a person selects two switches at random and tests them, find the probability that the second one is defective, given that the first one is defective.

5–97.   At the Avonlea Country Club, 73% of the members play bridge and swim, and 82% play bridge. If a member is selected at random, find the probability that the member swims, given that the member plays bridge.

5–98.   At a large university, the probability that a student takes calculus *and* is on the dean's list is 0.042. The probability that a student is on the dean's list is 0.21. Find the probability that the student is taking calculus, given that he or she is on the dean's list.

5–99.   In Rolling Acres Housing Plan, 42% of the houses have a deck and a garage; 60% have a deck. Find the probability that a home has a garage, given that it has a deck.

5–100.   In a pizza restaurant, 95% of the customers order pizza. If 65% of the customers order pizza and a salad, find the probability that a customer who orders pizza will also order a salad.

5–101.   At an exclusive country club, 68% of the members play bridge and drink champagne, and 83% play bridge. If a member is selected at random, find the probability that the member drinks champagne, given that he or she plays bridge.

5–102. Eighty students in a school cafeteria were asked if they favored a ban on smoking in the cafeteria. The results of the survey are shown in the table.

Class	Favor	Oppose	No opinion
Freshman	15	27	8
Sophomore	23	5	2

If a student is selected at random, find these probabilities.
a. Given that the student is a freshman, he or she opposes the ban.
b. Given that the student favors the ban, the student is a sophomore.

5–103. In a large shopping mall, a marketing agency conducted a survey on credit cards. The results are shown in the table.

Employment status	Owns a credit card	Does not own a credit card
Employed	18	29
Unemployed	28	34

If a person is selected at random, find these probabilities.
a. The person owns a credit card, given that the person is employed.
b. The person is unemployed, given that the person owns a credit card.

5–104. A study of graduates' average grades and degrees showed the following results.

Degree	Grade C	Grade B	Grade A
B.S.	5	8	15
B.A.	7	12	8

If a graduate is selected at random, find these probabilities.
a. The graduate has a B.S. degree, given that he or she has an A average.
b. Given that the graduate has a B.A. degree, the graduate has a C average.

5–105. In a lab there are eight technicians. Three are male and five are female. If three technicians are selected, find the probability that at least one is female.

5–106. There are five chemistry instructors and six physics instructors at a college. If a committee of four instructors is selected, find the probability of at least one of them being a physics instructor.

5–107. On a surprise quiz consisting of five true–false questions, an unprepared student guesses each answer. Find the probability that he gets at least one correct.

5–108. A lot of portable radios contains 15 good radios and three defective ones. If two are selected and tested, find the probability that at least one will be defective.

5–109. If a family has four children, find the probability that there is at least one girl among the children.

5–110. A carpool contains three kindergartners and five first-graders. If two children are ill, find the probability that at least one of them is a kindergartner.

5–111. If four cards are drawn from a deck and not replaced, find the probability of getting at least one club.

5–112. At a local clinic there are eight men, five women, and three children in the waiting room. If three patients are randomly selected, find the probability that there is at least one child among them.

5–113. It has been found that 6% of all automobiles on the road have defective brakes. If five automobiles are stopped and checked by the state police, find the probability that at least one will have defective brakes.

5–114. A medication is 75% effective against a bacterial infection. Find the probability that if 12 people take the medication, at least one person's infection will not improve.

5–115. A coin is tossed six times. Find the probability of getting at least one tail.

5–116. If three digits are randomly selected, find the probability of getting at least one 7. Digits can be used more than once.

5–117. If a die is rolled three times, find the probability of getting at least one 6.

5–118. At a teachers' conference, there were four English teachers, three mathematics teachers, and five science teachers. If four teachers are selected for a committee, find the probability that at least one is a science teacher.

5–119. If a die is rolled three times, find the probability of getting at least one even number.

5–120. At a faculty meeting, there were seven full professors, five associate professors, six assistant professors, and 12 instructors. If four people are selected at random to attend a conference, find the probability that at least one is a full professor.

5–5

Summary

In this chapter, the basic concepts and rules of probability are explained. The three types of probability are classical, empirical, and subjective. Classical probability uses sample spaces. Empirical probability uses frequency distributions and is based on observation. In subjective probability, the researcher makes an educated guess about the chance of an event occurring.

A probability event consists of one or more outcomes of a probability experiment. Two events are said to be mutually exclusive if they cannot occur at the same time. Events can also be classified as independent or dependent. If events are independent, whether or not the first event occurs does not affect the probability of the next event occurring. If the probability of the second event occurring is changed by the occurrence of the first event, then the events are dependent. The complement of an event is the set of outcomes in the sample space that are not included in the outcomes of the event itself. Complementary events are mutually exclusive.

Probability problems can be solved by using the addition rules, the multiplication rules, and the complementary event rules.

Important Terms

classical probability 169

complement of an event 171

compound event 169

conditional probability 188

dependent events 188

empirical probability 173

equally likely events 169

event 169

independent events 186

law of large numbers 176

mutually exclusive events 179

outcome 166

probability 166

probability experiment 166

sample space 166

simple event 169

subjective probability 176

Venn diagrams 172

Important Formulas

Formula for classical probability:

$$P(E) = \frac{\text{number of outcomes in } E}{\text{total number of outcomes in the sample space}} = \frac{n(E)}{n(S)}$$

Formula for empirical probability:

$$P(E) = \frac{\text{frequency for the class}}{\text{total frequencies in the distribution}} = \frac{f}{n}$$

Addition rule 1, for two mutually exclusive events:

$$P(A \text{ or } B) = P(A) + P(B)$$

Addition rule 2, for events that are not mutually exclusive:

$$P(A \text{ or } B) = P(A) + P(B) - P(A \text{ and } B)$$

Multiplication rule 1, for independent events:

$$P(A \text{ and } B) = P(A) \cdot P(B)$$

Multiplication rule 2, for dependent events:

$$P(A \text{ and } B) = P(A) \cdot P(B|A)$$

Formula for conditional probability:

$$P(B|A) = \frac{P(A \text{ and } B)}{P(A)}$$

Formula for complementary events:

$$P(\overline{E}) = 1 - P(E) \quad \text{or} \quad P(E) = 1 - P(\overline{E})$$
$$\text{or} \quad P(E) + P(\overline{E}) = 1$$

Review Exercises

5–121. When a die is rolled, find the probability of getting
a. A 5
b. A 6
c. A number less than 5

5–122. When a card is drawn from a deck, find the probability of getting
a. A heart
b. A 7 and a club
c. A 7 or a club
d. A jack
e. A black card

5–123. In a survey conducted at a local restaurant during breakfast hours, 20 people preferred orange juice, 16 preferred grapefruit juice, and 9 preferred apple juice with breakfast. If a person is selected at random, find the probability that he or she prefers grapefruit juice.

5–124. If a die is rolled one time, find these probabilities:
a. Getting a 5
b. Getting an odd number
c. Getting a number less than 3

5–125. A recent survey indicated that in a town of 1500 households, 850 had cordless telephones. If a household is randomly selected, find the probability that it has a cordless telephone.

5–126. During a sale at a men's store, 16 white sweaters, 3 red sweaters, 9 blue sweaters, and 7 yellow sweaters were purchased. If a customer is selected at random, find the probability that he bought the following.
a. A blue sweater
b. A yellow or a white sweater
c. A red, a blue, or a yellow sweater
d. A sweater that was not white

5–127. At a swimwear store, the managers found that 16 women bought white bathing suits, 4 bought red suits, 3 bought blue suits, and 7 bought yellow suits. If a customer is selected at random, find the probability that she bought the following.
a, A blue suit
b. A yellow or a red suit
c. A white or a yellow or a blue suit
d. A suit that was not red

5–128. When two dice are rolled, find the probability of getting
a. A sum of 5 or 6
b. A sum greater than 9
c. A sum less than 4 or greater than 9
d. A sum that is divisible by 4
e. A sum of 14
f. A sum less than 13

5–129. The probability that a person owns a car is 0.80, that a person owns a boat is 0.30, and that a person owns both a car and a boat is 0.12. Find the probability that a person owns either a boat or a car, but not both.

5–130. There is a 0.39 probability that John will purchase a new car, a 0.73 probability that Mary will purchase a new car, and a 0.36 probability that both will purchase a new car. Find the probability that neither will purchase a new car.

5–131. A Gallup Poll found that 78% of Americans worry about the quality and healthfulness of their diet. If five people are selected at random, find the probability that all five worry about the quality and healthfulness of their diet.
Source: *The Book of Odds* by Michael D. Shook and Robert C. Shook (New York: Penguin Putnam, Inc., 1991), p. 33.

5–132. Twenty-five percent of the engineering graduates of a university received a starting salary of $25,000 or more. If three of the graduates are selected at random, find the probability that all had a starting salary of $25,000 or more.

5–133. Three cards are drawn from an ordinary deck *without* replacement. Find the probability of getting
a. All black cards
b. All spades
c. All queens

5–134. A coin is tossed and a card is drawn from a deck. Find the probability of getting
a. A head and a 6
b. A tail and a red card
c. A head and a club

5–135. A box of candy contains six chocolate-covered cherries, three peppermint patties, two caramels, and two strawberry creams. If a piece of candy is selected, find the probability of getting a caramel or a peppermint patty.

5–136. A manufacturing company has three factories: X, Y, and Z. The daily output of each is shown below.

Product	Factory X	Factory Y	Factory Z
TVs	18	32	15
Stereos	6	20	13

If one item is selected at random, find these probabilities.

a. It was manufactured at factory X or is a stereo.

b. It was manufactured at factory Y or factory Z.

c. It is a TV or was manufactured at factory Z.

5–137. A vaccine has a 90% probability of being effective in preventing a certain disease. The probability of getting the disease if a person is not vaccinated is 50%. In a certain geographic region, 25% of the people get vaccinated. If a person is selected at random, find the probability that he or she will contract the disease.

5–138. A manufacturer makes three models of a television set, models A, B, and C. A store sells 40% of model A sets, 40% of model B sets, and 20% of model C sets. Of model A sets, 3% have stereo sound; of model B sets, 7% have stereo sound; and of model C sets, 9% have stereo sound. If a set is sold at random, find the probability that it has stereo sound.

5–139. The probability that Sue will live on campus and buy a new car is 0.37. If the probability that she will live on campus is 0.73, find the probability that she will buy a new car, given that she lives on campus.

5–140. The probability that a customer will buy a television set and buy an extended warranty is 0.03. If the probability that a customer will purchase a television set is 0.11, find the probability that the customer will also purchase the extended warranty.

5–141. Of the members of the Blue River Health Club, 43% have a lifetime membership and exercise regularly (three or more times a week). If 75% of the club members exercise regularly, find the probability that a randomly selected member is a life member, given that he or she exercises regularly.

5–142. The probability that it snows and the bus arrives late is 0.023. John hears the weather forecast, and there is a 40% chance of snow tomorrow. Find the probability that the bus will be late, given that it snows.

5–143. A number of students were grouped according to their reading ability and education. The table shows the results.

Education	Reading ability		
	Low	Average	High
Graduated high school	6	18	43
Did not graduate	27	16	7

If a student is selected at random, find these probabilities.

a. The student has a low reading ability, given that the student is a high school graduate.

b. The student has a high reading ability, given that the student did not graduate.

5–144. At a large factory, the employees were surveyed and classified according to their level of education and whether or not they smoked. The data are shown in the table.

Smoking habit	Educational level		
	Not high school graduate	High school graduate	College graduate
Smoke	6	14	19
Do not smoke	18	7	25

If an employee is selected at random, find these probabilities.

a. The employee smokes, given that he or she graduated from college.

b. Given that the employee did not graduate from high school, he or she is a smoker.

5–145. A survey done for *Prevention* magazine found that 77% of bike riders sometimes ride without a helmet. If four bike riders are randomly selected, find the probability that at least one of the riders does not wear a helmet all the time.

Souce: Snapshot, *USA Today,* May 26, 1995.

5–146. A coin is tossed five times. Find the probability of getting at least one tail.

5–147. The U.S. Department of Health and Human Services reports that 15% of Americans have chronic sinusitis. If five people are selected at random, find the probability that at least one has chronic sinusitis.

Source: *100% American* by Daniel Evans Weiss (New York: Poseidon Press, 1988).

Statistics Today

Would You Bet Your Life?   Revisited

In his book *Probabilities in Everyday Life,* John D. McGervey states that the chance of being killed on any given commercial airline flight is almost 1 in 1 million and that the chance of being killed during a transcontinental auto trip is about 1 in 8000. The corresponding probabilities are 1/1,000,0000 = 0.000001 as compared to 1/8000 = 0.000125. Since the second number is 125 times greater than the first number, you have a much higher risk driving than flying across the United States.

Quiz

Determine whether each statement is true or false. If the statement is false, explain why.

1. Subjective probability has little use in the real world.

2. Classical probability uses a frequency distribution to compute probabilities.

3. In classical probability, all outcomes in the sample space are equally likely.

4. When two events are not mutually exclusive, $P(A \text{ or } B) = P(A) + P(B)$.

5. If two events are dependent, they must have the same probability of occurring.

6. An event and its complement can occur at the same time.

Select the best answer.

7. The probability that an event happens is 0.42. What is the probability that the event won't happen?
a. −0.42
b. 0.58
c. 0
d. 1

8. When a meteorologist says that there is a 30% chance of showers, what type of probability is the person using?
a. Classical
b. Empirical
c. Relative
d. Subjective

9. The sample space for tossing three coins consists of how many outcomes?
a. 2
b. 4
c. 6
d. 8

10. The complement of guessing five correct answers on a five-question true–false exam is
a. Guessing five incorrect answers
b. Guessing at least one incorrect answer
c. Guessing at least one correct answer
d. Guessing no incorrect answers

11. When two dice are rolled, the sample space consists of how many events?
a. 6
b. 12
c. 36
d. 54

Complete the following statements with the best answer.

12. The set of all possible outcomes of a probability experiment is called the _____.

13. The probability of an event can be any number between and including _____ and _____.

14. If an event cannot occur, its probability is _____.

15. The sum of the probabilities of the events in the sample space is _____.

16. When two events cannot occur at the same time, they are said to be _____.

17. When a card is drawn, what is the probability of getting
a. A jack
b. A 4
c. A card less than 6 (an ace is considered above 6)

18. When a card is drawn from a deck, find the probability of getting
a. A diamond
b. A 5 or a heart
c. A 5 and a heart
d. A king
e. A red card

19. At a men's clothing store, 12 men purchased blue golf sweaters, 8 purchased green sweaters, 4 purchased gray sweaters, and 7 bought black sweaters. If a customer is selected at random, find the probability that he purchased
a. A blue sweater
b. A green or gray sweater
c. A green or black or blue sweater
d. A sweater that was not black

20. When two dice are rolled, find the probability of getting
a. A sum of 6 or 7
b. A sum greater than 8
c. A sum less than 3 or greater than 8
d. A sum that is divisible by 3
e. A sum of 16
f. A sum less than 11

21. The probability that a person owns a microwave oven is 0.75, that a person owns a compact disk player is 0.25, and that a person owns both a microwave and a CD player, 0.16. Find the probability that a person owns either a microwave or a CD player, but not both.

22. Of the physics graduates of a university, 30% received a starting salary of $30,000 or more. If five of the graduates are selected at random, find the probability that all had a starting salary of $30,000 or more.

23. Five cards are drawn from an ordinary deck *without* replacement. Find the probability of getting
a. All red cards
b. All diamonds
c. All aces

24. The probability that Sam will be accepted by the college of his choice *and* obtain a scholarship is 0.35. If the probability that he is accepted by the college is 0.65, find the probability that he will obtain a scholarship given that he is accepted by the college.

25. The probability that a customer will buy a car and an extended warranty is 0.16. If the probability that a customer will purchase a car is 0.30, find the probability that the customer will also purchase the extended warranty.

26. Of the members of the Spring Lake Bowling Lanes, 57% have a lifetime membership and bowl regularly (three or more times a week). If 70% of the club members bowl regularly, find the probability that a randomly selected member is a lifetime member given that he or she bowls regularly.

27. The probability that John has to work overtime and it rains is 0.028. John hears the weather forecast, and there is a 50% chance of rain. Find the probability that he will have to work overtime given that it rains.

28. At a large factory, the employees were surveyed and classified according to their level of education and whether or not they attend a sports event at least once a month. The data are shown in the table.

Sports event	Educational level		
	High school graduate	**Two-year college degree**	**Four-year college degree**
Attend	16	20	24
Do not attend	12	19	25

If an employee is selected at random, find the probability that
a. The employee attends sports events regularly given that he or she graduated from college (two- or four-year degree).
b. Given that the employee is a high school graduate, he or she does not attend sports events regularly.

29. In a certain high-risk group, the chances of a person suffering a heart attack are 55%. If six people are chosen, find the probability that at least one will have had a heart attack.

30. A single die is rolled four times. Find the probability of getting at least one 5.

31. If 85% of all people have brown eyes and six people are selected at random, find the probability that at least one of them has brown eyes.

Critical Thinking Challenges

1. Would You Take This Bet? Consider the following problem: A con man has three coins. One coin has been specially made and has a head on each side. A second coin has been specially made and on each side it has a tail. Finally, a third coin has a head and a tail on it. All coins are of the same denomination. The con man places the coins in his pocket, selects one, and shows you one side. It is heads. He is willing to bet you even money that it is the two-headed coin. His reasoning is that it can't be the two-tailed coin since a head is showing; therefore, there is a 50-50 chance of it being the two-headed coin. Would you take the bet? (*Hint:* See Exercise 1 in Data Projects.)

2. Gamblers' Run Chevalier de Méré won money when he bet unsuspecting patrons that in four rolls of a die, he could get at least one 6, but he lost money when he bet that in 24 rolls of two dice, he could get at least a double 6. Using the probability rules, find the probability of each event and explain why he won the majority of the time on the first game but lost the majority of the time when playing the second game. (*Hint:* Find the probabilities of losing each game and subtract from one.)

3. Classic Birthday Problem How many people do you think need to be in a room so that two people will have the same birthday (month and day)? You might think it is 366. This would, of course, guarantee it (excluding leap year), but how many people would need to be in a room so that there would be a 90% probability that two people would be born on the same day? What about a 50% probability?

Actually, the number is much smaller than you may think. For example, if you have 50 people in a room, the probability that two people will have the same birthday is 97%. If you have 23 people in a room, there is a 50% probability that two people will be born on the same day!

The problem can be solved by using the probability rules. It must be assumed that all birthdays are equally likely, but this assumption will have little effect on the answers. The way to find the answer is by using the complementary event rule as P (two people having the same birthday) $= 1 - P$ (all have different birthdays).

For example, suppose there were three people in the room. The probability that each had a different birthday would be

$$\left(\frac{365}{365}\right) \cdot \left(\frac{364}{365}\right) \cdot \left(\frac{363}{365}\right) = \frac{_{365}P_3}{365^3} = 0.992$$

Hence, the probability that at least two of the three people will have the same birthday will be

$$1 - 0.992 = 0.008$$

Hence, for k people, the formula is

$$P(\text{at least 2 people have the same birthday}) =$$

$$1 - \frac{_{365}P_k}{365^k}$$

Using your calculator, complete the following table and verify that for at least a 50% chance of two people having the same birthday, 23 or more people will be needed.

Number of people	Probability that at least two have the same birthday
1	0.000
2	0.003
5	0.027
10	
15	
20	
21	
22	
23	

4. 100% Contagious We know that if the probability of an event happening is 100%, then the event is a certainty. Can it be concluded that if there is a 50% chance of contracting a communicable disease through contact with an infected person, there would be a 100% chance of contracting the disease if two contacts were made with the infected person? Explain your answer.

Data Projects

1. Make a set of three cards—one with a red star on both sides, one with a black star on both sides, and one with a black star on one side and a red star on the other side. With a partner, play the game described in the critical thinking challenge "Would You Take This Bet?" 100 times and record the results of how many times you win and how many times your partner wins. (*Note:* Do not change options during the 100 trials.)

a. Do you think the game is fair (i.e., does one person win approximately 50% of the time)?
b. If you think the game is unfair, explain what the probabilities might be and why.

2. Take a coin and tape a small weight (e.g., part of a paper clip) to one side. Flip the coin 100 times and record the results. Do you think you have changed the probabilities of the results of flipping the coin? Explain.

3. This game is called "Diet Fractions." Roll two dice and use the numbers to make a fraction less than or equal to one. Player A wins if the fraction cannot be reduced; otherwise, player B wins.

a. Play the game 100 times and record the results.

b. Decide if the game is fair or not. Explain why or why not.

c. Using the sample space for two dice, compute the probabilities of player A winning and player B winning. Do these agree with the results obtained in part a?

Source: George W. Bright, John G. Harvey, and Margariete Montague Wheeler, "Fair Games, Unfair Games." Chapter 8, *Teaching Statistics and Probability*. NCTM 1981 Yearbook. Reston, Virginia: The National Council of Teachers of Mathematics, Inc., 1981, p. 49. Used with permission.

4. Often when playing gambling games or collecting items in cereal boxes, one wonders how long will it be before one achieves a success. For example, suppose there are six different types of toys with one toy packaged at random in a cereal box. If a person wanted a certain toy, about how many boxes would that person have to buy on average before obtaining that particular toy? Of course, there is a possibility that the particular toy would be in the first box opened or that the person might never obtain the particular toy. These are the extremes.

a. To find out, simulate the experiment using dice. Start rolling dice until a particular number, say 3, is obtained and keep track of how many rolls are necessary. Repeat 100 times. Then find the average.

b. You may decide to use another number, such as 10 different items. In this case, use 10 playing cards (ace through 10 of diamonds), select a particular card (say an ace), shuffle the deck each time, deal the cards, and count how many cards are turned over before the ace is obtained. Repeat 100 times, then find the average.

c. Summarize the findings for both experiments.

chapter

6

Probability Distributions

Outline

Objectives

After completing this chapter, you should be able to

1. Construct a probability distribution for a random variable.

2. Find the mean, variance, and expected value for a discrete random variable.

3. Find the exact probability for X successes in n trials of a binomial experiment.

4. Find the mean, variance, and standard deviation for the variable of a binomial distribution.

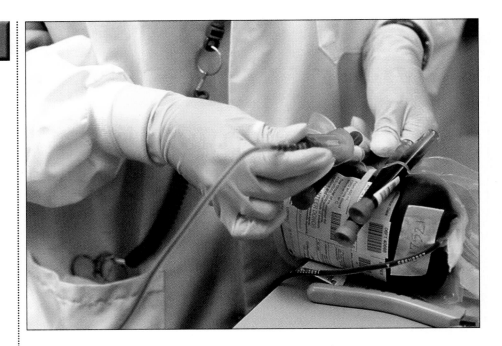

Is Pooling Worthwhile?

Blood samples are used to screen people for certain diseases. When the disease is rare, health care workers sometimes combine or pool the blood samples of a group of individuals into one batch and then test it. If the test result of the batch is negative, no further testing is needed since none of the individuals in the group has the disease. However, if the test result of the batch is positive, each individual in the group must be tested.

Consider this hypothetical example: Suppose the probability of a person having the disease is 0.05, and a pooled sample of 15 individuals is tested. What is the probability that no further testing would be needed for the individuals in the sample? The answer to this question can be found by using what is called the binomial distribution.

This chapter explains the probability distributions in general and a specific, often-used distribution called the binomial distribution.

6–1

Introduction

Many decisions in business, insurance, and other real-life situations are made by assigning probabilities to all possible outcomes pertaining to the situation and then evaluating the results. For example, a saleswoman can compute the probability that she will make 0, 1, 2, or 3 or more sales in a single day. An insurance company might be able to assign probabilities to the number of vehicles a family owns. A self-employed speaker might be able to compute the probabilities for giving 0, 1, 2, 3, or 4 or more speeches each week. Once these probabilities are assigned, statistics such as the mean, variance, and standard deviation can be computed for these events. With these statistics, various decisions can be made. The saleswoman will be able to compute the average number of sales she makes per week; and if she is working on commission, she will be able to approximate her weekly income over a period of time, say monthly. The public speaker will be able to plan ahead and approximate his average income and expenses. The insurance company can use its information to design special computer forms and programs to accommodate its customers' future needs.

This chapter explains the concepts and applications of what is called a *probability distribution*. In addition a special probability distribution, called the *binomial* distribution, is explained.

6–2

Probability Distributions

Objective 1. Construct a probability distribution for a random variable.

Before probability distribution is defined formally, the definition of variable is reviewed. In Chapter 1, a *variable* was defined as a characteristic or attribute that can assume different values. Various letters of the alphabet, such as X, Y, or Z, are used to represent variables. Since the variables in this chapter are associated with probability, they are called *random variables.*

For example, if a die is rolled, a letter such as X can be used to represent the outcomes. Then the value X can assume is 1, 2, 3, 4, 5, or 6, corresponding to the outcomes of rolling a single die. If two coins are tossed, a letter, say Y, can be used to represent the number of heads, in this case 0, 1, or 2. As another example, if the temperature at 8:00 A.M. is 43° and at noon it is 53°, then the values (T) the temperature assumes are said to be random, since they are due to various atmospheric conditions at the time the temperature was taken.

A **random variable** is a variable whose values are determined by chance.

Also recall from Chapter 1 that one can classify variables as discrete or continuous by observing the values the variable can assume. If a variable can assume only a specific number of values, such as the outcomes for the roll of a die or the outcomes for the toss of a coin, then the variable is called a *discrete variable. Discrete variables have values that can be counted.* For example, the number of joggers in Riverview Park each day and the number of phone calls received after a TV commercial airs are examples of discrete variables since they can be counted.

Variables that can assume all values in the interval between any two given values are called *continuous variables.* For example, if the temperature goes from 62° to 78° in a 24-hour period, it has passed through every possible number from 62 to 78. *Continuous random variables are obtained from data that can be measured rather than counted.* Examples of continuous variables are heights, weights, temperatures, and time. In this chapter only discrete random variables are used; Chapter 7 explains continuous random variables.

The procedure shown here for constructing a probability distribution for a discrete random variable uses the probability experiment of tossing three coins. Recall that when three coins are tossed, the sample space is represented as TTT, TTH, THT, HTT, HHT, HTH, THH, HHH, and if X is the random variable for the number of heads, then X assumes the value 0, 1, 2, or 3.

Probabilities for the values of X can be determined as follows:

No heads	One head			Two heads			Three heads
TTT	TTH	THT	HTT	HHT	HTH	THH	HHH
$\frac{1}{8}$	$\frac{1}{8}$	$\frac{1}{8}$	$\frac{1}{8}$	$\frac{1}{8}$	$\frac{1}{8}$	$\frac{1}{8}$	$\frac{1}{8}$
$\frac{1}{8}$	$\frac{3}{8}$			$\frac{3}{8}$			$\frac{1}{8}$

Hence, the probability of getting no heads is $\frac{1}{8}$, one head is $\frac{3}{8}$, two heads is $\frac{3}{8}$, and three heads is $\frac{1}{8}$. From these values, a probability distribution can be constructed by listing the outcomes and assigning the probability of each outcome, as shown.

Number of heads, X	0	1	2	3
Probability, $P(X)$	$\frac{1}{8}$	$\frac{3}{8}$	$\frac{3}{8}$	$\frac{1}{8}$

A **probability distribution** consists of the values a random variable can assume and the corresponding probabilities of the values. The probabilities are determined theoretically or by observation.

Example 6–1

Construct a probability distribution for rolling a single die.

Solution

Since the sample space is 1, 2, 3, 4, 5, 6, and each outcome has a probability of $\frac{1}{6}$, the distribution is as follows:

Outcome, X	1	2	3	4	5	6
Probability, $P(X)$	$\frac{1}{6}$	$\frac{1}{6}$	$\frac{1}{6}$	$\frac{1}{6}$	$\frac{1}{6}$	$\frac{1}{6}$

Probability distributions can be shown graphically by representing the values of X on the x axis and the probabilities $P(X)$ on the y axis.

Example 6–2

Represent graphically the probability distribution for the sample space for tossing three coins.

Number of heads, X	0	1	2	3
Probability, $P(X)$	$\frac{1}{8}$	$\frac{3}{8}$	$\frac{3}{8}$	$\frac{1}{8}$

Solution

The values X assumes are located on the x axis, and the values for $P(X)$ are located on the y axis. The graph is shown in Figure 6–1.

Note that for visual appearances, it is not necessary to start with 0 at the origin.

The preceding examples are illustrations of *theoretical* probability distributions. One did not need to actually perform the experiments to compute the probabilities. In contrast, to construct actual probability distributions, one must observe the variable over a period of time. They are empirical, as shown in the next example.

Figure 6-1

Probability Distribution
for Example 6-2

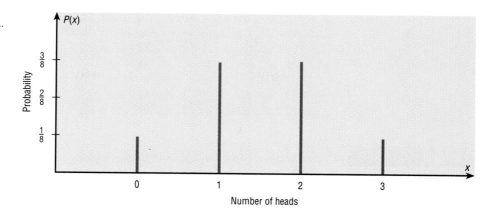

Example 6-3

During the summer months, a rental agency keeps track of the number of chain saws it rents each day for a period of 90 days. The number of saws rented per day is represented by the variable X. The results are shown here. Compute the probability $P(X)$ for each X, and construct a probability distribution and graph for the data.

X	Number of days
0	45
1	30
2	15
Total	90

Solution

The probability $P(X)$ can be computed for each X by dividing the number of days X saws were rented by total days.

for 0 saws: $\frac{45}{90} = 0.50$ for 1 saw: $\frac{30}{90} = 0.33$ for 2 saws: $\frac{15}{90} = 0.17$

The distribution is as follows:

Number of saws rented, X	0	1	2
Probability, $P(X)$	0.50	0.33	0.17

The graph is shown in Figure 6-2.

Figure 6-2

Probability Distribution
for Example 6-3

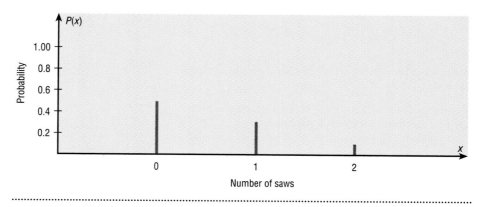

	Two Requirements for a Probability Distribution
1.	The sum of the probabilities of all the events in the sample space must equal 1; that is, $\Sigma P(X) = 1$.
2.	The probability of each event in the sample space must be between or equal to 0 and 1. That is, $0 \le P(X) \le 1$.

Example 6–4

Determine whether each distribution is a probability distribution.

a.

X	0	5	10	15	20
$P(X)$	$\frac{1}{5}$	$\frac{1}{5}$	$\frac{1}{5}$	$\frac{1}{5}$	$\frac{1}{5}$

c.

X	1	2	3	4
$P(X)$	$\frac{1}{4}$	$\frac{1}{8}$	$\frac{1}{16}$	$\frac{9}{16}$

b.

X	0	2	4	6
$P(X)$	-1.0	1.5	0.3	0.2

d.

X	2	3	7
$P(X)$	0.5	0.3	0.4

Solution

a. Yes, it is a probability distribution.

b. No, it is not a probability distribution, since $P(X)$ cannot be 1.5 or -1.0.

c. Yes, it is a probability distribution.

d. No, it is not, since $\Sigma P(X) = 1.2$.

Many variables in business, education, engineering, and other areas can be analyzed by using probability distributions. The next section shows methods for finding the mean and standard deviation for a probability distribution.

Exercises

6–1. (**W**) Define and give three examples of a random variable.

6–2. (**W**) Explain the difference between a discrete and a continuous random variable.

6–3. (**W**) Give three examples of a discrete random variable.

6–4. (**W**) Give three examples of a continuous random variable.

6–5. (**W**) What is a probability distribution? Give an example.

For Exercises 6–6 through 6–11, determine whether the distribution represents a probability distribution. If it is not, state why.

6–6.

X	1	3	5	7	9	11
$P(X)$	$\frac{1}{6}$	$\frac{1}{6}$	$\frac{1}{6}$	$\frac{1}{6}$	$\frac{1}{6}$	$\frac{1}{6}$

6–7.

X	1	3	5	7	9
$P(X)$	$\frac{1}{8}$	$\frac{1}{8}$	$\frac{3}{8}$	$\frac{2}{8}$	$\frac{1}{8}$

6–8.

X	3	6	8
$P(X)$	-0.3	0.6	0.7

6–9.

X	1	2	3	4	5
$P(X)$	$\frac{3}{10}$	$\frac{1}{10}$	$\frac{1}{10}$	$\frac{2}{10}$	$\frac{3}{10}$

6–10.

X	20	30	40	50
$P(X)$	1.1	0.2	0.9	0.3

6–11.

X	5	10	15
$P(X)$	1.2	0.3	0.5

For Exercises 6–12 through 6–18, state whether the variable is discrete or continuous.

6–12. The speed of a race car.

6–13.   The number of cups of coffee a fast-food restaurant serves each day.

6–14.   The number of people who play the state lottery each day.

6–15.   The weight of an elephant.

6–16.   The time it takes to complete an exercise session.

6–17.   The number of English majors in your school.

6–18.   The blood pressures of all patients admitted to a hospital on a specific day.

For Exercises 6–19 through 6–26, construct a probability distribution for the data and draw a graph for the distribution.

6–19.   The probabilities that a patient will have 0, 1, 2, or 3 medical tests performed upon entering a hospital are $\frac{6}{15}$, $\frac{5}{15}$, $\frac{3}{15}$, and $\frac{1}{15}$, respectively.

6–20.   The probabilities of a return on an investment of $1000, $2000, and $3000 are $\frac{1}{2}$, $\frac{1}{4}$, and $\frac{1}{4}$, respectively.

6–21.   The probabilities of a machine manufacturing 0, 1, 2, 3, 4, or 5 defective parts in one day are 0.75, 0.17, 0.04, 0.025, 0.01, and 0.005, respectively.

6–22.   The probabilities that a shopper will purchase 0, 1, 2, or 3 items are 0.20, 0.40, 0.20, and 0.20, respectively.

6–23.   A die is loaded in such a way that the probabilities of getting 1, 2, 3, 4, 5, and 6 are $\frac{1}{2}$, $\frac{1}{6}$, $\frac{1}{12}$, $\frac{1}{12}$, $\frac{1}{12}$, and $\frac{1}{12}$, respectively.

6–24.   The probabilities that a customer selects 1, 2, 3, 4, or 5 items at a convenience store are 0.32, 0.12, 0.23, 0.18, and 0.15, respectively.

6–25.   The probabilities that a tutor sees 1, 2, 3, 4, or 5 students in any one day are 0.10, 0.25, 0.25, 0.20, and 0.20, respectively.

6–26.   Three patients are given a headache relief tablet. The probabilities for 0, 1, 2, or 3 successes are 0.18, 0.52, 0.21, and 0.09, respectively.

6–27.   A box contains three $1 bills, two $5 bills, one $10 bill, and one $20 bill. Construct a probability distribution for the data.

6–28.   Construct a probability distribution for a family of three children. Let X represent the number of boys.

6–29.   Construct a probability distribution for drawing a card from a deck of 40 cards consisting of 10 cards numbered 1, 10 cards numbered 2, 15 cards numbered 3, and 5 cards numbered 4.

6–30.   Using the sample space for tossing two dice, construct a probability distribution for the sums 2 through 12.

A probability distribution can be written in formula notation as $P(X) = 1/X$, where $X = 2, 3, 6$. The distribution is shown as follows:

X	2	3	6
$P(X)$	$\frac{1}{2}$	$\frac{1}{3}$	$\frac{1}{6}$

For Exercises 6–31 through 6–36, write the distribution for the formula and determine whether it is a probability distribution.

*6–31. $P(X) = X/6$ for $X = 1, 2, 3$

*6–32. $P(X) = X$ for $X = 0.2, 0.3, 0.5$

*6–33. $P(X) = X/6$ for $X = 3, 4, 7$

*6–34. $P(X) = X + 0.1$ for $X = 0.1, 0.02, 0.04$

*6–35. $P(X) = X/7$ for $X = 1, 2, 4$

*6–36. $P(X) = X/(X + 2)$ for $X = 0, 1, 2$

6–3

Mean, Variance, and Expectation

The mean, variance, and standard deviation for a probability distribution are computed differently from the mean, variance, and standard deviation for samples. This section explains how these measures—as well as a new measure called the *expectation*—are calculated for probability distributions.

Mean

In Chapter 3, the means for a sample or population were computed by adding the values and dividing by the total number of values, as shown in the formulas:

$$\overline{X} = \frac{\Sigma X}{n} \qquad \mu = \frac{\Sigma X}{N}$$

But how would one compute the mean of the number of spots that show on top when a die is rolled? One could try rolling the die, say, 10 times, recording the number of spots, and finding the mean; however, this answer would only approximate the true mean. What about 50 rolls or 100 rolls? Actually, the more times the die is rolled, the better the approximation. One might ask then, "How many times must the die be rolled to get the exact answer?" *It must be rolled an infinite number of times.* Since this task is impossible, the above formulas cannot be used because the denominators would be infinity. Hence, a new method of computing the mean is necessary. This method gives the exact theoretical value of the mean as if it were possible to roll the die an infinite number of times.

Before the formula is stated, an example will be used to explain the concept. Suppose two coins are tossed repeatedly, and the number of heads that occurred is recorded. What will the mean of the number of heads be? The sample space is

HH, HT, TH, TT

and each outcome has a probability of $\frac{1}{4}$. Now, in the long run, one would *expect* two heads (HH) to occur approximately $\frac{1}{4}$ of the time, one head to occur approximately $\frac{1}{2}$ of the time (HT or TH), and no heads (TT) to occur approximately $\frac{1}{4}$ of the time. Hence, on average, one would expect the number of heads to be

$$\tfrac{1}{4} \cdot 2 + \tfrac{1}{2} \cdot 1 + \tfrac{1}{4} \cdot 0 = 1$$

That is, if it were possible to toss the coins many times or an infinite number of times, the *average* of the number of heads would be 1.

Hence, in order to find the mean for a probability distribution, one must multiply each possible outcome by its corresponding probability and find the sum of the products.

Formula for the Mean of a Probability Distribution

The mean of the random variable of a probability distribution is

$$\mu = X_1 \cdot P(X_1) + X_2 \cdot P(X_2) + X_3 \cdot P(X_3) + \cdots + X_n \cdot P(X_n)$$
$$= \Sigma X \cdot P(X)$$

where $X_1, X_2, X_3, \ldots, X_n$ are the outcomes and $P(X_1), P(X_2), P(X_3), \ldots, P(X_n)$ are the corresponding probabilities.

Note: $\Sigma X \cdot P(X)$ means to sum the products.

Rounding Rule

The rounding rule for the mean, variance, and standard deviation for variables of a probability distribution is this: The mean, variance, and standard deviation should be rounded to one more decimal place than the outcome, X. When fractions are used, they should be reduced to the lowest terms.

The next four examples illustrate the use of the formula.

Example 6–5

Find the mean of the number of spots that appear when a die is tossed.

Solution

In the toss of a die, the mean can be computed as follows.

Outcome, X	1	2	3	4	5	6
Probability, $P(X)$	$\frac{1}{6}$	$\frac{1}{6}$	$\frac{1}{6}$	$\frac{1}{6}$	$\frac{1}{6}$	$\frac{1}{6}$

$$\mu = \Sigma X \cdot P(X) = 1 \cdot \tfrac{1}{6} + 2 \cdot \tfrac{1}{6} + 3 \cdot \tfrac{1}{6} + 4 \cdot \tfrac{1}{6} + 5 \cdot \tfrac{1}{6} + 6 \cdot \tfrac{1}{6}$$
$$= \tfrac{21}{6} = 3\tfrac{1}{2} \text{ or } 3.5$$

That is, when a die is tossed many times, the theoretical mean will be 3.5. Note that even though the die cannot show a 3.5, the theoretical average is 3.5.

The reason why this formula gives the theoretical mean is that in the long run, each outcome would occur approximately $\frac{1}{6}$ of the time. Hence, multiplying the outcome by its corresponding probability and finding the sum would yield the theoretical mean. In other words, the outcome 1 would occur approximately $\frac{1}{6}$ of the time, the outcome 2 would occur approximately $\frac{1}{6}$ of the time, etc.

Also, since a mean is not a probability, it can be a number greater than one.

Example 6–6

In a family with two children, find the mean of the number of children who will be girls.

Solution

The probability distribution is as follows:

Number of girls, X	0	1	2
Probability, $P(X)$	$\frac{1}{4}$	$\frac{1}{2}$	$\frac{1}{4}$

Hence, the mean is

$$\mu = \Sigma X \cdot P(X) = 0 \cdot \tfrac{1}{4} + 1 \cdot \tfrac{1}{2} + 2 \cdot \tfrac{1}{4} = 1$$

Example 6–7

If three coins are tossed, find the mean number of heads that will occur.

Solution

The probability distribution is as follows:

Number of heads, X	0	1	2	3
Probability, $P(X)$	$\frac{1}{8}$	$\frac{3}{8}$	$\frac{3}{8}$	$\frac{1}{8}$

The mean is

$$\mu = \Sigma X \cdot P(X) = 0 \cdot \tfrac{1}{8} + 1 \cdot \tfrac{3}{8} + 2 \cdot \tfrac{3}{8} + 3 \cdot \tfrac{1}{8} = \tfrac{12}{8} = 1\tfrac{1}{2} = 1.5$$

The value 1.5 cannot occur as an outcome. Nevertheless, it is the long-run or theoretical average.

Example 6–8

In a restaurant, the following probability distribution was obtained for the number of items a person ordered for a large pizza.

Number of items, X	0	1	2	3	4
Probability, $P(X)$	0.3	0.4	0.2	0.06	0.04

Find the mean for the distribution.

Solution

$$\mu = \Sigma X \cdot P(X)$$
$$= (0)(0.3) + (1)(0.4) + (2)(0.2) + (3)(0.06) + (4)(0.04) = 1.14 \approx 1.1$$

The probability distribution in Example 6–8 is the same as a relative frequency distribution for discrete variables.

Variance

For a probability distribution, the mean of the random variable describes the measure of the so-called long-run or theoretical average, but it does not tell anything about the spread of the distribution. Recall from Chapter 3 that in order to measure this spread or variability, statisticians use the variance and standard deviation. The following formulas were used:

$$\sigma^2 = \frac{\Sigma (X - \mu)^2}{N} \qquad \text{or} \qquad \sigma = \sqrt{\frac{\Sigma (X - \mu)^2}{N}}$$

These formulas cannot be used for a random variable of a probability distribution since N is infinite, so the variance and standard deviation must be computed differently.

To find the variance for the random variable of a probability distribution, subtract the theoretical mean of the random variable from each outcome and square the difference. Then multiply each difference by its corresponding probability and add the products. The formula is

$$\sigma^2 = \Sigma [(X - \mu)^2 \cdot P(X)]$$

Finding the variance by using this formula is somewhat tedious. So for simplified computations, a shortcut formula can be used. This formula is algebraically equivalent to the longer one and will be used in the examples that follow.

Formula for the Variance of a Probability Distribution

Find the variance of a probability distribution by multiplying the square of each outcome by its corresponding probability, summing those products, and subtracting the square of the mean. The formula for the variance of a probability distribution is

$$\sigma^2 = \Sigma [X^2 \cdot P(X)] - \mu^2$$

The standard deviation of a probability distribution is

$$\sigma = \sqrt{\sigma^2}$$

Remember that the variance and standard deviation cannot be negative.

Example 6–9

Compute the variance and standard deviation for the data in Example 6–5.

Historical Note

The Fey Manufacturing Co., located in San Francisco, invented the first three-reel, automatic payout slot machine in 1895.

Solution

Recall that the mean is $\mu = 3.5$, as computed in Example 6–5. Square each outcome and multiply by the corresponding probability, sum those products, and then subtract the square of the mean.

$$\sigma^2 = [1^2 \cdot \tfrac{1}{6} + 2^2 \cdot \tfrac{1}{6} + 3^2 \cdot \tfrac{1}{6} + 4^2 \cdot \tfrac{1}{6} + 5^2 \cdot \tfrac{1}{6} + 6^2 \cdot \tfrac{1}{6}] - (3.5)^2 = 2.9$$

To get the standard deviation, find the square root of the variance.

$$\sigma = \sqrt{2.9} = 1.7$$

Example 6–10

Five balls numbered 0, 2, 4, 6, and 8 are placed in a bag. After the balls are mixed, one is selected, its number is noted, and then it is replaced. If this experiment is repeated many times, find the variance and standard deviation of the numbers on the balls.

Solution

Let X be the number on each ball. The probability distribution is as follows:

Number on ball, X	0	2	4	6	8
Probability, $P(X)$	$\tfrac{1}{5}$	$\tfrac{1}{5}$	$\tfrac{1}{5}$	$\tfrac{1}{5}$	$\tfrac{1}{5}$

The mean is

$$\mu = \Sigma X \cdot P(X) = 0 \cdot \tfrac{1}{5} + 2 \cdot \tfrac{1}{5} + 4 \cdot \tfrac{1}{5} + 6 \cdot \tfrac{1}{5} + 8 \cdot \tfrac{1}{5} = 4.0$$

The variance is

$$\sigma^2 = \Sigma [X^2 \cdot P(X)] - \mu^2$$
$$= [0^2 \cdot (\tfrac{1}{5}) + 2^2 \cdot (\tfrac{1}{5}) + 4^2 \cdot (\tfrac{1}{5}) + 6^2 \cdot (\tfrac{1}{5}) + 8^2 \cdot (\tfrac{1}{5})] - 4^2$$
$$= [0 + \tfrac{4}{5} + \tfrac{16}{5} + \tfrac{36}{5} + \tfrac{64}{5}] - 16$$
$$= \tfrac{120}{5} - 16$$
$$= 24 - 16 = 8$$

The standard deviation is $\sigma = \sqrt{8} = 2.8$.

The mean, variance, and standard deviation can also be found by using vertical columns, as shown. (0.2 is used for $P(X)$ since $\tfrac{1}{5} = 0.2$.)

X	$P(X)$	$X \cdot P(X)$	$X^2 \cdot P(X)$
0	0.2	0	0
2	0.2	0.4	0.8
4	0.2	0.8	3.2
6	0.2	1.2	7.2
8	0.2	1.6	12.8
		$\Sigma X \cdot P(X) = 4$	$\Sigma X^2 \cdot P(X) = 24$

Find the mean by summing the $X \cdot P(X)$ column and the variance by summing the $X^2 \cdot P(X)$ column and subtracting the square of the mean:

$$\sigma^2 = 24 - 4^2 = 8$$

Example 6–11

Historical Note

The probability that 0, 1, 2, 3, or 4 people will be placed on hold when they call a radio talk show is shown in the distribution. Find the variance and standard deviation for the data. The radio station has four phone lines.

X	0	1	2	3	4
$P(X)$	0.18	0.34	0.23	0.21	0.04

Solution

The mean is

$$\mu = \Sigma X \cdot P(X)$$
$$= 0 \cdot (0.18) + 1 \cdot (0.34) + 2 \cdot (0.23) + 3 \cdot (0.21) + 4 \cdot (0.04)$$
$$= 1.6$$

The variance is

$$\sigma^2 = [\Sigma X^2 \cdot P(X)] - \mu^2$$
$$= [0^2 \cdot (0.18) + 1^2 \cdot (0.34) + 2^2 \cdot (0.23)$$
$$+ 3^2 \cdot (0.21) + 4^2 \cdot (0.04)] - 1.6^2$$
$$= [0 + 0.34 + 0.92 + 1.89 + 0.64] - 2.56$$
$$= 3.79 - 2.56$$
$$= 1.2$$

The standard deviation is $\sigma = \sqrt{\sigma^2}$, or $\sigma = \sqrt{1.2} = 1.1$.

Expectation

Related to the concept of the mean for a probability distribution is the concept of expected value or expectation. Expected value is used in various types of games of chance, in insurance, and in other areas, such as decision theory.

The **expected value** of a discrete random variable of a probability distribution is the theoretical average of the variable. The formula is

$$\mu = E(X) = \Sigma X \cdot P(X)$$

The symbol $E(X)$ is used for the expected value.

The formula for the expected value is the same as the formula for the theoretical mean. The expected value, then, is the theoretical mean of the probability distribution. That is, $E(X) = \mu$.

When expected value problems involve money, it is customary to round the answer to the nearest cent.

Example 6–12

One thousand tickets are sold at $1 each for a color television valued at $350. What is the expected value of the gain if a person purchases one ticket?

Solution

The problem can be set up as follows:

	Win	Lose
Gain, X	$349	$-$1
Probability, $P(X)$	$\dfrac{1}{1000}$	$\dfrac{999}{1000}$

Two things should be noted. First, for a win, the net gain is $349, since the person does not get the cost of the ticket ($1) back. Second, for a loss, the gain is represented by a negative number, in this case $-$1. The solution, then, is

$$E(X) = \$349 \cdot \frac{1}{1000} + (-\$1) \cdot \frac{999}{1000} = -\$0.65$$

Expected value problems of this type can also be solved by finding the overall gain and subtracting the cost of the tickets, as shown:

$$E(X) = \$350 \cdot \frac{1}{1000} - \$1 = -\$0.65$$

Here, the overall gain ($350) must be used.

Note that the expectation is $-$0.65. This does not mean that a person loses $0.65, since the person can only win a television set valued at $350 or lose $1 on the ticket. What this expectation means is that the average of the losses is $0.65 for each of the 1000 ticket holders. Here is another way of looking at this situation: If a person purchased one ticket each week over a long period of time, the average loss would be $0.65 per ticket, since theoretically, on average, that person would win the set once for each 1000 tickets purchased.

Example 6–13

A ski resort loses $70,000 per season when it does not snow very much and makes $250,000 profit when it does snow a lot. The probability of it snowing at least 75 inches (i.e., a good season) is 40%. Find the expectation for the profit.

Solution

	$250,000	$-$70,000
Profit, X	$250,000	$-$70,000
Probability, $P(X)$	0.40	0.60

$$E(X) = (\$250{,}000)(0.40) + (-\$70{,}000)(0.60) = \$58{,}000$$

Example 6–14

One thousand tickets are sold at $1 each for four prizes of $100, $50, $25, and $10. What is the expected value if a person purchases two tickets?

	$98	$48	$23	$8	$-$2
Gain, X	$98	$48	$23	$8	$-$2
Probability, $P(X)$	$\dfrac{2}{1000}$	$\dfrac{2}{1000}$	$\dfrac{2}{1000}$	$\dfrac{2}{1000}$	$\dfrac{992}{1000}$

Solution

$$E(X) = \$98 \cdot \frac{2}{1000} + \$48 \cdot \frac{2}{1000} + \$23 \cdot \frac{2}{1000} + \$8 \cdot \frac{2}{1000} + (-\$2) \cdot \frac{992}{1000}$$
$$= -\$1.63$$

An alternative solution is

$$E(X) = \$100 \cdot \frac{2}{1000} + \$50 \cdot \frac{2}{1000} + \$25 \cdot \frac{2}{1000} + \$10 \cdot \frac{2}{1000} - \$2$$
$$= -\$1.63$$

In gambling games, if the expected value of the gain is zero, the game is said to be fair. If the expected value of the gain of a game is positive, then the game is in favor of the player. That is, the player has a better-than-even chance of winning. If the expected value of the gain is negative, then the game is said to be in favor of the house. That is, in the long run, the players will lose money.

Exercises

6–37. From past experience, a company has found that in cartons of transistors, 92% contain no defective transistors, 3% contain one defective transistor, 3% contain two defective transistors and 2% contain three defective transistors. Find the mean, variance, and standard deviation for the defective transistors.

6–38. The number of suits sold per day at a retail store is shown in the table, with the corresponding probabilities. Find the mean, variance, and standard deviation of the distribution.

Number of suits sold, X	19	20	21	22	23
Probability, $P(X)$	0.2	0.2	0.3	0.2	0.1

6–39. A shoe-store manager computes the probabilities of selling shoes for each size, as shown here. Find the mean, variance, and standard deviation for the distribution.

Size, X	8	9	10	11	12	13
Probability, $P(X)$	0.13	0.39	0.23	0.14	0.08	0.03

6–40. The probability distribution for the number of customers per day at the Sunrise Coffee Shop is shown below. Find the mean, variance, and standard deviation of the distribution.

Number of customers, X	50	51	52	53	54
Probability, $P(X)$	0.10	0.20	0.37	0.21	0.12

6–41. A public speaker computes the probabilities for the number of speeches she gives each week. Compute the mean, variance, and standard deviation of the distribution shown.

Number of speeches, X	0	1	2	3	4	5
Probability, $P(X)$	0.06	0.42	0.22	0.12	0.15	0.03

6–42. A recent survey by an insurance company showed the following probabilities for the number of automobiles each policyholder owned. Find the mean, variance, and standard deviation for the distribution.

Number of automobiles, X	1	2	3	4
Probability, $P(X)$	0.4	0.3	0.2	0.1

6–43. A concerned parents group determined the number of commercials shown in each of five children's programs over a period of time. Find the mean, variance, and standard deviation for the distribution shown.

Number of commercials, X	5	6	7	8	9
Probability, $P(X)$	0.2	0.25	0.38	0.10	0.07

6–44. A study conducted by a TV station showed the number of televisions per household and the corresponding probabilities for each. Find the mean, variance, and standard deviation.

Number of televisions, X	1	2	3	4
Probability, $P(X)$	0.32	0.51	0.12	0.05

6–45. The following distribution shows the number of students enrolled in CPR classes offered by the local fire department. Find the mean, variance, and standard deviation for the distribution.

Number of students, X	12	13	14	15	16
Probability, $P(X)$	0.15	0.20	0.38	0.18	0.09

6–46. A florist determines the probabilities for the number of flower arrangements she delivers each day. Find the mean, variance, and standard deviation for the distribution shown.

Number of arrangements, X	6	7	8	9	10
Probability, $P(X)$	0.2	0.2	0.3	0.2	0.1

6–47. A cash prize of $5000 is to be awarded by Lincoln Fire Department. If 2500 tickets are sold at $5 each, find the expected value of the gain.

6–48. A box contains ten $1 bills, five $2 bills, three $5 bills, one $10 bill, and one $100 bill. A person is charged $20 to select one bill. Find the expectation. Is the game fair?

6–49. If a person rolls doubles when he tosses two dice, he wins $5. For the game to be fair, how much should the person pay to play the game?

6–50. If a player rolls two dice and gets a sum of 2 or 12, she wins $20. If the person gets a 7, she wins $5. The cost to play the game is $3. Find the expectation of the game.

6–51. A lottery offers one $1000 prize, one $500 prize, and five $100 prizes. One thousand tickets are sold at $3 each. Find the expectation if a person buys one ticket.

6–52. In Exercise 6–51, find the expectation if a person buys two tickets. Assume that the player's ticket is replaced after each draw and the same ticket can win more than one prize.

6–53. For a daily lottery, a person selects a three-digit number. If the person plays for $1, she can win $500. Find the expectation. In the same daily lottery, if a person boxes a number, she will win $80. Find the expectation if the number 123 is played for $1 and boxed. (When a number is "boxed," it can win when the digits occur in any order.)

6–54. If a 60-year-old buys a $1000 life insurance policy at a cost of $60 and has a probability of 0.972 of living to age 61, find the expectation of the policy.

6–55. A roulette wheel has 38 numbers: 1 through 36, 0, and 00. Half of the numbers from 1 through 36 are red and half are black. A ball is rolled, and it falls into one of the 38 slots, giving a number and a color. Green is the color for 0 and 00. The payoffs for a $1 bet are as follows:

Red or black	$1	0		$35
Odd or even	$1	00		$35
1–18	$1	Any single number	$35
19–36	$1	0 or 00		$17

If a person bets $1, find the expected value for each.
a. Red
b. Even (exclude 0 and 00)
c. 0
d. Any single value
e. 0 or 00

*6–56. Construct a probability distribution for the sum shown on the faces when two dice are rolled. Find the mean, variance, and standard deviation of the distribution.

*6–57. When one die is rolled, the expected value of the number of spots is 3.5. In Exercise 6–56, the mean number of spots was found for rolling two dice. What is the mean number of spots if three dice are rolled?

*6–58. Another formula for finding the variance for a probability distribution is

$$\sigma^2 = \Sigma (X - \mu)^2 \cdot P(X)$$

Verify algebraically that this formula gives the same result as the formula shown in this section.

*6–59. Roll a die 100 times. Compute the mean and standard deviation. How does the result compare with the theoretical results of Example 6–5?

*6–60. Roll two dice 100 times and find the mean, variance, and standard deviation of the sum of the spots. Compare the result with the theoretical results obtained in Exercise 6–56.

*6–61. Conduct a survey of the number of books your classmates have brought to class today. Construct a probability distribution and find the mean, variance, and standard deviation.

*6–62. In a recent promotional campaign, a company offered the following prizes and the corresponding probabilities. Find the expected value of winning. The tickets are free.

Number of Prizes	Amount	Probability
1	$100,000	$\dfrac{1}{1,000,000}$
2	$10,000	$\dfrac{1}{50,000}$
5	$1,000	$\dfrac{1}{10,000}$
10	$100	$\dfrac{1}{1,000}$

If the winner has to mail in the winning ticket to claim the prize, what will be the expectation if the cost of the stamp is considered? Assume a stamp costs $0.33.

6–4

The Binomial Distribution

Objective 3. Find the exact probability for *X* successes in *n* trials of a binomial experiment.

Many types of probability problems have only two outcomes, or they can be reduced to two outcomes. For example, when a coin is tossed, it can land heads or tails. When a baby is born, it will be either male or female. In a basketball game, a team either wins or loses. A true–false item can be answered in only two ways, true or false. Other situations can be reduced to two outcomes. For example, a medical treatment can be classified as effective or ineffective, depending on the results. A person can be classified as having normal or abnormal blood pressure, depending on the measure of the blood pressure gauge. A multiple-choice question, even though there are four or five answer choices, can be classified as correct or incorrect. Situations like these are called *binomial experiments*.

Historical Note

In 1653, Blaise Pascal created a triangle of numbers called Pascal's triangle that can be used in the binomial distribution.

A **binomial experiment** is a probability experiment that satisfies the following four requirements:

1. Each trial can have only two outcomes or outcomes that can be reduced to two outcomes. These outcomes can be considered as either success or failure.
2. There must be a fixed number of trials.
3. The outcomes of each trial must be independent of each other.
4. The probability of a success must remain the same for each trial.

A binomial experiment and its results give rise to a special probability distribution called the *binomial distribution*.

The outcomes of a binomial experiment and the corresponding probabilities of these outcomes are called a **binomial distribution**.

In binomial experiments, the outcomes are usually classified as successes or failures. For example, the correct answer to a multiple-choice item can be classified as a success, but any of the other choices would be incorrect and hence classified as a failure. The notation that is commonly used for binomial experiments and the binomial distribution is defined next.

Notation for the Binomial Distribution

$P(S)$	The symbol for the probability of success.
$P(F)$	The symbol for the probability of failure.
p	The numerical probability of success.
q	The numerical probability of a failure.

$$P(S) = p \quad \text{and} \quad P(F) = 1 - p = q$$

n	The number of trials.
X	The number of successes.

Note that $q = 1 - p$ and $0 \leq X \leq n$.

The probability of a success in a binomial experiment can be computed with the following formula.

Binomial Probability Formula

In a binomial experiment, the probability of exactly X successes in n trials is

$$P(X) = \frac{n!}{(n - X)!X!} \cdot p^X \cdot q^{n-X}$$

An explanation of why the formula works will be given following Example 6–15.

Example 6–15

A coin is tossed three times. Find the probability of getting exactly two heads.

Solution

This problem can be solved by looking at the sample space. There are three ways to get two heads.

HHH, <u>HHT, HTH, THH,</u> TTH, THT, HTT, TTT

The answer is $\frac{3}{8}$, or 0.375.

Looking at the problem in Example 6–15 from the standpoint of a binomial experiment, one can show that it meets the four requirements.

1. There are only two outcomes for each trial, heads or tails.
2. There is a fixed number of trials (three).
3. The outcomes are independent of each other (the outcome of one toss in no way affects the outcome of another toss).
4. The probability of a success (heads) is $\frac{1}{2}$ in each case.

In this case, $n = 3$, $X = 2$, $p = \frac{1}{2}$, and $q = \frac{1}{2}$. Hence, substituting in the formula gives

$$P(2 \text{ heads}) = \frac{3!}{(3-2)!2!} \cdot \left(\frac{1}{2}\right)^2 \left(\frac{1}{2}\right)^1 = \frac{3}{8} = 0.375$$

which is the same answer obtained by using the sample space.

The same example can be used to explain the formula. First, note that there are three ways to get exactly two heads and one tail from a possible eight ways. They are HHT, HTH, and THH. In this case, then, the number of ways of obtaining two heads from three coin tosses is $_3C_2$, or 3, as shown in Chapter 4. In general, the number of ways to get X successes from n trials without regard to order is

$$_nC_X = \frac{n!}{(n-X)!X!}$$

This is the first part of the binomial formula. (Some calculators can be used for this.)

Next, each success has a probability of $\frac{1}{2}$, and can occur twice. Likewise, each failure has a probability of $\frac{1}{2}$ and can occur once, giving the $(\frac{1}{2})^2(\frac{1}{2})^1$ part of the formula. Generalizing, then, each success has a probability of p and can occur X times, and each failure has a probability of q and can occur $(n - X)$ times. Putting it all together yields the binomial probability formula.

Example 6–16

If a student randomly guesses at five multiple-choice questions, find the probability that the student gets exactly three correct. Each question has five possible choices.

Solution

In this case $n = 5$, $X = 3$, and $p = \frac{1}{5}$, since there is one chance in five of guessing a correct answer. Then,

$$P(3) = \frac{5!}{(5-3)!3!} \cdot \left(\frac{1}{5}\right)^3 \left(\frac{4}{5}\right)^2 = 0.05$$

Example 6–17

A survey from Teenage Research Unlimited (Northbrook, Ill.) found that 30% of teenage consumers received their spending money from part-time jobs. If five teenagers are selected at random, find the probability that at least three of them will have part-time jobs.

Solution

To find the probability that at least three have a part-time job, it is necessary to find the individual probabilities for either 3, or 4, or 5, and then add them to get the total probability.

$$P(3) = \frac{5!}{(5-3)!(3!)} \cdot (0.3)^3(0.7)^2 = 0.132$$

$$P(4) = \frac{5!}{(5-4)!(4)!} \cdot (0.3)^4(0.7)^1 = 0.028$$

$$P(5) = \frac{5!}{(5-5)!(5)!} \cdot (0.3)^5(0.7)^0 = 0.002$$

Hence,

P(at least three teenagers have part-time jobs)
$= 0.1323 + 0.02835 + 0.00243 = 0.16308$
$\approx 0.132 + 0.028 + 0.002 = 0.162$

Computing probabilities using the binomial probability formula can be quite tedious at times, so tables have been developed for selected values of n and p. Table B in Appendix C gives the probabilities for individual events. The next example shows how to use Table B to compute probabilities for binomial experiments.

Example 6–18

Solve the problem in Example 6–15 by using Table B.

Solution

Since $n = 3$, $X = 2$, and $p = 0.5$, the value 0.375 is found as shown in Figure 6–3.

Figure 6–3

Using Table B for Example 6–18

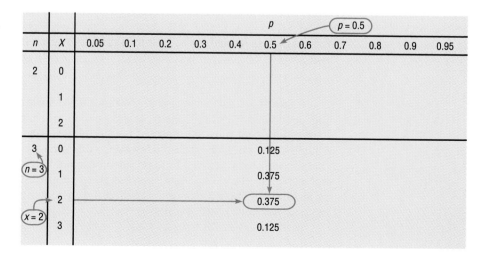

| **Example 6–19** | |

Public Opinion reported that 5% of Americans are afraid of being alone in a house at night. If a random sample of 20 Americans is selected, find these probabilities using the binomial table:

a. There are exactly five people in the sample who are afraid of being alone at night.

b. There are at most three people in the sample who are afraid of being alone at night.

c. There are at least three people in the sample who are afraid of being alone at night.

Source: *100% American* by Daniel Evan Weiss (New York: Poseidon Press, 1988).

Solution

a. $n = 20$, $p = 0.05$, and $X = 5$. From the table, one gets 0.002.

b. $n = 20$ and $p = 0.05$. "At most three people" means 0, or 1, or 2, or 3. Hence, the solution is

$$P(0) + P(1) + P(2) + P(3) = 0.358 + 0.377 + 0.189 + 0.060$$
$$= 0.984$$

c. $n = 20$ and $p = 0.05$. "At least three people" means 3, 4, 5, . . . , 20. This problem can best be solved by finding $P(0) + P(1) + P(2)$ and subtracting from 1.

$$P(0) + P(1) + P(2) = 0.358 + 0.377 + 0.189 = 0.924$$
$$1 - 0.924 = 0.076$$

| **Example 6–20** | |

A report from the Secretary of Health and Human Services stated that 70% of single-vehicle traffic fatalities that occur at night on weekends involve an intoxicated driver. If a sample of 15 single-vehicle traffic fatalities that occur at night on a weekend is selected, find the probability that exactly 12 involve a driver who is intoxicated.

Source: *100% American* by Daniel Evan Weiss (New York: Poseidon Press, 1988).

Solution

$n = 15$, $p = 0.70$, and $X = 12$. From Table B, $P(12) = 0.170$. Hence, the probability is 0.17.

Remember that in the use of the binomial distribution, the outcomes must be independent. For example, in the selection of components from a batch to be tested, each component must be replaced before the next one is selected. Otherwise, the outcomes are not independent. However, a dilemma arises because there is a chance that the same component could be selected again. This situation can be avoided by not replacing the component, but by using another distribution to calculate the probabilities. The distribution is called the hypergeometric distribution, but it is beyond the scope of this book.

Objective 4. Find the mean, variance, and standard deviation for the variable of a binomial distribution.

Mean, Variance, and Standard Deviation for the Binomial Distribution

The mean, variance, and standard deviation of a variable that has the **binomial distribution** can be found by using the following formulas.

$$\text{mean} \quad \mu = n \cdot p$$
$$\text{variance} \quad \sigma^2 = n \cdot p \cdot q$$
$$\text{standard deviation} \quad \sigma = \sqrt{n \cdot p \cdot q}$$

These formulas are algebraically equivalent to the formulas for the mean, variance, and standard deviation of the variables for probability distributions, but because they are for variables of the binomial distribution, they have been simplified using algebra. The algebraic derivation is omitted here, but their equivalence is shown in the next example.

Example 6–21

A coin is tossed four times. Find the mean, variance, and standard deviation of the number of heads that will be obtained.

Solution

With the formulas for the binomial distribution and $n = 4$, $p = \frac{1}{2}$, and $q = \frac{1}{2}$, the results are

$$\mu = n \cdot p = 4 \cdot \tfrac{1}{2} = 2$$
$$\sigma^2 = n \cdot p \cdot q = 4 \cdot \tfrac{1}{2} \cdot \tfrac{1}{2} = 1$$
$$\sigma = \sqrt{1} = 1$$

From Example 6–21, when four coins are tossed many, many times, the average of the number of heads that appear is two, and the standard deviation of the number of heads is one. Note that these are theoretical values.

As stated previously, this problem can be solved by using the expected value formulas. The distribution is shown as follows:

No of heads, X	0	1	2	3	4
Probability, $P(X)$	$\frac{1}{16}$	$\frac{4}{16}$	$\frac{6}{16}$	$\frac{4}{16}$	$\frac{1}{16}$

$$\mu = E(X) = \Sigma X \cdot P(X) = 0 \cdot \tfrac{1}{16} + 1 \cdot \tfrac{4}{16} + 2 \cdot \tfrac{6}{16} + 3 \cdot \tfrac{4}{16} + 4 \cdot \tfrac{1}{16} = \tfrac{32}{16} = 2$$
$$\sigma^2 = \Sigma X^2 \cdot P(X) - \mu^2$$
$$= 0^2 \cdot \tfrac{1}{16} + 1^2 \cdot \tfrac{4}{16} + 2^2 \cdot \tfrac{6}{16} + 3^2 \cdot \tfrac{4}{16} + 4^2 \cdot \tfrac{1}{16} - 2^2 = \tfrac{80}{16} - 4 = 1$$
$$\sigma = \sqrt{1} = 1$$

Hence, the simplified binomial formulas give the same results.

Example 6–22

A die is rolled 240 times. Find the mean, variance, and standard deviation for the number of threes that will be rolled.

Solution

$n = 240$ and $p = \frac{1}{6}$, since this is a binomial situation where getting a 3 is a success and *not* getting a 3 is a failure.

$$\mu = n \cdot p = 240 \cdot (\tfrac{1}{6}) = 40$$

$$\sigma^2 = n \cdot p \cdot q = 240 \, (\tfrac{1}{6}) \, (\tfrac{5}{6}) = 33.3$$

$$\sigma = \sqrt{n \cdot p \cdot q} = \sqrt{33.3} = 5.8$$

On average, there will be 40 threes. The standard deviation is 5.8.

Example 6–23

The *Statistical Bulletin* published by Metropolitan Life Insurance Co. reported that 2% of all American births result in twins. If a random sample of 8000 births is taken, find the mean, variance, and standard deviation of the number of births that would result in twins.

Source: *100% American* by Daniel Evan Weiss (New York: Poseidon Press, 1988).

Solution

This is a binomial situation, since a birth can result in either twins or not twins (i.e., two outcomes).

$$\mu = n \cdot p = (8000)(0.02) = 160$$
$$\sigma^2 = n \cdot p \cdot q = (8000)(0.02)(0.98) = 156.8$$
$$\sigma = \sqrt{n \cdot p \cdot q} = \sqrt{156.8} = 12.5$$

For the sample, the average number of births that would result in twins is 160, the variance is 156.8, or 157, and the standard deviation is 12.5, or 13 if rounded.

Exercises

6–63. Which of the following are binomial experiments or can be reduced to binomial experiments?
a. Surveying 100 people to determine if they like Sudsy Soap.
b. Tossing a coin 100 times to see how many heads occur.
c. Drawing a card from a deck and getting a heart.
d. Asking 1000 people which brand of cigarettes they smoke.
e. Testing four different brands of aspirin to see which brands are effective.
f. Testing one brand of aspirin using 10 people to determine whether it is effective.
g. Asking 100 people if they smoke.

h. Checking 1000 applicants to see whether they were admitted to White Oak College.
i. Surveying 300 prisoners to see how many different crimes they were convicted of.
j. Surveying 300 prisoners to see whether this is their first offense.

6–64. (ans) Compute the probability of X successes, using Table B in Appendix C.
a. $n = 2, p = 0.30, X = 1$
b. $n = 4, p = 0.60, X = 3$
c. $n = 5, p = 0.10, X = 0$
d. $n = 10, p = 0.40, X = 4$
e. $n = 12, p = 0.90, X = 2$
f. $n = 15, p = 0.80, X = 12$
g. $n = 17, p = 0.05, X = 0$

h. $n = 20, p = 0.50, X = 10$
i. $n = 16, p = 0.20, X = 3$

6–65. Compute the probability of X successes, using the binomial formula.
a. $n = 6, X = 3, p = 0.03$
b. $n = 4, X = 2, p = 0.18$
c. $n = 5, X = 3, p = 0.63$
d. $n = 9, X = 0, p = 0.42$
e. $n = 10, X = 5, p = 0.37$

For Exercises 6–66 through 6–75, assume all variables are binomial. (*Note:* **If values are not found in Table B (Appendix C), use the binomial formula.**)

6–66. A burglar alarm system has six fail-safe components. The probability of each failing is 0.05. Find these probabilities.
a. Exactly three will fail.
b. Fewer than two will fail.
c. None will fail.
d. Compare the answers for parts *a, b,* and *c,* and explain why the results are reasonable.

6–67. A student takes a 10-question, true–false exam and guesses on each question. Find the probability of passing if the lowest passing grade is 6 correct out of 10.

6–68. If the quiz in Exercise 6–67 was a multiple-choice quiz with five choices for each question, find the probability of guessing at least 6 correct out of 10 correctly.

6–69. In a survey, 30% of the people interviewed said that they bought most of their books during the last three months of the year (October, November, December). If nine people are selected at random, find the probability that exactly three of these people buy most of their books during October, November, and December.
Source: USA Snapshot, *USA Today,* May 24, 1995.

6–70. In a Gallup Survey, 90% of the people interviewed were unaware that maintaining a healthy weight could reduce the risk of stroke. If 15 people are selected at random, find the probability that at least 9 are unaware that maintaining a proper weight could reduce the risk of stroke.
Source: USA Snapshot, *USA Today,* June 24, 1995.

6–71. In a survey, three of four students said the courts show "too much concern" for criminals. Find the probability that at most three out of seven randomly selected students will agree with this statement.
Source: *Harper's Index* 290, no. 1739 (April 1995), p. 13.

6–72. It was found that 60% of American victims of health care fraud are senior citizens. If 10 victims are randomly selected, find the probability that exactly 3 are senior citizens.
Source: *100% American* by Daniel Evan Weiss (New York: Poseidon Press, 1988).

6–73. R. H. Bruskin Associates Market Research found that 40% of Americans do not think having a college education is important to succeed in the business world. If a random sample of five Americans is selected, find these probabilities.
a. Exactly two people will agree with that statement.
b. At most three people will agree with that statement.
c. At least two people will agree with that statement.
d. Fewer than three people will agree with that statement.
Source: *100% American* by Daniel Evans Weiss (New York: Poseidon Press, 1988).

6–74. Find these probabilities for a sample of nine children if 60% of them had German measles by the time they were 12 years old.
a. At least five have had German measles.
b. Exactly seven have had German measles.
c. More than three have had German measles.

6–75. If 20% of the people in a community use the emergency room at a hospital in one year, find these probabilities for a sample of 10 people.
a. At most three used the emergency room.
b. Exactly three used the emergency room.
c. At least five used the emergency room.

6–76. (**ans**) Find the mean, variance, and standard deviation for each of the values of n and p when the conditions for the binomial distribution are met.
a. $n = 100, p = 0.75$
b. $n = 300, p = 0.3$
c. $n = 20, p = 0.5$
d. $n = 10, p = 0.8$
e. $n = 1000, p = 0.1$
f. $n = 500, p = 0.25$
g. $n = 50, p = \frac{2}{5}$
h. $n = 36, p = \frac{1}{6}$

6–77. A study found that 1% of Social Security recipients are too young to vote. If 800 Social Security recipients are randomly selected, find the mean, variance, and standard deviation of the number of recipients who are too young to vote.
Source: *Harper's Index* 289, no. 1737 (February 1995), p. 11.

6–78. Find the mean, variance, and standard deviation for the number of heads when 10 coins are tossed.

6–79. If 2% of automobile carburetors are defective, find the mean, variance, and standard deviation of a lot of 500 carburetors.

6–80. It has been reported that 83% of federal government employees use e-mail. If a sample of 200 federal government employees is selected, find the mean, variance, and standard deviation of the number who use e-mail.
Source: *USA Today,* June 14, 1995.

6–81. A survey found that 21% of Americans watch fireworks on television on July 4. Find the mean, variance, and standard deviation of the number of individuals who watch fireworks on television if a random sample of 1000 Americans is selected.
Source: USA Snapshot, *USA Today,* July 3–4, 1995.

6–82. In a restaurant, a study found that 42% of all patrons smoked. If the seating capacity of the restaurant is 80 people, find the mean, variance, and standard deviation of the number of smokers.

6–83. A survey found that 25% of pet owners had their pets bathed professionally rather than doing it themselves. If 18 pet owners are randomly selected, find the probability that exactly five people have their pets bathed professionally.
Source: USA Snapshot, *USA Today,* June 15, 1995.

6–84. In a survey, 63% of Americans said they own an answering machine. If 14 Americans are selected at random, find the probability that exactly 9 own an answering machine.
Source: USA Snapshot, *USA Today,* May 22, 1995.

6–85. One out of every three Americans believes that the U.S. government should take "primary responsibility" for eliminating poverty in the United States. If 10 Americans are selected, find the probability that at most 3 will believe that the U.S. government should take primary responsibility for eliminating poverty.
Source: *Harper's Index* 289, no. 1735 (December 1994), p. 13.

6–86. In a survey, 58% of American adults said they had never heard of the Internet. If 20 American adults are selected at random, find the probability that exactly 12 will say they have never heard of the Internet.
Source: *Harper's Index* 289, no. 1737 (February 1995), p. 11.

6–87. In the past year, 13% of businesses have eliminated jobs. If five businesses are selected at random, find the probability that at least three have eliminated jobs during the last year.
Source: *USA Today,* June 15, 1995.

6–88. Of graduating high school seniors, 14% said that their generation will be remembered for their social concerns. If seven graduating seniors are selected at random, find the probability that either two or three will agree with that statement.
Source: *USA Today,* June 8, 1995.

6–89. A survey found that 86% of Americans have never been a victim of violent crime. If a sample of 12 Americans is selected at random, find the probability that 10 or more have never been victims of violent crime.
Source: *Harper's Index* 290, no 1741 (June 1995), p. 11.

*6–90. The graph shown here represents the probability distribution for the number of girls in a family of three children. From this graph, construct a probability distribution.

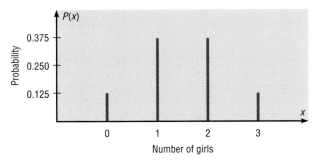

*6–91. Construct a binomial distribution graph for the number of defective computer chips in a lot of five if $p = 0.2$.

Summary

Many variables have special probability distributions. This chapter presented one of the most common probability distributions, the binomial distribution.

The binomial distribution is used when there are only two outcomes for an experiment, there is a fixed number of trials, the probability is the same for each trial, and the outcomes are independent of each other.

A probability distribution can be graphed, and the mean, variance, and standard deviation can be found. The mathematical expectation can also be calculated for a probability distribution. Expectation is used in insurance and games of chance.

Important Terms

binomial distribution 221

binomial experiment 221

expected value 216

probability distribution 208

random variable 207

Important Formulas

Formula for the mean of a probability distribution:

$$\mu = \Sigma\, X \cdot P(X)$$

Formula for the variance of a probability distribution:

$$\sigma^2 = \Sigma\, [X^2 \cdot P(X)] - \mu^2$$

Formula for expected value:

$$E(X) = \Sigma\, X \cdot P(X)$$

Binomial probability formula:

$$P(X) = \frac{n!}{(n-X)!X!} \cdot p^X \cdot q^{n-X}$$

Formula for the mean of the binomial distribution:

$$\mu = n \cdot p$$

Formulas for the variance and standard deviation of the binomial distribution:

$$\sigma^2 = n \cdot p \cdot q \qquad \sigma = \sqrt{n \cdot p \cdot q}$$

Review Exercises

For Exercises 6–92 through 6–95, determine whether the distribution represents a probability distribution. If it does not, state why.

6–92.

X	1	2	3	4	5
$P(X)$	$\frac{3}{10}$	$\frac{1}{10}$	$\frac{2}{10}$	$\frac{3}{10}$	$\frac{1}{10}$

6–93.

X	5	10	15	20
$P(X)$	0.5	0.3	0.1	0.4

6–94.

X	100	200	300
$P(X)$	0.3	0.3	0.1

6–95.

X	8	12	16	20
$P(X)$	$\frac{5}{6}$	$\frac{1}{12}$	$\frac{1}{12}$	$\frac{1}{12}$

6–96. The number of rescue calls a helicopter ambulance service receives per 24-hour period is distributed as shown here. Construct a graph for the data.

Number of calls, X	6	7	8	9	10
Probability, $P(X)$	0.12	0.31	0.40	0.15	0.02

6–97. A study was conducted to determine the number of radios each household has. The data are shown here. Construct a probability distribution and draw a graph for the data.

Number of radios	Probability
0	0.05
1	0.30
2	0.45
3	0.12
4	0.08

6–98. A box contains five pennies, three dimes, one quarter, and one half-dollar. Construct a probability distribution and draw a graph for the data.

6–99. At Tyler's Tie Shop, Tyler found the probabilities that a customer will buy 0, 1, 2, 3, or 4 ties, as shown. Construct a graph for the distribution.

Number of ties, X	0	1	2	3	4
Probability, P(X)	0.30	0.50	0.10	0.08	0.02

6–100. A bank has a drive-through service. The number of customers arriving during a 15-minute period is distributed as shown. Find the mean, variance, and standard deviation for the distribution.

Number of customers, X	0	1	2	3	4
Probability, P(X)	0.12	0.20	0.31	0.25	0.12

6–101. At a small community library, the number of visitors per hour during the day has the distribution shown. Find the mean, variance, and standard deviation for the data.

Number of visitors, X	8	9	10	11	12
Probability, P(X)	0.15	0.25	0.29	0.19	0.12

6–102. During a recent paint sale at Corner Hardware, the number of cans of paint purchased was distributed as shown. Find the mean, variance, and standard deviation of the distribution.

Number of cans, X	1	2	3	4	5
Probability, P(X)	0.42	0.27	0.15	0.10	0.06

6–103. The number of inquiries received per day for a college catalog is distributed as shown. Find the mean, variance, and standard deviation for the data.

Number of inquiries, X	22	23	24	25	26	27
Probability, P(X)	0.08	0.19	0.36	0.25	0.07	0.05

6–104. There are five envelopes in a box. One envelope contains a penny, one a nickel, one a dime, one a quarter, and one a half-dollar. A person selects an envelope. Find the expected value of the draw.

6–105. A person selects a card from a deck. If it is a red card, he wins $1. If it is a black card between or including 2 and 10, he wins $5. If it is a black face card, he wins $10, and if it is a black ace, he wins $100. Find the expectation of the game.

6–106. If 30% of all commuters ride the train to work, find the probability that if 10 workers are selected, 5 will ride the train.

6–107. If 90% of all people between the ages of 30 and 50 drive a car, find these probabilities for a sample of 20 people in that age group.
a. Exactly 20 drive a car.
b. At least 15 drive a car.
c. At most 15 drive a car.

6–108. If 10% of the people who are given a certain drug experience dizziness, find these probabilities for a sample of 15 people who take the drug.
a. At least two people will become dizzy.
b. Exactly three people will become dizzy.
c. At most four people will become dizzy.

6–109. If 75% of nursing students are able to pass a drug calculation test, find the mean, variance, and standard deviation of the number of students who pass the test in a sample of 180 nursing students.

6–110. A club has 50 members. If there is a 10% absentee rate per meeting, find the mean, variance, and standard deviation of the number of people who will be absent from each meeting.

6–111. The chance that a U.S. police chief believes the death penalty "significantly reduces the number of homicides" is one in four. If a random sample of eight police chiefs is selected, find the probability that at most three believe that the death penalty significantly reduces the number of homicides.
Source: *Harper's Index* 290, no. 1741 (June 1995), p. 11.

6–112. *American Energy Review* reported that 27% of American households burn wood. If a random sample of 500 American households is selected, find the mean, variance, and standard deviation of the number of households that burn wood.
Source: *100% American* by Daniel Evan Weiss (New York: Poseidon Press, 1988).

6–113. Three out of four American adults under 35 have eaten pizza for breakfast. If a random sample of 20 adults under 35 is selected, find the probability that exactly 16 have eaten pizza for breakfast.
Source: *Harper's Index* 290, no. 1738 (March 1995), p. 9.

6–114. One out of four Americans over 55 has eaten pizza for breakfast. If a sample of 10 Americans over 55 is selected at random, find the probability that at most three have eaten pizza for breakfast.
Source: *Harper's Index* 290, no. 1738 (March 1995), p. 9.

Statistics Today

Is Pooling Worthwhile?   Revisited

In the case of the pooled sample, the probability that only one test will be needed can be determined by using the binomial distribution. The question being asked is, "In a sample of 15 individuals, what is the probability that no individual will have the disease?" Hence, $n = 15$, $p = 0.05$, and $X = 0$. From Table B in Appendix C, the probability is 0.463, or 46% of the time, only one test will be needed. For screening purposes, then, pooling samples in this case would save considerable time, money, and effort as opposed to testing every individual in the population.

Quiz

Determine whether each statement is true or false. If the statement is false, explain why.

1. The expected value of a random variable can be thought of as a long-run average.

2. The number of courses a student is taking this semester is an example of a continuous random variable.

3. When the binomial distribution is used, the outcomes must be dependent.

4. The binomial distribution can be used to represent discrete random variables.

Complete the following statements with the best answer.

5. Random variable values are determined by _____.

6. The mean for a binomial variable can be found by using the formula _____.

7. One requirement for a probability distribution is that the sum of all the events in the sample space must equal _____.

Select the best answer.

8. What is the sum of the probabilities of all outcomes in a probability distribution?
a. 0
b. 1/2
c. 1
d. It cannot be determined.

9. How many outcomes are there in a binomial experiment?
a. 0
b. 1
c. 2
d. It varies.

For questions 10 through 13, determine if the distribution represents a probability distribution. If not, state why.

10.

X	1	2	3	4	5
$P(X)$	$\frac{1}{7}$	$\frac{2}{7}$	$\frac{2}{7}$	$\frac{3}{7}$	$\frac{2}{7}$

11.

X	3	6	9	12	15
$P(X)$	0.3	0.5	0.1	0.08	0.02

12.

X	50	75	100
$P(X)$	0.5	0.2	0.3

13.

X	4	8	12	16
$P(X)$	$\frac{1}{6}$	$\frac{3}{12}$	$\frac{1}{2}$	$\frac{1}{12}$

14. The number of fire calls the Conestoga Valley Fire Company receives per day is distributed as follows:

Number X	5	6	7	8	9
Probability $P(X)$	0.28	0.32	0.09	0.21	0.10

Construct a graph for the data.

15. A study was conducted to determine the number of telephones each household has. The data are shown here.

Number of telephones	Frequency
0	2
1	30
2	48
3	13
4	7

Construct a probability distribution and draw a graph for the data.

16. During a recent cassette sale at Matt's Music Store, the number of tapes customers purchased was distributed as follows:

Number X	0	1	2	3	4
Probability $P(X)$	0.10	0.23	0.31	0.27	0.09

Find the mean, variance, and standard deviation of the distribution.

17. The number of calls received per day at a crisis hot line is distributed as follows:

Number X	30	31	32	33	34
Probability $P(X)$	0.05	0.21	0.38	0.25	0.11

Find the mean, variance, and standard deviation of the distribution.

18. There are six playing cards placed face down in a box. They are the 4 of diamonds, the 5 of hearts, the 2 of clubs, the 10 of spades, the 3 of diamonds, and the 7 of hearts. A person selects a card. Find the expected value of the draw.

19. A person selects a card from an ordinary deck of cards. If it is a black card, she wins $2. If it is a red card between or including 3 and 7, she wins $10. If it is a red face card, she wins $25, and if it is a black jack, she wins $100. Find the expectation of the game.

20. If 40% of all commuters ride to work in carpools, find the probability that if eight workers are selected, five will ride in carpools.

21. If 60% of all women are employed outside the home, find the probability that in a sample of 20 women,
a. Exactly 15 are employed.
b. At least 10 are employed.
c. At most five are not employed outside the home.

22. If 80% of the applicants are able to pass a driver's proficiency road test, find the mean, variance, and standard deviation of the number of people who pass the test in a sample of 300 applicants.

23. A history class has 75 members. If there is a 12% absentee rate per class meeting, find the mean, variance, and standard deviation of the number of students who will be absent from each class.

Critical Thinking Challenges

1. Simulation Simulation techniques can be used to compute the probability of an event when mathematical calculations are difficult. Simulation uses random numbers to create conditions similar to real life problems. Simulation techniques go back to ancient times when chess was invented to simulate warfare. Modern techniques date to the mid-1940's when two physicists, John Von Neuman and Stanislaw Ulam, developed simulation techniques to study the behavior of neutrons in the design of atomic reactors. The following example illustrates how simulation is used.

Suppose that there are six different toys packaged in cereal boxes, one toy per box. If you wanted to collect the entire set, on average, how many boxes would you have to buy?

Since the probability of getting a specific toy is ⅙, a die can be rolled until one gets all the numbers, one through six, keeping track of the number of rolls it takes. The experiment is then repeated a specific number of times, say 10 or 20. (The more times the better.) Finally, the average of the number of rolls is found. This will give an approximate number. Try it and see how many boxes, on average, need to be purchased.

2. Simulation Try this one. To win a certain lottery, a person must spell the word BIG. One-half of the tickets sold contain the letter B, ⅓ of the tickets sold contain the letter I, and ⅙ of the tickets sold contain the letter G. Using a die, simulate the experiment and determine the average number of tickets that need to be purchased to win.

For more information on simulation, see "Speaking of Statistics" on page 234.

Data Project

Probability Distributions Roll three dice 100 times, recording the sum of the spots on the faces as you roll. Then find the average of the spots. How close is this to the theoretical average?

TI-83 Calculator

The Binomial Distribution

A. To find the probability for a binomial variable:

Example: $n = 5$ $x = 3$ $p = 0.2$

1. Press **2nd [DISTR]** then 0 to get `binompdf(`
2. Enter 5, .2, {3}) then press **ENTER**.

The calculator will display {.0512}.

B. To find the probability for several values of a binomial variable:

Example: $n = 5$ $x = 1, 2, 3,$ $p = 0.2$

1. Press **2nd [DISTR] 0** to get `binompdf(`
2. Enter 5, .2, {1, 2, 3}) then press **ENTER**.

The calculator will display { .4096   .2048   .0512}
Use the ◄ and ► keys to scroll the line from left to right.

Games such as Monopoly use simulation to create the real life situations of buying, owning, and managing real estate. The game of Risk simulates global warfare. Simulation can also be used to play sports games. To see how this is accomplished, play the simulated bowling game shown here.

For more examples of simulation, see the "Critical Thinking Challenge" section on page 232.

Simulated Bowling Game

Let's use the random digit table to simulate a bowling game. Our game is much simpler than commercial simulation games.

First Ball		Second Ball					
		2-Pin Split			No split		
Digit	Results	Digit	Results		Digit	Results	
1–3	Strike	1	Spare		1–3	Spare	
4–5	2-pin split	2–8	Leave one pin		4–6	Leave 1 pin	
6–7	9 pins down	9–0	Miss both pins		7–8	*Leave 2 pins	
8	8 pins down				9	+Leave 3 pins	
9	7 pins down				0	Leave all pins	
0	6 pins down						

*If there are fewer than 2 pins, result is a spare.
+If there are fewer than 3 pins, those pins are left.

Here's how to score bowling:
1. There are 10 frames to a **game** or **line**.
2. You roll two balls for each game, unless you knock all the pins down with the first ball (a **strike**).
3. Your score for a frame is the sum of the pins knocked down by the two balls, if you don't knock down all 10.
4. If you knock all 10 pins down with two balls (a **spare**, shown as ▱), your score is 10 pins plus the number knocked down with the next ball.
5. If you knock all 10 pins down with the first ball (a **strike**, shown as ⊠), your score is 10 pins plus the number knocked down by the next **two** balls.
6. A **split** (shown as 0) is when there is a big space between the remaining pins. Place in the circle the number of pins remaining after the second ball.
7. A **miss** is shown as —.

Here is how one person simulated a bowling game using the random digits 7 2 7 4 8 2 2 3 6 1 6 0 4 6 1 5 5, chosen in that order from the table.

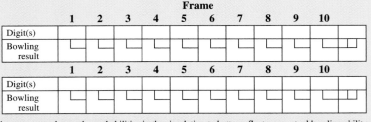

Frame

	1	2	3	4	5	6	7	8	9	10
Digit(s)	7/2	7/4	8/2	2	3	6/1	6/0	4/6	1	5/5
Bowling result	9▱ 19	9— 28	8▱ 48	⊠ 77	⊠ 97	9▱ 116	9— 125	8① 134	⊠ 153	8① 162	162

Now you try several.

Frame

	1	2	3	4	5	6	7	8	9	10
Digit(s)										
Bowling result										

	1	2	3	4	5	6	7	8	9	10
Digit(s)										
Bowling result										

If you wish to, you can change the probabilities in the simulation to better reflect *your* actual bowling ability.

Source: Albert Shulte, "Simulated Bowling Game," Student Math Notes, March 1986. Published by the National Council of Teachers of Mathematics. Reprinted with permission.

c h a p t e r

7

The Normal Distribution

Objectives

After completing this chapter, you should be able to

1. Identify distributions as symmetrical or skewed.

2. Identify the properties of the normal distribution.

3. Find the area under the standard normal distribution, given various z values.

4. Find probabilities for a normally distributed variable by transforming it into a standard normal variable.

5. Find specific data values for given percentages using the standard normal distribution.

6. Use the central limit theorem to solve problems involving sample means.

7. Use the normal approximation to compute probabilities for a binomial variable.

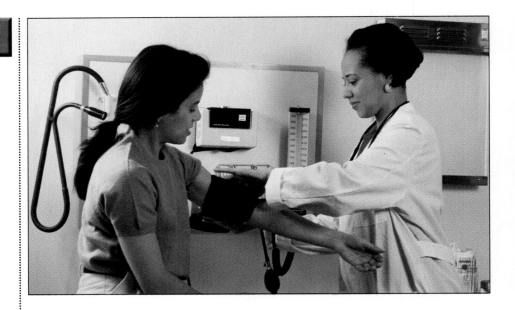

What Is Normal?

Medical researchers have determined so-called normal intervals for a person's blood pressure, cholesterol, triglycerides, and the like. For example, the normal range of systolic blood pressure is 110 to 140. The normal interval for a person's triglycerides is from 30 to 200 milligrams per deciliter (mg/dl). By measuring these variables, a physician can determine if a patient's vital statistics are within the normal interval, or if some type of treatment is needed to correct a condition and avoid future illnesses. The question then is, "How does one determine the so-called normal intervals?"

In this chapter, you will learn how researchers determine normal intervals for specific medical tests using the normal distribution. You will see how the same methods are used to determine the lifetimes of batteries, the strength of ropes, and many other traits.

7–1

Introduction

Random variables can be either discrete or continuous. Discrete variables and their distributions were explained in Chapter 6. Recall that a discrete variable cannot assume all values between any two given values of the variables. On the other hand, a continuous variable can assume all values between any two given values of the variables. Examples of continuous variables are the heights of adult men, body temperatures of rats, and cholesterol levels of adults. Many continuous variables, such as the examples just mentioned, have distributions that are bell-shaped and are called *approximately normally distributed variables*. For example, if a researcher selects a random sample of 100 adult women, measures their heights, and constructs a histogram, the researcher gets a graph similar to the one shown in Figure 7–1(a). Now, if the researcher increases the sample size and decreases the width of the classes, the histograms will look like the ones shown in Figure 7–1(b)

Figure 7–1

Histograms for the
Distribution of Heights
of Adult Women

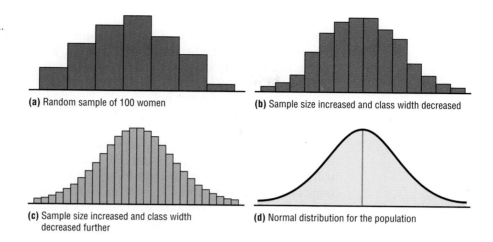

(a) Random sample of 100 women

(b) Sample size increased and class width decreased

(c) Sample size increased and class width decreased further

(d) Normal distribution for the population

and 7–1(c). Finally, if it were possible to measure exactly the heights of all adult females in the United States and plot them, the histogram would approach what is called the *normal distribution,* shown in Figure 7–1(d). This distribution is also known as the *bell curve* or the *Gaussian distribution,* named for the German mathematician Carl Friedrich Gauss (1777–1855), who derived its equation.

No variable fits the normal distribution perfectly, since the normal distribution is a theoretical distribution. However, the normal distribution can be used to describe many variables, because the deviations from the normal distribution are very small. This concept will be explained further in the next section.

Objective 1. Identify distributions as symmetrical or skewed.

When the data values are evenly distributed about the mean, the distribution is said to be **symmetrical.** Figure 7–2(a) shows a symmetrical distribution. When the majority of the data values fall to the left or right of the mean, the distribution is said to be *skewed.* When the majority of the data values fall to the right of the mean, the distribution is said to be **negatively skewed.** The mean is to the left of

Figure 7–2

Normal and Skewed Distributions

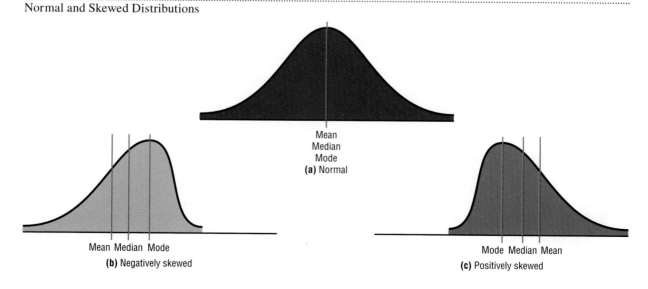

Mean
Median
Mode
(a) Normal

Mean Median Mode
(b) Negatively skewed

Mode Median Mean
(c) Positively skewed

the median, and the mean and the median are to the left of the mode. See Figure 7–2(b). When the majority of the data values fall to the left of the mean, the distribution is said to be **positively skewed.** The mean falls to the right of the median and both the mean and the median fall to the right of the mode. See Figure 7–2(c).

The "tail" of the curve indicates the direction of skewness (right is positive, left negative). This distribution can be compared with the one in Figure 3–2. Both types follow the same principles.

This chapter will present the properties of the normal distribution and discuss its applications. Then a very important theorem called the *central limit theorem* will be explained. Finally, the chapter will explain how the normal curve distribution can be used as an approximation to other distributions, such as the binomial distribution. Since the binomial distribution is a discrete distribution, a correction for continuity may be employed when the normal distribution is used for its approximation.

7–2

Properties of the Normal Distribution

Objective 2. Identify the properties of the normal distribution.

In mathematics, curves can be represented by equations. For example, the equation of the circle shown in Figure 7–3 is $x^2 + y^2 = r^2$, where r is the radius. The circle can be used to represent many physical objects, such as a wheel or a gear. Even though it is not possible to manufacture a wheel that is perfectly round, the equation and the properties of the circle can be used to study the many aspects of the wheel, such as area, velocity, and acceleration. In a similar manner, the theoretical curve, called the *normal distribution curve,* can be used to study many variables that are not perfectly normally distributed but are nevertheless approximately normal.

The mathematical equation for the normal distribution is

$$y = \frac{e^{-(X-\mu)^2/2\sigma^2}}{\sigma\sqrt{2\pi}}$$

where

$e \approx 2.718$ ($\approx$ means "is approximately equal to")

$\pi \approx 3.14$

μ = population mean

σ = population standard deviation

Figure 7–3

Graph of a Circle and an Application

Circle

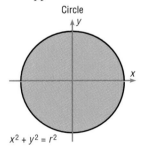

$x^2 + y^2 = r^2$

Wheel

This equation may look formidable, but in applied statistics, tables are used for specific problems instead of the equation.

Another important aspect in applied statistics is that *the area under the normal distribution curve is more important than the frequencies.* Therefore, when the normal distribution is pictured, the y axis, which indicates the frequencies, is sometimes omitted.

The shape and position of the normal distribution curve depend on two parameters, the mean and the standard deviation. Each normally distributed variable has its own normal distribution curve, which depends on the values of the variable's mean and standard deviation. Figure 7–4(a) shows two normal distributions with the same mean values but different standard deviations. The larger the standard deviation, the more dispersed, or spread out, the distribution is. Figure 7–4(b) shows two normal distributions with the same standard deviation but with different

Figure 7–4

Shapes of Normal Distributions

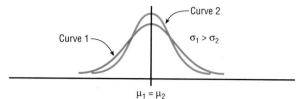

(a) Same means but different standard deviations

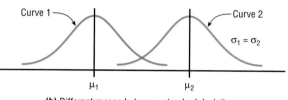

(b) Different means but same standard deviations

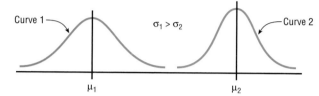

(c) Different means and different standard deviations

The discovery of the equation for the normal distribution can be traced to three mathematicians. In 1733, the French mathematician Abraham DeMoivre derived an equation for the normal distribution based on the random variation of the number of heads appearing when a large number of coins were tossed. Not realizing any connection with the naturally occurring variables, he showed this formula to only a few friends. About 100 years later, two mathematicians, Pierre Laplace in France and Carl Gauss in Germany, derived the equation of the normal curve independently and without any knowledge of DeMoivre's work. In 1924, Karl Pearson found that DeMoivre had discovered the formula before Laplace or Gauss.

means. These curves have the same shapes but are located at different positions on the x axis. Figure 7–4(c) shows two normal distributions with different means and different standard deviations.

The **normal distribution** is a continuous, symmetric, bell-shaped distribution of a variable.

The properties of the normal distribution, including those mentioned in the definition, are explained next.

Summary of the Properties of the Theoretical Normal Distribution

1. The normal distribution curve is bell-shaped.
2. The mean, median, and mode are equal and located at the center of the distribution.
3. The normal distribution curve is unimodal (i.e., it has only one mode).
4. The curve is symmetrical about the mean, which is equivalent to saying that its shape is the same on both sides of a vertical line passing through the center.
5. The curve is continuous—i.e., there are no gaps or holes. For each value of X, there is a corresponding value of Y.
6. The curve never touches the x axis. Theoretically, no matter how far in either direction the curve extends, it never meets the x axis—but it gets increasingly closer.
7. The total area under the normal distribution curve is equal to 1.00, or 100%. This fact may seem unusual, since the curve never touches the x axis, but one can prove it mathematically by using calculus. (The proof is beyond the scope of this textbook.)
8. The area under the normal curve that lies within one standard deviation of the mean is approximately 0.68, or 68%; within two standard deviations, about 0.95, or 95%; and within three standard deviations, about 0.997, or 99.7%. See Figure 7–5, which also shows the area in each region.

Figure 7–5

Areas under the Normal
Distribution Curve

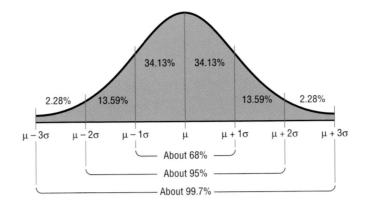

These values follow the *empirical rule* for data given in Section 3–3.

One must know these properties in order to solve problems using applications involving distributions that are approximately normal.

7–3

The Standard Normal Distribution

Objective 3. Find the area under the standard normal distribution, given various *z* values.

Since each normally distributed variable has its own mean and standard deviation, as stated earlier, the shape and location of these curves will vary. In practical applications, then, one would have to have a table of areas under the curve for each variable. To simplify this situation, statisticians use what is called the *standard normal distribution*.

The **standard normal distribution** is a normal distribution with a mean of 0 and a standard deviation of 1.

The standard normal distribution is shown in Figure 7–6.

Figure 7–6

Standard Normal
Distribution

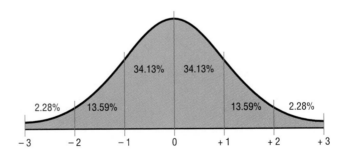

The values under the curve indicate the proportion of area in each section. For example, the area between the mean and one standard deviation above or below the mean is about 0.3413, or 34.13%.

The formula for the standard normal distribution is

$$y = \frac{e^{-\frac{X^2}{2}}}{\sqrt{2\pi}}$$

All normally distributed variables can be transformed into the standard normally distributed variable by using the formula for the standard score:

$$z = \frac{\text{value} - \text{mean}}{\text{standard deviation}} \quad \text{or} \quad z = \frac{X - \mu}{\sigma}$$

This is the same formula used in Section 3–4. The use of this formula will be explained in the next section.

As stated earlier, the area under the normal distribution curve is used to solve practical application problems, such as finding the percentage of adult women whose height is between 5 feet 4 inches and 5 feet 7 inches, or finding the probability that a new battery will last longer than four years. Hence, the major emphasis of this section will be to show the procedure for finding the area under the normal distribution curve for any z value. The applications will be shown in the next section. Once the X values are transformed using the preceding formula, they are called z values. The **z value** is actually the number of standard deviations that a particular X value is away from the mean. Table E in Appendix C gives the area (to four decimal places) under the standard normal curve for any z value from 0 to 3.09.

Finding Areas under the Normal Distribution Curve

For the solution of problems using the normal distribution, a four-step procedure is recommended with the use of Procedure Table 6.

STEP 1   Draw a picture.

STEP 2   Shade the area desired.

STEP 3   Find the correct figure in Procedure Table 6 (the figure that is similar to the one you've drawn).

STEP 4   Follow the directions given in the appropriate block of Procedure Table 6 to get the desired area.

There are seven basic types of problems and all seven are summarized in Procedure Table 6. Note that this table is presented as an aid in understanding how to use the normal distribution table and in visualizing the problems. After learning the procedures, one should *not* find it necessary to refer to the procedure table for every problem.

Procedure Table 6

Finding the Area under the Normal Distribution Curve

1. Between 0 and any z value:
   Look up the z value in the table to get the area.

2. In any tail:
   a. Look up the z value to get the area.
   b. Subtract the area from 0.5000.

(continued)

Procedure Table 6 (concluded)

Finding the Area under the Normal Distribution Curve

3. Between two z values on the same side of the mean:
   a. Look up both z values to get the areas.
   b. Subtract the smaller area from the larger area.

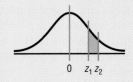 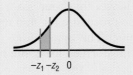

4. Between two z values on opposite sides of the mean:
   a. Look up both z values to get the areas.
   b. Add the areas.

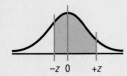

5. Less than any z value to the right of the mean:

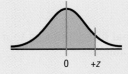

   a. Look up the z value to get the area.
   b. Add 0.5000 to the area

6. Greater than any z value to the left of the mean:

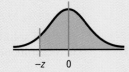

   a. Look up the z value in the table to get the area.
   b. Add 0.5000 to the area.

7. In any two tails:

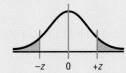

   a. Look up the z values in the table to get the areas.
   b. Subtract both areas from 0.5000.
   c. Add the answers.

Procedure

1. Draw the picture.
2. Shade the area desired.
3. Find the correct figure.
4. Follow the directions.
Note: Table E gives the area between 0 and any z value to the right of 0.

Situation 1 Finding the area under the normal curve between 0 and any z value.

Example 7–1

Find the area under the normal distribution curve between $z = 0$ and $z = 2.34$.

Solution

Draw the figure and represent the area as shown in Figure 7–7.

Figure 7–7
Area under the
Standard Normal Curve
for Example 7–1

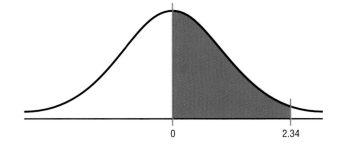

Since Table E gives the area between 0 and any z value to the right of 0, one need only look up the z value in the table. Find 2.3 in the left column and 0.04 in the top row. The value where the column and row meet in the table is the answer, 0.4904. See Figure 7–8. Hence, the area is 0.4904, or 49.04%.

Figure 7–8

Using Table E in the Appendix for Example 7–1

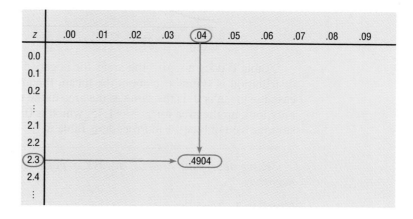

Example 7–2

Find the area between $z = 0$ and $z = 1.5$.

Solution

Draw the figure and represent the area as shown in Figure 7–9.

Figure 7–9

Area under the Standard Normal Curve for Example 7–2

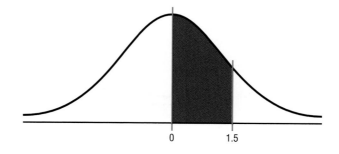

Find the area in Table E by finding 1.5 in the left column and 0.00 in the top row. The value is 0.4332, or 43.32%.

Next, one must be able to find the areas for values that are not in Table E. This is done by using the properties of the normal distribution described in Section 7–2.

Example 7–3

Find the area between $z = 0$ and $z = -1.75$.

Solution

Represent the area as shown in Figure 7–10.

Figure 7–10
....
Area under the
Standard Normal Curve
for Example 7–3

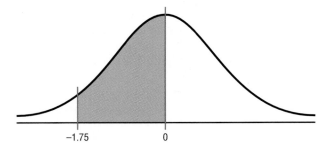

Table E does not give the areas for negative values of z. But since the normal distribution is symmetric about the mean, the area to the left of the mean (in this case the mean is 0) is the same as the area to the right of the mean. Hence one need only look up the area for $z = +1.75$, which is 0.4599, or 45.99%. This solution is summarized in block 1 in Procedure Table 6.

....

Remember that area is always a positive number, even if the z value is negative.

Situation 2   Find the area under the curve in either tail.

....

Example 7–4

Find the area to the right of $z = 1.11$.

Solution

Draw the figure and represent the area as shown in Figure 7–11.

Figure 7–11
....
Area under the
Standard Normal Curve
for Example 7–4

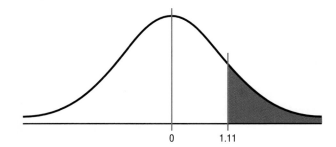

The required area is in the tail of the curve. Since Table E gives the area between $z = 0$ and $z = 1.11$, first find that area. Then subtract this value from 0.5000, since half of the area under the curve is to the right of $z = 0$. See Figure 7–12.

Figure 7–12
....
Finding the Area in the
Tail of the Curve
(Example 7–4)

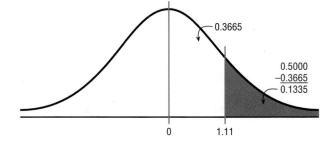

The area between $z = 0$ and $z = 1.11$ is 0.3665, and the area to the right of $z = 1.11$ is 0.1335, or 13.35%, obtained by subtracting 0.3665 from 0.5000.

| **Example 7–5** | Find the area to the left of $z = -1.93$. |

Solution

The desired area is shown in Figure 7–13.

Figure 7–13

Area under the
Standard Normal Curve
for Example 7–5

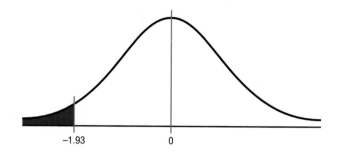

Again, Table E gives the area for positive z values. But from the symmetric property of the normal distribution, the area to the left of -1.93 is the same as the area to the right of $z = +1.93$, as shown in Figure 7–14.

Figure 7–14

Comparison of Areas to
the Right of $+1.93$ and
to the Left of -1.93
(Example 7–5)

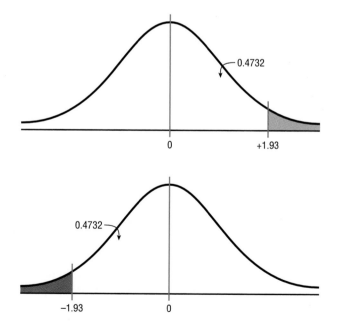

Now, one need only find the area between 0 and $+1.93$ and subtract it from 0.5000, as shown:

$$\begin{array}{r} 0.5000 \\ -0.4732 \\ \hline 0.0268, \text{ or } 2.68\% \end{array}$$

This procedure was summarized in block 2 of Procedure Table 6.

Situation 3   Find the area under the curve between any two z values on the same side of the mean.

| Example 7–6 | Find the area between $z = 2.00$ and $z = 2.47$. |

Solution

The desired area is shown in Figure 7–15.

Figure 7–15

Area under the Curve for Example 7–6

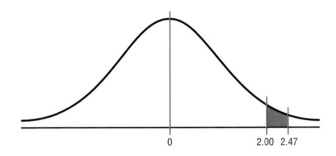

For this situation, look up the area from $z = 0$ to $z = 2.47$ and the area from $z = 0$ to $z = 2.00$. Then subtract the two areas, as shown in Figure 7–16.

Figure 7–16

Finding the Area under the Curve for Example 7–6

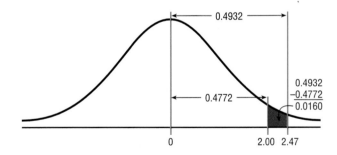

The area between $z = 0$ and $z = 2.47$ is 0.4932. The area between $z = 0$ and $z = 2.00$ is 0.4772. Hence, the desired area is $0.4932 - 0.4772 = 0.0160$, or 1.60%. This procedure is summarized in block 3 of Procedure Table 6.

Two things should be noted here. First, the *areas,* not the z values, are subtracted. Subtracting the z values will yield an incorrect answer. Second, the procedure in Example 7–6 is used when both z values are on the same side of the mean.

| Example 7–7 | Find the area between $z = -2.48$ and $z = -0.83$. |

Solution

The desired area is shown in Figure 7–17.

Figure 7–17

Area under the Curve
for Example 7–7

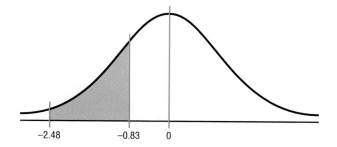

-2.48   -0.83   0

The area between $z = 0$ and $z = -2.48$ is 0.4934. The area between $z = 0$ and $z = -0.83$ is 0.2967. Subtracting yields $0.4934 - 0.2967 = 0.1967$, or 19.67%. This solution is summarized in block 3 of Procedure Table 6.

Situation 4   Find the area under the curve between any two z values on opposite sides of the mean.

Example 7–8   Find the area betwen $z = +1.68$ and $z = -1.37$.

Solution

The desired area is shown in Figure 7–18.

Figure 7–18

Area under the Curve
for Example 7–8

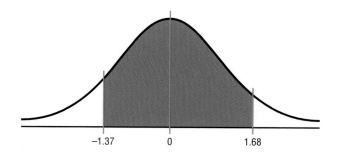

-1.37   0   1.68

Now, since the two areas are on opposite sides of $z = 0$, one must find both areas and add them. The area between $z = 0$ and $z = 1.68$ is 0.4535. The area between $z = 0$ and $z = -1.37$ is 0.4147. Hence, the total area between $z = -1.37$ and $z = +1.68$ is $0.4535 + 0.4147 = 0.8682$, or 86.82%.

This type of problem is summarized in block 4 of Procedure Table 6.

Situation 5   Find the area under the curve less than any z value to the right of the mean.

Example 7–9   Find the area to the left of $z = 1.99$.

Solution

The desired area is shown in Figure 7–19.

Figure 7–19

Area under the Curve for Example 7–9

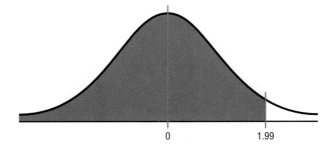

Since Table E gives only the area between $z = 0$ and $z = 1.99$, one must add 0.5000 to the table area, since 0.500 (half) of the total area lies to the left of $z = 0$. The area between $z = 0$ and $z = 1.99$ is 0.4767, and the total area is 0.4767 + 0.5000 = 0.9767, or 97.67%.

This solution is summarized in block 5 of Procedure Table 6.

The same procedure is used when the z value is to the left of the mean, as shown in the next example.

Situation 6   Find the area under the curve greater than any z value to the left of the mean.

Example 7–10

Find the area to the right of $z = -1.16$.

Solution

The desired area is shown in Figure 7–20.

Figure 7–20

Area under the Curve for Example 7–10

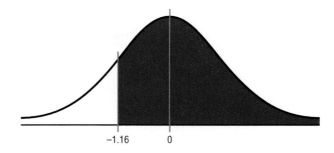

The area between $z = 0$ and $z = -1.16$ is 0.3770. Hence, the total area is 0.3770 + 0.5000 = 0.8770, or 87.70%.

This type of problem is summarized in block 6 of Procedure Table 6.

The final type of problem is that of finding the area in two tails. To solve it, find the area in each tail and add them, as shown in the next example.

Situation 7   Find the total area under the curve in any two tails.

Example 7–11

Find the area to the right of $z = +2.43$ and to the left of $z = -3.01$.

Solution

The desired area is shown in Figure 7–21.

Figure 7–21

Area under the Curve
for Example 7–11

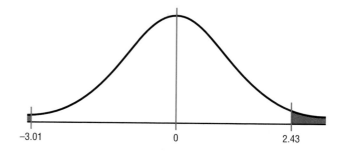

-3.01   0   2.43

The area to the right of 2.43 is $0.5000 - 0.4925 = 0.0075$. The area to the left of $z = -3.01$ is $0.5000 - 0.4987 = 0.0013$. The total area, then, is $0.0075 + 0.0013 = 0.0088$, or 0.88%.

This solution is summarized in block 7 of Procedure Table 6.

The Normal Distribution Curve as a Probability Distribution Curve

The normal distribution curve can be used as a probability distribution curve for normally distributed variables. Recall that the normal distribution is a *continuous distribution,* as opposed to a discrete probability distribution, as explained in Chapter 6. The fact that it is continuous means that there are no gaps in the curve. In other words, for every z value on the x axis, there is a corresponding height, or frequency value.

However, as stated earlier, the area under the curve is more important than the frequencies. *This area corresponds to a probability.* That is, if it were possible to select any z value at random, the probability of choosing one, say, between 0 and 2.00 would be the same as the area under the curve between 0 and 2.00. In this case, the area is 0.4772. Therefore, the probability of selecting any z value between 0 and 2.00 is 0.4772. The problems involving probability are solved in the same manner as the previous examples involving areas in this section. For example, if the problem is to find the probability of selecting a z value between 2.25 and 2.94, solve it by using the method shown in block 3 of Procedure Table 6.

For probabilities, a special notation is used. For example, if the problem is to find the probability of any z value between 0 and 2.32, this probability is written as $P(0 < z < 2.32)$.

Example 7–12

Find the probability for each.

a. $P(0 < z < 2.32)$
b. $P(z < 1.65)$
c. $P(z > 1.91)$

Solution

a. $P(0 < z < 2.32)$ means to find the area under the normal distribution curve between 0 and 2.32. Look up the area in Table E corresponding to $z = 2.32$. It is 0.4898, or 48.98%. The area is shown in Figure 7–22.

Figure 7–22

Area under the Curve for Part *a* of Example 7–12

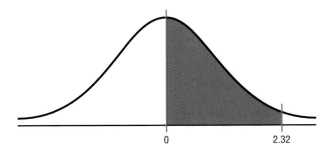

b. $P(z < 1.65)$ is represented in Figure 7–23.

Figure 7–23

Area under the Curve for Part *b* of Example 7–12

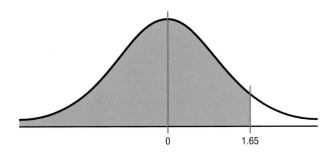

First, find the area between 0 and 1.65 in Table E. Then add it to 0.5000 to get 0.4505 + 0.5000 = 0.9505, or 95.05%.

c. $P(z > 1.91)$ is shown in Figure 7–24.

Figure 7–24

Area under the Curve for Part *c* of Example 7–12

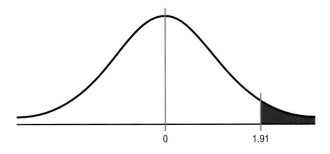

Since this area is a tail area, find the area between 0 and 1.91 and subtract it from 0.5000. Hence, 0.5000 − 0.4719 = 0.0281.

Sometimes, one must find a specific z value for a given area under the normal distribution. The procedure is to work backward, using Table E.

Example 7–13	Find the z value such that the area under the normal distribution curve between 0 and the z value is 0.2123.

Solution

Draw the figure. The area is shown in Figure 7–25.

Figure 7–25

Area under the Curve for Example 7–13

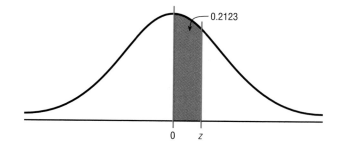

Next, find the area in Table E, as shown in Figure 7–26. Then read the correct z value in the left column as 0.5 and in the top row as 0.06 and add these two values to get 0.56.

Figure 7–26

Finding the z Value from Table E (Example 7–13)

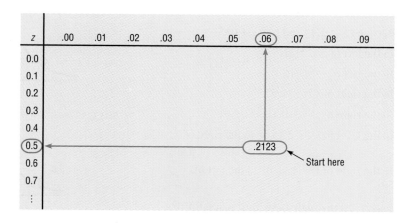

Finding the area under the standard normal distribution curve is the first step in solving a wide variety of practical applications in which the variables are normally distributed. Some of these applications will be presented in Section 7–4.

Exercises

7–1. (**W**) What are the characteristics of the normal distribution?

7–2. (**W**) Why is the normal distribution important in statistical analysis?

7–3. (**W**) What is the total area under the normal distribution curve?

7–4. (**W**) What percentage of the area falls below the mean? Above the mean?

7–5. (**W**) What percentage of the area under the normal distribution curve falls within one standard deviation above and below the mean? Two standard deviations? Three standard deviations?

For Exercises 7–6 through 7–25, find the area under the normal distribution curve.

7–6. Between $z = 0$ and $z = 1.97$.

7–7. Between $z = 0$ and $z = 0.56$.

7–8. Between $z = 0$ and $z = -0.48$.

7–9. Between $z = 0$ and $z = -2.07$.

7–10. To the right of $z = 1.02$.

7–11. To the right of $z = 0.23$.

7–12. To the left of $z = -0.42$.

7–13. To the left of $z = -1.43$.

7–14. Between $z = 1.23$ and $z = 1.90$.

7–15. Between $z = 0.79$ and $z = 1.28$.

7–16. Between $z = -0.87$ and $z = -0.21$.

7–17. Between $z = -1.56$ and $z = -1.83$.

7–18. Between $z = 0.24$ and $z = -1.12$.

7–19. Between $z = 2.47$ and $z = -1.03$.

7–20. To the left of $z = 1.22$.

7–21. To the left of $z = 2.16$.

7–22. To the right of $z = -1.92$.

7–23. To the right of $z = -0.18$.

7–24. To the left of $z = -2.15$ or to the right of $z = 1.62$.

7–25. To the right of $z = 1.98$ or to the left of $z = -0.59$.

In Exercises 7–26 through 7–39, find probabilities for each, using the standard normal distribution.

7–26. $P(0 < z < 1.69)$

7–27. $P(0 < z < 0.67)$

7–28. $P(-1.23 < z < 0)$

7–29. $P(-1.57 < z < 0)$

7–30. $P(z > 2.59)$

7–31. $P(z > 2.83)$

7–32. $P(z < -1.77)$

7–33. $P(z < -1.51)$

7–34. $P(-0.05 < z < 1.10)$

7–35. $P(-2.46 < z < 1.74)$

7–36. $P(1.32 < z < 1.51)$

7–37. $P(1.46 < z < 2.97)$

7–38. $P(z > -1.39)$

7–39. $P(z < 1.42)$

For Exercises 7–40 through 7–45, find the z value that corresponds to the given area.

7–40.

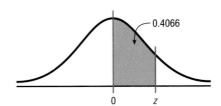

7–41.

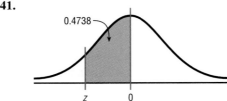

7–42.

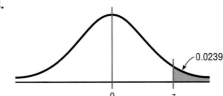

7–43.

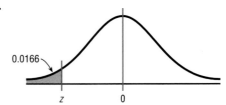

7–44.

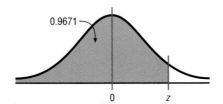

7–45.

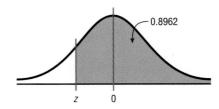

*7–46. Find a z value to the right of the mean so that 53.98% of the distribution lies to the left of it.

*7–47. Find a z value to the left of the mean so that 96.86% of the area lies to the right of it.

*7–48. Find two z values so that 40% of the middle area is bounded by them.

*7–49. Find two z values, one positive and one negative, so that the areas in the two tails total the following values.
a. 5%
b. 10%
c. 1%

*7–50. Find the z values that correspond to the 90th percentile, 80th percentile, 50th percentile, and 5th percentile.

*7–51. Draw a normal distribution with a mean of 100 and a standard deviation of 15.

*7–52. Find the equation for the standard normal distribution by substituting 0 for μ and 1 for σ in the equation

$$y = \frac{e^{-(X-\mu)^2/2\sigma^2}}{\sigma\sqrt{2\pi}}$$

*7–53. Graph the standard normal distribution by using the formula derived in Exercise 7–52. Let $\pi \approx 3.14$ and $e \approx 2.718$. Use X values of $-2, -1.5, -1, -0.5, 0, 0.5, 1, 1.5, 2$.

7–4

Applications of the Normal Distribution

Objective 4. Find probabilities for a normally distributed variable by transforming it into a standard normal variable.

The standard normal distribution curve can be used to solve a wide variety of practical problems. The only requirement is that the variable be normally or approximately normally distributed. There are several mathematical tests to determine whether a variable is normally distributed. However, those tests are not included here; for all the problems presented in this chapter, one can assume that the variable is normally or approximately normally distributed.

To solve problems by using the standard normal distribution, transform the original variable into a standard normal distribution variable by using the formula

$$z = \frac{\text{value} - \text{mean}}{\text{standard deviation}} \qquad \text{or} \qquad z = \frac{X - \mu}{\sigma}$$

This is the same formula presented in Section 3–4. This formula transforms the values of the variable into standard units or z values. Once the variable is transformed, then Procedure Table 6 and Table E in Appendix C can be used to solve problems.

For example, suppose that the scores for a standardized test are normally distributed, have a mean of 100, and have a standard deviation of 15. When the scores are transformed into z values, the two distributions coincide, as shown in Figure 7–27. (Recall that the z distribution has a mean of 0 and a standard deviation of 1.)

Figure 7–27

Test Scores and Their Corresponding z Values

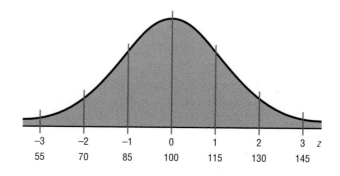

z	−3	−2	−1	0	1	2	3
	55	70	85	100	115	130	145

To solve the application problems in this section, transform the values of the variable into z values and then use Procedure Table 6 and Table E, as shown in the next examples.

Example 7–14

If the scores for the test have a mean of 100 and a standard deviation of 15, find the percentage of scores that will fall below 112.

Solution

STEP 1   Draw the figure and represent the area, as shown in Figure 7–28.

Figure 7–28

Area under the Curve for Example 7–14

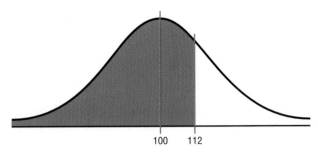

100   112

STEP 2   Find the z value corresponding to a score of 112.

$$z = \frac{X - \mu}{\sigma} = \frac{112 - 100}{15} = \frac{12}{15} = 0.8$$

Hence, 112 is 0.8 standard deviation above the mean of 100, as shown for the z distribution in Figure 7–29.

Figure 7–29

Area and z Values for Example 7–14

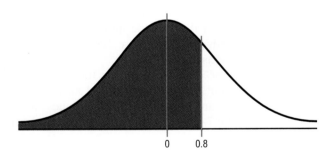

0   0.8

STEP 3   Find the area using Table E. The area between $z = 0$ and $z = 0.8$ is 0.2881. Since the area under the curve to the left of $z = 0.8$ is desired, add 0.5000 to 0.2881 ($0.5000 + 0.2881 = 0.7881$). Therefore, 78.81% of the scores fall below 112.

Example 7–15

Each month, an American household generates an average of 28 pounds of newspaper for garbage or recycling. Assume the standard deviation is two pounds. If a household is selected at random, find the probability of its generating

a.   Between 27 and 31 pounds per month.

Historical Note

Astronomers in the late 1700s and 1800s used the principles underlying the normal distribution to correct measurement errors that occurred in charting the positions of the planets.

b.   More than 30.2 pounds per month.
Assume the variable is approximately normally distributed.

Source: Michael D. Shook and Robert L. Shook, *The Book of Odds* (New York: Plume, 1991).

Solution *a*

STEP 1   Draw the figure and represent the area. See Figure 7–30.

Figure 7–30

Area under the Curve for Part *a* of Example 7–15

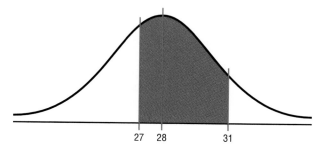

STEP 2   Find the two *z* values.

$$z_1 = \frac{X - \mu}{\sigma} = \frac{27 - 28}{2} = -\frac{1}{2} = -0.5$$

$$z_2 = \frac{X - \mu}{\sigma} = \frac{31 - 28}{2} = \frac{3}{2} = 1.5$$

STEP 3   Find the appropriate area, using Table E. The area between $z = 0$ and $z = -0.5$ is 0.1915. The area between $z = 0$ and $z = 1.5$ is 0.4332. Add 0.1915 and 0.4332 (0.1915 + 0.4332 = 0.6247). Thus, the total area is 62.47%. See Figure 7–31.

Figure 7–31

Area and *z* Values for Part *a* of Example 7–15

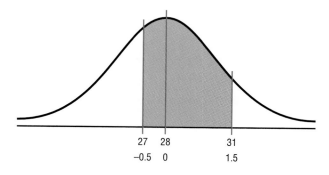

Hence, the probability that a randomly selected household generates between 27 and 31 pounds of newspapers per month is 62.47%.

Solution *b*

STEP 1   Draw the figure and represent the area, as shown in Figure 7–32.

Figure 7–32

Area under the Curve
for Part *b* of
Example 7–15

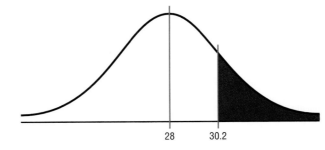

28   30.2

STEP 2   Find the *z* value for 30.2.

$$z = \frac{X - \mu}{\sigma} = \frac{30.2 - 28}{2} = \frac{2.2}{2} = 1.1$$

STEP 3   Find the appropriate area. The area between $z = 0$ and $z = 1.1$ obtained from Table E is 0.3643. Since the desired area is in the right tail, subtract 0.3643 from 0.5000.

$$0.5000 - 0.3643 = 0.1357$$

Hence, the probability that a randomly selected household will accumulate more than 30.2 pounds of newspapers is 0.1357, or 13.57%.

The normal distribution can also be used to answer questions of "How many?" This application is shown in the next example.

Example 7–16

The American Automobile Association reports that the average time it takes to respond to an emergency call is 25 minutes. Assume the variable is approximately normally distributed and the standard deviation is 4.5 minutes. If 80 calls are randomly selected, approximately how many will be responded to in less than 15 minutes?

Source: Michael D. Shook and Robert L. Shook, *The Book of Odds* (New York: Penguin Putnam, Inc., 1991), p. 70.

Solution

To solve the problem, find the area under the normal distribution curve to the left of 15.

STEP 1   Draw a figure and represent the area as shown in Figure 7–33.

Figure 7–33

Area under the Curve
for Example 7–16

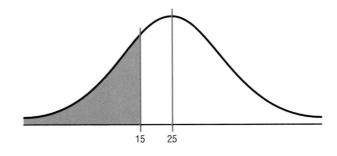

15   25

STEP 2   Find the z value for 15.

$$z = \frac{X - \mu}{\sigma} = \frac{15 - 25}{4.5} = -2.22$$

STEP 3   Find the appropriate area. The area obtained from Table E is 0.4868, which corresponds to the area between $z = 0$ and $z = -2.22$ (Use +2.22).

STEP 4   Subtract 0.4868 from 0.5000 to get 0.0132.

STEP 5   To find how many calls will be made in less than 15 minutes, multiply the sample size (80) by the area (0.0132) to get 1.056. Hence, 1.056, or approximately one, call will be responded to in under 15 minutes.

....

Note: For problems using percentage, be sure to change the percentage to a decimal before multiplying. Also, round the answer to the nearest whole number, since it is not possible to have 1.056 call.

Finding Data Values Given Specific Probabilities

The normal distribution can also be used to find specific data values for given percentages. This application is shown in the next example.

....

Example 7–17

Objective 5. Find specific data values for given percentages using the standard normal distribution.

An exclusive college desires to accept only the top 10% of all graduating seniors based on the results of a national placement test. This test has a mean of 500 and a standard deviation of 100. Find the cutoff score for the exam. Assume the variable is normally distributed.

Solution

Since the test scores are normally distributed, the test value (X) that cuts off the upper 10% of the area under the normal distribution curve is desired. This area is shown in Figure 7–34.

Figure 7–34

Area under the Curve for Example 7–17

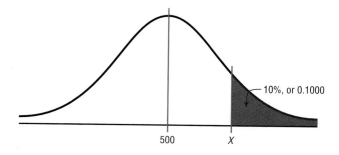

Work backward to solve this problem

STEP 1   Subtract 0.1000 from 0.5000 to get the area under the normal distribution between 500 and X: $0.5000 - 0.1000 = 0.4000$.

STEP 2   Find the z value that corresponds to an area of 0.4000 by looking up 0.4000 in the area portion of Table E. If the specific value cannot be found, use the closest value—in this case, 0.3997, as shown in Figure 7–35. The corresponding z value is 1.28. (If the area falls exactly halfway between two z values, use the larger of the two z values. For example, the area 0.4500 falls halfway between 0.4495 and 0.4505. In this case use 1.65 rather than 1.64 for the z value.)

Figure 7–35

Finding the z Value from Table E (Example 7–17)

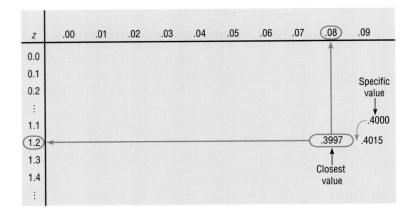

STEP 3   Substitute in the formula $z = (X - \mu)/\sigma$ and solve for X.

$$1.28 = \frac{X - 500}{100}$$

$$X = (1.28)(100) + 500 = 628$$

The score of 628 should be used as a cutoff. Anybody scoring below 628 should not be admitted.

Instead of using the formula shown in Step 3 one can use the formula $X = z \cdot \sigma + \mu$. This is obtained by solving

$$z = \frac{(X - \mu)}{\sigma} \text{ for } X$$

as follows:

$z \cdot \sigma = X - \mu$ \qquad Multiply both sides by σ.

$z \cdot \sigma + \mu = X$ \qquad Add μ to both sides.

$X = z \cdot \sigma + \mu$ \qquad Exchange both sides of the equation.

When one must find the value of X, the following formula can be used:

$$X = z \cdot \sigma + \mu$$

Example 7–18

For a medical study, a researcher wishes to select people in the middle 60% of the population based on blood pressure. If the mean systolic blood pressure is 120 and the standard deviation is 8, find the upper and lower readings that would qualify people to participate in the study.

Solution

Assuming that blood pressure readings are normally distributed, the cutoff points are as shown in Figure 7–36.

Figure 7–36
Area under the Curve for Example 7–18

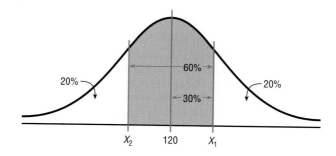

Note that two values are needed, one above the mean and one below the mean. Find the value to the right of the mean first. The closest z value for an area of 0.3000 is 0.84. Substituting in the formula $X = z\sigma + \mu$, one gets

$$X = z\sigma + \mu = (0.84)(8) + 120 = 126.72$$

On the the other side, $z = -0.84$; hence,

$$X = (-0.84)(8) + 120 = 113.28$$

Therefore, the middle 60% will have blood pressure readings of $113.28 < X < 126.72$.

As shown in this section, the normal distribution is a useful tool in answering many questions about variables that are normally or approximately normally distributed.

Exercises

7–54. Explain why the standard normal distribution can be used to solve many real-life problems.

7–55. The average hourly wage of production workers in manufacturing is $11.76. Assume the variable is normally distributed. If the standard deviation of earnings is $2.72, find these probabilities for a randomly selected production worker.
a. The production worker earns more than $12.55.
b. The production worker earns less than $8.00.
Source: *Statistical Abstract of the United States 1994* (Washington, DC: U.S. Bureau of the Census).

7–56. The Speedmaster IV automobile gets an average 22.0 miles per gallon in the city. The standard deviation is 3 miles per gallon. Assume the variable is normally distributed. Find the probability that on any given day, the car will get more than 26 miles per gallon when driven in the city.

7–57. If the mean salary of high school teachers in the United States is $29,835, and the standard deviation is $3,000, find these probabilities for a randomly selected teacher. Assume the variable is normally distributed.

a. The teacher earns more than $35,000.
b. The teacher earns less than $25,000.

7–58. For a specific year, Americans spent an average of $71.12 for books. Assume the variable is normally distributed. If the standard deviation of the amount spent on books is $8.42, find these probabilities for a randomly selected American.
a. He or she spent more than $60 per year on books.
b. He or she spent less than $80 per year on books.
Source: *Statistical Abstract of the United States 1994.*

7–59. A survey found that people keep their television sets an average of 4.8 years. The standard deviation is 0.89 year. If a person decides to buy a new TV set, find the probability that he or she has owned the old set for the following amount of time. Assume the variable is normally distributed.
a. Less than 2.5 years
b. Between 3 and 4 years
c. More than 4.2 years

7–60. The average age of CEOs is 56 years. Assume the variable is normally distributed. If the standard deviation is four years, find the probability that the age of a randomly selected CEO will be in the following range.
a. Between 53 and 59 years old
b. Between 58 and 63 years old
c. Between 50 and 55 years old
Source: Michael D. Shook and Robert L. Shook, *The Book of Odds* (New York: Penguin Putnam Inc., 1993), p. 49.

7–61. The average life of a brand of automobile tires is 30,000 miles, with a standard deviation of 2,000 miles. If a tire is selected and tested, find the probability that it will have the following lifetime. Assume the variable is normally distributed.
a. Between 25,000 and 28,000 miles
b. Between 27,000 and 32,000 miles
c. Between 31,500 and 33,500 miles

7–62. The average time a person spends at the West Newton Zoo is 62 minutes. The standard deviation is 12 minutes. If a visitor is selected at random, find the probability that he or she will spend the following time at the zoo. Assume the variable is normally distributed.
a. At least 82 minutes
b. At most 50 minutes

7–63. The average time for a courier to travel from Pittsburgh to Harrisburg is 200 minutes, and the standard deviation is 10 minutes. If one of these trips is selected at random, find the probability that the courier will have the following travel time. Assume the variable is normally distributed.
a. At least 180 minutes
b. At most 205 minutes

7–64. The average amount of snow per season in Trafford is 44 inches. The standard deviation is 6 inches. Find the probability that next year Trafford will receive the following amount of snowfall. Assume the variable is normally distributed.
a. At most 50 inches of snow
b. At least 53 inches of snow

7–65. The average waiting time for a drive-in window at a local bank is 9.2 minutes, with a standard deviation of 2.6 minutes. When a customer arrives at the bank, find the probability that the customer will have to wait the following time. Assume the variable is normally distributed.
a. Between 5 and 10 minutes
b. Less than 6 minutes or more than 9 minutes

7–66. The average time it takes college freshmen to complete the Mason Basic Reasoning Test is 24.6 minutes. The standard deviation is 5.8 minutes. Find these probabilities. Assume the variable is normally distributed.
a. It will take a student between 15 and 30 minutes to complete the test.
b. It will take a student less than 18 minutes or more than 28 minutes to complete the test.

7–67. A brisk walk at 4 miles per hour burns an average of 300 calories per hour. If the standard deviation of the distribution is 8 calories, find the probability that a person who walks one hour at the rate of 4 miles per hour will burn the following calories. Assume the variable is normally distributed.
a. More than 280 calories
b. Less than 293 calories
c. Between 285 and 320 calories

7–68. During September, the average temperature of Laurel Lake is 64.2° and the standard deviation is 3.2°. Assume the variable is normally distributed. For a randomly selected day, find the probability that the temperature will be as follows:
a. Above 62°   *c.* Between 65° and 68°
b. Below 67°

7–69. If the systolic blood pressure for a certain group of obese people has a mean of 132 and a standard deviation of 8, find the probability that a randomly selected person will have the following blood pressure. Assume the variable is normally distributed.
a. Above 130   *c.* Between 131 and 136
b. Below 140

7–70. In order to qualify for letter sorting, applicants are given a speed-reading test. The scores are normally distributed, with a mean of 80 and a standard deviation of 8. The variable is also normally distributed. If only the top 15% of the applicants are selected, find the cutoff score.

7–71. The scores on a test have a mean of 100 and a standard deviation of 15. If a personnel manager wishes to select from the top 75% of applicants who take the test, find the cutoff score. Assume the variable is normally distributed.

7–72. For an educational study, a volunteer must place in the middle 50% on a test. If the mean for the population is 100 and the standard deviation is 15, find the two limits (upper and lower) for the scores that would enable a volunteer to participate in the study. Assume the variable is normally distributed.

7–73. A contractor decided to build homes that will include the middle 80% of the market. If the average size (in square feet) of homes built is 1,810, find the maximum and minimum sizes of the homes the contractor should build. Assume that the standard deviation is 92 square feet and the variable is normally distributed.
Source: Congressional Research Service, in *The Book of Odds,* 1991, p. 15.

7–74. If the average price of a new home is $145,500, find the maximum and minimum prices of the houses a contractor will build to include the middle 80% of the market. Assume that the standard deviation of prices is $1,500 and the variable is normally distributed.
Source: Congressional Research Service, in *The Book of Odds,* 1991, p. 15.

7–75. An athletic association wants to sponsor a footrace. The average time it takes to run the course is 58.6 minutes, with a standard deviation of 4.3 minutes. If the association decides to include only the top 20% of the racers, what should the cutoff time be in the tryout run? Assume the variable is normally distributed.

7–76. In order to help students improve their reading, a school district decides to implement a reading program. It is to be administered to the bottom 5% of the students in the district, based on the scores on a reading achievement exam. If the average score for the students in the district is 122.6, find the cutoff score that will make a student eligible for the program. The standard deviation is 18. Assume the variable is normally distributed.

7–77. An automobile dealer finds that the average price of a previously owned vehicle is $8,256. He decides to sell cars that will appeal to the middle 60% of the market in terms of price. Find the maximum and

minimum prices of the cars the dealer will sell. The standard deviation is $1,150 and the variable is normally distributed.

7–78. A small publisher wishes to publish self-improvement books. After a survey of the market, the publisher finds that the average cost of the type of book that she wishes to publish is $12.80. If she wants to price her books to sell in the middle 70% range, what should the maximum and minimum prices of the books be? The standard deviation is $0.83 and the variable is normally distributed.

7–79. A special enrichment program in mathematics is to be offered to the top 12% of students in a school district. A standardized mathematics achievement test given to all students has a mean of 57.3 and a standard deviation of 16. Find the cutoff score. Assume the variable is normally distributed.

7–80. A pet-shop owner decides to sell tropical fish that will appeal to the middle 60% of customers. The owner reads in a study that the mean price of tropical fish sold is $9.52, with a standard deviation of $1.02. Find the maximum and minimum prices of tropical fish the owner should sell. Assume the variable is normally distributed.

7–81. An advertising company plans to market a product to low-income families. A study states that for a particular area, the average income per family is $24,596 and the standard deviation is $6,256. If the company plans to target the bottom 18% of the families based on income, find the cutoff income. Assume the variable is normally distributed.

7–82. If a one-person household spends an average of $40 per week on groceries, find the maximum and minimum dollar amount spent per week for the middle 50% of one-person households. Assume that the standard deviation is $5 and the variable is normally distributed.
Source: Food Marketing Institute, in *The Book of Odds,* 1991, p. 192.

7–83. The mean lifetime of a wristwatch is 25 months, with a standard deviation of 5 months. If the distribution is normal, for how many months should a guarantee be if the manufacturer does not want to exchange more than 10% of the watches? Assume the variable is normally distributed.

7–84. In order to qualify for police academy training, recruits are tested for stress tolerance. The scores are normally distributed, with a mean of 60 and a standard deviation of 10. If only the top 20% of recruits are selected, find the cutoff score.

7–85. In the distributions shown, state the mean and standard deviation for each. Hint: See Figures 7–5 and 7–6. Hint: The vertical lines are one standard deviation apart.

a.

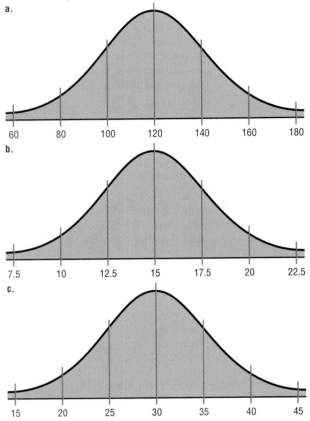

b.

c.

7–86. Suppose that the mathematics SAT scores for high school seniors for a specific year have a mean of 456 and a standard deviation of 100 and are approximately normally distributed. If a subgroup of these high school seniors, those who are in the National Honor Society, is selected, would you expect the distribution of scores to have the same mean and standard deviation? Explain your answer.

7–87. Given a data set, how could you decide if the distribution of the data was approximately normal?

7–88. If a distribution of raw scores were plotted and then the scores were transformed into z scores, would the shape of the distribution change? Explain your answer.

7–89. In a normal distribution, find σ when $\mu = 100$ and 2.68% of the area lies to the right of 105.

7–90. In a normal distribution, find μ when σ is 6 and 3.75% of the area lies to the left of 85.

7–91. In a certain normal distribution, 1.25% of the area lies to the left of 42 and 1.25% of the area lies to the right of 48. Find μ and σ.

7–92. An instructor gives a 100-point examination in which the grades are normally distributed. The mean is 60 and the standard deviation is 10. If there are 5% A's and 5% F's, 15% B's and 15% D's, and 60% C's, find the scores that divide the distribution into those categories.

7–5

The Central Limit Theorem

In addition to knowing how individual data values vary about the mean for a population, statisticians are also interested in knowing about the distribution of the means of samples taken from a population. This topic is discussed in the subsections that follow.

Distribution of Sample Means

Objective 6. Use the central limit theorem to solve problems involving sample means.

Suppose a researcher selects 100 samples of a specific size from a large population and computes the mean of the same variable for each of the 100 samples. These sample means, $\overline{X}_1, \overline{X}_2, \overline{X}_3, \ldots, \overline{X}_{100}$, constitute a sampling distribution of sample means.

A **sampling distribution of sample means** is a distribution obtained by using the means computed from random samples of a specific size taken from a population.

If the samples are randomly selected, the sample means, for the most part, will be somewhat different from the population mean μ. These differences are caused by sampling error.

Sampling error is the difference between the sample measure and the corresponding population measure due to the fact that the sample is not a perfect representation of the population.

When all possible samples of a specific size are selected from a population, the distribution of the sample means for a variable has two important properties, which are explained next.

Properties of the Distribution of Sample Means

1. The mean of the sample means will be the same as the population mean.
2. The standard deviation of the sample means will be smaller than the standard deviation of the population, and it will be equal to the population standard deviation divided by the square root of the sample size.

The following example illustrates these two properties. Suppose a professor gave an eight-point quiz to a small class of four students. The results of the quiz were 2, 6, 4, and 8. For the sake of discussion, assume that the four students constitute the population. The mean of the population is

$$\mu = \frac{2 + 6 + 4 + 8}{4} = 5$$

The standard deviation of the population is

$$\sigma = \sqrt{\frac{(2-5)^2 + (6-5)^2 + (4-5)^2 + (8-5)^2}{4}} = 2.236$$

The graph of the original distribution is shown in Figure 7–37. This is called a *uniform distribution*.

Figure 7–37

Distribution of Quiz Scores

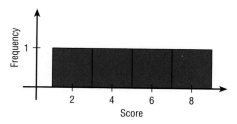

Now, if all samples of size 2 are taken with replacement, and the mean of each sample is found, the distribution is as shown next.

Sample	Mean	Sample	Mean
2, 2	2	6, 2	4
2, 4	3	6, 4	5
2, 6	4	6, 6	6
2, 8	5	6, 8	7
4, 2	3	8, 2	5
4, 4	4	8, 4	6
4, 6	5	8, 6	7
4, 8	6	8, 8	8

A frequency distribution of sample means is as follows.

$\overline{X}$	f
2	1
3	2
4	3
5	4
6	3
7	2
8	1

For the data from the example just discussed, Figure 7–38 shows the graph of the sample means. The graph appears to be somewhat normal, even though it is a histogram.

Figure 7–38

Distribution of
Sample Means

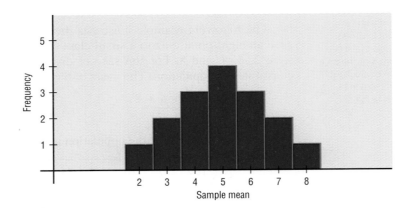

The mean of the sample means, denoted by $\mu_{\overline{X}}$, is

$$\mu_{\overline{X}} = \frac{2 + 3 + \cdots + 8}{16} = \frac{80}{16} = 5$$

which is the same as the population mean. Hence,

$$\mu_{\overline{X}} = \mu$$

The standard deviation of sample means, denoted by $\sigma_{\overline{X}}$, is

$$\sigma_{\overline{X}} = \sqrt{\frac{(2 - 5)^2 + (3 - 5)^2 + \cdots + (8 - 5)^2}{16}} = 1.581$$

which is the same as the population standard deviation divided by $\sqrt{2}$:

$$\sigma_{\overline{X}} = \frac{2.236}{\sqrt{2}} = 1.581$$

(*Note:* Rounding rules were not used here in order to show that the answers coincide.)

In summary, if all possible samples of size n are taken from the same population, the mean of the sample means, denoted by $\mu_{\overline{X}}$, equals the population mean μ; and the standard deviation of the sample means, denoted by $\sigma_{\overline{X}}$, equals $\sigma/\sqrt{n}$.

The standard deviation of the sample means is called the **standard error of the mean.** Hence,

$$\sigma_{\overline{X}} = \frac{\sigma}{\sqrt{n}}$$

A third property of the sampling distribution of sample means pertains to the shape of the distribution and is explained by the **central limit theorem.**

The Central Limit Theorem

As the sample size n increases, the shape of the distribution of the sample means taken from a population with mean μ and standard deviation σ will approach a normal distribution. As previously shown, this distribution will have a mean μ and a standard deviation $\sigma/\sqrt{n}$.

The central limit theorem can be used to answer questions about sample means in the same manner that the normal distribution can be used to answer questions about individual values. The only difference is that a new formula must be used for the z values. It is

$$z = \frac{\overline{X} - \mu}{\sigma/\sqrt{n}}$$

Notice that $\overline{X}$ is the sample mean, and the denominator is the standard error of the mean.

If a large number of samples of a given size were selected from a large population, and the sample means computed, the distribution of sample means would look like the one shown in Figure 7–39. The percentages indicate the areas of the regions.

Figure 7–39
Distribution of Sample Means for Large Number of Samples

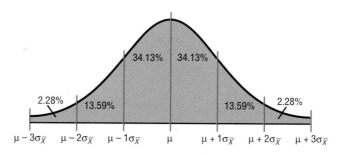

It's important to remember two things when using the central limit theorem:

1. When the original variable is normally distributed, the distribution of the sample means will be normally distributed, for any sample size n.
2. When the distribution of the original variable departs from normality, a sample size of 30 or more is needed to use the normal distribution to approximate the distribution of the sample means. The larger the sample, the better the approximation will be.

The next several examples show how the standard normal distribution can be used to answer questions about sample means.

Example 7–19

A. C. Neilsen reported that children between the ages of 2 and 5 watch an average of 25 hours of television per week. Assume the variable is normally distributed and the standard deviation is 3 hours. If 20 children between the ages of 2 and 5 are randomly selected, find the probability that the mean of the number of hours they watch television will be greater than 26.3 hours.

Source: Michael D. Shook and Robert L. Shook, *The Book of Odds* (New York: Penguin Putnam Inc., 1991), p. 161.

Solution

Since the variable is approximately normally distributed, the distribution of sample means will be approximately normal, with a mean of 25. The standard deviation of the sample means is

$$\sigma_{\overline{X}} = \frac{\sigma}{\sqrt{n}} = \frac{3}{\sqrt{20}} = 0.671$$

The distribution of the means is shown in Figure 7–40, with the appropriate area shaded.

Figure 7–40

Distribution of the Means for Example 7–19

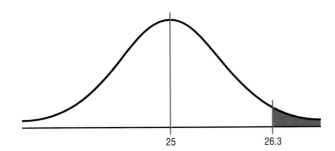

The z value is

$$z = \frac{\overline{X} - \mu}{\sigma/\sqrt{n}} = \frac{26.3 - 25}{3/\sqrt{20}} = \frac{1.3}{0.671} = 1.94$$

The area between 0 and 1.94 is 0.4738. Since the desired area is in the tail, subtract 0.4738 from 0.5000. Hence, 0.5000 − 0.4738 = 0.0262, or 2.62%.

One can conclude that the probability of obtaining a sample mean larger than 26.3 hours is 2.62% (i.e., $P[\overline{X} > 26.3] = 2.62\%$).

Example 7–20

The average age of a vehicle registered in the United States is 8 years, or 96 months. Assume the standard deviation is 16 months. If a random sample of 36 cars is selected, find the probability that the mean of their age is between 90 and 100 months.

Source: *Harper's Index* 290, no. 1740 (May 1995), p. 11.

Solution

The desired area is shown in Figure 7–41.

Figure 7–41

Area under the Curve
for Example 7–20

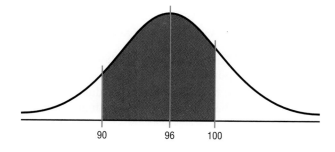

90   96   100

The two z values are

$$z_1 = \frac{90 - 96}{16/\sqrt{36}} = -2.25$$

$$z_2 = \frac{100 - 96}{16/\sqrt{36}} = 1.50$$

The two areas corresponding to the z values of -2.25 and 1.50, respectively, are 0.4878 and 0.4332. Since the z values are on opposite sides of the mean, find the probability by adding the areas: $0.4878 + 0.4332 = 0.921$, or 92.1%.

Hence, the probability of obtaining a sample mean between 90 and 100 months is 92.1% (i.e., $P[90 < \overline{X} < 100] = 92.1\%$).

Since the sample size is 30 or larger, the normality assumption is not necessary, as shown in Example 7–20.

Students sometimes have difficulty deciding whether to use

$$z = \frac{\overline{X} - \mu}{\sigma/\sqrt{n}} \qquad \text{or} \qquad z = \frac{X - \mu}{\sigma}$$

The formula

$$z = \frac{\overline{X} - \mu}{\sigma/\sqrt{n}}$$

should be used to gain information about a sample mean, as shown in this section. The formula

$$z = \frac{X - \mu}{\sigma}$$

is used to gain information about an individual data value obtained from the population. Notice that the first formula contains $\overline{X}$, the symbol for the sample mean, while the second formula contains X, the symbol for an individual data value. The next example illustrates the uses of the two formulas.

Example 7–21

The average number of pounds of meat a person consumes a year is 218.4 pounds. Assume that the standard deviation is 25 pounds and the distribution is approximately normal.

Source: American Dietetic Association, *The Book of Odds*, 1991, p. 164.

a. Find the probability that a person selected at random consumes less than 224 pounds per year.

b. If a sample of 40 individuals is selected, find the probability that the mean of the sample will be less than 224 pounds per year.

Solution

a. Since the question asks about an individual person, the formula $z = (X - \mu)/\sigma$ is used.
   The distribution is shown in Figure 7–42.

Figure 7–42

Area under the Curve for Part *a* of Example 7–21

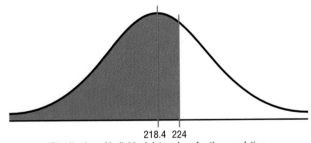

218.4 224
Distribution of individual data values for the population

The z value is

$$z = \frac{X - \mu}{\sigma} = \frac{224 - 218.4}{25} = 0.22$$

The area between 0 and 0.22 is 0.0871; this area must be added to 0.5000 to get the total area to the left of $z = 0.22$.

$$0.0871 + 0.5000 = 0.5871$$

Hence, the probability of selecting an individual who consumes less than 224 pounds of meat per year is 0.5871, or 58.71% (i.e., $P[X < 224] = 0.5871$).

b. Since the question concerns the mean of a sample with a size of 40, the formula $z = (\overline{X} - \mu)/(\sigma/\sqrt{n})$ is used.

The area is shown in Figure 7–43.

Figure 7–43

Area under the Curve
for Part *b* of
Example 7–21

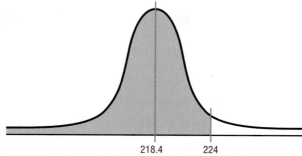

218.4   224

Distribution of means for all samples of size 40 taken from the population

The *z* value is

$$z = \frac{\overline{X} - \mu}{\sigma/\sqrt{n}} = \frac{224 - 218.4}{25/\sqrt{40}} = 1.42$$

The area between $z = 0$ and $z = 1.42$ is 0.4222; this value must be added to 0.5000 to get the total area.

$$0.4222 + 0.5000 = 0.9222$$

Hence, the probability that the mean of a sample of 40 individuals is less than 224 pounds per year is 0.9222, or 92.22%. That is, $P(\overline{X} < 224) = 0.9222$.

Comparing the two probabilities, one can see that the probability of selecting an individual who consumes less than 224 pounds of meat per year is 58.71%, but the probability of selecting a sample of 40 people with a mean consumption of meat that is less than 224 pounds per year is 92.22%. This rather large difference is due to the fact that the distribution of sample means is much less variable than the distribution of individual data values.

As this section has shown, the central limit theorem enables the researcher to determine probabilities associated with sample means of sampling distributions. The formulas and their uses are summarized in Table 7–1.

Table 7–1 Summary of Formulas and Their Uses	
Formula	**Use**
1. $z = \dfrac{X - \mu}{\sigma}$	Used to gain information about an individual data value when the variable is normally distributed.
2. $z = \dfrac{\overline{X} - \mu}{\sigma/\sqrt{n}}$	Used to gain information when applying the central limit theorem about a sample mean when the variable is normally distributed or when the sample size is 30 or more.

Exercises

7–93. (W) If samples of a specific size are selected from a population and the means are computed, what is this distribution of means called?

7–94. (W) Why do most of the sample means differ somewhat from the population mean? What is this difference called?

7–95. (W) What is the mean of the sample means?

7–96. (W) What is the standard deviation of the sample means called? What is the formula for this standard deviation?

7–97. (W) What does the central limit theorem say about the shape of the distribution of sample means?

7–98. What formula is used to gain information about an individual data value when the variable is normally distributed?

7–99. What formula is used to gain information about a sample mean when the variable is normally distributed or when the sample size is 30 or more?

For Exercises 7–100 through 7–117, assume that the sample is taken from a large population.

7–100. A survey found that Americans generate an average of 17.2 pounds of glass garbage each year. Assume the standard deviation of the distribution is 2.5 pounds. Find the probability that the mean of a sample of 55 families will be between 17 and 18 pounds.
Source: Michael D. Shook and Robert L. Shook, *The Book of Odds* (New York: Penguin Putnam Inc., 1991), p. 14.

7–101. The mean serum cholesterol of a large population of overweight adults is 220 mg/dl and the standard deviation is 16.3 mg/dl. If a sample of 30 adults is selected, find the probability that the mean will be between 220 and 222 mg/dl.

7–102. For a certain large group of individuals, the mean hemoglobin level in the blood is 21.0 grams per milliliter (g/ml). The standard deviation is 2 g/ml. If a sample of 25 individuals is selected, find the probability that the mean will be greater than 21.3 g/ml. Assume the variable is normally distributed.

7–103. The mean weight of 18-year-old females is 126 pounds, and the standard deviation is 15.7. If a sample of 25 females is selected, find the probability that the mean of the sample will be greater than 128.3 pounds. Assume the variable is normally distributed.

7–104. The mean grade point average of the engineering majors at a large university is 3.23, with a standard deviation of 0.72. In a class of 48 students, find the probability that the mean grade point average of the students is less than 3.15.

7–105. The average price of a pound of sliced bacon is $2.02. Assume the standard deviation is $0.08. If a random sample of 40 one-pound packages is selected, find the probability that the mean of the sample will be less than $2.00.
Source: *Statistical Abstract of the United States 1994.*

7–106. The average hourly wage of fast-food workers employed by a nationwide chain is $5.55. The standard deviation is $1.15. If a sample of 50 workers is selected, find the probability that the mean of the sample will be between $5.25 and $5.90.

7–107. The mean score on a dexterity test for 12-year-olds is 30. The standard deviation is 5. If a psychologist administers the test to a class of 22 students, find the probability that the mean of the sample will be between 27 and 31. Assume the variable is normally distributed.

7–108. A recent study of the life span of portable radios found the average to be 3.1 years, with a standard deviation of 0.9 year. If the number of radios owned by the students in one dormitory is 47, find the probability that the mean lifetime of these radios will be less than 2.7 years.

7–109. The average age of accountants is 43 years, with a standard deviation of 5 years. If an accounting firm employs 30 accountants, find the probability that the average age of the group is greater than 44.2 years old.

7–110. The average annual precipitation for Des Moines is 30.83 inches, with a standard deviation of 5 inches. If a random sample of 10 years is selected, find the probability that the mean will be between 32 and 33 inches. Assume the variable is normally distributed.

7–111. Procter & Gamble reported that an American family of 4 washes an average of one ton (2,000 pounds) of clothes each year. If the standard deviation of the distribution is 187.5 pounds, find the probability that the mean of a randomly selected sample of 50 families of four will be between 1,980 and 1,990 pounds.
Source: Lewis H. Lapham, Michael Pollan, and Eric Etheridge, *The Harper's Index Book* (New York: Henry Holt & Co., 1987).

7–112. The average annual salary in Pennsylvania was $24,393 in 1992. Assume that salaries were normally distributed for a certain group of wage earners, and the standard deviation of this group was $4,362.
a. Find the probability that a randomly selected individual earned less than $26,000.
b. Find the probability that for a randomly selected sample of 25 individuals, the mean salary was less than $26,000.
c. Why is the probability for Part *b* higher than the probability for Part *a*?

Associated Press, December 23, 1992.

7–113. The average time it takes a group of adults to complete a certain achievement test is 46.2 minutes. The standard deviation is 8 minutes. Assume the variable is normally distributed.
a. Find the probability that a randomly selected adult will complete the test in less than 43 minutes.
b. Find the probability that if 50 randomly selected adults take the test, the mean time it takes the group to complete the test will be less than 43 minutes.
c. Does it seem reasonable that an adult would finish the test in less than 43 minutes? Explain.
d. Does it seem reasonable that the mean of the 50 adults could be less than 43 minutes?

7–114. Assume that the mean systolic blood pressure of normal adults is 120 millimeters of mercury (mmHg) and the standard deviation is 5.6. Assume the variable is normally distributed.
a. If an individual is selected, find the probability that the individual's pressure will be between 120 and 121.8 mmHg.
b. If a sample of 30 adults is randomly selected, find the probability that the sample mean will be between 120 and 121.8 mmHg.
c. Why is the answer to Part *a* so much smaller than the answer to Part *b*?

7–115. The average cholesterol content of a certain brand of eggs is 215 milligrams and the standard deviation is 15 milligrams. Assume the variable is normally distributed.

a. If a single egg is selected, find the probability that the cholesterol content will be more than 220 milligrams.
b. If a sample of 25 eggs is selected, find the probability that the mean of the sample will be larger than 220 milligrams.

Source: *Living Fit* (Englewood Cliffs, NJ: Best Foods, CPC International, Inc., 1991).

7–116. At a large university, the mean age of graduate students who are majoring in psychology is 32.6 years, and the standard deviation is 3 years. Assume the variable is normally distributed.
a. If an individual from the department is randomly selected, find the probability that his or her age will be between 31.0 and 33.2 years.
b. If a random sample of 15 individuals is selected, find the probability that the mean age of the students in the sample will be between 31.0 and 33.2 years.

7–117. The average labor cost for car repairs for a large chain of car repair shops is $48.25. The standard deviation is $4.20. Assume the variable is normally distributed.
a. If a store is selected at random, find the probability that the labor cost will range between $46 and $48.
b. If 20 stores are selected at random, find the probability that the mean of the sample will be between $46 and $48.
c. Which answer is larger? Explain why.

***7–118.** The average breaking strength of a certain brand of steel cable is 2000 pounds, with a standard deviation of 100 pounds. A sample of 20 cables is selected and tested. Find the sample mean that will cut off the upper 95% of all of the samples of size 20 taken from the population. Assume the variable is normally distributed.

***7–119.** The standard deviation of a variable is 15. If a sample of 100 individuals is selected, compute the standard error of the mean. What size sample is necessary to double the standard error of the mean?

***7–120.** In Exercise 7–119, what size sample is needed to cut the standard error of the mean in half?

7–6

The Normal Approximation to the Binomial Distribution

The normal distribution is often used to solve problems that involve the binomial distribution since when *n* is large (say, 100), the calculations are too difficult to do by hand using the binomial distribution. Recall from Chapter 6 that a binomial distribution has the following characteristics:

1. There must be a fixed number of trials.
2. The outcome of each trial must be independent.
3. Each experiment can have only two outcomes or be reduced to two outcomes.
4. The probability of a success must remain the same for each trial.

Also, recall that a binomial distribution is determined by n (the number of trials) and p (the probability of a success). When p is approximately 0.5, and as n increases, the shape of the binomial distribution becomes similar to the normal distribution. The larger n and the closer p is to 0.5, the more similar the shape of binomial distribution is to the normal distribution.

Objective 7. Use the normal approximation to compute probabilities for a binomial variable.

But when p is close to 0 or 1 and n is relatively small, the normal approximation is inaccurate. As a rule of thumb, statisticians generally agree that the normal approximation should be used only when $n \cdot p$ and $n \cdot q$ are both greater than or equal to 5. (*Note;* $q = 1 - p$.) For example, if p is 0.3 and n is 10, then $np = (10)(0.3) = 3$, and the normal distribution should not be used as an approximation. On the other hand, if $p = 0.5$ and $n = 10$, then $np = (10)(0.5) = 5$ and $nq = (10)(0.5) = 5$, and the normal distribution can be used as an approximation. See Figure 7–44.

Figure 7–44
....

Comparison of the Binomial Distribution and the Normal Distribution

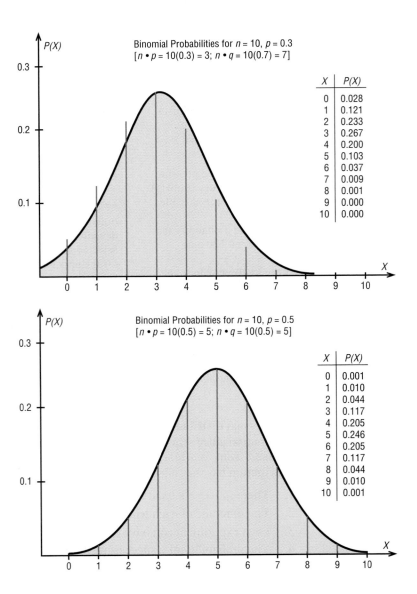

Binomial Probabilities for $n = 10$, $p = 0.3$
[$n \cdot p = 10(0.3) = 3$; $n \cdot q = 10(0.7) = 7$]

X	P(X)
0	0.028
1	0.121
2	0.233
3	0.267
4	0.200
5	0.103
6	0.037
7	0.009
8	0.001
9	0.000
10	0.000

Binomial Probabilities for $n = 10$, $p = 0.5$
[$n \cdot p = 10(0.5) = 5$; $n \cdot q = 10(0.5) = 5$]

X	P(X)
0	0.001
1	0.010
2	0.044
3	0.117
4	0.205
5	0.246
6	0.205
7	0.117
8	0.044
9	0.010
10	0.001

The formulas for the mean and standard deviation for the binomial distribution are necessary for calculations. They are

$$\mu = n \cdot p \text{ and } \sigma = \sqrt{n \cdot p \cdot q}$$

Procedure for the Normal Approximation to the Binomial Distribution

STEP 1   Check to see whether the normal approximation can be used.

STEP 2   Find the mean μ and the standard deviation σ.

STEP 3   Write the problem in probability notation, using X.

STEP 4   Rewrite the problem by using the continuity correction factor, and show the corresponding area under the normal distribution.

STEP 5   Find the corresponding z values.

STEP 6   Find the solution.

In addition to the previous condition of $np \geq 5$ and $nq \geq 5$, a correction for continuity may be used in the normal approximation.

A **correction for continuity** is a correction employed when a continuous distribution is used to approximate a discrete distribution.

The continuity correction means that for any specific value of X, say 8, the boundaries of X in the binomial distribution (in this case, 7.5 to 8.5) must be used. (See Chapter 1, Section 1–3.) Hence, when one employs the normal distribution to approximate the binomial, the boundaries of any specific value X must be used as they are shown in the binomial distribution. For example, for $P(X = 8)$, the correction is $P(7.5 < X < 8.5)$. For $P(X \leq 7)$, the correction is $P(X < 7.5)$. For $P(X \geq 3)$, the correction is $P(X > 2.5)$.

··

Example 7–22

A magazine reported that 6% of American drivers read the newspaper while driving. If 300 drivers are selected at random, find the probability that exactly 25 say they read the newspaper while driving.

Source: USA Snapshot, *USA Today,* June 26, 1995.

Solution

Here, $p = 0.06$, $q = 0.94$, and $n = 300$.

STEP 1   Check to see whether the normal approximation can be used.

$$np = (300)(0.06) = 18 \qquad nq = (300)(0.94) = 282$$

Since $np \geq 5$ and $nq \geq 5$, the normal distribution can be used.

STEP 2   Find the mean and standard deviation.

$$\mu = np = (300)(0.06) = 18$$
$$\sigma = \sqrt{npq} = \sqrt{(300)(0.06)(0.94)} = \sqrt{16.92} = 4.11$$

STEP 3   Write the problem in probability notation: $P(X = 25)$.

STEP 4   Rewrite the problem by using the continuity correction factor: $P(24.5 < X < 25.5)$. Show the corresponding area under the normal distribution curve (see Figure 7–45).

Figure 7–45

Area under the Curve
and X Values for
Example 7–22

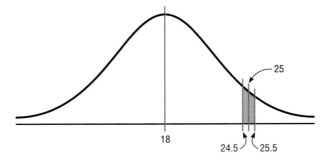

STEP 5 Find the corresponding z values. Since 25 represents any value between 24.5 and 25.5, find both z values.

$$z_1 = \frac{25.5 - 18}{4.11} = 1.82 \qquad z_2 = \frac{24.5 - 18}{4.11} = 1.58$$

STEP 6 Find the solution. Find the corresponding areas in the table: The area for $z = 1.82$ is 0.4656, and the area for $z = 1.58$ is 0.4429. Subtract the areas to get the approximate value: $0.4656 - 0.4429 = 0.0227$, or 2.27%.

Hence, the probability that exactly 25 people read the newspaper while driving is 2.27%.

Example 7–23

Of the members of a bowling league, 10% are widowed. If 200 bowling league members are selected at random, find the probability that 10 or more will be widowed.

Solution

Here, $p = 0.10$, $q = 0.90$, and $n = 200$.

STEP 1 Since np is $(200)(0.10) = 20$ and nq is $(200)(0.90) = 180$, the normal approximation can be used.

STEP 2 $\mu = np = (200)(0.10) = 20$
$\sigma = \sqrt{npq} = \sqrt{(200)(0.10)(0.90)} = \sqrt{18} = 4.24$

STEP 3 $P(X \geq 10)$.

STEP 4 $P(X > 9.5)$. The desired area is shown in Figure 7–46.

Figure 7–46

Area under the Curve
and X Value for
Example 7–23

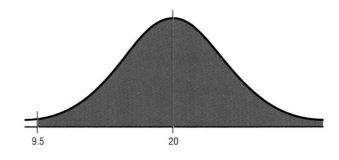

STEP 5   Since the problem is to find the probability of 10 or more positive responses, the normal distribution graph is as shown in Figure 7–46. Hence, the area between 9.5 and 20 must be added to 0.5000 to get the correct approximation.

The z value is

$$z = \frac{9.5 - 20}{4.24} = -2.48$$

STEP 6   The area between 20 and 9.5 is 0.4934. Thus, the probability of getting 10 or more responses is $0.5000 + 0.4934 = 0.9934$, or 99.34%.

It can be concluded, then, that the probability of 10 or more widowed people in a random sample of 200 bowling league members is 99.34%.

Example 7–24

If a baseball player's batting average is 0.320 (32%), find the probability that the player will get at most 26 hits in 100 times at bat.

Solution

Here, $p = 0.32$, $q = 0.68$, and $n = 100$.

STEP 1   Since $np = (100)(0.320) = 32$ and $nq = (100)(0.680) = 68$, the normal distribution can be used to approximate the binomial distribution.

STEP 2 $\mu = np = (100)(0.320) = 32$
$\sigma = \sqrt{npq} = \sqrt{(100)(0.32)(0.68)} = \sqrt{21.76} = 4.66$

STEP 3 $P(X \leq 26)$.

STEP 4 $P(X < 26.5)$. The desired area is shown in Figure 7–47.

Figure 7–47

Area under the Curve
for Example 7–24

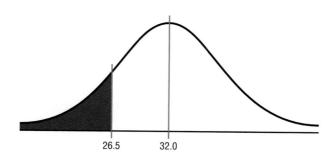

26.5   32.0

STEP 5   The z value is

$$z = \frac{26.5 - 32}{4.66} = -1.18$$

STEP 6   The area between the mean and 26.5 is 0.3810. Since the area in the left tail is desired, 0.3810 must be subtracted from 0.5000. So the probability is $0.5000 - 0.3810 = 0.1190$, or 11.9%.

The closeness of the normal approximation is shown in the next example.

Example 7–25

When $n = 10$ and $p = 0.5$, use the binomial distribution table (Table B in Appendix C) to find the probability that $X = 6$. Then use the normal approximation to find the probability that $X = 6$.

Solution

From Table B, for $n = 10$, $p = 0.5$, and $X = 6$, the probability is 0.205.

For the normal approximation,

$$\mu = np = (10)(0.5) = 5$$
$$\sigma = \sqrt{npq} = \sqrt{(10)(0.5)(0.5)} = 1.58$$

Now, $X = 6$ is represented by the boundaries 5.5 and 6.5. So the z values are

$$z_1 = \frac{6.5 - 5}{1.58} = 0.95 \qquad z_2 = \frac{5.5 - 5}{1.58} = 0.32$$

The corresponding area for 0.95 is 0.3289, and the corresponding area for 0.32 is 0.1255.

The solution is $0.3289 - 0.1255 = 0.2034$, which is very close to the binomial table value of 0.205. The desired area is shown in Figure 7–48.

Figure 7–48

Area under the Curve for Example 7–25

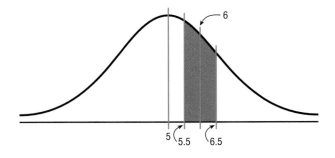

Students sometimes have difficulty deciding whether to add 0.5 or subtract 0.5 from the data value for the correction factor. Table 7–2 summarizes the different situations.

Table 7–2 Summary of the Normal Approximation to the Binomial Distribution	
Binomial	**Normal**
When finding	Use
1. $P(X = a)$	$P(a - 0.5 < X < a + 0.5)$
2. $P(X \geq a)$	$P(X > a - 0.5)$
3. $P(X > a)$	$P(X > a + 0.5)$
4. $P(X \leq a)$	$P(X < a + 0.5)$
5. $P(X < a)$	$P(X < a - 0.5)$
For all cases, $\mu = n \cdot p$, $\sigma = \sqrt{n \cdot p \cdot q}$, $n \cdot p \geq 5$, and $n \cdot q \geq 5$.	

Exercises

7–121.   (**W**) Explain why the normal distribution can be used as an approximation to the binomial distribution. What conditions must be met to use the normal distribution to approximate the binomial distribution? Why is a correction for continuity necessary?

7–122.   (**ans**) Use the normal approximation to the binomial to find the probabilities for the specific value(s) of X.

a. $n = 30, p = 0.5, X = 18$
b. $n = 50, p = 0.8, X = 44$
c. $n = 100, p = 0.1, X = 12$
d. $n = 10, p = 0.5, X \geq 7$
e. $n = 20, p = 0.7, X \leq 12$
f. $n = 50, p = 0.6, X \leq 40$

7–123.   Check each binomial distribution to see whether it can be approximated by the normal distribution (i.e., are $np \geq 5$ and $nq \geq 5$?).

a. $n = 20, p = 0.5$ d. $n = 50, p = 0.2$
b. $n = 10, p = 0.6$ e. $n = 30, p = 0.8$
c. $n = 40, p = 0.9$ f. $n = 20, p = 0.85$

7–124.   Of all 3- to 5-year-old children, 56% are enrolled in school. If a sample of 500 such children is randomly selected, find the probability that at least 250 will be enrolled in school.
Source: *Statistical Abstract of the United States 1994*.

7–125.   Two out of five adult smokers acquired the habit by age 14. If 400 smokers are randomly selected, find the probability that 170 or more acquired the habit by age 14.
Source: *Harper's Index* 289, no. 1735 (December 1994).

7–126.   An airline company has found that 5% of its passengers do not show up for their scheduled flights. If the plane has 100 seats, find the probability that six or more people will not show up for the fully booked flight.

7–127.   In a survey, 15% of Americans said they believe that they will eventually get cancer. In a random sample of 80 Americans used in the survey, find these probabilities.
a.   At least six people in the sample believe they will get cancer.

b.   Fewer than five people in the sample believe they will get cancer.
Source: *Harper's Index* (December 1994), p. 13.

7–128.   If the probability that a newborn child will be female is 50%, find the probability that in 100 births, 55 or more will be females.

7–129.   *Prevention* magazine reports that 18% of drivers use a car phone while driving. Find the probability that in a random sample of 100 drivers, exactly 18 will use a car phone while driving.
Source: USA Snapshot, *USA Today,* June 26, 1995.

7–130.   A dealer states that 90% of all automobiles sold have air conditioning. If the dealer sells 250 cars, find the probability that fewer than 5 of them will not have air conditioning.

7–131.   In 1993, only 3% of elementary and secondary schools in the United States did not have microcomputers. If a random sample of 180 elementary and secondary schools is selected, find the probability that in 1993 6 or fewer schools did not have microcomputers.
Source: *Statistical Abstract of the United States 1994.*

7–132.   Last year, 17% of American workers carpooled to work. If a random sample of 25 workers is selected, find the probability that exactly 5 people will carpool to work.
Source: USA Snapshot, *USA Today,* July 6, 1995.

7–133.   A political candidate estimates that 30% of the voters in his party favor his proposed tax reform bill. If there are 400 people at a rally, find the probability that at least 100 favor his tax bill.

*7–134.   Recall that for use of the normal distribution as an approximation to the binomial distribution, the condition $np \geq 5$ and $nq \geq 5$ must be met. For each given probability, compute the minimum sample size needed for use of the normal approximation.
a. $p = 0.1$ d. $p = 0.8$
b. $p = 0.3$ e. $p = 0.9$
c. $p = 0.5$

7–7

Summary

The normal distribution can be used to describe a variety of variables, such as heights, weights, and temperatures. The normal distribution is bell-shaped, unimodal, symmetric, and continuous; its mean, median, and mode are equal. Since each variable has its own distribution with mean μ and standard deviation σ, mathematicians use the standard normal distribution, which has a mean of 0 and a standard deviation of 1. Other approximately normally distributed variables can be transformed into the standard normal distribution with the formula $z = (X - \mu)/\sigma$.

The normal distribution can also be used to describe a sampling distribution of sample means. These samples must be of the same size and randomly selected from the population. The means of the samples will differ somewhat from the population mean, since samples are generally not perfect representations of the population from which they came. The mean of the sample means will be equal to the population mean, and the standard deviation of the sample means will be equal to the population standard deviation divided by the square root of the sample size. The central limit theorem states that as the size of the samples increases, the distribution of sample means will be approximately normal.

The normal distribution can be used to approximate other distributions, such as the binomial distribution. For the normal distribution to be used as an approximation, the conditions $np \geq 5$ and $nq \geq 5$ must be met. Also, a correction for continuity may be used for more accurate results.

Important Terms

central limit theorem 265

correction for continuity 273

negatively skewed distribution 237

normal distribution 239

positively skewed distribution 238

sampling distribution of sample means 262

sampling error 263

standard error of the mean 265

standard normal distribution 240

symmetrical distribution 237

z value 241

Important Formulas

Formula for the z value (or standard score):

$$z = \frac{X - \mu}{\sigma}$$

Formula for finding a specific data value:

$$X = z \cdot \sigma + \mu$$

Formula for the mean of the sample means:

$$\mu_{\overline{X}} = \mu$$

Formula for the standard error of the mean:

$$\sigma_{\overline{X}} = \frac{\sigma}{\sqrt{n}}$$

Formula for the z value for the central limit theorem:

$$z = \frac{\overline{X} - \mu}{\sigma/\sqrt{n}}$$

Formulas for the mean and standard deviation for the binomial distribution:

$$\mu = n \cdot p \qquad \sigma = \sqrt{n \cdot p \cdot q}$$

Review Exercises

7–135.   Find the area under the standard normal distribution curve for each.
a.   Between $z = 0$ and $z = 1.95$
b.   Between $z = 0$ and $z = 0.37$
c.   Between $z = 1.32$ and $z = 1.82$
d.   Between $z = -1.05$ and $z = 2.05$
e.   Between $z = -0.03$ and $z = 0.53$
f.   Between $z = +1.10$ and $z = -1.80$
g.   To the right of $z = 1.99$
h.   To the right of $z = -1.36$

i. To the left of $z = -2.09$
j. To the left of $z = 1.68$

7–136. Using the standard normal distribution, find each probability.
a. $P(0 < z < 2.07)$
b. $P(-1.83 < z < 0)$
c. $P(-1.59 < z < +2.01)$
d. $P(1.33 < z < 1.88)$
e. $P(-2.56 < z < 0.37)$
f. $P(z > 1.66)$
g. $P(z < -2.03)$
h. $P(z > -1.19)$
i. $P(z < 1.93)$
j. $P(z > -1.77)$

7–137. The average reaction time for a laboratory animal to respond to a stimulus is 1.5 seconds, with a standard deviation of 0.3 second. Find the probability that it will take the animal the following time to react. Assume the variable is normally distributed.
a. Between 1.0 and 1.3 seconds
b. More than 2.2 seconds
c. Less than 1.1 seconds

7–138. The average diastolic blood pressure of a certain age group of people is 85 mmHg. The standard deviation is 6. If an individual from this age group is selected, find the probability that his or her pressure will be the following. Assume the variable is normally distributed.
a. Greater than 90
b. Below 80
c. Between 85 and 95
d. Between 88 and 92

7–139. The average number of miles a mail carrier walks on a typical route in a certain city is 5.6. The standard deviation is 0.8 mile. Find the probability that a carrier chosen at random walks the following miles. Assume the variable is normally distributed.
a. Between 5 and 6 miles
b. Less than 4 miles
c. More than 6.3 miles

7–140. The average number of years a person lives after being diagnosed with a certain disease is 4 years. The standard deviation is 6 months. Find the probability that an individual diagnosed with the disease will live the following numbers of years. Assume the variable is normally distributed.
a. More than 5 years
b. Less than 2 years
c. Between 3.5 and 5.2 years
d. Between 4.2 and 4.8 years

7–141. Americans spend on average $617 per year on their insured vehicles. Assume the variable is normally distributed. If the standard deviation of the distribution is $52, find the probability that a randomly selected insured-vehicle owner will spend the following amounts on his or her vehicle per year.
a. Between $600 and $700
b. Less than $575
c. More than $624
Source: *Statistical Abstract of the United States 1994.*

7–142. The average weight of an airline passenger's suitcase is 45 pounds. The standard deviation is 2 pounds. If 15% of the suitcases are overweight, find the maximum weight allowed by the airline. Assume the variable is normally distributed.

7–143. An educational study to be conducted requires a test score in the middle 40% range. If $\mu = 100$ and $\sigma = 15$, find the highest and lowest acceptable test scores that would enable a candidate to participate in the study. Assume the variable is normally distributed.

7–144. The average repair cost for automatic washers is $73, with a standard deviation of $8. The costs are normally distributed. If 9 washers are repaired, find the probability that the mean of the repair bills will be less than $70. Assume the variable is normally distributed.

7–145. Americans spend on average 12.2 minutes in the shower. If the standard deviation of the variable is 2.3 minutes and the variable is normally distributed, find the probability that the mean time of a sample of 12 Americans who shower will be less than 11 minutes.
Source: USA Snapshot, *USA Today,* July 11, 1995.

7–146. The probability of winning on a slot machine is 5%. If a person plays the machine 500 times, find the probability of winning 30 times. Use the normal approximation to the binomial distribution.

7–147. Of the total population of older Americans, 18% live in Florida. For a randomly selected sample of 200 older Americans, find the probability that more than 40 live in Florida.
Source: Elizabeth Vierck, *Fact Book on Aging* (Santa Barbara, CA: ABC-CLIO 1990).

7–148. In a corporation, 30% of the people elect to enroll in the financial investment program offered by the company. Find the probability that of 800 randomly selected people, at least 260 have enrolled in the program.

7–149. Of the total population of the United States, 20% live in the Northeast. If 200 residents of the United States are selected at random, find the probability that at least 50 live in the Northeast.
Source: *Statistical Abstract of the United States 1994.*

Statistics Today	**What Is Normal? Revisited**

Many of the variables measured in medical tests—blood pressure, triglyceride level, etc.—are approximately normally distributed for the majority of the population in the United States. Thus, researchers can find the mean and standard deviation of these variables. Then, using these two measures along with the z values, they can find normal intervals for healthy individuals. For example, 95% of the systolic blood pressures of healthy individuals fall within two standard deviations of the mean. If an individual's pressure is outside the determined normal range (either above or below), the physician will look for a possible cause and prescribe treatment if necessary.

Quiz

Determine whether each statement is true or false. If the statement is false, explain why.

1. The total area under the normal distribution is infinite.

2. The standard normal distribution is a continuous distribution.

3. All variables that are approximately normally distributed can be transformed into standard normal variables.

4. The z value corresponding to a number below the mean is always negative.

5. The area under the standard normal distribution to the left of $z = 0$ is negative.

6. The central limit theorem applies to means of samples selected from different populations.

Select the best answer.

7. The mean of the standard normal distribution is
a. 0
b. 1
c. 100
d. variable

8. Approximately what percentage of normally distributed data values will fall within one standard deviation above or below the mean?
a. 68%
b. 95%
c. 99.7%
d. variable

9. Which is not a property of the standard normal distribution?
a. It's symmetric about the mean.
b. It's uniform.

c. It's bell-shaped.
d. It's unimodal.

10. When a distribution is positively skewed, the relationship of the mean, median, and mode from left to right will be
a. Mean, median, mode
b. Mode, median, mean
c. Median, mode, mean
d. Mean, mode, median

11. The standard deviation of the sample means equals
a. The population standard deviation
b. The population standard deviation divided by the population mean
c. The population standard deviation divided by the square root of the sample size
d. The square root of the population standard deviation

Complete the following statements with the best answer.

12. When using the standard normal distribution, $P(z < 0) = $ _____.

13. The difference between a sample mean and a population mean is due to _____.

14. The mean of the sample means equals _____.

15. The standard deviation of the sample means is called _____.

16. The normal distribution can be used to approximate the binomial distribution when $n \cdot p$ and $n \cdot q$ are both greater than or equal to _____.

17. When a distribution is negatively skewed, the mean will fall to the _____ of the median.

18. Find the area under the standard normal distribution for each.
a. Between 0 and 1.50
b. Between 0 and -1.25
c. Between 1.56 and 1.96
d. Between -1.20 and -2.25
e. Between -0.06 and 0.73
f. Between 1.10 and -1.80
g. To the right of $z = 1.75$
h. To the right of $z = -1.28$
i. To the left of $z = -2.12$
j. To the left of $z = 1.36$

19. Using the standard normal distribution, find each probability.
a. $P(0 < z < 2.16)$
b. $P(-1.87 < z < 0)$
c. $P(-1.63 < z < 2.17)$
d. $P(1.72 < z < 1.98)$
e. $P(-2.17 < z < 0.71)$
f. $P(z > 1.77)$
g. $P(z < -2.37)$
h. $P(z > -1.73)$
i. $P(z < 2.03)$
j. $P(z > -1.02)$

20. The time it takes for a certain pain reliever to begin to reduce symptoms is 30 minutes, with a standard deviation of 4 minutes. Assuming the variable is normally distributed, find the probability that it will take the medication
a. Between 34 and 35 minutes to begin to work
b. More than 35 minutes to begin to work
c. Less than 25 minutes to begin to work
d. Between 35 and 40 minutes to begin to work

21. The average height of a certain age group of people is 53 inches. The standard deviation is 4 inches. If the variable is normally distributed, find the probability that a selected individual's height will be
a. Greater than 59 inches
b. Less than 45 inches
c. Between 50 and 55 inches
d. Between 58 and 62 inches

22. The average number of gallons of lemonade consumed by the football team during a game is 20, with a standard deviation of 3 gallons. Assume the variable is normally distributed. When a game is played, find the probability of using
a. Between 20 and 25 gallons
b. Less than 19 gallons
c. More than 21 gallons
d. Between 26 and 28 gallons

23. The average number of years a person takes to complete a graduate degree program is 3. The standard deviation is 4 months. Assume the variable is

normally distributed. If an individual enrolls in the program, find the probability that it will take
a. More than 4 years to complete the program
b. Less than 3 years to complete the program
c. Between 3.8 and 4.5 years to complete the program
d. Between 2.5 and 3.1 years to complete the program

24. On the daily run of an express bus, the average number of passengers is 48. The standard deviation is 3. Assume the variable is normally distributed. Find the probability that the bus will have
a. Between 36 and 40 passengers
b. Less than 42 passengers
c. More than 48 passengers
d. Between 43 and 47 passengers

25. The average height of books on a library shelf is 8.3 centimeters. The standard deviation is 0.6 centimeter. If 20% of the books are oversized, find the minimum height of the oversized books on the library shelf. Assume the variable is normally distributed.

26. Membership in an elite organization requires a test score in the upper 30% range. If $\mu = 115$ and $\sigma = 12$, find the lowest acceptable score that would enable a candidate to apply for membership. Assume the variable is normally distributed.

27. The average repair cost of a microwave oven is $55, with a standard deviation of $8. The costs are normally distributed. If 12 ovens are repaired, find the probability that the mean of the repair bills will be greater than $60.

28. The average electric bill in a residential area is $72 for the month of April. The standard deviation is $6. If the amounts of the electric bills are normally distributed, find the probability that the mean of the bill for 15 residents will be less than $75.

29. The probability of winning on a slot machine is 8%. If a person plays the machine 300 times, find the probability of winning at most 20 times. Use the normal approximation to the binomial distribution.

30. If 10% of the people in a certain factory are members of a union, find the possibility that in a sample of 2000, fewer than 180 people are union members.

31. In a corporation, 30% of the employees contributed to a retirement program offered by the company. Find the probability that of 300 randomly selected people, at least 105 are enrolled in the program.

32. A company that installs swimming pools for residential customers sells a pool package to 18% of the customers its representatives visit. If the sales reps visit 180 customers, find the probability of their making at least 40 sales.

Critical Thinking Challenge

When Is a Distribution Normal?   Sometimes a researcher must decide whether or not a variable is normally distributed. There are several ways to do this. One simple but very subjective method uses special graph paper, which is called *normal probability paper*. For the distribution of systolic blood pressure readings given in Chapter 3 of the textbook, the following method can be used:

1.   Make a table, as shown.

Boundaries	Frequency	Cumulative frequency	Cumulative percent frequency
89.5–104.5	24		
104.5–119.5	62		
119.5–134.5	72		
134.5–149.5	26		
149.5–164.5	12		
164.5–179.5	4		
	200		

2.   Find the cumulative frequencies for each class and place the results in the third column.

3.   Find the cumulative percents for each class by dividing each cumulative frequency by 200 (the total frequencies) and multiplying by 100. (For the first

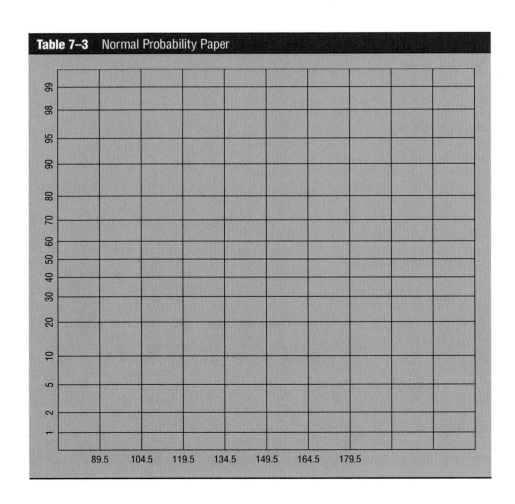

Table 7–3   Normal Probability Paper

class, it would be 24/200 × 100 = 12%.) Place these values in the last column.

4. Using the normal probability paper, label the x axis with the class boundaries as shown and plot the percents.

5. If the points fall approximately in a straight line, it can be concluded that the distribution is normal. Do you feel that this distribution is approximately normal? Explain your answer. (Another method for determining if a variable is approximately normally distributed is shown in Chapter 12.)

6. To find an approximation of the mean or median, draw a horizontal line from the 50% point on the y axis

over to the curve and then a vertical line down to the x axis. Compare this approximation of the mean with the computed mean.

7. To find an approximation of the standard deviation, locate the values on the x axis that correspond to the 16% and 84% values on the y axis. Subtract these two values and divide the result by 2. Compare this approximate standard deviation to the computed standard deviation.

8. Explain why the method used in Step 7 works.

Data Project

The Normal Distribution   Select a variable (interval or ratio) and collect 30 data values.

a.   Contruct a frequency distribution for the variable.

b.   Use the procedure described in the critical thinking challenge "When Is a Distribution Normal?" to

graph the distribution on normal probability paper.

c.   Can you conclude that the data are approximately normally distributed? Explain your answer.

TI-83 Calculator

The Normal Distribution

A.   To find the area under the standard normal distribution curve between any two z values:
Example: Find the area between $z = 2.00$ and $z = 2.47$.

1.   Press **2nd [DISTR]** then **2** to get `normalcdf(`

2.   Enter 2, 2.47) then press **ENTER**
The calculator will display .0159944012.
Use −1E99 and 1E99 to specify infinity
Example: Find the area to the left of −1.93

1.   Press **2nd [DISTR]** then **2** to get `normalcdf(`

2.   Enter −1E99, −1.93) then press **ENTER**
The calculator will display .0268033499
Note: Use **2nd [EE]** to obtain E

B.   To find the area under a normal distribution curve between any two z values given a specific mean and standard deviation:
Example: Find the area between 13 and 17 when $\overline{X} = 14$ and $\sigma = 2$.

1.   Press **2nd [DISTR]** then **2** to get `normalcdf(`.

2.   Enter 13, 17, 14, 2) then press **ENTER**
The calculator will display .6246552391

chapter

8

Confidence Intervals and Sample Size

Outline

Objectives

After completing this chapter, you should be able to

1. Find the confidence interval for the mean when σ is known or $n \geq 30$.

2. Determine the minimum sample size for finding a confidence interval for the mean.

3. Find the confidence interval for the mean when σ is unknown and $n < 30$.

4. Find the confidence interval for a proportion.

5. Determine the minimum sample size for finding a confidence interval for a proportion.

6. Find a confidence interval for a variance and a standard deviation.

Statistics Today

Would You Change the Channel?

In the article shown on the next page, a survey by the Roper Organization found that 45% of the people who were offended by a television program would change the channel, while 15% would turn off their television sets. The article further states that the margin of error is 3 percentage points, and 4000 adults were interviewed.

 Several questions arise:

1. How do these estimates compare with the true population percentages?
2. What is meant by a margin of error of 3 percentage points?
3. Is the sample of 4000 large enough to represent the population of all adults who watch television in the United States?

 After reading this chapter, you will be able to answer these questions, since this chapter explains how statisticians can use statistics to make estimates of parameters.

TV Survey Eyes On-Off Habits

NEW YORK (AP)—You're watching TV and the show offends you. Do you sit there and take it or do you get up and go?

A survey released yesterday on the public's attitudes toward television said 12 percent of the respondents do nothing, 45 percent change channels and 15 percent turn off their sets.

National Association of Broadcasters and the Network Television Association, an alliance of the Big Three broadcast networks, co-sponsored the survey, which has been taken every two years since 1959.

The Roper Organization polled 4,000 adults nationwide in personal interviews between November and December last year. The results had a margin of sampling error of plus or minus 3 percentage points.

The study, "America's Watching," said that even in an age of channel "grazing" about two in three viewers regularly make a special effort to watch shows they particularly like.

And despite increasing cable alternatives, about three-quarters of these "appointment" viewers with cable said that ABC, CBS and NBC still offer most of their favorites.

Viewers find TV news highly credible, the survey also said. In the event of conflicting news reports, 56 percent of respondents said they would believe television's account above others.

A hefty 69 percent said they get most of their news from TV. Newspapers were second, with 43 percent. Radio, word-of-mouth and magazines fell far behind.

A majority of the respondents reported having seen something on television that they found "either personally offensive or morally objectionable" in the past few weeks. But only 42 percent could specify whether it was profanity, sex or violence.

Source: Associated Press. Reprinted with permission.

8–1

Introduction

One aspect of inferential statistics is **estimation,** which is the process of estimating the value of a parameter from information obtained from a sample. For example, *The Book of Odds,* by Michael D. Shook and Robert L. Shook (New York: Penguin Putnam Inc., 1991), contains the following statements:

> *"One out of 4 Americans is currently dieting." (Calorie Control Council. Used with permission.)*
>
> *"Seventy-two percent of Americans have flown in commercial airlines." ("The Bristol Meyers Report: Medicine in the Next Century." Used with permission.)*
>
> *"The average kindergarten student has seen more than 5000 hours of television." (U.S. Department of Education.)*
>
> *"The average school nurse makes $32,786 a year." (National Association of School Nurses. Used with permission.)*
>
> *"The average amount of life insurance is $108,000 per household with life insurance." (American Council of Life Insurance. Used with permission.)*

Since the populations from which these values were obtained are large, these values are only *estimates* of the true parameters and are derived from data collected from samples.

The statistical procedures for estimating the population mean, proportion, variance, and standard deviation will be explained in this chapter.

An important question in estimation is that of sample size. How large should the sample be in order to make an accurate estimate? This question is not easy to answer since the size of the sample depends on several factors, such as the accuracy desired and the probability of making a correct estimate. The question of sample size will be explained in this chapter also.

8–2

Confidence Intervals for the Mean (σ Known or $n \geq 30$) and Sample Size

Suppose a college president wishes to estimate the average age of the students attending classes this semester. The president could select a random sample of 100 students and find the average age of these students, say 22.3 years. From the sample mean, the president could infer that the average age of all the students is 22.3 years. This type of estimate is called a *point estimate*.

Objective 1. Find the confidence interval for the mean when σ is known or $n \geq 30$.

A **point estimate** is a specific numerical value estimate of a parameter. The best point estimate of the population mean μ is the sample mean $\overline{X}$.

One might ask why other measures of central tendency, such as the median and mode, are not used to estimate the population mean. The reason is that the means of samples vary less than other statistics (such as medians and modes) when many samples are selected from the same population. Therefore, the sample mean is the best estimate of the population mean.

Sample measures (i.e., statistics) are used to estimate population measures (i.e., parameters). These statistics are called **estimators.** As previously stated, the sample mean is a better estimator of the population mean than the sample median or sample mode.

A good estimator must satisfy the three properties described next.

Three Properties of a Good Estimator

1. The estimator must be an **unbiased estimator.** That is, the expected value or the mean of the estimates obtained from samples of a given size is equal to the parameter being estimated.
2. The estimator must be consistent. For a **consistent estimator,** as sample size increases, the value of the estimator approaches the value of the parameter estimated.
3. The estimator must be a **relatively efficient estimator.** That is, of all the statistics that can be used to estimate a parameter, the relatively efficient estimator has the smallest variance.

Confidence Intervals

As stated in Chapter 7, the sample mean will be, for the most part, somewhat different from the population mean due to sampling error. Therefore, one might ask a second question: How good is a point estimate? The answer is that there is no way of knowing how close the point estimate is to the population mean.

This answer places some doubt on the accuracy of point estimates. For this reason, statisticians prefer another type of estimate called an *interval estimate*.

An **interval estimate** of a parameter is an interval or a range of values used to estimate the parameter. This estimate may or may not contain the value of the parameter being estimated.

In an interval estimate, the parameter is specified as being between two values. For example, an interval estimate for the average age of all students might be $26.9 < \mu < 27.7$, or 27.3 ± 0.4 years.

Either the interval contains the parameter or it does not. A degree of confidence (usually a percent) can be assigned before an interval estimate is made. For instance, one may wish to be 95% confident that the interval contains the true population mean. Another question then arises. Why 95%? Why not 99% or 99.5%?

If one desires to be more confident, such as 99% or 99.5% confident, then the interval must be larger. For example, a 99% confidence interval for the mean age of college students might be $26.7 < \mu < 27.9$, or 27.3 ± 0.6. Hence, a trade-off occurs. To be more confident that the interval contains the true population mean, one must make the interval wider.

A **confidence interval** is a specific interval estimate of a parameter determined by using data obtained from a sample and the specific confidence level of the estimate.

The **confidence level** of an interval estimate of a parameter is the probability that the interval estimate will contain the parameter.

Intervals constructed in this way are called *confidence intervals*. Three common confidence intervals are used: the 90%, the 95%, and the 99% confidence intervals.

The algebraic derivation of the formula for determining a confidence interval for a mean will be shown later. A brief intuitive explanation will be given first.

The central limit theorem states that when the sample size is large, approximately 95% of the sample means will fall within 1.96 standard errors of the population mean. That is,

$$\mu \pm 1.96 \left(\frac{\sigma}{\sqrt{n}} \right)$$

Now, if a specific sample mean is selected, say $\overline{X}$, there is a 95% probability that it falls within the range of $\mu \pm 1.96(\sigma/\sqrt{n})$. Likewise, there is a 95% probability that the interval specified by

$$\overline{X} \pm 1.96 \left(\frac{\sigma}{\sqrt{n}} \right)$$

will contain μ, as will be shown later. Stated another way,

$$\overline{X} - 1.96 \left(\frac{\sigma}{\sqrt{n}} \right) < \mu < \overline{X} + 1.96 \left(\frac{\sigma}{\sqrt{n}} \right)$$

Hence, one can be 95% confident that the population mean is contained within that interval when the values of the variable are normally distributed in the population.

The value used for the 95% confidence interval, 1.96, is obtained from Table E in Appendix C. For a 99% confidence interval, the value 2.58 is used instead of 1.96 in the formula. This value is also obtained from Table E and is based on the standard normal distribution. Since other confidence intervals are used in statistics, the symbol $z_{\alpha/2}$ (read "zee sub alpha over two") is used in the general formula for confidence intervals. The Greek letter α (alpha) represents the total area in both of the tails of the standard normal distribution curve. $\alpha/2$ represents the area in each one of the tails. More will be said after Examples 8–1 and 8–2 about finding other values for $z_{\alpha/2}$.

The relationship between α and the confidence level is that the stated confidence level is the percentage equivalent to the decimal value of $1 - \alpha$, and vice versa. When the 95% confidence interval is to be found, $\alpha = 0.05$, since $1 - 0.05 = 0.95$, or 95%. When $\alpha = 0.01$, then $1 - \alpha = 1 - 0.01 = 0.99$, and the 99% confidence interval is being calculated.

Formula for the Confidence Interval of the Mean for a Specific α

$$\overline{X} - z_{\alpha/2}\left(\frac{\sigma}{\sqrt{n}}\right) < \mu < \overline{X} + z_{\alpha/2}\left(\frac{\sigma}{\sqrt{n}}\right)$$

For a 95% confidence interval, $z_{\alpha/2} = 1.96$; and for a 99% confidence interval, $z_{\alpha/2} = 2.58$.

The term $z_{\alpha/2}(\sigma/\sqrt{n})$ is called the *maximum error of estimate*. For a specific value, say $\alpha = 0.05$, 95% of the sample means will fall within this error value on either side of the population mean, as previously explained. See Figure 8–1.

Figure 8–1

95% Confidence Interval

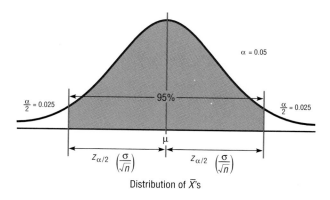

Distribution of $\overline{X}$'s

The **maximum error of estimate** is the maximum difference between the point estimate of a parameter and the actual value of the parameter.

A more detailed explanation of the error of estimate follows the examples illustrating the computation of confidence intervals.

Rounding Rule for a Confidence Interval for a Mean   When computing a confidence interval for a population mean using *raw data,* round off to one more decimal place than the number of decimal places in the original data. When computing a confidence interval for a population mean using a sample mean and a standard deviation, round off to the same number of decimal places as given for the mean.

....

Example 8–1

The president of a large university wishes to estimate the average age of the students presently enrolled. From past studies, the standard deviation is known to be 2 years. A sample of 50 students is selected, and the mean is found to be 23.2 years. Find the 95% confidence interval of the population mean.

Solution

Since the 95% confidence interval is desired, $z_{\alpha/2} = 1.96$. Hence, substituting in the formula

$$\overline{X} - z_{\alpha/2}\left(\frac{\sigma}{\sqrt{n}}\right) < \mu < \overline{X} + z_{\alpha/2}\left(\frac{\sigma}{\sqrt{n}}\right)$$

one gets

$$23.2 - (1.96)\left(\frac{2}{\sqrt{50}}\right) < \mu < 23.2 + (1.96)\left(\frac{2}{\sqrt{50}}\right)$$

$$23.2 - 0.6 < \mu < 23.2 + 0.6$$

$$22.6 < \mu < 23.8$$

or 23.2 ± 0.6 years. Hence, the president can say, with 95% confidence, that the average age of the students is between 22.6 and 23.8 years, based on 50 students.

....

Example 8–2

A certain medication is known to increase the pulse rate of its users. The standard deviation of the pulse rate is known to be 5 beats per minute. A sample of 30 users had an average pulse rate of 104 beats per minute. Find the 99% confidence interval of the true mean.

Solution

Since the 99% confidence interval is desired, $z_{\alpha/2} = 2.58$. Hence, substituting in the formula

$$\overline{X} - z_{\alpha/2}\left(\frac{\sigma}{\sqrt{n}}\right) < \mu < \overline{X} + z_{\alpha/2}\left(\frac{\sigma}{\sqrt{n}}\right)$$

one gets

$$104 - (2.58)\left(\frac{5}{\sqrt{30}}\right) < \mu < 104 + (2.58)\left(\frac{5}{\sqrt{30}}\right)$$

$$104 - 2.4 < \mu < 104 + 2.4$$

$$101.6 < \mu < 106.4$$

$$\text{rounded to } 102 < \mu < 106 \text{ or } 104 \pm 2$$

....

Another way of looking at a confidence interval is shown in Figure 8–2. According to the central limit theorem, approximately 95% of the sample means fall within 1.96 standard deviations of the population mean if the sample size is 30 or more or if σ is known when n is less than 30. If it were possible to build a confidence interval about each sample mean, as was done in the previous examples, 95% of these intervals would contain the population mean, as shown in Figure 8–3. Hence, one can be 95% confident that a confidence interval built around a specific sample mean would contain the population mean.

Figure 8–2

95% Confidence Interval for Sample Means

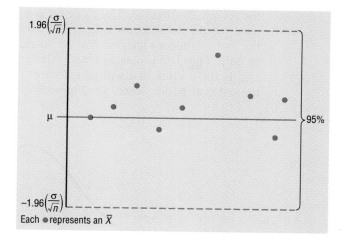

Each ● represents an $\overline{X}$

Figure 8–3

95% Confidence Intervals for Each Sample Mean

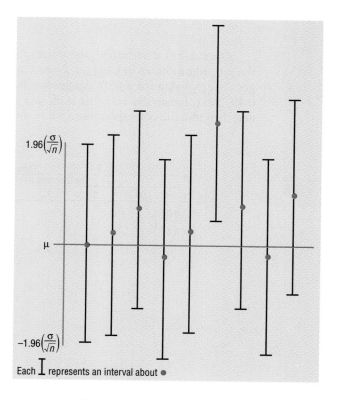

Each ⊥ represents an interval about ●

If one desires to be 99% confident, the confidence intervals must be enlarged so that 99 out of every 100 intervals contain the population mean.

Since other confidence intervals (besides 90%, 95%, and 99%) are sometimes used in statistics, an explanation of how to find the values for $z_{\alpha/2}$ is necessary. As stated previously, the Greek letter α represents the total of the areas in both tails of the normal distribution. The value for α is found by subtracting the decimal equivalent for the desired confidence level from 1. For example, if one wanted to find the 98% confidence interval, one would change 98% to 0.98 and find $\alpha = 1 - 0.98$, or 0.02. Then $\alpha/2$ is obtained by dividing α by 2. So $\alpha/2$ is 0.02/2, or 0.01. Finally, $z_{0.01}$ is the z value that will give an area of 0.01 in the right tail of the standard normal distribution curve. See Figure 8–4.

Figure 8–4

Finding $\alpha/2$ for a 98% Confidence Interval

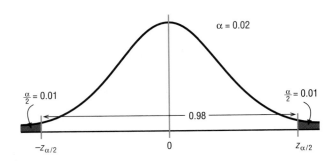

Once $\alpha/2$ is determined, the corresponding $z_{\alpha/2}$ value can be found by using the procedure shown in Chapter 7 (see Example 7–17), which is reviewed here. To get the $z_{\alpha/2}$ value for a 98% confidence interval, subtract 0.01 from 0.5000 to get 0.49. Next, locate the area that is closest to 0.49 (in this case, 0.4901) in Table E, and then find the corresponding z value. In this example, it is 2.33. See Figure 8–5.

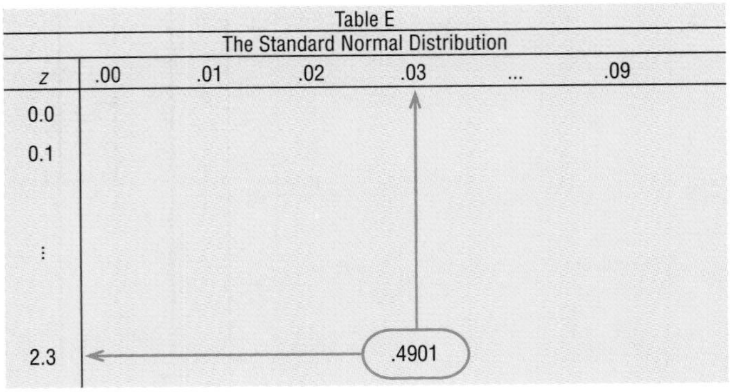

Figure 8–5

Finding $z_{\alpha/2}$ for a 98% Confidence Interval

			Table E			
			The Standard Normal Distribution			
z	.00	.01	.02	.03	...	.09
0.0						
0.1						
⋮						
2.3				.4901		

For confidence intervals, only the positive value is used in the formula.

When the original variable is normally distributed and σ is known, the standard normal distribution can be used to find confidence intervals regardless of the size of the sample. When $n \geq 30$, the distribution of means will be approximately normal even if the original distribution of the variable departs from normality. Also, if $n \geq 30$ (some authors use $n > 30$), s can be substituted for σ in the formula for confidence intervals; and the standard normal distribution can be used to find confidence intervals for means, as shown in the next example.

Example 8–3

A sample of 50 days showed that a fast-food restaurant served an average of 182 customers during lunchtime (11:00 A.M. to 2:00 P.M.). The standard deviation of the sample was 8. Find the 90% confidence interval for the mean.

Solution

STEP 1   Find $\alpha/2$. Since the 90% confidence interval is to be used, $\alpha = 1 - 0.90 = 0.10$, and

$$\frac{\alpha}{2} = \frac{0.10}{2} = 0.05$$

STEP 2   Find $z_{\alpha/2}$. Subtract 0.05 from 0.5000 to get 0.4500. The corresponding z value obtained from Table E is 1.65. (*Note:* This value is found by using the z value for an area between 0.4495 and 0.4505. A more precise z value obtained mathematically is 1.645 and is sometimes used; however, 1.65 will be used in this textbook.)

STEP 3   Substitute in the formula

$$\overline{X} - z_{\alpha/2} \left(\frac{s}{\sqrt{n}} \right) < \mu < \overline{X} + z_{\alpha/2} \left(\frac{s}{\sqrt{n}} \right)$$

(s is used in place of σ when σ is unknown, since $n \geq 30$.)

$$182 - 1.65 \left(\frac{8}{\sqrt{50}} \right) < \mu < 182 + 1.65 \left(\frac{8}{\sqrt{50}} \right)$$

$$180.1 < \mu < 183.9 \text{ or}$$

$$180 < \mu < 184$$

Hence, one can be 90% confident that the true population mean is between 180 and 184, or 182 ± 2.

Computer Applications for z Confidence Interval for a Mean

MINITAB z confidence interval for mean:

1. Enter the data into C1.
2. Click Stat > Basic Statistics > 1-Sample Z.
3. Highlight C1 in the dialog box and click on Select.
4. Click on Confidence interval. Press Tab.
5. Type in the desired level (e.g., 90, 95, 99). Then press Tab.
6. Type in the value for sigma.
7. Click on OK.

Example: Find the 95% confidence interval of the mean when $\sigma = 11$ using the following data.

Data:

C1	43	52	18	20	25	45	43	21	42	32
	24	32	19	25	26	44	42	41	41	53
	22	25	23	21	27	33	36	47	19	20

MINITAB printout for this example:

Z Confidence Intervals

```
The assumed sigma = 11.0
Variable     N      Mean    StDev    SE Mean        95.0 % CI
C1          30     32.03    11.01      2.01    (    28.10,      35.97)
```

Summary: The 95% confidence interval is $28.10 < \mu < 35.97$. The output also shows the sample size (N), the mean, the standard deviation, and the standard error of the mean.

Sample Size

Objective 2. Determine the minimum sample size for finding a confidence interval for the mean.

Sample size determination is closely related to statistical estimation. Quite often, one asks, "How large a sample is necessary to make an accurate estimate?" The answer is not simple, since it depends on three things: the maximum error of estimate, the population standard deviation, and the degree of confidence. For example, how close to the true mean does one want to be (2 units, 5 units, etc.), and how confident does one wish to be (90%, 95%, 99%, etc.)? For the purpose of this chapter, it will be assumed that the population standard deviation of the variable is known or has been estimated from a previous study.

The formula for sample size is derived from the maximum error of estimate formula,

$$E = z_{\alpha/2}\left(\frac{\sigma}{\sqrt{n}}\right)$$

and this formula is solved for n as follows:

$$E\sqrt{n} = z_{\alpha/2}(\sigma)$$

$$\sqrt{n} = \frac{z_{\alpha/2}\cdot\sigma}{E}$$

Hence, $n = \left(\dfrac{z_{\alpha/2}\cdot\sigma}{E}\right)^2$

Formula for the Minimum Sample Size Needed for an Interval Estimate of the Population Mean

$$n = \left(\frac{z_{\alpha/2}\cdot\sigma}{E}\right)^2$$

where E is the maximum error of estimate. If necessary, round the answer up to obtain a whole number. That is, if there is any fraction or decimal portion in the answer, use the next whole number for sample size, n.

Example 8–4

The college president asks the statistics teacher to estimate the average age of the students at their college. How large a sample is necessary? The statistics teacher decides the estimate should be accurate within 1 year and be 99% confident. From a previous study, the standard deviation of the ages is known to be 3 years.

Solution

Since $\alpha = 0.01$ (or $1 - 0.99$), $z_{\alpha/2} = 2.58$, and $E = 1$, substituting in the formula, one gets

$$n = \left(\frac{z_{\alpha/2} \cdot \sigma}{E} \right)^2 = \left[\frac{(2.58)(3)}{1} \right]^2 = 59.9$$

which is rounded up to 60. Therefore, in order to be 99% confident that the estimate is within 1 year of the true mean age, the teacher needs a sample size of at least 60 students.

....

Notice that when one is finding the sample size, the size of the population is irrelevant when the population is large or infinite or when sampling is done with replacement. In other cases, an adjustment is made in the formula for computing sample size. This adjustment is beyond the scope of this book.

The formula for determining sample size requires the use of the population standard deviation. What then happens when σ is unknown? In this case, an attempt is made to estimate σ. One such way is to use the standard deviation, s, obtained from a sample taken previously as an estimate for σ.

Sometimes, interval estimates rather than point estimates are reported. For instance, one may read a statement such as "On the basis of a sample of 200 families, the survey estimates that an American family of two spends an average of $84 per week for groceries. One can be 95% confident that this estimate is accurate within $3 of the true mean." This statement means that the 95% confidence interval of the true mean is

$$\$81 < \mu < \$87$$

The algebraic derivation of the formula for a confidence interval is shown next. As explained in Chapter 7, the sampling distribution of the mean is approximately normal when large samples ($n \geq 30$) are taken from a population. Also,

$$z = \frac{\overline{X} - \mu}{\sigma/\sqrt{n}}$$

Furthermore, there is a probability of $1 - \alpha$ that a z will have a value between $-z_{\alpha/2}$ and $+z_{\alpha/2}$. Hence,

$$-z_{\alpha/2} < \frac{\overline{X} - \mu}{\sigma/\sqrt{n}} < z_{\alpha/2}$$

Using algebra,

$$-z_{\alpha/2} \cdot \frac{\sigma}{\sqrt{n}} < \overline{X} - \mu < z_{\alpha/2} \cdot \frac{\sigma}{\sqrt{n}}$$

Subtracting $\overline{X}$ from both sides and the middle, one gets

$$-\overline{X} - z_{\alpha/2} \cdot \frac{\sigma}{\sqrt{n}} < -\mu < -\overline{X} + z_{\alpha/2} \cdot \frac{\sigma}{\sqrt{n}}$$

Multiplying by -1, one gets

$$+\overline{X} + z_{\alpha/2} \cdot \frac{\sigma}{\sqrt{n}} > \mu > \overline{X} - z_{\alpha/2} \cdot \frac{\sigma}{\sqrt{n}}$$

Reversing the inequality, one gets the formula for the confidence interval:

$$\overline{X} - z_{\alpha/2} \cdot \frac{\sigma}{\sqrt{n}} < \mu < \overline{X} + z_{\alpha/2} \cdot \frac{\sigma}{\sqrt{n}}$$

Exercises

8–1. (**W**) What is the difference between a point estimate and an interval estimate of a parameter? Which is better? Why?

8–2. (**W**) What information is necessary to calculate a confidence interval?

8–3. (**W**) What is the maximum error of estimate?

8–4. (**W**) What is meant by the 95% confidence interval of the mean?

8–5. (**W**) What are three properties of a good estimator?

8–6. (**W**) What statistic best estimates μ?

8–7. (**W**) What is necessary to determine sample size?

8–8. (**W**) When one is determining the sample size for a confidence interval, is the size of the population relevant?

8–9. Find the critical values for each.
  a. $z_{\alpha/2}$ for the 99% confidence interval
  b. $z_{\alpha/2}$ for the 98% confidence interval
  c. $z_{\alpha/2}$ for the 95% confidence interval
  d. $z_{\alpha/2}$ for the 90% confidence interval
  e. $z_{\alpha/2}$ for the 94% confidence interval

8–10. A study of 36 camels showed that they could walk at an average rate of 2.6 miles per hour. The sample standard deviation is 0.4. Find the 95% confidence interval of the mean for all camels.

8–11. A sample of the reading scores of 35 fifth-graders has a mean of 82. The standard deviation of the sample is 15.
  a. Find the 95% confidence interval of the mean reading scores of all fifth-graders.

  b. Find the 99% confidence interval of the mean reading scores of all fifth-graders.
  c. Which interval is larger? Explain why.

8–12. The average annual wind speed in Rochester, Minnesota, is 13.1 miles per hour. If a sample of 100 days was used to determine the average wind speed, find the 98% confidence interval of the mean. Assume the standard deviation of the sample was 2.8 miles per hour.
Source: Snapshot, *USA Today,* July 26, 1995.

8–13. A study of 40 English composition professors showed that they spent, on average, 12.6 minutes correcting a student's term paper.
  a. Find the 90% confidence interval of the mean time for all composition papers when $\sigma = 2.5$ minutes.
  b. If a professor stated that he spent, on average, 30 minutes correcting a term paper, what would be your reaction?

8–14. A study of 40 bowlers showed that their average score was 186. The standard deviation of the population is 6.
  a. Find the 95% confidence of the mean score for all bowlers.
  b. Find the 95% confidence interval of the mean score if a sample of 100 bowlers is used instead of a sample of 40.
  c. Which interval is smaller? Explain why.

8–15. A study found that 8- to 12-year-olds spend an average of $18.50 per trip to a mall. If a sample of 49 children was used, find the 90% confidence interval of the mean. Assume the standard deviation of the sample is $1.56.
Source: *USA Today,* July 25, 1995.

8–16.   A study found that the average time it took a person to find a new job was 5.9 months. If a sample of 36 job seekers was surveyed, find the 95% confidence interval of the mean. Assume the standard deviation of the sample is 0.8 month.

Source: *The Book of Odds* by Michael D. Shook and Robert Shook (New York: Penguin Putnam Inc., 1991) p. 46.

8–17.   A random sample of 48 days taken at a large hospital shows that an average of 38 patients were treated in the emergency room per day. The standard deviation of the population is 4.

a.   Find the 99% confidence interval of the mean number of ER patients treated each day at the hospital.

b.   Find the 99% confidence interval of the mean number of ER patients treated each day if the standard deviation were 8 instead of 4.

c.   Why is the confidence interval for part *b* wider than the one for part *a*?

8–18.   A random sample of 49 female shoppers showed that they spent an average of $23.45 per visit at a grocery store. The standard deviation of the sample was $2.80. Find the 90% confidence interval of the true mean.

8–19.   Noise levels at various area urban hospitals were measured in decibels. The mean of the noise levels in 84 corridors was 61.2 decibels, and the standard deviation was 7.9. Find the 95% confidence interval of the true mean.

Source: M. Bayo, A. Garcia, and A. Garcia, "Noise Levels in an Urban Hospital and Workers' Subjective Responses," *Archives of Environmental Health* 50, no. 3 (May/June 1995).

8–20.   A researcher is interested in estimating the average salary of police officers in a large city. She wants to be 95% confident that her estimate is correct. If the standard deviation is $1050, how large a sample is needed to get the desired information and to be accurate within $200?

8–21.   A university dean wishes to estimate the average number of hours his part-time instructors teach per week. The standard deviation from a previous study is 2.6 hours. How large a sample must be selected if he wants to be 99% confident of finding whether the true mean differs from the sample mean by 1 hour?

8–22.   In the hospital study cited in Exercise 8–19, the mean noise level in the 171 ward areas was 58.0 decibels, and the standard deviation was 4.8. Find the 90% confidence interval of the true mean.

Source: M. Bayo, A. Garcia, and A. Garcia, "Noise Levels in an Urban Hospital and Workers' Subjective Responses," *Archives of Environmental Health* 50, no. 3 (May/June 1995).

8–23.   An insurance company is trying to estimate the average number of sick days that full-time food-service workers use per year. A pilot study found the standard deviation to be 2.5 days. How large a sample must be selected if the company wants to be 95% confident of getting an interval that contains the true mean with a maximum error of 1 day?

8–24.   A restaurant owner wishes to find the 99% confidence interval of the true mean cost of a dry martini. How large should the sample be if she wishes to be accurate within $0.10? A previous study showed that the standard deviation of the price was $0.12.

8–25.   A health care professional wishes to estimate the birth weights of infants. How large a sample must she select if she desires to be 90% confident that the true mean is within 6 ounces of the sample mean? The standard deviation of the birth weights is known to be 8 ounces.

8–3

Confidence Intervals for the Mean (σ Unknown and $n < 30$)

Objective 3. Find the confidence interval for the mean when σ is unknown and $n < 30$.

When σ is known and the variable is normally distributed or when σ is unknown and $n \geq 30$, the standard normal distribution is used to find confidence intervals for the mean. However, in many situations, the population standard deviation is not known and the sample size is less than 30. In such situations, the standard deviation from the sample can be used in place of the population standard deviation for confidence intervals. But a somewhat different distribution, called the ***t* distribution,** must be used when the sample size is less than 30 and the variable is normally or approximately normally distributed.

Some important characteristics of the *t* distribution are described next.

Characteristics of the *t* Distribution

The *t* distribution shares some characteristics of the normal distribution and differs from it in others. The *t* distribution is similar to the standard normal distribution in the following ways.

1. It is bell-shaped.
2. It is symmetrical about the mean.
3. The mean, median, and mode are equal to 0 and are located at the center of the distribution.
4. The curve never touches the *x* axis.

The *t* distribution differs from the standard normal distribution in the following ways.

1. The variance is greater than 1.
2. The *t* distribution is actually a family of curves based on the concept of *degrees of freedom*, which is related to sample size.
3. As the sample size increases, the *t* distribution approaches the standard normal distribution. See Figure 8–6.

Figure 8–6

The *t* Family of Curves

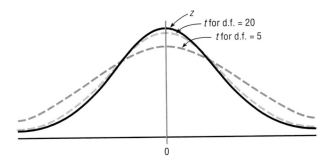

Many statistical distributions use the concept of degrees of freedom, and the formulas for finding the degrees of freedom vary for different statistical tests. The **degrees of freedom** are the number of values that are free to vary after a sample statistic has been computed, and they tell the researcher which specific curve to use when a distribution consists of a family of curves.

For example, if the mean of 5 values is 10, then 4 of the 5 values are free to vary. But once 4 values are selected, the fifth value must be a specific number to get a sum of 50, since $50 \div 5 = 10$. Hence, the degrees of freedom are $5 - 1 = 4$, and this value tells the researcher which *t* curve to use.

The symbol d.f. will be used for degrees of freedom. The degrees of freedom for a confidence interval for the mean are found by subtracting 1 from the sample size. That is, d.f. $= n - 1$. *Note:* For other statistical tests used later in this book, the degrees of freedom are not equal to $n - 1$.

The formula for finding a confidence interval about the mean using the *t* distribution is given next.

> **Formula for a Specific Confidence Interval for the Mean When σ Is Unknown and $n < 30$**
>
> $$\overline{X} - t_{\alpha/2}\left(\frac{s}{\sqrt{n}}\right) < \mu < \overline{X} + t_{\alpha/2}\left(\frac{s}{\sqrt{n}}\right)$$
>
> The degrees of freedom are $n - 1$.

The values for $t_{\alpha/2}$ are found in Table F in Appendix C. The top row of Table F, labeled "Confidence Intervals," is used to get these values. The other two rows, labeled "One Tail" and "Two Tails," will be explained in the next chapter and should not be used here.

The next two examples show how to find the value in Table F for $t_{\alpha/2}$.

Example 8–5

Find the $t_{\alpha/2}$ value for a 95% confidence interval when the sample size is 22.

Solution

d.f. = 22 − 1, or 21. Find 21 in the left column and 95% in the row labeled "Confidence Intervals." The intersection where the two meet gives the value for $t_{\alpha/2}$, which is 2.080. See Figure 8–7.

Figure 8–7

Finding $t_{\alpha/2}$ for Example 8–5

	Confidence Intervals	50%	80%	90%	95%	98%	99%
d.f.	One Tail α	0.25	0.10	0.05	0.025	0.01	0.005
	Two Tails α	0.50	0.20	0.10	0.05	0.02	0.01
1							
2							
3							
⋮							
21						2.080	2.518	2.831
⋮							
z ∞			.674	1.282[a]	1.645[b]	1.960	2.326[c]	2.576[d]

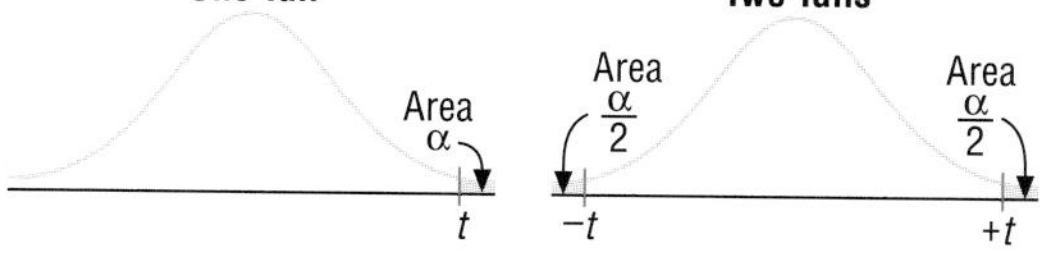

Table F — The *t* Distribution

One Tail   Two Tails

Source: W. H. Beyer, *Handbook of Tables for Probability and Statistics,* 2nd edition, CRC Press, Boca Raton, Florida, 1986. With permission.

[a]This value has been rounded to 1.28 in the textbook.

[b]This value has been rounded to 1.65 in the textbook.

[c]This value has been rounded to 2.33 in the textbook.

[d]This value has been rounded to 2.58 in the textbook.

Example 8–6

Find the $t_{\alpha/2}$ value for a 99% confidence interval when the sample size is 14.

Solution

d.f. $= n - 1 = 14 - 1 = 13$. Look up 13 in the column labeled "d.f." and look up 99% in the row labeled "Confidence Intervals." The value where the two meet is 3.012.

Note: At the bottom of Table F where d.f. $= \infty$, the $z_{\alpha/2}$ values can be found for specific confidence intervals. The reason is that as the degrees of freedom increase, the *t* distribution approaches the standard normal distribution.

The next example shows how to find the confidence interval when one is using the *t* distribution.

Example 8–7

Ten randomly selected automobiles were stopped, and the tread depth of the right front tire was measured. The mean was 0.32 inch, and the standard deviation was 0.08 inch. Find the 95% confidence interval of the mean depth. Assume that the variable is approximately normally distributed.

Solution

Since σ is unknown and *s* must replace it, the *t* distribution (Table F) must be used with $\alpha = 0.05$. Hence, with 9 degrees of freedom, $t_{\alpha/2} = 2.262$.

The 95% confidence interval of the population mean is found by substituting in the formula

$$\overline{X} - t_{\alpha/2}\left(\frac{s}{\sqrt{n}}\right) < \mu < \overline{X} + t_{\alpha/2}\left(\frac{s}{\sqrt{n}}\right)$$

Hence, $\quad 0.32 - (2.262)\left(\frac{0.08}{\sqrt{10}}\right) < \mu < 0.32 + (2.262)\left(\frac{0.08}{\sqrt{10}}\right)$

$$0.32 - 0.057 < \mu < 0.32 + 0.057$$
$$0.26 < \mu < 0.38$$

Therefore, one can be 95% confident that the population mean is between 0.26 and 0.38.

Students sometimes have difficulty deciding whether to use $z_{\alpha/2}$ or $t_{\alpha/2}$ values when finding confidence intervals for the mean. As stated previously, when the σ is known, $z_{\alpha/2}$ values can be used *no matter what the sample size is,* as long as the variable is normally distributed or $n \geq 30$. When σ is unknown and $n \geq 30$, *s* can be used in the formula and $z_{\alpha/2}$ values can be used. Finally, when σ is unknown and $n < 30$, *s* is used in the formula and $t_{\alpha/2}$ values are used, as long as the variable is approximately normally distributed. These rules are summarized in Figure 8–8.

Figure 8–8
....
When to Use the z or t
Distribution

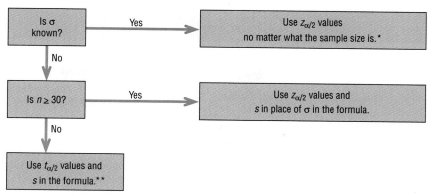

*Variable must be normally distributed when $n < 30$.
**Variable must be approximately normally distributed.

It should be pointed out that some statisticians have a different point of view. They use $z_{\alpha/2}$ values when σ is known and $t_{\alpha/2}$ values when σ is unknown. In these circumstances, a t table that contains t values for sample sizes greater than or equal to 30 would be needed. The procedure shown in Figure 8–8 is the one used throughout this textbook.

Computer Applications for *t* Confidence Interval for a Mean

MINITAB t confidence interval for mean:

1. Enter the data into C1.
2. Click Stat > Basic Statistics > 1-Sample t.
3. Highlight C1 in the dialog box and click on Select.
4. Click on Confidence Interval. Press Tab.
5. Type in the desired level (e.g., 90, 95, 99).
6. Click on OK.

Example: Find the 95% confidence interval for the following data.

Data:

C1	625	675	535	406	512
	680	483	522	619	575

MINITAB printout for this example:

```
T Confidence Intervals
Variable    N     Mean    StDev    SE Mean       95.0 % CI
C1          10    563.2   87.9     27.8      (   500.4,    626.0)
```

Summary: The 95% confidence interval is $500.4 < \mu < 626.0$. The output also shows the sample size (N), the mean, the standard deviation, and the standard error of the mean.

Exercises

8-26. (W) What are the properties of the *t* distribution?

8-27. (W) Who developed the *t* distribution?

8-28. (W) What is meant by degrees of freedom?

8-29. (W) When should the *t* distribution be used to find a confidence interval for the mean?

8-30. (ans) Find the values for each.

a. $t_{\alpha/2}$ and $n = 18$ for the 99% confidence interval for the mean

b. $t_{\alpha/2}$ and $n = 23$ for the 95% confidence interval for the mean

c. $t_{\alpha/2}$ and $n = 15$ for the 98% confidence interval for the mean

d. $t_{\alpha/2}$ and $n = 10$ for the 90% confidence interval for the mean

e. $t_{\alpha/2}$ and $n = 20$ for the 95% confidence interval for the mean

For Exercises 8-31 through 8-46, assume that all variables are approximately normally distributed.

8-31. The average hemoglobin reading for a sample of 20 teachers was 16 grams per 100 milliliters, with a sample standard deviation of 2 grams. Find the 99% confidence interval of the true mean.

8-32. A meteorologist who sampled 15 cold weather fronts found that the average speed at which they traveled across a certain state was 18 miles per hour. The standard deviation of the sample was 2 miles per hour. Find the 95% confidence interval of the mean.

8-33. A sample of 25 two-year-old chickens shows that they lay an average of 21 eggs per month. The standard deviation of the sample was 2 eggs. Find the 99% confidence interval of the true mean.

8-34. A sample of 20 tuna showed that they swim an average of 8.6 miles per hour. The standard deviation for the sample was 1.6. Find the 90% confidence interval of the true mean.

8-35. A sample of 6 adult elephants had an average weight of 12,200 pounds, with a sample standard deviation of 200 pounds. Find the 95% confidence interval of the true mean.

8-36. It has been reported that the average daily intake of calories for young women is 1667. To see if this is valid for nurses, a researcher sampled 15 nurses and found their average daily intake of calories was

1593. The standard deviation of the sample was 36 calories. Find the 90% confidence interval of the mean. Can the researcher conclude that the mean for nurses is the same as the mean for all young women?

8-37. A recent study of 28 city residents showed that the mean of the time they had lived at their present address was 9.3 years. The standard deviation of the sample was 2 years. Find the 90% confidence interval of the true mean.

8-38. An automobile shop manager timed six employees and found that the average time it took them to change a water pump was 18 minutes. The standard deviation of the sample was 3 minutes. Find the 99% confidence interval of the true mean.

8-39. A recent study of 25 students showed that they spent an average of $18.53 for gasoline per week. The standard deviation of the sample was $3.00. Find the 95% confidence interval of the true mean.

8-40. For a group of 10 men subjected to a stress situation, the mean number of heartbeats per minute was 126, and the standard deviation was 4. Find the 95% confidence interval of the true mean.

8-41. For the stress test described in Exercise 8-40, six women had an average heart rate of 115 beats per minute. The standard deviation of the sample was 6 beats. Find the 95% confidence interval of the true mean for the women.

8-42. For a sample of 24 operating rooms taken in the hospital study mentioned in Exercise 8-19, the mean noise level was 41.6 decibels, and the standard deviation was 7.5. Find the 95% confidence interval of the true mean of the noise levels in the operating rooms.
Source: M. Bayo, A. Garcia, and A. Garcia, "Noise Levels in an Urban Hospital and Workers' Subjective Responses," *Archives of Environmental Health* 50, no. 3 (May/June 1995).

8-43. A sample of 15 food servers showed an average weekly income of $320.20. The standard deviation of the sample was $12. Find the 98% confidence interval of the true mean.

8-44. For a group of 20 students taking a final exam, the mean heart rate was 96 beats per minute, and the standard deviation was 5. Find the 95% confidence interval of the true mean.

8-45. The average yearly income for 28 married couples living in city C is $58,219. The standard deviation of the sample is $56. Find the 95% confidence interval of the true mean.

*8–46.   A random sample of 20 parking meters in a large municipality showed the following incomes for a day.

$2.60	$1.05	$2.45	$2.90
$1.30	$3.10	$2.35	$2.00
$2.40	$2.35	$2.40	$1.95
$2.80	$2.50	$2.10	$1.75
$1.00	$2.75	$1.80	$1.95

Find the 95% confidence interval of the true mean.

THE MAN IN THE STREET
(SUBJECT TO A SAMPLING ERROR OF PLUS OR MINUS THREE PERCENTAGE POINTS)

Source: Drawing by Nurit Karlin:
© 1988 *The New Yorker* Magazine, Inc.

8–4

Confidence Intervals and Sample Size for Proportions

Objective 4. Find the confidence interval for a proportion.

A *USA Today* Snapshots feature (July 2, 1993) stated that 12% of the pleasure boats in the United States were named *Serenity*. The parameter 12% is called a **proportion.** It means that of all the pleasure boats in the United States, 12 out of every 100 are named *Serenity*. A proportion represents a part of a whole. It can be expressed as a fraction, decimal, or percentage. In this case, $12\% = 0.12 = \frac{12}{100}$ or $\frac{3}{25}$. Proportions can also represent probabilities. In this case, if a pleasure boat is selected at random, the probability that it is called *Serenity* is 0.12.

Proportions can be obtained from samples or populations. The following symbols will be used.

Symbols Used in Proportion Notation

p = symbol for the population proportion
$\hat{p}$ (read "*p* hat") = symbol for the sample proportion

For a sample proportion,

$$\hat{p} = \frac{X}{n} \quad \text{and} \quad \hat{q} = \frac{n - X}{n} \quad \text{or} \quad 1 - \hat{p}$$

where X = number of sample units that possess the characteristics of interest and n = sample size.

For example, in a study, 200 people were asked if they were satisfied with their job or profession; 162 said that they were. In this case, $n = 200$, $X = 162$, and $\hat{p} = X/n = 162/200 = 0.81$. It can be said that for this sample, 0.81 or 81% of those surveyed were satisfied with their job or profession. The sample proportion is $\hat{p} = 0.81$.

The proportion of people who did not respond favorably when asked if they were satisfied with their job or profession constituted $\hat{q}$, where $\hat{q} = (n - X)/n$. For a survey, $\hat{q} = (200 - 162)/200 = 38/200$, or 0.19, or 19%.

When $\hat{p}$ and $\hat{q}$ are given in decimals or fractions, $\hat{p} + \hat{q} = 1$. When $\hat{p}$ and $\hat{q}$ are given in percentages, $\hat{p} + \hat{q} = 100\%$. It follows, then, that $\hat{q} = 1 - \hat{p}$, or $\hat{p} = 1 - \hat{q}$, when $\hat{p}$ and $\hat{q}$ are in decimal or fraction form. For the sample survey on job satisfaction, $\hat{q}$ can also be found by using $\hat{q} = 1 - \hat{p}$, or $1 - 0.81 = 0.19$.

Similar reasoning applies to population proportions; i.e., $p = 1 - q$, $q = 1 - p$, and $p + q = 1$, when p and q are expressed in decimal or fraction form. When p and q are expressed as percentages, $p + q = 100\%$, $p = 100\% - q$, and $q = 100\% - p$.

Example 8–8

In a recent survey of 150 households, 54 had central air conditioning. Find $\hat{p}$ and $\hat{q}$.

Solution

Since $X = 54$ and $n = 150$,

$$\hat{p} = \frac{X}{n} = \frac{54}{150} = 0.36 = 36\%$$

$$\hat{q} = \frac{n - X}{n} = \frac{150 - 54}{150} = \frac{96}{150} = 0.64 = 64\%$$

One can also find $\hat{q}$ by using the formula $\hat{q} = 1 - \hat{p}$. In this case, $\hat{q} = 1 - 0.36 = 0.64$.

As with means, the statistician, given the sample population, tries to estimate the population proportion. Point and interval estimates for a population proportion can be made by using the sample proportion. For a point estimate of p (the population proportion), $\hat{p}$ (the sample proportion) is used. On the basis of the three properties of a good estimator, $\hat{p}$ is unbiased, consistent, and relatively efficient. But as with means, one is not able to decide how good the point estimate of p is. Therefore, statisticians also use an interval estimate for a proportion, and they can assign a probability that the interval will contain the population proportion.

Confidence Intervals

To construct a confidence interval about a proportion, one must use the maximum error of estimate, which is

$$E = z_{\alpha/2} \sqrt{\frac{\hat{p}\,\hat{q}}{n}}$$

Confidence intervals about proportions must meet the criteria that $np \geq 5$ and $nq \geq 5$.

Speaking of | **STATISTICS**

This study uses proportions. The sample size was 200,000. Find the actual number of boats in each group.

USA SNAPSHOTS®

A look at statistics that shape the sports world

Most popular boat names

Out of the millions of possibilities, just five names are used on more than one-third of pleasure boats in the USA:

Serenity — 12%
Obsession — 7%
Osprey — 6%
Fantasea — 5%
Liquid Asset — 4%

Source: Boat/U.S. (sample of 200,000 boat names)   By Sam Ward, USA TODAY

Source: Based on a survey by Boat/U.S. Copyright 1993, *USA TODAY.* Reprinted with permission.

Formula for a Specific Confidence Interval for a Proportion

$$\hat{p} - (z_{\alpha/2})\sqrt{\frac{\hat{p}\,\hat{q}}{n}} < p < \hat{p} + (z_{\alpha/2})\sqrt{\frac{\hat{p}\,\hat{q}}{n}}$$

when np and nq are each greater than or equal to 5.

Rounding Rule for a Confidence Interval for a Proportion   Round off to three decimal places.

Example 8–9

A sample of 500 nursing applications included 60 from men. Find the 90% confidence interval of the true proportion of men who applied to the nursing program.

Solution

Since $\alpha = 1 - 0.90 = 0.10$, and $z_{\alpha/2} = 1.65$, substituting in the formula

$$\hat{p} - (z_{\alpha/2})\sqrt{\frac{\hat{p}\,\hat{q}}{n}} < p < \hat{p} + (z_{\alpha/2})\sqrt{\frac{\hat{p}\,\hat{q}}{n}}$$

when $\hat{p} = 60/500 = 0.12$ and $\hat{q} = 1 - 0.12 = 0.88$, one gets

$$0.12 - (1.65)\sqrt{\frac{(0.12)(0.88)}{500}} < p < 0.12 + (1.65)\sqrt{\frac{(0.12)(0.88)}{500}}$$

$$0.12 - 0.024 < p < 0.12 + 0.024$$
$$0.096 < p < 0.144$$
or $$9.6\% < p < 14.4\%$$

Hence, one can be 90% confident that the percentage of men who applied is between 9.6% and 14.4%.

When a specific percentage is given, the percentage becomes $\hat{p}$ when it is changed to a decimal. For example, if the problem states that 12% of the applicants were men, then $\hat{p} = 0.12$.

Example 8–10

Using the information in the *USA Today* Snapshot about boat names, find the 95% confidence interval of the true proportion of the boats named *Serenity*.

Solution

From the Snapshot, $\hat{p} = 0.12$ (i.e., 12%), and $n = 200{,}000$ is shown below the figure. Since $Z_{\alpha/2} = 1.96$, substituting in the formula

$$\hat{p} - z_{\alpha/2}\sqrt{\frac{\hat{p}\,\hat{q}}{n}} < p < \hat{p} + z_{\alpha/2}\sqrt{\frac{\hat{p}\,\hat{q}}{n}} \quad \text{yields}$$

$$0.12 - (1.96)\sqrt{\frac{(0.12)(0.88)}{200{,}000}} < p < 0.12 + (1.96)\sqrt{\frac{(0.12)(0.88)}{200{,}000}}$$
$$.119 < p < 0.121$$

Hence, one can say with 95% confidence that the true percentage of boats named *Serenity* is between 11.9% and 12.1%.

Sample Size for Proportions

In order to find the sample size needed to determine a confidence interval about a proportion, use the following formula.

Objective 5. Determine the minimum sample size for finding a confidence interval for a proportion.

Formula for Minimum Sample Size Needed for Interval Estimate of a Population Proportion

$$n = \hat{p}\,\hat{q}\left(\frac{z_{\alpha/2}}{E}\right)^2$$

If necessary, round up to obtain a whole number.

This formula can be found by solving the maximum error of estimate value for n:

$$E = z_{\alpha/2}\sqrt{\frac{\hat{p}\,\hat{q}}{n}}$$

There are two situations to consider. First, if some approximation of $\hat{p}$ is known (e.g., from a previous study), that value can be used in the formula.

Second, if no approximation of $\hat{p}$ is known, one should use $\hat{p} = 0.5$. This value will give a sample size sufficiently large to guarantee an accurate prediction, given the confidence interval and the error of estimate. The reason is that when $\hat{p}$ and $\hat{q}$ are each 0.5, the product $\hat{p} \cdot \hat{q}$ is at maximum, as shown here.

$\hat{p}$	$\hat{q}$	$\hat{p}\,\hat{q}$
0.1	0.9	0.09
0.2	0.8	0.16
0.3	0.7	0.21
0.4	0.6	0.24
0.5	**0.5**	**0.25**
0.6	0.4	0.24
0.7	0.3	0.21
0.8	0.2	0.16
0.9	0.1	0.09

Example 8–11

A researcher wishes to estimate, with 95% confidence, the number of people who own a home computer. A previous study shows that 40% of those interviewed had a computer at home. The researcher wishes to be accurate within 2% of the true proportion. Find the minimum sample size necessary.

Solution

Since $\alpha = 0.05$, $z_{\alpha/2} = 1.96$, $E = 0.02$, $\hat{p} = 0.40$, and $\hat{q} = 0.60$, then

$$n = \hat{p}\,\hat{q}\left(\frac{z_{\alpha/2}}{E}\right)^2 = (0.40)(0.60)\left(\frac{1.96}{0.02}\right)^2 = 2304.96$$

which, when rounded up, is 2305 people to interview.

In this survey, two groups of people were used. Identify the groups. What sample sizes were used? Do you think the sample sizes were adequate to draw the conclusions in the article? Suggest several ways that

patients might be made more aware of their need to ask questions about the prescription medicine they are about to take.

The Right R$_X$: One Minute to More Effective Medicines

There's a simple question that could spare you a hospital stay. Next time you buy a prescription, ask if there's anything else you need to know about what's inside that bottle.

Currently, about 10 percent of all hospital admissions are due to problems with medications. And many could be avoided if patients knew more about the drugs they took. But you don't have to become a pharmacologist or study the microscopic print on the drug package (if there is one) to learn more. Instead, ask your pharmacist. Fielding questions is what pharmacists are there for—and is why they spend five to seven years in school studying pharmacological medicine.

Pharmacists are now required by law to provide information about prescriptions to patients who they

think need it. However, you have to remember that your pharmacist is much more likely to dispense information when you ask for it, says Jon C. Schommer, Ph.D., assistant professor at the College of Pharmacy, Ohio State University, Columbus. His survey of 697 pharmacists showed that the number-one reason they gave information about prescription drugs was based on their perception that patients wanted or needed the information (*American Journal of Hospital Pharmacists*, February 1994.) Very often pharmacists base this perception on whether or not their patients ask questions. Yet only about a third of 360 people in a drugstore survey asked for more facts about their medicines.

Even what seems to be spelled

out on the label may need further clarification. Take the warning against alcohol use, for instance. The amount of alcohol that's in cough syrup can cause extra sleepiness when taken in combination with some drugs and can even cause hospitalization with others. In addition, some drugs can't do their jobs when taken with food or milk. The National Council on Patient Information and Education estimates that up to half of all medications are not taken correctly.

Also, don't let your pharmacist do all the talking. Volunteer information about all medications you're taking, even prescriptions from another pharmacy, aspirin or other over-the-counter remedies.

Example 8–12

The same researcher wishes to estimate the proportion of executives who own a car phone. She wants to be 90% confident and be accurate within 5% of the true proportion. Find the minimum sample size necessary.

Solution

Since no prior knowledge of p is known, statisticians assign the values $\hat{p} = 0.5$ and $\hat{q} = 0.5$. The sample size obtained by using these values will be large enough to ensure the specified degree of confidence. Hence,

$$n = \hat{p}\,\hat{q}\left(\frac{z_{\alpha/2}}{E}\right)^2 = (0.5)(0.5)\left(\frac{1.65}{0.05}\right)^2 = 272.25$$

which, when rounded up, is 273 executives to ask.

In determining the sample size, the size of the population is irrelevant. Only the degree of confidence and the maximum error are necessary to make the determination.

Exercises

8–47. In each case, find $\hat{p}$ and $\hat{q}$.
a. $n = 80$ and $X = 40$
b. $n = 200$ and $X = 90$
c. $n = 130$ and $X = 60$
d. $n = 60$ and $X = 35$
e. $n = 95$ and $X = 43$

8–48. (ans) Find $\hat{p}$ and $\hat{q}$ for each percentage. (Use each percentage for $\hat{p}$.)
a. 12%
b. 29%
c. 65%
d. 53%
e. 67%

8–49. A U.S. Travel Data Center survey conducted for *Better Homes and Gardens* of 1500 adults found that 39% said that they would take more vacations this year than last year. Find the 95% confidence interval of the true proportion of adults who said that they will travel more this year.
Source: *USA Today,* April 20, 1995.

8–50. In a recent study of 100 people, 85 said that they were satisfied with their present home. Find the 90% confidence interval of the true proportion of individuals who are satisfied with their present home.

8–51. An employment counselor found that in a sample of 100 unemployed workers, 65% were not interested in returning to work. Find the 95% confidence interval of the true proportion of workers who do not wish to return to work.

8–52. A *Today*/CNN/Gallup Poll of 1015 adults found that 13% approved of the job Congress was doing in 1995. Find the 95% confidence interval of the true proportion of adults who feel this way.
Source: *USA Today,* March 31, 1995.

8–53. A nutritionist found that in a sample of 80 families, 22% said they ate apples at least once a week. Find the 90% confidence interval of the true proportion of families who eat apples at least once a week.

8–54. A survey of 120 female freshmen showed that 18 did not wish to work after marriage. Find the 95% confidence interval of the true proportion of females who do not wish to work after marriage.

8–55. A study by the University of Michigan found that one in five 13- and 14-year-olds is a sometime smoker. To see how the smoking rate of the students at a large school district compared to the national rate, the superintendent surveyed 200 13- and 14-year-old students and found that 23% said they were sometime smokers. Find the 99% confidence interval of the true proportion and compare this with the University of Michigan's study.
Source: *USA Today,* July 20, 1995.

8–56. A survey of 80 recent fatal traffic accidents showed that 46 were alcohol-related. Find the 95% confidence interval of the true proportion of people who die in alcohol-related accidents.

8–57. A survey of 90 families showed that 40 owned at least one gun. Find the 95% confidence interval of the true proportion of families who own at least one gun.

8–58. In a certain state, a survey of 500 workers showed that 45% belonged to a union. Find the 90% confidence interval of the true proportion of workers who belong to a union.

8–59. For a certain age group, a study of 100 people who died showed that 25% had died of cancer. Find the true proportion of individuals who die of cancer in that age group. Use the 98% confidence interval.

8–60. A researcher wishes to be 95% confident that her estimate of the true proportion of individuals who travel overseas is within 0.04 of the true proportion. Find the sample size necessary. In a prior study, a sample of 200 people showed that 80 traveled overseas last year.

8–61. A medical researcher wishes to determine the percentage of females who take vitamins. He wishes to be 99% confident that the estimate is within 2 percentage points of the true proportion. How large should the sample size be? A recent study of 180 females showed that 25% took vitamins.

8–62. How large a sample must one take to be 90% confident that the estimate is within 0.05 of the true proportion of women over 55 who are widows? A recent study indicated that 29% of the 100 women over 55 in the study were widows.

8–63. A researcher wishes to estimate the proportion of adult males who are under 5 feet 5 inches tall. She wants to be 90% confident that her estimate is within 5% of the true proportion. How large a sample should be taken if in a sample of 300 males, 30 were under 5 feet 5 inches tall?

8–64. An educator desires to estimate, within 0.03, the true proportion of high school students who study at least 1 hour each school night. He wants to be 98% confident. How large a sample is necessary? Previously,

he conducted a study and found that 60% of the 250 students surveyed spent at least 1 hour each school night studying.

*8–65. If a sample of 600 people is chosen and the researcher desires to have a maximum error of estimate of 4% on the specific proportion who favor gun control, find the degree of confidence. A recent study showed that 50% were in favor of some form of gun control.

*8–66. In a study, 68% of 1015 adults said they believe the Republicans favor the rich. If the margin of error was 3 percentage points, what was the confidence interval used for the true proportion?
Source: *USA Today,* March 31, 1995.

8–5

Confidence Intervals for Variances and Standard Deviations

Objective 6. Find a confidence interval for a variance and a standard deviation.

In the previous section, confidence intervals were calculated for means and proportions. This section will explain how to find confidence intervals for variances and standard deviations. In statistics, the variance and standard deviation of a variable are as important as the mean. For example, when products that fit together (such as pipes) are manufactured, it is important to keep the variations of the diameters of the products as small as possible; otherwise, they will not fit together properly and have to be scrapped. When medicines are manufactured, the variance and standard deviation of the medication in the pills play an important role in making sure patients receive the proper dosage. For these reasons, confidence intervals for variances and standard deviations are necessary.

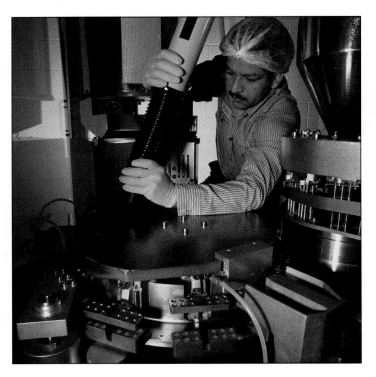

In order to calculate these confidence intervals, a new statistical distribution is needed. It is called the **chi-square distribution.**

The chi-square variable is similar to the t variable in that its distribution is a family of curves based on the number of degrees of freedom. The symbol for chi-square is χ^2 (Greek letter chi, pronounced "ki"). Several of the distributions are shown in Figure 8–9, along with the corresponding degrees of freedom. The chi-square distribution is obtained from the values of $(n - 1)s^2/\sigma^2$ when random samples are selected from a normally distributed population whose variance is σ^2.

Figure 8–9

The Chi-Square Family of Curves

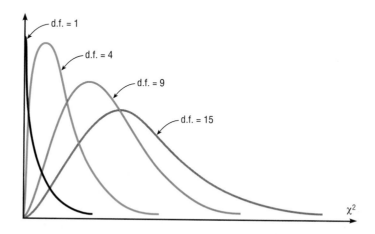

Historical Note

The χ^2 distribution with 2 degrees of freedom was formulated by a mathematician named Hershel in 1869 while he was studying the accuracy of shooting arrows at a target. Many other mathematicians have since contributed to its development.

A chi-square variable cannot be negative, and the distributions are positively skewed. At about 100 degrees of freedom, the chi-square distribution becomes somewhat symmetrical. The area under each chi-square distribution is equal to 1.00 or 100%.

Table G in Appendix C gives the values for the chi-square distribution. These values are used in the denominators of the formulas for confidence intervals. Two different values are used in the formula. One value is found on the left side of the table and the other is on the right. For example, to find the table values corresponding to the 95% confidence interval, one must first change 95% to a decimal and subtract it from 1 to find the value of α ($\alpha = 1 - 0.95 = 0.05$). Then divide the answer by 2 ($\alpha/2 = 0.05/2 = 0.025$). This is the column on the right side of the table, used to get the values for χ^2_{right}. To get the value for χ^2_{left}, subtract the value of $\alpha/2$ from the 1 ($1 - 0.05/2 = 0.975$). Finally, find the appropriate row corresponding to the degrees of freedom, $n - 1$. A similar procedure is used to find the values for a 90% or 99% confidence interval.

Example 8–13

Find the values for χ^2_{right} and χ^2_{left} for a 90% confidence interval when $n = 25$.

Solution

To find χ^2_{right}, subtract $1 - 0.90 = 0.10$ and divide by 2 to get 0.05.

To find χ^2_{left}, subtract $1 - 0.05$ to get 0.95. Hence, use the 0.95 and 0.05 columns and the row corresponding to 24 d.f. See Figure 8–10.

Figure 8–10

χ^2 Table for Example 8–13

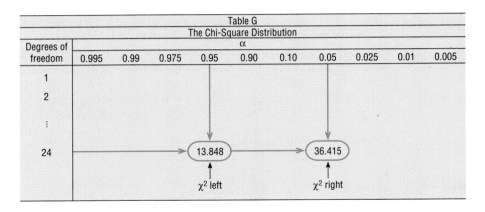

Degrees of freedom	0.995	0.99	0.975	0.95	0.90	0.10	0.05	0.025	0.01	0.005
1										
2										
⋮										
24				13.848			36.415			

Table G
The Chi-Square Distribution
α

χ^2 left χ^2 right

The answers are

$$\chi^2_{\text{right}} = 36.415$$
$$\chi^2_{\text{left}} = 13.848$$

The best estimators of σ^2 and σ are s^2 and s respectively.

In order to find confidence intervals for variances and standard deviations, one must assume that the variable is normally distributed.

The formulas for the confidence intervals are shown next.

Formula for the confidence interval for a variance

$$\frac{(n-1)s^2}{\chi^2_{\text{right}}} < \sigma^2 < \frac{(n-1)s^2}{\chi^2_{\text{left}}}$$

d.f. = $n - 1$

Formula for the confidence interval for a standard deviation

$$\sqrt{\frac{(n-1)s^2}{\chi^2_{\text{right}}}} < \sigma < \sqrt{\frac{(n-1)s^2}{\chi^2_{\text{left}}}}$$

d.f. = $n - 1$

Recall that s^2 is the symbol for the sample variance and s is the symbol for the sample standard deviation. If the problem gives the sample standard deviation (s), *square* it when substituting into the formula. But if the problem gives the sample variance (s^2), substitute it directly into the formula.

Rounding Rule for a Confidence Interval for a Variance or Standard Deviation   When computing a confidence interval for a population variance or standard deviation using raw data, round off to one more decimal than the number of decimal places in the original data.

When computing a confidence interval for a population variance or standard deviation using a sample variance or standard deviation, round off to the same number of decimal places as given for the sample variance or standard deviation.

The next example shows how to find a confidence interval for a variance and standard deviation.

| **Example 8–14** | Find the 95% confidence interval for the variance and standard deviation of the nicotine content of cigarettes manufactured if a sample of 20 cigarettes has a standard deviation of 1.6 milligrams. |

Solution

Since $\alpha = 0.05$, the two critical values for the 0.025 and 0.975 levels for 19 degrees of freedom are 32.852 and 8.907. The 95% confidence interval for the variance is found by substituting in the formula:

$$\frac{(n-1)s^2}{\chi^2_{\text{right}}} < \sigma^2 < \frac{(n-1)s^2}{\chi^2_{\text{left}}}$$

$$\frac{(20-1)(1.6)^2}{32.852} < \sigma^2 < \frac{(20-1)(1.6)^2}{8.907}$$

$$1.5 < \sigma^2 < 5.5$$

Hence, one can be 95% confident that the true variance for the nicotine content is between 1.5 and 5.5.

For the standard deviation, the confidence interval is

$$\sqrt{1.5} < \sigma < \sqrt{5.5}$$

$$1.2 < \sigma < 2.3$$

Hence, one can be 95% confident that the true standard deviation is between 1.2 and 2.3.

Exercises

8–67. What distribution must be used when computing confidence intervals for variances and standard deviations?

8–68. What assumption must be made when computing confidence intervals for variances and standard deviations?

8–69. Using Table G, find the values for χ^2_{left} and χ^2_{right}.

a. $\alpha = 0.05$ $n = 16$
b. $\alpha = 0.10$ $n = 5$
c. $\alpha = 0.01$ $n = 23$
d. $\alpha = 0.05$ $n = 29$
e. $\alpha = 0.10$ $n = 14$

8–70. Find the 95% confidence interval for the variance and standard deviation for the lifetime of batteries if a sample of 20 batteries has a standard deviation of 1.7 months. Assume the variable is normally distributed.

8–71. Find the 90% confidence interval for the variance and standard deviation for the time it takes an inspector to check a bus for safety if a sample of 27 buses has a standard deviation of 6.8 minutes. Assume the variable is normally distributed.

8–72. Find the 99% confidence interval for the variance and standard deviation of the weights of 25 gallon containers of motor oil if a sample of 14 containers has a variance of 3.2. The weights are given in ounces. Assume the variable is normally distributed.

8–73. Find the 99% confidence interval for the variance and standard deviation for the sugar content in sherbet (in milligrams) if a sample of nine servings has a variance of 36. Assume the variable is normally distributed.

8–74. Find the 90% confidence interval for the variance and standard deviation of the ages of seniors at Oak Park College if a sample of 24 students has a standard deviation of 2.3 years. Assume the variable is normally distributed.

8–75. Find the 98% confidence interval for the variance and standard deviation for the time it takes a telephone company to transfer a call to the correct office. A sample of 15 calls has a standard deviation of 1.6 minutes. Assume the variable is normally distributed.

8–76. A random sample of 15 television sets was selected, and the lifetimes (in months) of the picture tubes were measured. The variance of the sample was 8.6. Find the 90% confidence interval of the true variance.

8–77. A service station advertises that customers will have to wait no more than 30 minutes for an oil change. A sample of 28 oil changes has a standard deviation of 5.2 minutes. Find the 95% confidence interval of the population standard deviation of the time spent waiting for an oil change.

*8–78. Find the 95% confidence interval for the variance and standard deviation of the ounces of coffee that a machine dispenses in 12-ounce cups. Assume the variable is normally distributed. The data are given here.
12.03, 12.10, 12.02, 11.98, 12.00, 12.05, 11.97, 11.99.

*8–79. Find the 99% confidence interval for the variance and standard deviation for the breaking strength of cables (in pounds). Assume the variable is normally distributed. A sample of 12 cables is given here.
2001, 1998, 2002, 2000, 1998, 1999, 1997, 2005, 2003, 2001, 1999, 2006.

Summary

An important aspect of inferential statistics is estimation. Estimations of parameters of populations are accomplished by selecting a random sample from that population and choosing and computing a statistic that is the best estimator of the parameter. A good estimator must be unbiased, consistent, and relatively efficient. The best estimators of μ and p are $\overline{X}$ and $\hat{p}$, respectively. The best estimators of σ^2 and σ are s^2 and s respectively.

There are two type of estimates of a parameter: point estimates and interval estimates. A point estimate is a specific value. For example, if a researcher wishes to estimate the average length of a certain adult fish, a sample of the fish is selected and measured. The mean of this sample is computed—e.g., 3.2 centimeters. From this sample mean, the researcher estimates the population mean to be 3.2 centimeters.

The problem with point estimates is that the accuracy of the estimate cannot be determined. For this reason, statisticians prefer to use the interval estimate. By computing an interval about the sample value, statisticians can be 95% or 99% (or some other percentage) confident that their estimate contains the true parameter. The confidence level is determined by the researcher. The higher the confidence level, the wider the interval of the estimate must be. For example, a 95% confidence interval of the true mean length of a certain species of fish might be

$$3.17 < \mu < 3.23$$

whereas the 99% confidence interval might be

$$3.15 < \mu < 3.25$$

When the confidence interval of the mean is computed, the z or t values are used, depending on whether or not the population standard deviation is known and depending on the size of the sample. If σ is known or $n \geq 30$, the z values can be used. If σ is not known, the t values must be used when the sample size is less than 30, and the population is normally distributed.

Closely related to computing confidence intervals is determining the sample size to make an estimate of the mean. The following information is needed to determine the minimum sample size necessary.

1. The degree of confidence must be stated.
2. The population standard deviation must be known or be able to be estimated.
3. The maximum error of estimate must be stated.

Confidence intervals and sample sizes can also be computed for proportions, and confidence intervals for variances and standard deviations can be computed using the chi-square distribution.

Important Terms

Chi-square distribution 311	Degrees of freedom 298	Maximum error of estimate 289	t distribution 297
Confidence interval 288	Estimation 286	Point estimate 287	Unbiased estimator 287
Confidence level 288	Estimator 287	Proportion 303	
Consistent estimator 287	Interval estimate 287	Relatively efficient estimator 287	

Important Formulas

Formula for the confidence interval of the mean when σ is known (when $n \geq 30$, s can be used if σ is unknown):

$$\bar{X} - z_{\alpha/2}\left(\frac{\sigma}{\sqrt{n}}\right) < \mu < \bar{X} + z_{\alpha/2}\left(\frac{\sigma}{\sqrt{n}}\right)$$

Formula for the sample size for means:

$$n = \left(\frac{z_{\alpha/2} \cdot \sigma}{E}\right)^2$$

where E is the maximum error.

Formula for the confidence interval of the mean when σ is unknown and $n < 30$:

$$\bar{X} - t_{\alpha/2}\left(\frac{s}{\sqrt{n}}\right) < \mu < \bar{X} + t_{\alpha/2}\left(\frac{s}{\sqrt{n}}\right)$$

Formula for the confidence interval for a proportion:

$$\hat{p} - (z_{\alpha/2})\sqrt{\frac{\hat{p}\,\hat{q}}{n}} < p < \hat{p} + (z_{\alpha/2})\sqrt{\frac{\hat{p}\,\hat{q}}{n}}$$

where $\hat{p} = X/n$ **and** $\hat{q} = 1 - \hat{p}$

Formula for the sample size for proportions:

$$n = \hat{p}\,\hat{q}\left(\frac{z_{\alpha/2}}{E}\right)^2$$

Formula for the confidence interval for a variance:

$$\frac{(n-1)s^2}{\chi^2_{right}} < \sigma^2 < \frac{(n-1)s^2}{\chi^2_{left}}$$

Formula for confidence interval for a standard deviation:

$$\sqrt{\frac{(n-1)s^2}{\chi^2_{right}}} < \sigma < \sqrt{\frac{(n-1)s^2}{\chi^2_{left}}}$$

Review Exercises

8–80. *TV Guide* reported that Americans have their television sets turned on an average of 54 hours per week. A researcher surveyed 50 local households and found they had their sets on an average of 47.3 hours. The standard deviation of the sample was 6.2 hours. Find the 90% confidence interval of the true mean. How does this compare with the *TV Guide* findings?
Source: *TV Guide* 43, no. 30 (1995), p. 26.

8–81. The owner of a small business complained that the rent was too high. He randomly surveyed 32 other businesses in his locality and found that the mean rent on their small stores was $1250 per month. The standard deviation of the sample was $33. Find the 90% confidence interval of the population mean.

8–82. The average weight of 60 randomly selected compact automobiles was 2627 pounds. The sample standard deviation was 400 pounds. Find the 99% confidence interval of a true mean weight of the automobiles.

8–83. A U.S. Travel Data Center survey reported that Americans stayed an average of 7.5 nights when they went on vacation. The sample size was 1500. Find the 95% confidence interval of the true mean. Assume the standard deviation was 0.8 day.
Source: *USA Today,* April 20, 1995.

8–84. In a hospital, a sample of 10 weeks was selected, and the researcher found that an average of 12 babies were born each week. The standard deviation of the sample was 2. Find the 99% confidence interval of the true mean.

8–85. For a certain urban area, in a sample of five months, an average of 28 mail carriers were bitten by dogs each month. The standard deviation of the sample was 3. Find the 90% confidence interval of the true mean number of mail carriers who are bitten by dogs each month.

8–86. How large a sample is necessary to be 95% sure that the estimate of the mean income of graphic artists is within $200 of the true mean? The standard deviation of income is $800.

8–87. A researcher wishes to estimate, within $25, the true average amount of postage a community college spends each year. If she wishes to be 90% confident, how large a sample is necessary? The standard deviation is known to be $80.

8–88. A U.S. Travel Data Center's survey of 1500 adults found that 42% of respondents stated that they favor historical sites as vacations. Find the 95% confidence interval of the true proportion of all adults who favor visiting historical sites as vacations.
Source: *USA Today,* April 20, 1995.

8–89. In a study of 200 accidents that required treatment in an emergency room, 40% occurred at home. Find the 90% confidence interval of the true proportion of accidents that occur at home.

8–90. A political analyst found that 60% of 300 Republican voters believe that the federal government has too much power. Find the 95% confidence interval of the population proportion of Republican voters who feel this way.

8–91. A nutritionist wishes to determine, within 2%, the true proportion of adults who snack before bedtime. If she wishes to be 95% confident that her estimate contains the population proportion, how large a sample will she need? A previous study found that 18% of the 100 people surveyed said they did snack before bedtime.

8–92. A survey of 200 adults showed that 15% played basketball for regular exercise. If a researcher desires to find the 99% confidence interval of the true proportion of adults who play basketball and be within 1% of the true population, how large a sample should be selected?

8–93. The standard deviation of the diameter of 28 oranges was 0.34 inch. Find the 99% confidence interval of the true standard deviation of the diameters of the oranges.

8–94. A random sample of 22 lawn mowers was selected, and the motors were tested to see how many miles per gallon of gasoline each one obtained. The variance of the measurements was 2.6. Find the 95% confidence interval of the true variance.

8–95. A flu vaccine is tested to see how long (in months) it prevents patients from contracting the flu. The variance obtained from five tests was 1.6. Find the 98% confidence interval of the true variance.

8–96. The heights of 28 police officers from a large-city police force were measured. The standard deviation of the sample was 1.83 inches. Find the 95% confidence interval of the standard deviation of the heights of the officers.

Statistics Today

Would You Change the Channel?   Revisited

The estimates given in the article are point estimates. However, since the margin of error is stated to be 3 percentage points, an interval estimate can easily be obtained. For example, if 45% of the people changed the channel, then the confidence interval of the true percentages of people who changed channels would be $42\% < p < 48\%$. The article fails to state whether a 90%, 95%, or some other percentage was used for the confidence interval.

Using the formula given in Section 8–4, a minimum sample size of 1068 would be needed to obtain a 95% confidence interval for p, as shown next. Use $\hat{p}$ and $\hat{q}$ as 0.5, since no value is known for $\hat{p}$.

$$n = \hat{p}\,\hat{q}\left(\frac{z_{\alpha/2}}{E}\right)^2$$

$$n = (0.5)\,(0.5)\left(\frac{1.96}{0.03}\right)^2$$

$$n = 1068$$

Data Analysis

1. From the Data Bank in Appendix D, choose a variable, find the mean, and construct the 95% and 99% confidence intervals of the population mean. Use a sample of at least 30 subjects. Find the mean of the population and determine whether it falls within the confidence interval.

2. Repeat Exercise 1 using a different variable and a sample of 15.

3. Repeat Exercise 1 using a proportion. For example, construct a confidence interval for the proportion of individuals who did not complete high school.

Quiz

Determine whether each statement is true or false. If the statement is false, explain why.

1. Interval estimates are preferred over point estimates since a confidence level can be specified.

2. For a specific confidence interval, the larger the sample size, the smaller the maximum error of estimate will be.

3. An estimator is consistent if, as the sample size decreases, the value of the estimator approaches the value of the parameter estimated.

4. In order to determine the sample size needed to estimate a parameter, one must know the maximum error of estimate.

Select the best answer.

5. When a 99% confidence interval is calculated instead of a 95% confidence interval with n being the same, the maximum error of estimate will be
a. Smaller
b. Larger
c. The same
d. It cannot be determined.

6. The best point estimate of the population mean is
a. The sample mean
b. The sample median
c. The sample mode
d. The sample midrange

7. When the population standard deviation is unknown and sample size is less than 30, what table value should be used in computing a confidence interval for a mean?

a. z
b. t
c. chi-square
d. None of the above

Complete the following statements with the best answer.

8. A good estimator should be _____, _____, and _____.

9. The maximum difference between the point estimate of a parameter and the actual value of the parameter is called _____.

10. The statement "The average height of an adult male is 5 feet 10 inches" is an example of a(n) _____ estimate.

11. The three confidence intervals used most often are the _____%, _____%, and _____%.

12. A study of 35 adult females from a certain population showed the mean systolic blood pressure to be 116 mmHg and the standard deviation to be 8.4. Find the 90% confidence interval of the true mean of the population.

13. An irate patient complained that the cost of a doctor's visit was too high. She randomly surveyed 20 other patients and found that the mean amount of money they spent on each doctor's visit was $44.80. The standard deviation of the sample was $3.53. Find the 95% confidence interval of the population mean.

14. The average weight of 40 randomly selected school buses was 4150 pounds. The standard deviation was 480 pounds. Find the 99% confidence interval of the true mean weight of the buses.

15. In a study of 10 insurance sales reps from a certain large city, the average of the group was 48.6 years old, and the standard deviation was 4.1 years. Find the 95% confidence interval of the population mean of all insurance sales reps in that city.

16. In a hospital, a sample of 8 weeks was selected, and it was found that an average of 438 patients were treated in the emergency room each week. The standard deviation was 16. Find the 99% confidence interval of the true mean.

17. For a certain urban area, it was found that in a sample of 4 months, an average of 31 burglaries occurred each month. The standard deviation was 4. Find the 90% confidence interval of the true mean number of burglaries each month.

18. How large a sample is needed to be 95% confident that the estimate of the mean income of entry-level accountants is within $150 of the true mean? The standard deviation of income is $950.

19. A researcher wishes to estimate within $300 the true average amount of money a county spends on road repairs each year. If she wants to be 90% confident, how large a sample is necessary? The standard deviation is known to be $900.

20. A recent study of 75 workers found that 53 people rode the bus to work each day. Find the 95% confidence interval of the proportion of all workers who rode the bus to work.

21. In a study of 150 accidents that required treatment in an emergency room, 36% involved children under 6 years of age. Find the 90% confidence interval of the true proportion of accidents that involve children under the age of 6.

22. A political analyst found that 50% of 280 Democratic voters think the minimum wage should be increased. Find the 95% confidence interval of the population proportion of Democratic voters who feel this way.

23. A nutritionist wishes to determine, within 3%, the true proportion of adults who do not eat any lunch. If he wishes to be 95% confident that it contains the population proportion, how large a sample will be necessary? A previous study found that 15% of the 125 people surveyed said they did not eat lunch.

24. A sample of 25 novels has a standard deviation of nine pages. Find the 95% confidence interval of the population standard deviation.

25. The variance of the breaking strengths of a sample of 10 chords is nine. Find the 99% confidence interval of the population variance.

26. A sample of 20 automobiles has a pollution by-product release standard deviation of 2.3 ounces when one gallon of gasoline is used. Find the 90% confidence interval of the population standard deviation.

Critical Thinking Challenge

Hazardous Train Wrecks   There were 351 hazardous material rail accidents in 1987 and 482 in 1992. In 1987, 53% of the rail accidents involved a rail car carrying hazardous materials, and in 1992, 48% of the rail accidents involved a rail car carrying hazardous materials. Find the 95% confidence interval for the proportions for both years and compare the results. Answer the following questions.

1.   Do the intervals overlap?

2.   If so, can you conclude that the percentages, 53% and 48%, are the same? Explain your answer.

3.   If not, can you conclude the percentages, 53% and 48%, are different? Explain your answer.

Source: *Pittsburgh Tribune Review*, June 5, 1994. Used with permission.

Data Projects

Use MINITAB, the TI-83, or a computer program of your choice to complete the following exercises.

1.   Select several variables, such as the number of points a football team scored in each game of a specific season, the number of passes completed, or the number of yards gained. Using confidence intervals for the mean, determine the 90%, 95%, and 99% confidence intervals. (Use z or t, whichever is relevant.) Decide which you think is most appropriate. When this is completed, write a summary of your findings by answering the following questions.

a.   What was the purpose of the study?

b.   What was the population?
c.   How was the sample selected?
d.   What were the results obtained using confidence intervals?
e.   Did you use z or t? Why?

2.   Using the same data or different data, construct a confidence interval for a proportion. For example, you might want to find the proportion of passes completed by the quarterback or the proportion of passes that were intercepted. Write a short paragraph summarizing the results.

TI-83 Calculator

Note: For confidence intervals, the calculator will accept either raw data or summary statistics.

z Confidence Interval: One Mean

A.   z Confidence Interval One Mean (Data)
1.   Enter the data into L$_1$.
2.   Press **STAT** and move the cursor to TESTS.
3.   Press **7**.
4.   Select Data, and press **ENTER**, and move cursor to σ.
5.   Enter the values for σ. Make sure L$_1$ is selected and Freq is 1.
6.   Type in the correct confidence level.
7.   Select Calculate and press **ENTER**.

Example 7: Find the 99% confidence interval for the mean when σ = 6 using the following data:

27,   16,   9,   14,   32,   15,   16,   18,   16,   13

Input

```
ZInterval
 Inpt:DATA Stats
 σ:6
 List:L₁
 Freq:1
 C-Level:.99
 Calculate
```

Output

```
ZInterval
 (12.713,22.487)
 x̄=17.6
 Sx=6.818276094
 n=10
■
```

The confidence interval is 12.713 < μ < 22.487. The values of the sample mean and standard deviation are also given.

B. Confidence Interval One Mean (Stats)
1. Press **STAT** and move the cursor to TESTS.
2. Press **7**.
3. Select Stats and press **ENTER**. Move cursor to σ.
4. Enter the value for the population standard deviation, σ.
5. Enter the sample mean, $\bar{X}$.
6. Enter the sample size.
7. Type in the correct confidence level.
8. Select Calculate and press **ENTER**.

Example 8: Find the 90% confidence interval of the mean when $\bar{X} = 182$, $\sigma = 8$, and $n = 50$.

Input

```
ZInterval
 Inpt:Data Stats
 σ:8
 x̄:182
 n:50
 C-Level:.9
 Calculate
```

Output

```
ZInterval
 (180.14,183.86)
 x̄=182
 n=50



■
```

The 90% confidence interval of the mean is 180.14 < μ < 183.86.

***t* Confidence Interval: One Mean**

A. *t* Confidence Interval One Mean (Data)
1. Enter the data into L_1.
2. Press **STAT** and move the cursor to TESTS.
3. Press **8**.
4. Select Data and press **ENTER**. Move cursor to c-level.
5. Type in the correct confidence level. Make sure L_1 is selected and Freq is 1.
6. Select Calculate and press **ENTER**.

Example 9: Find the 95% confidence interval for
62, 81, 86, 79, 73, 88, 90, 98, 78,
93, 87, 82, 78, 59, 63, 97, 93, 84.

Input

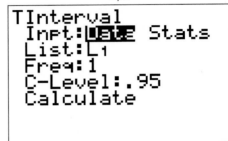

```
TInterval
 Inpt:Data Stats
 List:L₁
 Freq:1
 C-Level:.95
 Calculate
```

Output

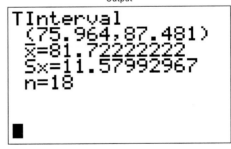

```
TInterval
 (75.964,87.481)
 x̄=81.72222222
 Sx=11.57992967
 n=18
■
```

The 95% confidence interval of the mean is 75.964 < μ < 87.481. The values of the sample mean and standard deviation are also given.

B. *t* Confidence Interval One Mean (Stats)
1. Press **STAT** and move the cursor to TESTS.
2. Press **8**.
3. Select Stats and press **ENTER**. Move cursor to $\bar{X}$.
4. Enter the value for the sample mean, $\bar{X}$.
5. Enter the value for the sample standard deviation, s_X.
6. Enter the sample size, n.
7. Type in the correct confidence interval.
8. Select Calculate and press **ENTER**.

Example 10: Find the 95% confidence level when $\bar{X} = 0.32$, $s = 0.08$, and $n = 10$.

Input

```
TInterval
 Inpt:Data Stats
 x̄:.32
 Sx:.08
 n:10
 C-Level:.95
 Calculate
```

Output

```
TInterval
 (.26277,.37723)
 x̄=.32
 Sx=.08
 n=10
```

The 95% confidence interval is $0.26277 < \mu < 0.37723$.

z Confidence Interval: One Proportion

1. Press **STAT** and move the cursor to TESTS.
2. Press **A** (**ALPHA**, **MATH**).
3. Enter the value for X.
4. Enter the value for n.
5. Type in the correct confidence interval.
6. Select Calculate and press **ENTER**.

Example 11: Find the 95% confidence interval of p when $X = 60$ and $n = 500$.

Input

```
1-PropZInt
 x:60
 n:500
 C-Level:.95
 Calculate
```

Output

```
1-PropZInt
 (.09152,.14848)
 p̂=.12
 n=500
```

The 95% confidence level for p is $0.09152 < p < 0.14848$. $\hat{p}$ is also given.

chapter

9

Hypothesis Testing

Objectives

After completing this chapter, you should be able to

1. Understand the definitions used in hypothesis testing.

2. State the null and alternative hypotheses.

3. Find critical values for the z test.

4. State the five steps used in hypothesis testing.

5. Test means for large samples using the z test.

6. Test means for small samples using the t test.

7. Test proportions using the z test.

8. Test variances or standard deviation using the chi-square test.

9. Test hypotheses using confidence intervals.

Statistics Today

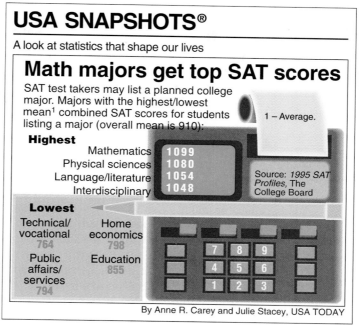

USA SNAPSHOTS®

A look at statistics that shape our lives

Math majors get top SAT scores

SAT test takers may list a planned college major. Majors with the highest/lowest mean[1] combined SAT scores for students listing a major (overall mean is 910):

1 – Average.

Highest

Mathematics	1099
Physical sciences	1080
Language/literature	1054
Interdisciplinary	1048

Source: *1995 SAT Profiles*, The College Board

Lowest

Technical/vocational	Home economics
764	798
Public affairs/services	Education
794	855

By Anne R. Carey and Julie Stacey, USA TODAY

Source: *USA Today,* October 12, 1995. Copyright *USA TODAY.* Used with permission.

How Much Better Is Better?

Suppose a school superintendent reads the *USA Today* Snapshot shown here, which states that the overall mean score for the SAT test is 910. Furthermore, suppose that, for a sample of students, the average of the SAT scores in the superintendent's school district is 960. Can the superintendent conclude that the students in his school district scored higher than average? At first glance, you might be inclined to say yes, since 960 is higher than 910. But recall that the means of samples vary about the population mean when samples are selected from a specific population. So the question arises, "Is there a real difference in the means, or is the difference simply due to chance (i.e., sampling error)?" In this chapter, you will learn how to answer that question using statistics that explain hypothesis testing.

9–1

Introduction

Researchers are interested in answering many types of questions. For example, a scientist might want to know whether the earth is warming up. A physician might want to know whether a new medication will lower a person's blood pressure. An educator might wish to see whether a new teaching technique is better than a traditional one. A retail merchant might want to know whether the public prefers a certain color in a new line of fashion. Automobile manufacturers are interested in determining whether seat belts will reduce the severity of injuries caused by accidents. These types of questions can be addressed through statistical **hypothesis testing,** which is a decision-making process for evaluating claims about a population. In hypothesis testing, the researcher must define the population under study, state the particular hypotheses that will be investigated, give the significance level, select a sample from the population, collect the data, perform the calculations required for the statistical test, and reach a conclusion.

Hypotheses concerning parameters such as means and proportions can be investigated. There are two specific statistical tests used for hypotheses concerning means: the *z test* and the *t test*. This chapter will explain in detail the hypothesis-testing procedure along with the *z* test and the *t* test. In addition, a hypothesis-testing procedure for testing a single variation or standard deviation using the chi-square distribution is explained in Section 9–6.

9–2

Steps in Hypothesis Testing

Objective 1. Understand the definitions used in hypothesis testing.

Every hypothesis-testing situation begins with the statement of a hypothesis.

A **statistical hypothesis** is a conjecture about a population parameter. This conjecture may or may not be true.

There are two types of statistical hypotheses for each situation: the null hypothesis and the alternative hypothesis.

The **null hypothesis,** symbolized by H_0, is a statistical hypothesis that states that there is no difference between a parameter and a specific value or that there is no difference between two parameters.

The **alternative hypothesis,** symbolized by H_1, is a statistical hypothesis that states a specific difference between a parameter and a specific value or states that there is a difference between two parameters.

As an illustration of how hypotheses should be stated, three different statistical studies will be used as examples.

Situation A A medical researcher is interested in finding out whether a new medication will have any undesirable side effects. The researcher is particularly concerned with the pulse rate of the patients who take the medication. Will the pulse rate increase, decrease, or remain unchanged after a patient takes the medication?

Since the researcher knows that the mean pulse rate for the population under study is 82 beats per minute, the hypotheses for this situation are

$$H_0: \mu = 82$$
$$H_1: \mu \neq 82$$

The null hypothesis specifies that the mean will remain unchanged, and the alternative hypothesis states that it will be different. This test is called a *two-tailed test* (a term that will be formally defined later in this section), since the possible side effects of the medicine could be to raise or lower the pulse rate.

Situation B   A chemist invents an additive to increase the life of an automobile battery. If the mean lifetime of the automobile battery is 36 months, then his hypotheses are

$$H_0: \mu \leq 36$$
$$H_1: \mu > 36$$

In this situation, the chemist is interested only in increasing the lifetime of the batteries, so his alternative hypothesis is that the mean is greater than 36 months. The null hypothesis is that the mean is less than or equal to 36 months. This test is called right-tailed, since the interest is in an increase only.

Situation C   A contractor wishes to lower heating bills by using a special type of insulation in houses. If the average of the monthly heating bills is $78, her hypotheses about heating costs with the use of insulation are

$$H_0: \mu \geq \$78$$
$$H_1: \mu < \$78$$

This test is a left-tailed test, since the contractor is interested only in lowering heating costs.

In order to state hypotheses correctly, researchers must translate the *conjecture* or *claim* from words into mathematical symbols. The basic symbols used are as follows:

Equal	$=$	Less than	$<$
Not equal	$\neq$	Greater than or equal to	$\geq$
Greater than	$>$	Less than or equal to	$\leq$

The null and alternative hypotheses are stated together, and the null hypothesis contains the equal sign, as shown.

Two-tailed test	Right-tailed test	Left-tailed test
$H_0: =$	$H_0: \leq$	$H_0: \geq$
$H_1: \neq$	$H_1: >$	$H_1: <$

The formal definitions of the different types of tests are given later in this section.

Table 9–1 shows some common phrases that are used in hypotheses conjectures and the corresponding symbols. This table should be helpful in translating verbal conjectures into mathematical symbols.

Objective 2. State the null and alternative hypotheses.

Table 9–1 Hypothesis-Testing Common Phrases

>	<
Is greater than	Is less than
Is more than	Is below
Is larger than	Is lower than
Is longer than	Is shorter than
Is bigger than	Is smaller than
Is better than	Is reduced from

$\geq$	$\leq$
Is greater than or equal to	Is less than or equal to
Is at least	Is not more than
Is not less than	Is at most

=	$\neq$
Is equal to	Is not equal to
Is exactly the same as	Is different from
Has not changed from	Has changed from
Is the same as	Is not the same as

Example 9–1

State the null and alternative hypotheses for each conjecture.

a. A researcher thinks that if expectant mothers use vitamin pills, the birth weight of the babies will increase. The average of the birth weights of the population is 8.6 pounds.

b. An engineer hypothesizes that the mean number of defects can be decreased in a manufacturing process of compact discs by using robots instead of humans for certain tasks. The mean number of defective discs per 1000 is 18.

c. A psychologist feels that playing soft music during a test will change the results of the test. The psychologist is not sure whether the grades will be higher or lower. In the past, the mean of the scores was 73.

Solution

a. H_0: $\mu \leq 8.6$ and H_1: $\mu > 8.6$.   c. H_0: $\mu = 73$ and H_1: $\mu \neq 73$.

b. H_0: $\mu \geq 18$ and H_1: $\mu < 18$.

After stating the hypothesis, the researcher's next step is to design the study. The researcher selects the correct *statistical test*, chooses an appropriate *level of significance*, and formulates a plan for conducting the study. In situation A, for instance, the researcher will select a sample of patients who will be given the drug. After allowing a suitable period of time for the drug to be absorbed, the researcher will measure each person's pulse rate. Recall that the sample means vary about the population mean. So even if the null hypothesis is true, the mean of the sample will not, in most cases, be exactly equal to the population mean of 82.

Now a question arises. If the mean of the sample is not exactly equal to the population mean, how does one know whether the medication affects the pulse rate? That is, is the difference due to chance, or is it due to the effects of the medication?

If the mean pulse rate of the sample were, say, 83, the researcher would probably conclude that this difference was due to chance and would not reject the null hypothesis. But if the sample mean were, say, 90, then in all likelihood the re-

Unusual Stat

Researchers studying lying found that people lie more often on the phone than in face-to-face chats. One lie was discovered for every seven statements. About 10% of all lies told were merely exaggerations, but 60% were outright deceptions. The rest were subtle omissions. Finally, 70% of liars would tell the same lies again. Reprinted with permission from *Psychology Today* magazine Copyright © 1996 (Sussex Publishers, Inc.).

searcher would conclude that the medication increased the pulse rate of the users and would reject the null hypothesis. The question is, "Where does the researcher draw the line?" This decision is not made on feelings or intuition; it is made statistically. That is, the difference must be significant and in all likelihood not due to chance. Here is where the concepts of statistical test and level of significance are used.

A **statistical test** uses the data obtained from a sample to make a decision about whether or not the null hypothesis should be rejected.

The numerical value obtained from a statistical test is called the **test value.**

In this type of statistical test, the mean is computed for the data obtained from the sample and is compared with the population mean. Then a decision is made to reject or not reject the null hypothesis on the basis of the value obtained from the statistical test. If the difference is significant, the null hypothesis is rejected. If it is not, then the null hypothesis is not rejected.

In the hypothesis-testing situation, there are four possible outcomes. In reality, the null hypothesis may or may not be true, and a decision is made to reject or not reject it on the basis of the data obtained from a sample. The four possible outcomes are shown in Figure 9–1. Notice that there are two possibilities for a correct decision and two possibilities for an incorrect decision.

Figure 9–1

Possible Outcomes of a Hypothesis Test

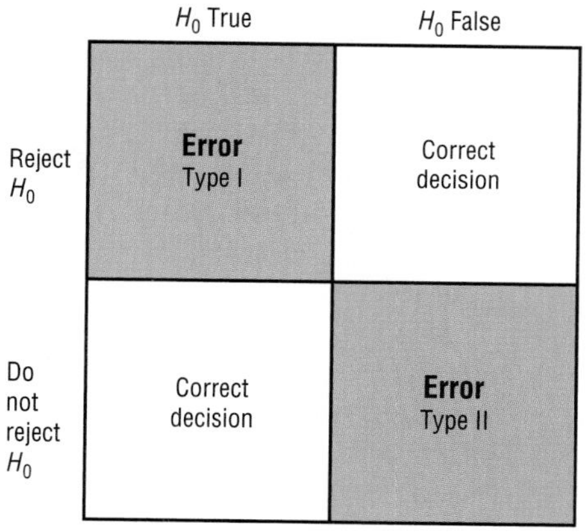

If a null hypothesis is true and it is rejected, then a *type I error* is made. In situation A, for instance, the medication might not significantly change the pulse rate of all the users in the population; but it might change the rate, by chance, of the subjects in the sample. In this case, the researcher will reject the null hypothesis when it is really true, thus committing a type I error.

On the other hand, the medication might not change the pulse rate of the subjects in the sample; but when it is given to the general population, it might cause a significant increase or decrease in the pulse rate of the users. The researcher, on the basis of the data obtained from the sample, will not reject the null hypothesis, thus committing a *type II error.*

In situation B, the additive might not significantly increase the lifetimes of automobile batteries in the population, but it might increase the lifetimes of the batteries in the sample. In this case, the null hypothesis would be rejected when it is not really true. This would be a type I error. On the other hand, the additive might not work on the batteries selected for the sample, but if it were to be used in the general population of batteries, it might significantly increase their lifetimes. The researcher, on the basis of information obtained from the sample, would not reject the null hypothesis, thus committing a type II error.

A **type I error** occurs if one rejects the null hypothesis when it is true.

A **type II error** occurs if one does not reject the null hypothesis when it is false.

The decision to reject or not reject the null hypothesis does not prove anything. *The only way to prove anything statistically is to use the entire population,* which, in most cases, is not possible. The decision, then, is made on the basis of probabilities. That is, when there is a large difference between the mean obtained from the sample and the hypothesized mean, the null hypothesis is probably not true. The question is, "How large a difference is necessary to reject the null hypothesis?" Here is where the level of significance is used.

The **level of significance** is the maximum probability of committing a type I error. This probability is symbolized by α (Greek letter **alpha**). That is, P (type I error) $= \alpha$.

The probability of a type II error is symbolized by β (Greek letter **beta**). That is, P (type II error) $= \beta$. In most hypothesis-testing situations, β cannot easily be computed.

Statisticians generally agree on using three arbitrary significance levels: the 0.10, 0.05, and 0.01 level. That is, if the null hypothesis is rejected, the probability of a type I error will be 10%, 5%, or 1%, and the probability of a correct decision will be 90%, 95%, or 99%, depending on which level of significance is used. Here is another way of putting it: When $\alpha = 0.10$, there is a 10% chance of rejecting a true null hypothesis; when $\alpha = 0.05$, there is a 5% chance of rejecting a true null hypothesis; and when $\alpha = 0.01$, there is a 1% chance of rejecting a true null hypothesis.

In a hypothesis-testing situation, the researcher decides what level of significance to use. It does not have to be the 0.10, 0.05, or 0.01 level. It can be any level, depending on the seriousness of the type I error. After a significance level is chosen, a *critical value* is selected from a table for the appropriate test. If a z test is used, for example, the z table (Table E in Appendix C) is consulted to find the critical value. The critical value determines the critical and noncritical regions.

The **critical value(s)** separates the critical region from the noncritical region.

The symbol for critical value is C.V.

The **critical** or **rejection region** is the range of values of the test value that indicates that there is a significant difference and that the null hypothesis should be rejected.

The **noncritical** or **nonrejection region** is the range of values of the test value that indicates that the difference was probably due to chance and that the null hypothesis should not be rejected.

The critical value can be on the right side of the mean or on the left side of the mean for a one-tailed test. Its location depends on the inequality sign of the alternative hypothesis. For example, in situation B, where the chemist is interested in increasing the average lifetime of automobile batteries, the alternative hypothesis is $H_1: \mu > 36$. Since the inequality sign is $>$, the null hypothesis will be rejected only when the sample mean is significantly greater than 36. Hence, the critical value must be on the right side of the mean. Therefore, this test is called a *one-tailed right test.*

A **one-tailed test** indicates that the null hypothesis should be rejected when the test value is in the critical region on one side of the mean. A one-tailed test is either **right-tailed** or **left-tailed,** depending on the direction of the inequality of the alternative hypothesis.

Objective 3. Find critical values for the *z* test.

To obtain the critical value, the researcher must choose an alpha level. In situation B, suppose the researcher chose $\alpha = 0.01$. Then, the researcher must find a *z* value such that 1% of the area falls to the right of the *z* value and 99% falls to the left of the *z* value, as shown in Figure 9–2(a).

Next, the researcher must find the value in Table E closest to 0.4900. Note that because the table gives the area between 0 and the *z*, 0.5000 must be subtracted from 0.9900 to get 0.4900. The critical *z* value is 2.33, since that value gives the area closest to 0.4900, as shown in Figure 9–2(b).

Figure 9–2

Finding the Critical Value for $\alpha = 0.01$ (Right-Tailed Test)

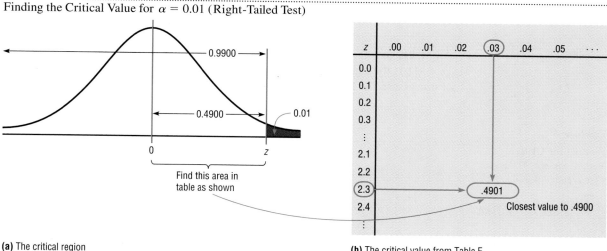

(a) The critical region

(b) The critical value from Table E

The critical and noncritical regions and the critical value are shown in Figure 9–3.

Figure 9–3

Critical and Noncritical Regions for $\alpha = 0.01$ (Right-Tailed Test)

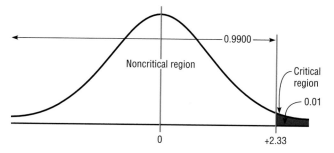

Now, move on to situation C, where the contractor is interested in lowering the heating bills. The alternative hypothesis is $H_1: \mu < \$78$. Hence, the critical value falls to the left of the mean. This test is thus a left-tailed test. At $\alpha = 0.01$, the critical value is -2.33, as shown in Figure 9–4.

Figure 9–4
Critical and Noncritical
Regions for $\alpha = 0.01$
(Left-Tailed Test)

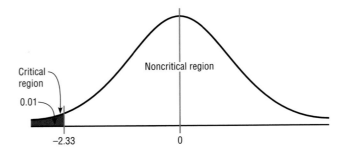

When a researcher conducts a two-tailed test, as in situation A, the null hypothesis can be rejected when there is a significant difference in either direction, above or below the mean.

In a **two-tailed test,** the null hypothesis should be rejected when the test value is in either of the two critical regions.

For a two-tailed test, then, the critical region must be split into two equal parts. If $\alpha = 0.01$, then half of the area, or 0.005, must be to the right of the mean and half must be to the left of the mean, as shown in Figure 9–5.

Figure 9–5
Finding the Critical
Values for $\alpha = 0.01$
(Two-Tailed Test)

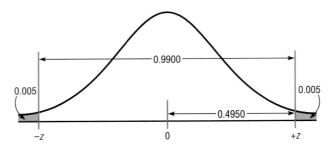

In this case, the area to be found in Table E is 0.4950. The critical values are $+2.58$ and -2.58, as shown in Figure 9–6.

Figure 9–6
Critical and Noncritical
Regions for $\alpha = 0.01$
(Two-Tailed Test)

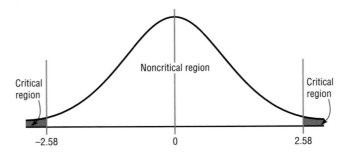

A similar procedure is used for other values of α.

Figure 9–7 shows the critical values (C.V.) for the three situations discussed in this section for α values of $\alpha = 0.10$, $\alpha = 0.05$, and $\alpha = 0.01$. The procedure for finding critical values is outlined next.

Figure 9–7

Summary of Hypothesis Testing and Critical Values

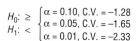

$H_0: \geq$
$H_1: <$
$\begin{cases} \alpha = 0.10,\ \text{C.V.} = -1.28 \\ \alpha = 0.05,\ \text{C.V.} = -1.65 \\ \alpha = 0.01,\ \text{C.V.} = -2.33 \end{cases}$

(a) Left-tailed

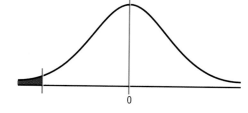

$H_0: \leq$
$H_1: >$
$\begin{cases} \alpha = 0.10,\ \text{C.V.} = +1.28 \\ \alpha = 0.05,\ \text{C.V.} = +1.65 \\ \alpha = 0.01,\ \text{C.V.} = +2.33 \end{cases}$

(b) Right-tailed

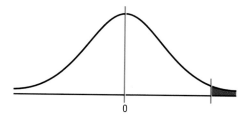

$H_0: =$
$H_1: \neq$
$\begin{cases} \alpha = 0.10,\ \text{C.V.} = \pm 1.65 \\ \alpha = 0.05,\ \text{C.V.} = \pm 1.96 \\ \alpha = 0.01,\ \text{C.V.} = \pm 2.58 \end{cases}$

(c) Two–tailed

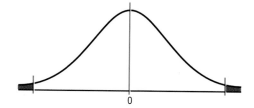

Procedure for Finding the Critical Values for Specific α Values, Using Table E

1. Draw the figure and indicate the appropriate area.
   a. If the test is left-tailed, the critical region, with an area equal to α, will be on the left side of the mean.
   b. If the test is right-tailed, the critical region, with an area equal to α, will be on the right side of the mean.
   c. If the test is two-tailed, α must be divided by 2; half of the area will be to the right of the mean, and the other half will be to the left of the mean.
2. For a one-tailed test, subtract the area (equivalent to α) in the critical region from 0.5000, since Table E gives the area under the standard normal distribution curve between 0 and any z to the right of 0. For a two-tailed test, subtract the area (equivalent to $\alpha/2$) from 0.5000.
3. Find the area in Table E corresponding to the value obtained in step 2. If the exact value cannot be found in the table, use the closest value.
4. Find the z value that corresponds to the area. This will be the critical value.
5. Determine the sign of the critical value for a one-tailed test.
   a. If the test is left-tailed, the critical value will be negative.
   b. If the test is right-tailed, the critical value will be positive.
   For a two-tailed test, one value will be positive and the other negative.

Example 9–2	Using Table E in Appendix C, find the critical value(s) for each situation and draw the appropriate figure, showing the critical region.

  a. A left-tailed test with $\alpha = 0.10$.

  b. A two-tailed test with $\alpha = 0.02$.

  c. A right-tailed test with $\alpha = 0.005$.

Solution *a*

STEP 1 Draw the figure and indicate the appropriate area. Since this is a left-tailed test, the area of 0.10 is located in the left tail, as shown in Figure 9–8.

STEP 2 Subtract 0.10 from 0.5000 to get 0.4000.

STEP 3 In Table E, find the area that is closest to 0.4000; in this case, it is 0.3997.

STEP 4 Find the z value that corresponds to this area. It is 1.28.

STEP 5 Determine the sign of the critical value (i.e., the z value). Since this is a left-tailed test, the sign of the critical value is negative. Hence, the critical value is -1.28. See Figure 9–8.

Figure 9–8

Critical Value and
Critical Region for
Part *a* of Example 9–2

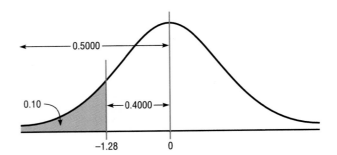

Solution *b*

STEP 1 Draw the figure and indicate the appropriate area. In this case, there are two areas equivalent to $\alpha/2$, or $0.02/2 = 0.01$.

STEP 2 Subtract 0.01 from 0.5000 to get 0.4900.

STEP 3 Find the area in Table E closest to 0.4900. In this case, it is 0.4901.

STEP 4 Find the z value that corresponds to this area. It is 2.33.

STEP 5 Determine the sign of the critical value. Since this test is a two-tailed test, there are two critical values: one is positive and the other is negative. They are $+2.33$ and -2.33. See Figure 9–9.

Figure 9–9

Critical Values and
Critical Regions for
Part *b* of Example 9–2

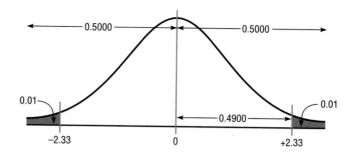

Solution *c*

STEP 1   Draw the figure and indicate the appropriate area. Since this is a right-tailed test, the area 0.005 is located in the right tail, as shown in Figure 9–10.

Figure 9–10

Critical Value and
Critical Region for
Part *c* of Example 9–2

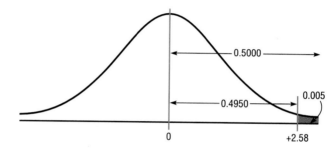

STEP 2   Subtract 0.005 from 0.5000 to get 0.4950.

STEP 3   Find the area in Table E equal to 0.4950.

STEP 4   Find the *z* value that corresponds to this area. It is 2.58.

STEP 5   Determine the sign of the critical value. Since this is a right-tailed test, the sign is positive; hence, the critical value is +2.58.

Objective 4. State the five steps used in hypothesis testing.

In hypothesis testing, the following steps are recommended.

1. State the hypotheses. Be sure to state both the null and the alternative hypotheses.
2. Design the study. This step includes selecting the correct statistical test, choosing a level of significance, and formulating a plan to carry out the study. The plan should include information such as the definition of the population, the way the sample will be selected, and the methods that will be used to collect the data.
3. Conduct the study and collect the data.
4. Evaluate the data. The data should be tabulated in this step, and the statistical test should be conducted. Finally, decide whether to reject or not reject the null hypothesis.
5. Summarize the results.

For the purposes of this chapter, a simplified version of the hypothesis-testing procedure will be used, since designing the study and collecting the data will be omitted. This procedure is summarized in Procedure Table 7.

Procedure Table 7

Procedure for Solving Hypothesis-Testing Problems

STEP 1 State the hypotheses, and identify the claim.

STEP 2 Find the critical value(s) from Table E in Appendix C.

STEP 3 Compute the test value.

STEP 4 Make the decision to reject or not reject the null hypothesis.

STEP 5 Summarize the results.

Exercises

9–1. (**W**) Define *null* and *alternative hypothesis,* and give an example of each.

9–2. (**W**) What is meant by a type I error? A type II error? How are they related?

9–3. (**W**) What is meant by a statistical test?

9–4. (**W**) Explain the difference between a one-tailed and a two-tailed test.

9–5. (**W**) What is meant by the critical region? The noncritical region?

9–6. (**W**) What symbols are used to represent the null hypothesis and the alternative hypothesis?

9–7. (**W**) What symbols are used to represent the probabilities of type I and type II errors?

9–8. (**W**) Explain what is meant by a significant difference.

9–9. (**W**) When should a one-tailed test be used? A two-tailed test?

9–10. (**W**) List the steps in hypothesis testing.

9–11. (**W**) In hypothesis testing, why can't the hypothesis be proved true?

9–12. (**ans**) Using the z table (Table E), find the critical value (or values) for each.
a. $\alpha = 0.01$, two-tailed test
b. $\alpha = 0.05$, right-tailed test
c. $\alpha = 0.005$, left-tailed test
d. $\alpha = 0.10$, left-tailed test
e. $\alpha = 0.05$, two-tailed test
f. $\alpha = 0.04$, right-tailed test
g. $\alpha = 0.01$, left-tailed test
h. $\alpha = 0.10$, two-tailed test
i. $\alpha = 0.02$, right-tailed test
j. $\alpha = 0.02$, two-tailed test

9–13. (**W**) For each conjecture, state the null and alternative hypotheses.
a. The average age of taxi drivers in New York City is 36.3 years.
b. The average income of nurses is $36,250.
c. The average age of disc jockeys is greater than 27.6 years.
d. The average pulse rate of female joggers is less than 72 beats per minute.
e. The average bowling score of people who enrolled in a basic bowling class is less than 100.
f. The average cost of a VCR is $297.75.
g. The average electric bill for residents of White Pine Estates exceeds $52.98 per month.
h. The average number of calories of brand A's low-calorie meals is at most 300.
i. The average weight loss of people who use brand A's low-calorie meals for six weeks is at least 3.6 pounds.

9–3

Large Sample Mean Test

Objective 5. Test means for large samples using the z test.

In this chapter, two statistical tests will be explained: the z test, used to test for the mean of a large sample, and the t test, used for the mean of a small sample. This section explains the z test, and Section 9–4 explains the t test.

Many hypotheses are tested using a statistical test based on the following general formula:

$$\text{test value} = \frac{(\text{observed value}) - (\text{expected value})}{\text{standard error}}$$

The observed value is the statistic (such as the mean) that is computed from the sample data. The expected value is the parameter (such as the mean) that one would expect to obtain if the null hypothesis were true—in other words, the hypothesized value. The denominator is the standard error of the statistic being tested (in this case, the standard error of the mean).

The z test is defined formally as follows.

The **z test** is a statistical test for the mean of a population. It can be used when $n \geq 30$, or when the population is normally distributed and σ is known.

The formula for the z test is

$$z = \frac{\bar{X} - \mu}{\sigma/\sqrt{n}}$$

where

$\bar{X}$ = sample mean

μ = hypothesized population mean

σ = population deviation

n = sample size

For the z test, the observed value is the value of the sample mean. The expected value is the value of the population mean, assuming that the null hypothesis is true. The denominator $\sigma/\sqrt{n}$ is the standard error of the mean.

The formula for the z test is the same formula shown in Chapter 7 for the situation where one is using a distribution of sample means. Recall that the central limit theorem allows one to use the standard normal distribution to approximate the distribution of sample means when $n \geq 30$. If σ is unknown, s can be used when $n \geq 30$.

Note: The student's first encounter with hypothesis testing can be somewhat challenging and confusing, since there are many new concepts being introduced at the same time. *In order to understand all the concepts, the student must carefully follow each step in the examples and try each exercise that is assigned.* Only after careful study and patience will these concepts become clear.

As stated in the previous section, there are five steps for solving *hypothesis-testing* problems:

STEP 1   State the hypothesis and identify the claim.

STEP 2   Find the critical value(s).

STEP 3   Compute the test value.

STEP 4   Make the decision to reject or not reject the null hypothesis.

STEP 5   Summarize the results.

In this study on margarine consumption, many conclusions are stated. State hypotheses that may have been used to test these conclusions. Define the population and the sample used. Do you think the sample would be representative of all adults? Explain your answer. Comment on the sample size.

Stick Margarine May Boost Risk of Heart Attacks

By Nancy Hellmich
USA TODAY

The bad news about margarine continues to mount.

Eating stick margarine increases the risk of heart attack, according to new findings from the Harvard Nurses' Health Study, an ongoing analysis of the diets of 90,000 nurses.

This new report adds to growing evidence that trans fatty acids—the fats that form when liquid vegetable oils are processed or hydrogenated raise blood cholesterol in much the same way that saturated fat does.

Other culprits besides margarine: any solid vegetable shortening, some fried foods at chains such as McDonald's and Burger King and processed foods made with partially hydrogenated vegetable oils (check the ingredient list).

The new study reported in Saturday's *Lancet* shows:

• Women who frequently use margarine have more than a 50% higher risk of heart disease than those who infrequently use margarine.

• Those who eat a couple of cookies a day (cookies often contain partially hydrogenated oils) are at a 50% higher risk.

• Women who eat lots of foods high in trans fatty acids have a 70% higher risk of heart disease than those who don't.

In recent years, many people have switched to margarine from butter, which is high in saturated fat.

But don't go back to butter just because margarine is unhealthy, says Dr. Walter Willett, study author. Instead, consider switching to liquid oils instead of solid margarine or shortening.

He's a strong supporter of olive oil because "there's a strong indication that olive oil is at least safe." And it's a monounsaturated fat that lowers bad cholesterol (LDL) without lowering good cholesterol (HDL).

Also, some tub margarines contain more water and air and therefore less fat and fewer trans fatty acids than stick margarines.

Source: Copyright 1993, *USA TODAY*. Reprinted with permission.

The next example illustrates these steps.

Example 9–3
A researcher reports that the average salary of assistant professors is more than $42,000. A sample of 30 assistant professors has a mean salary of $43,260. At $\alpha = 0.05$, test the claim that assistant professors earn more than $42,000 a year. The standard deviation of the population is $5230.

Solution

STEP 1 State the hypotheses and identify the claim.

$$H_0: \mu \leq \$42,000 \qquad H_1; \mu > \$42,000 \text{ (claim)}$$

STEP 2 Find the critical value. Since $\alpha = 0.05$ and the test is a right-tailed test, the critical value is $z = +1.65$.

STEP 3 Compute the test value.

$$z = \frac{\overline{X} - \mu}{\sigma/\sqrt{n}} = \frac{\$43,260 - 42,000}{5230/\sqrt{30}} = 1.32$$

STEP 4 Make the decision. Since the test value, $+1.32$, is less than the critical value, $+1.65$, and not in the critical region, the decision is, "Do not reject the null hypothesis." This test is summarized in Figure 9–11.

Figure 9–11

Summary of the z Test of Example 9–3

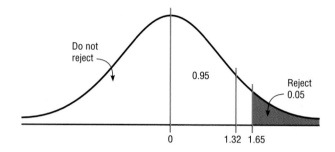

STEP 5 Summarize the results. There is not enough evidence to support the claim that assistant professors earn more on average than $42,000 a year.

Comment: Even though in Example 9–3 the sample mean, $43,260, is higher than the hypothesized population mean of $42,000, it is not *significantly* higher. Hence, the difference may be due to chance. When the null hypothesis is not rejected, there is still a probability of a type II error—i.e, of not rejecting the null hypothesis when it is false.

The probability of a type II error is not easily ascertained. For now, it is only necessary to realize that the type II error exists.

It should also be noted that when the null hypothesis is not rejected, it cannot be accepted as true. There is merely not enough evidence to say that it is false. This guideline may sound a little confusing, but the situation is analogous to a jury trial. The verdict is either guilty or not guilty and is based on the evidence presented. If a person is judged not guilty, it does not mean that the person is proved innocent; it only means that there was not enough evidence to reach the guilty verdict.

Example 9–4

A national magazine claims that the average college student watches less television than the general public. The national average is 29.4 hours per week, with a standard deviation of 2 hours. A sample of 30 college students has a mean of 27 hours. Is there enough evidence to support the claim at $\alpha = 0.01$?

Solution

STEP 1 State the hypotheses and identify the claim.

$$H_0: \mu \geq 29.4 \quad \text{and} \quad H_1: \mu < 29.4 \text{ (claim)}$$

STEP 2   Find the critical value. Since $\alpha = 0.01$ and the test is a left-tailed test, the critical value is -2.33.

STEP 3   Compute the test value.

$$z = \frac{\overline{X} - \mu}{\sigma/\sqrt{n}} = \frac{27 - 29.4}{2/\sqrt{30}} = -6.57$$

STEP 4   Make the decision. Since the test value, -6.57, falls in the critical region, the decision is to reject the null hypothesis. The test is summarized in Figure 9–12.

Figure 9–12

Summary of the z Test of Example 9–4

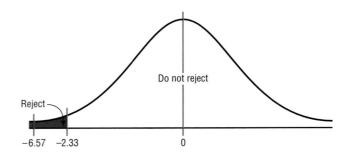

STEP 5   Summarize the results. There is enough evidence to support the claim that college students watch less television than the general public.

Comment: In Example 9–4, the difference is said to be significant. However, when the null hypothesis is rejected, there is always a chance of a type I error. In this case, the probability of a type I error is at most 0.01, or 1%.

Example 9–5

The Medical Rehabilitation Education Foundation reports that the average cost of rehabilitation for stroke victims is $24,672. To see if the average cost of rehabilitation is different at a large hospital, a researcher selected a random sample of 35 stroke victims and found that the average cost of their rehabilitation is $25,226. The standard deviation of the population is $3,251. At $\alpha = 0.01$, can it be concluded that the average cost at a large hospital is different from $24,672?
Source: Snapshot, *USA Today,* September 18, 1995.

Solution

STEP 1   State the hypotheses and identify the claim.

$$H_0: \mu = \$24{,}672 \qquad \text{and} \qquad H_1: \mu \neq \$24{,}672 \text{ (claim)}$$

STEP 2   Find the critical values. Since $\alpha = 0.01$ and the test is a two-tailed test, the critical values are $+2.58$ and -2.58.

STEP 3   Compute the test value.

$$z = \frac{\overline{X} - \mu}{\sigma/\sqrt{n}} = \frac{25{,}226 - 24{,}672}{3{,}251/\sqrt{35}} = 1.01$$

STEP 4   Make the decision. Do not reject the null hypothesis, since the test value falls in the noncritical region, as shown in Figure 9–13.

Figure 9–13

Critical and Test Values for Example 9–5

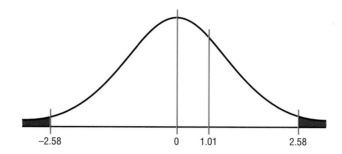

STEP 5   Summarize the results. There is not enough evidence to support the claim that the average cost of rehabilitation at the large hospital is different from $24,672.

As with confidence intervals, the central limit theorem states that when the population standard deviation σ is unknown, the sample standard deviation s can be used in the formula as long as the sample size is 30 or more. The formula for the z test in this case is

$$z = \frac{\overline{X} - \mu}{s/\sqrt{n}}$$

When n is less than 30 and σ is unknown, the t test must be used. The t test will be explained in the next section.

Students sometimes have difficulty summarizing the results of a hypothesis test. Figure 9–14 shows the four possible outcomes and the summary statement for each situation.

Figure 9–14

Outcomes of a Hypothesis-Testing Situation

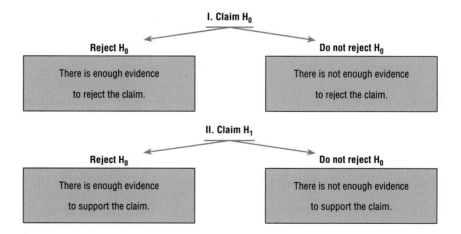

First of all, the claim can be either the null or alternative hypothesis, and one should identify which it is. Second, after the study is completed, the null hypothesis is either rejected or not rejected. From these two facts, the decision can be identified in the appropriate block of Figure 9–14.

For example, suppose a researcher claims that the mean weight of an adult animal of a particular species is 42 pounds. In this case, the claim would be the null hypothesis, H_0: $\mu = 42$. If the null hypothesis is rejected, the conclusion would be that there is enough evidence to reject the claim that the mean weight of the adult animal is 42 pounds. See Figure 9–15(a).

Figure 9–15

Outcomes of a
Hypothesis-Testing
Situation for Two
Specific Cases

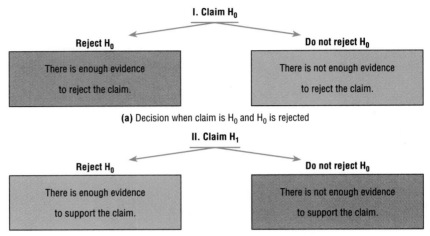

(a) Decision when claim is H_0 and H_0 is rejected

(b) Decision when claim is H_1 and H_0 is not rejected

On the other hand, suppose the researcher claims that the mean weight of the adult animals is not 42 pounds. The claim would be the alternative hypothesis, H_1: $\mu \neq 42$. Furthermore, suppose that the null hypothesis is not rejected. The conclusion, then, would be that there is not enough evidence to support the claim that the mean weight of the adult animals is not 42 pounds. See Figure 9–15(b).

Again, remember that nothing is being proven true or false. The statistician is only stating that there is or is not enough evidence to say that a claim is *probably* true or false. As noted previously, the only way to prove something would be to use the entire population under study, and usually this cannot be done, especially when the population is large.

P-Values

Statisticians usually test hypotheses at the common α levels of 0.05 or 0.01 and sometimes at 0.10. Recall that the choice of the level depends on the seriousness of the type I error. Besides listing an α value, many computer statistical packages give a P-value for hypothesis tests. The **P-value** is the actual probability of getting the sample mean value or a more extreme sample mean value in the direction of the alternative hypothesis ($<$ or $>$) if the null hypothesis is true. In other words, the P-value is the actual area under the standard normal distribution curve (or other curve, depending on what statistical test is being used) representing the probability of a particular sample mean or a more extreme sample mean occurring if the null hypothesis is true.

For example, suppose that a null hypothesis is H_0: $\mu \leq 50$ and the mean of a sample is $\overline{X} = 52$. If the computer printed a P-value of 0.0356 for a statistical test, then the probability of getting a sample mean of 52 or greater is 0.0356 if the true population mean is 50. The relationship between the P-value and the α value can be explained in this manner. For $P = 0.0356$, the null hypothesis would be rejected at $\alpha = 0.05$ but not at $\alpha = 0.01$. See Figure 9–16.

Figure 9–16

Comparison of α Values and P-Values

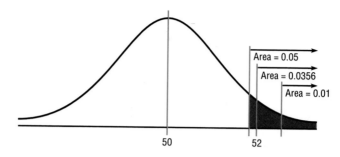

When the hypothesis test is two-tailed, the area corresponding to the P-value must be doubled. For a two-tailed test, if α is 0.05 and the P-value given is 0.0356, the actual P-value would be $2(0.0356) = 0.0712$. That is, the null hypothesis should not be rejected at $\alpha = 0.05$, since 0.0712 is greater than 0.05.

The next example shows how to find a P-value for use with the z test.

Example 9–6

A researcher wishes to test the claim that the average age of lifeguards in Ocean City is greater than 24 years. She selects a sample of 36 guards and finds the mean of the sample to be 24.7 years, with a standard deviation of 2 years. Is there evidence to support the claim at $\alpha = 0.05$? Find the P-value.

Solution

STEP 1   State the hypotheses and identify the claim.

$$H_0: \mu \leq 24 \quad \text{and} \quad H_1: \mu > 24 \text{ (claim)}$$

STEP 2   Compute the test value.

$$z = \frac{24.7 - 24}{2/\sqrt{36}} = 2.10$$

STEP 3   Using Table E in Appendix C, find the corresponding area under the normal distribution for $z = 2.10$. It is 0.4821.

STEP 4   Subtract this value for the area from 0.5000 to find the area in the right tail.

$$0.5000 - 0.4821 = 0.0179$$

Hence, the P-value is 0.0179.

STEP 5   Make the decision. Since the P-value is less than 0.05, the decision is to reject the null hypothesis.

STEP 6   Summarize the results. There is enough evidence to support the claim that the average age of lifeguards in Ocean City is greater than 24 years.

Interesting Fact

Nasty fights are hazardous not only to your relationship but also to your health. Researchers found that when people fight, stress hormones like cortisol and norepinephrine not only cause the body to react to danger but also reduce its supply of disease-fighting immune cells. It was originally thought that men experience more stress during arguments than women. However, researchers found that hormone levels and blood pressure rise far more in women because they tend to remember and dwell on disputes long afterward. Reprinted with permission from *Psychology Today* magazine Copyright © 1996 (Sussex Publishers, Inc.).

Note: Had the researcher chosen $\alpha = 0.01$, the null hypothesis would not have been rejected, since the *P*-value (0.0179) is greater than 0.01.

....

Example 9–7

A researcher claims that the average wind speed in a certain city is 8 miles per hour. A sample of 32 days has an average wind speed of 8.2 miles per hour. The standard deviation of the sample is 0.6 mile per hour. At $\alpha = 0.05$, is there enough evidence to reject the claim? Use the *P*-value method.

Solution

STEP 1   State the hypotheses and identify the claim.

$$H_0: \mu = 8 \text{ (claim)} \qquad \text{and} \qquad H_1: \mu \neq 8$$

STEP 2   Compute the test value.

$$z = \frac{8.2 - 8}{0.6/\sqrt{32}} = 1.89$$

STEP 3   Using Table E, find the corresponding area for $z = 1.89$. It is 0.4706.

STEP 4   Subtract the value from 0.5000.

$$0.5000 - 0.4706 = 0.0294.$$

STEP 5   Make the decision: Since this test is two-tailed, the value 0.0294 must be doubled; $2(0.0294) = 0.0588$. Hence, the decision is not to reject the null hypothesis, since the *P*-value is greater than 0.05.

STEP 6   Summarize the results. There is not enough evidence to reject the claim that the average wind speed is 8 miles per hour.

....

A clear distinction between the α value and the *P*-value should be made. The α value is chosen by the researcher before the statistical test is conducted. The *P*-value is computed after the sample mean has been found.

There are two schools of thought on *P*-values. Some researchers do not choose an α value but report the *P*-value and allow the reader to decide whether the null hypothesis should be rejected. Others decide on the α value in advance and use the *P*-value to make the decision, as shown in the previous two examples. A note of caution is needed here: If a researcher selects $\alpha = 0.01$ and the *P*-value is 0.03, the researcher may decide to change the α value from 0.01 to 0.05 so that the null hypothesis will be rejected. This, of course, should not be done. If the α level is selected in advance, it should be used in making the decision. If the researcher is using the *P*-value, then no α value should be selected and the reader should make the decision about whether the null hypothesis should be rejected.

P-values given on calculators and computers are slightly different from those found in Table E. This is due to the fact that *z*-values and the values in Table E have been rounded. Also, most calculators and computers give the exact *P*-value for two-tailed tests, so it should not be doubled (as it should when the area found in Table E is used).

One additional note on hypothesis testing is that the researcher should distinguish between *statistical significance* and *practical significance*. When the null hypothesis is rejected at a specific significance level, it can be concluded that the difference is probably not due to chance and thus is statistically significant. However, the results may not have any practical significance. For example, suppose that a new fuel additive increases the miles per gallon a car can get by $\frac{1}{4}$ mile for a sample of 1000 automobiles. The results may be statistically significant at the 0.05 level, but it would hardly be worthwhile to market the product for such a small increase. Hence, there is no practical significance to the results. It is up to the researcher to use common sense when interpreting the results of a statistical test.

Computer Application for a z Test for a Single Mean

MINITAB z test for a single mean:

1. Enter the data into C1.
2. Click Stat > Basic Statistics > 1-Sample Z.
3. Highlight C1 in the dialog box and click on Select.
4. Click on Test mean. Press Tab.
5. Type in the value of the hypothesized mean. Press Tab.
6. If the test is two-tailed, press Tab. If the test is one-tailed, click on the arrow in the box labeled Alternative and select the appropriate hypothesis. Then press Tab.
7. Type in the value for sigma.
8. Click on OK.

Example: If $\sigma = 3$, test the claim H_0: $\mu = 32$ using the following data: (Use $\alpha = 0.05$)

Data:

C1	32	28	24	18	35	41	27	28	28	31
	25	31	35	40	36	32	29	28	19	21
	26	24	20	19	27	26	23	21	30	20

MINITAB printout for this example:

Z-Test

```
Test of mu = 32.000 vs mu not = 32.000
The assumed sigma = 3.00

Variable     N      Mean     StDev     SE Mean        Z          P
C1          30    27.467     6.118       0.548     -8.28     0.0000
```

Summary: The test value is -8.28, which is significant beyond the 0.01 level since the P-value is 0.000.

Exercises

For Exercises 9–14 through 9–27, perform each of the following steps.
a. State the hypotheses and identify the claim.
b. Find the critical value(s).
c. Compute the test value.
d. Make the decision.
e. Summarize the results.

Use diagrams to show the critical region (or regions).

9–14. A shoe-store manager claims that the average cost of a pair of tennis shoes is $89.95. A sample of 68 pairs of tennis shoes has an average cost of $90.26. The standard deviation of the sample is $3.00. At

$\alpha = 0.05$, is there enough evidence to reject the manager's claim?

9–15. A survey claims that the average cost of a hotel room in Atlanta is $69.21. In order to see if this is correct, a researcher selects a sample of 30 hotel rooms and finds that the average cost is $68.43. The standard deviation of the population is $3.72. At $\alpha = 0.05$, is there enough evidence to reject the claim?
Source: *USA Today,* August 11, 1995.

9–16. The manager of a large factory believes that the average hourly wage of the employees is below $9.78 per hour. A sample of 18 employees has a mean hourly wage of $9.60. The standard deviation of all salaries is $1.42. Assume the variable is normally distributed. At $\alpha = 0.10$, is there enough evidence to support the manager's claim?

9–17. The average SAT score in mathematics is 483, with a standard deviation of 100. A special preparation course states that it can increase scores. A sample of 32 students completed the course, and the average of their scores was 494. At $\alpha = 0.05$, does the course do what it claims?

9–18. A maker of frozen meals claims that the average caloric content of its meals is 800, and the standard deviation is 25. A researcher tested 12 meals and found that the average number of calories was 873. Is there enough evidence to reject the claim at $\alpha = 0.02$? Assume the variable is normally distributed.

9–19. A report in *USA Today* stated that the average age of commercial jets in the United States is 14 years. An executive of a large airline company selects a sample of 36 planes and finds the average age of the planes is 11.8 years. The standard deviation of the sample is 2.7 years. At $\alpha = 0.01$, can it be concluded that the average age of the planes in his company is less than the national average?
Source: *USA Today,* July 7, 1995.

9–20. A diet clinic states that there is an average loss of 24 pounds for patients who stay on the program for 20 weeks. The standard deviation is 5 pounds. The clinic tries a new diet, reducing the salt intake to see whether that strategy will produce a greater weight loss. A group of 40 volunteers loses an average of 16.3 pounds each over 20 weeks. Should the clinic change to the new diet? Use $\alpha = 0.05$.

9–21. A travel agent claims that the average cost of a three-day trip to Atlantic City is $915. The standard deviation is $35. Sixty people who scheduled the trip paid an average of $927 for the trip. At $\alpha = 0.05$, should the agent's claim be rejected?

9–22. A survey found that women over the age of 55 consume an average of 1660 calories a day. In order to see if the number of calories consumed by women over age 55 in assisted-living residences is the same, a researcher sampled 43 women over the age of 55 in a large assisted-living facility and found the mean number of calories consumed was 1446. The standard deviation of the sample is 56 calories. At $\alpha = 0.10$, can it be concluded that there is no difference between the number of calories consumed by the residents and that consumed by other women over 55?

9–23. A manufacturer states that the average lifetime of its lightbulbs is 3 years, or 36 months. The standard deviation is 8 months. Fifty bulbs are selected, and the average lifetime is found to be 32 months. Should the manufacturer's statement be rejected at $\alpha = 0.01$?

9–24. A real estate agent claims that the average price of a condominium in Naples, Florida, is at most $56,900. The standard deviation is $2500. A sample of 36 condos has an average selling price of $57,374. Does the evidence support the claim at $\alpha = 0.05$?

9–25. The average serum cholesterol level in a certain group of patients is 240 milligrams. The standard deviation is 18 milligrams. A new medication is designed to lower the cholesterol level if taken for one month. A sample of 40 people used the medication for 30 days, after which their average cholesterol level was 229 milligrams. At $\alpha = 0.01$, does the medication lower the cholesterol level of the patients?

9–26. The state's education secretary claims that the average cost of one year's tuition for all private high schools in the state is $2350. A sample of 30 private high schools is selected, and the average tuition is $2315. The standard deviation for the population is $38. At $\alpha = 0.05$, is there enough evidence to reject the claim that the average cost of tuition is equal to $2350?

9–27. To see if young men ages 8–17 spend more or less than the national average of $24.44 per shopping trip to a local mall, the manager surveyed 33 young men and found the average amount spent per visit was $22.97. The standard deviation of the sample was $3.70. At $\alpha = 0.02$, can it be concluded that the average amount spent at a local mall is not equal to the national average of $24.44?
Source: *USA Today,* July 25, 1995.

9–28. (**W**) What is meant by a *P*-value?

9–29. State whether or not the null hypothesis should be rejected on the basis of the given *P*-value.
a. *P*-value = 0.258, $\alpha = 0.05$, one-tailed test

b. *P*-value = 0.6841, α = 0.10, two-tailed test
c. *P*-value = 0.0153, α = 0.01, one-tailed test
d. *P*-value = 0.0232, α = 0.05, two-tailed test
e. *P*-value = 0.002, α = 0.01, one-tailed test

9–30. A college professor claims that the average cost of a paperback textbook is greater than $27.50. A sample of 50 books has an average cost of $29.30. The standard deviation of the sample is $5.00. Find the *P*-value for the test. On the basis of the *P*-value, should the null hypothesis be rejected at α = 0.05?

9–31. A study found that the average stopping distance of a school bus traveling 50 miles per hour was 264 feet (Snapshots, *USA Today,* March 12, 1992). A group of automotive engineers decided to conduct a study of its school buses and found that for 20 buses, the average stopping distance of buses traveling 50 miles per hour was 262.3 feet. The standard deviation of the population was 3 feet. Test the claim that the average stopping distance of the company's buses is actually less than 264 feet. Find the *P*-value. On the basis of the *P*-value, should the null hypothesis be rejected at α = 0.01? Assume that the variable is normally distributed.

9–32. For a certain group of individuals, the average cost of a trip to the Super Bowl was $875. The standard deviation of the population was $50. This year, 49 fans who scheduled the trip paid an average of $890 for the three-day trip. Test the claim that the average cost is greater than last year's cost. Find the *P*-value. On the basis of the *P*-value, should the null hypothesis be rejected at α = 0.01?

9–33. A manufacturer states that the average lifetime of its television sets is more than 84 months. The standard deviation of the population is 10 months. One hundred sets are randomly selected and tested. The average lifetime of the sample is 85.1 months. Test the claim that the average lifetime of the sets is more than 84 months, and find the *P*-value. On the basis of the *P*-value, should the null hypothesis be rejected at α = 0.01?

9–34. A special cable has a breaking strength of 800 pounds. The standard deviation of the population is 12 pounds. A researcher selects a sample of 20 cables and finds that the average breaking strength is 793 pounds. Can one reject the claim that the breaking strength is 800 pounds? Find the *P*-value. Should the null hypothesis be rejected at α = 0.01? Assume that the variable is normally distributed.

9–35. The average hourly wage last year for members of the hospital clerical staff in a large city was $6.32. The standard deviation of the population was $0.54. This year a sample of 50 workers had an average hourly wage of $6.51. Test the claim, at α = 0.05, that the average has not changed by finding the *P*-value for the test.

9–36. Ten years ago, the average acreage of farms in a certain geographic region was 65 acres. The standard deviation of the population was 7 acres. A recent study consisting of 22 farms showed that the average was 63.2 acres per farm. Test the claim, at α = 0.10, that the average has not changed by finding the *P*-value for the test. Assume that σ has not changed and the variable is normally distributed.

9–37. A car dealer recommends that transmissions should be serviced at 30,000 miles. In order to see whether her customers are adhering to this recommendation, the dealer selects a sample of 40 customers and finds that the average mileage of the automobiles serviced is 30,456. The standard deviation of the sample is 1684 miles. By finding the *P*-value, determine whether the owners are having their transmissions serviced at 30,000 miles, for α = 0.10. Do you think the α value of 0.10 is an appropriate significance level?

9–38. A motorist claims that the South Boro Police issue an average of 60 speeding tickets per day. The following data show the number of speeding tickets issued each day for a period of one month. Assume σ is 13.42. Is there enough evidence to reject the motorist's claim at α = 0.05?

72	45	36	68	69	71	57	60
83	26	60	72	58	87	48	59
60	56	64	68	42	57	57	
58	63	49	73	75	42	63	

*9–39. A manager states that in his factory, the average number of days per year missed by the employees due to illness is less than the national average of 10. The following data show the number of days missed by 40 employees last year. Is there sufficient evidence to believe the manager's statement at α = 0.05? (Use *s* to estimate σ.)

0	6	12	3	3	5	4	1
3	9	6	0	7	6	3	4
7	4	7	1	0	8	12	3
2	5	10	5	15	3	2	5
3	11	8	2	2	4	1	9

*9–40. Suppose a statistician chose to test a hypothesis at α = 0.01. The critical value for a right-tailed test is +2.33. If the test value was 1.97, what would the decision be? What would happen if, after seeing the test value, he decided to choose α = 0.05?

What would the decision be? Explain the contradiction, if there is one.

*9–41. The president of a company states that the average hourly wage of her employees is \$8.65. A sample of 50 employees has the distribution shown. At $\alpha = 0.05$, is the president's statement believable? (Use s to approximate σ.)

Class	Frequency
8.35–8.43	2
8.44–8.52	6
8.53–8.61	12
8.62–8.70	18
8.71–8.79	10
8.80–8.88	2

Small Sample Mean Test

Objective 6. Test means for small samples using the *t* test.

When the population standard deviation is unknown and the sample size is less than 30, the z test is inappropriate for testing hypotheses involving means. A different test, called the t test, is used. The t test is used when σ is unknown and $n < 30$. (Some authors use $n \le 30$.)

As stated in Chapter 8, the t distribution is similar to the standard normal distribution in the following ways.

1. It is bell-shaped.
2. It is symmetrical about the mean.
3. The mean, median, and mode are equal to 0 and are located at the center of the distribution.
4. The curve never touches the x axis.

The t distribution differs from the standard normal distribution in the following ways.

1. The variance is greater than 1.
2. The t distribution is a family of curves based on the **degrees of freedom,** which is a number related to sample size. (Recall that the symbol for degrees of freedom is d.f. See Section 8–3 for an explanation of degrees of freedom.)
3. As the sample size increases, the t distribution approaches the normal distribution.

The t test is defined formally next.

The **t test** is a statistical test for the mean of a population and is used when the population is normally or approximately normally distributed, σ is unknown, and $n < 30$.

The formula for the t test is

$$t = \frac{\bar{X} - \mu}{s/\sqrt{n}}$$

The degrees of freedom are d.f. = $n - 1$.

The formula for the t test is similar to the formula for the z test. But since the population standard deviation σ is unknown, the sample standard deviation s is used instead.

The critical values for the t test are given in Table F in Appendix C. For a one-tailed test, find the α level by looking at the top row of the table and finding the appropriate column. Find the degrees of freedom by looking down the left-hand column. Notice that degrees of freedom are given for values from 1 through

28. When the degrees of freedom are 29 or more, the row with ∞ (infinity) is used. Note that the values in this row are the same as the values for the z distribution, since as the sample size increases, the t distribution approaches the z distribution. When the sample size is 30 or more, statisticians generally agree that the two distributions can be considered identical, since the difference between their values is relatively small.

Example 9–8

Find the critical t value for $\alpha = 0.05$ with d.f. = 16 for a right-tailed t test.

Solution

Find the 0.05 column in the top row and 16 in the left-hand column. Where the row and column meet, the appropriate critical value is found; it is +1.746. See Figure 9–17.

Figure 9–17

Finding the Critical Value for the t Test in Table F (Example 9–8)

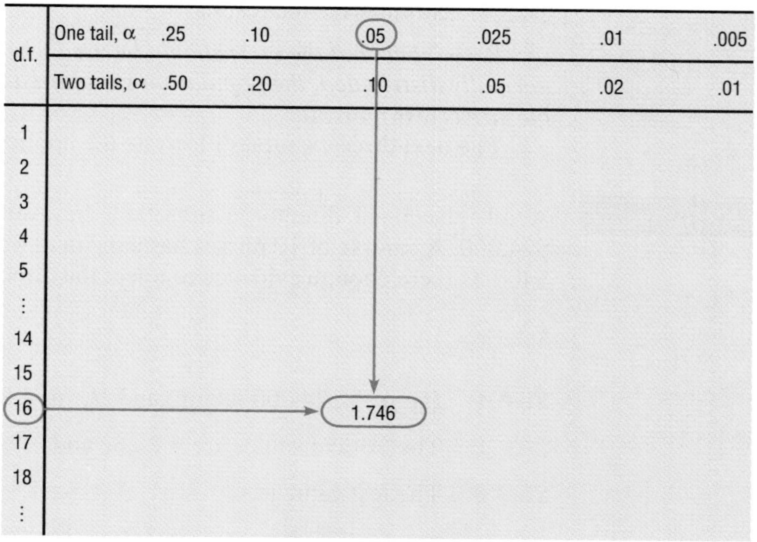

d.f.	One tail, α	.25	.10	.05	.025	.01	.005
	Two tails, α	.50	.20	.10	.05	.02	.01
1							
2							
3							
4							
5							
⋮							
14							
15							
16				1.746			
17							
18							
⋮							

Example 9–9

Find the critical t value for $\alpha = 0.01$ with d.f. = 22 for a left-tailed test.

Solution

Find the 0.01 column in the row labeled "One tail" and find 22 in the left column. The critical value is −2.508 since the test is a one-tailed left test.

Example 9–10

Find the critical values for $\alpha = 0.10$ with d.f. = 18 for a two-tailed t test.

Solution

Find the 0.10 column in the row labeled "Two tails" and find 18 in the column labeled "d.f." The critical values are +1.734 and −1.734.

Example 9–11

Find the critical value for $\alpha = 0.05$ with d.f. = 28 for a right-tailed t test.

Solution

Find the 0.05 column in the "One tail" row and 28 in the left column. The critical value is $+1.701$.

When testing hypotheses by using the t test, follow the same procedure as for the z test.

STEP 1   State the hypotheses and identify the claim.

STEP 2   Find the critical value(s) from Table F.

STEP 3   Compute the test value.

STEP 4   Make the decision to reject or not reject the null hypothesis.

STEP 5   Summarize the results.

Remember that the t *test should be used when the population is approximately normally distributed, the population standard deviation is unknown, and the sample size is less than 30.*

The next three examples illustrate the application of the t test.

Example 9–12

A job placement director claims that the average starting salary for nurses is $24,000. A sample of 10 nurses has a mean of $23,450 and a standard deviation of $400. Is there enough evidence to reject the director's claim at $\alpha = 0.05$?

Solution

STEP 1 H_0: $\mu = \$24,000$ (claim) and H_1: $\mu \neq \$24,000$.

STEP 2   The critical values are $+2.262$ and -2.262 for $\alpha = 0.05$ and d.f. = 9.

STEP 3   The test value is

$$t = \frac{\bar{X} - \mu}{s/\sqrt{n}} = \frac{23,450 - 24,000}{400/\sqrt{10}} = -4.35$$

STEP 4   Reject the null hypothesis, since $-4.35 < -2.262$, as shown in Figure 9–18.

Figure 9–18

Summary of the t Test of Example 9–12

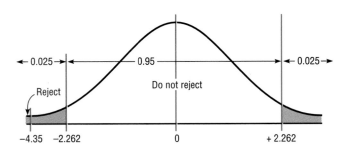

STEP 5   There is enough evidence to reject the claim that the starting salary of nurses is $24,000.

Example 9–13

A machine is designed to fill jars with 16 ounces of coffee. A consumer suspects that the machine is not filling the jars completely. A sample of 8 jars has a mean of 15.6 ounces and a standard deviation of 0.3 ounce. Is there enough evidence to support the consumer's conjecture at $\alpha = 0.10$?

Solution

STEP 1 H_0: $\mu \geq 16$ and H_1: $\mu < 16$ (claim).

STEP 2 At $\alpha = 0.10$ and d.f. = 7, the critical value is -1.415.

STEP 3 The test value is

$$t = \frac{\overline{X} - \mu}{s/\sqrt{n}} = \frac{15.6 - 16}{0.3/\sqrt{8}} = -3.77$$

STEP 4 Reject the null hypothesis, since the test value falls in the critical region, as shown in Figure 9–19.

Figure 9–19

Summary of the t Test of Example 9–13

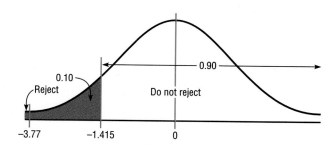

STEP 5 There is enough evidence to support the claim that the machine is filling the jars with less than 16 ounces of coffee.

Example 9–14

A physician claims that joggers' maximal volume oxygen uptake is greater than the average of all adults. A sample of 15 joggers has a mean of 43.6 milliliters per kilogram (ml/kg) and a standard deviation of 6 ml/kg. If the average of all adults is 36.7 ml/kg, is there enough evidence to support the physician's claim at $\alpha = 0.01$?

Solution

STEP 1 H_0: $\mu \leq 36.7$ and H_1: $\mu > 36.7$ (claim).

STEP 2 The critical value is $+2.624$, since $\alpha = 0.01$ and d.f. = 14.

STEP 3 The test value is

$$t = \frac{\overline{X} - \mu}{s/\sqrt{n}} = \frac{43.6 - 36.7}{6/\sqrt{15}} = 4.45$$

STEP 4 Reject the null hypothesis, since 4.45 falls in the critical region, as shown in Figure 9–20.

Figure 9–20

Summary of the *t* Test of
Example 9–14

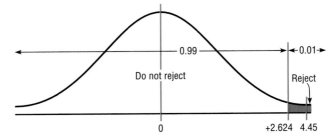

STEP 5   There is enough evidence to support the claim that the joggers' maximal
volume oxygen uptake is greater than 36.7 ml/kg.

P-values for the *t* test are computed the same way as they are for the *z* test.
There is a slight problem, however, when using Table F. Table F gives *t*-values for
only selected values of α (e.g., 0.25, 0.50). To compute exact *P*-values for a *t* test,
one would need a table similar to Table E for each degree of freedom. Since this is
not practical, only a specific interval can be found for a *P*-value.

Suppose the test value is 2.056 for a sample size of 11 and a right-tailed test.
To get the *P*-value, one would look in Table F across the row with 10 degrees of
freedom and find that 2.056 falls between 1.812 and 2.228. Since this is a one-
tailed right test, looking at the row labeled "One Tail, α," one would find that
2.056 falls between 0.05 and 0.025. Hence, the *P*-value would be contained in the
interval $0.025 < P\text{-value} < 0.05$.

Students sometimes have difficulty deciding whether to use the *z* test or *t* test.
The rules are the same as those pertaining to confidence intervals.

1.   If σ is known, use the *z* test.
2.   If σ is unknown but $n \geq 30$, use the *z* test and use *s* in place of σ in the
     formula.
3.   If σ is unknown and $n < 30$, use the *t* test. (The population must be
     approximately normally distributed.)

These rules are summarized in Figure 9–21.

Figure 9–21

Using the *z* or *t* Test

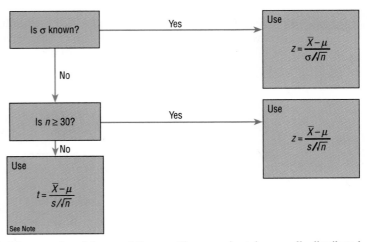

Note: With d.f. $= n - 1$, and the population must be approximately normally distributed.

Computer Application for a *t* Test for a Single Mean

MINITAB *t* test for a single mean:

1. Enter the data into C1.
2. Click Stat > Basic Statistics > 1-Sample t.
3. Highlight C1 in the dialog box and click on Select. Press Tab.
4. Click on Test mean.
5. Type in the value of the hypothesized mean. Press Tab.
6. If the test is two-tailed, press Tab. If the test is one-tailed, click on the arrow in the box labeled Alternative and select appropriate hypothesis.
7. Click on OK.

Example: Test the claim H_0: $\mu = 2$ at $\alpha = 0.05$, using the following data.

Data:

C1	6	8	3	2	0	0	1	5	4	3	3	2

MINITAB printout for this example:

```
T-Test of the Mean

Test of mu = 2.000 vs mu not = 2.000

Variable        N       Mean      StDev     SE Mean          T        P
C1             12      3.083      2.392       0.690       1.57     0.14
```

Summary: In this case, the test value is 1.57 and the *P*-value is 0.14. The decision is not to reject the null hypothesis at $\alpha = 0.05$.

Exercises

9–42. (W) In what ways is the *t* distribution similar to the standard normal distribution?

9–43. (W) In what ways is the *t* distribution different from the standard normal distribution?

9–44. (W) What are the degrees of freedom for the *t* test?

9–45. (W) How does the formula for the *t* test differ from the formula for the *z* test?

9–46. (ans) Find the critical value (or values) for the *t* test for each.

a. $n = 10$, $\alpha = 0.05$, right-tailed
b. $n = 18$, $\alpha = 0.10$, two-tailed
c. $n = 6$, $\alpha = 0.01$, left-tailed
d. $n = 9$, $\alpha = 0.025$, right-tailed
e. $n = 15$, $\alpha = 0.05$, two-tailed
f. $n = 23$, $\alpha = 0.005$, left-tailed
g. $n = 28$, $\alpha = 0.01$, two-tailed
h. $n = 17$, $\alpha = 0.02$, two-tailed

For Exercises 9–47 through 9–60, perform each of the following steps.

a. State the hypotheses and identify the claim.
b. Find the critical value(s).
c. Find the test value.
d. Make the decision.
e. Summarize the results.

Assume that the population is approximately normally distributed.

9–47. The average amount of rainfall during the summer months for the northeast part of the United States is 11.52 inches. A researcher selects a random sample of 10 cities in the northeast and finds that the average amount of rainfall for 1995 was 7.42 inches. The standard deviation of the sample is 1.3 inches. At $\alpha = 0.05$, can it be concluded that for 1995 the mean rainfall was below 11.52 inches?

Source: Based on information in *USA Today*, September 14, 1995.

9–48.   The manager of a car rental agency claims that the average mileage of cars rented is less than 8000. A sample of five automobiles has an average mileage of 7723, with a standard deviation of 500 miles. At $\alpha = 0.01$, is there enough evidence to reject the manager's claim?

9–49.   A rental agency claims that the average rent that small-business establishments pay in Eagle City is $800. A sample of 10 establishments shows an average rental rate of $863. The standard deviation of the sample is $20. At $\alpha = 0.05$, is there evidence to reject the agency's claim?

9–50.   A large hospital instituted a fitness program to reduce absenteeism. The director reported that the average number of working hours lost due to illness per employee is 48 hours per year. After one year, a sample of 18 employees showed an average of 41 hours of work lost, with a standard deviation of 5. Did the program reduce absenteeism? Use $\alpha = 0.10$.

9–51.   In order to increase customer service, a muffler repair shop claims its mechanics can replace a muffler in less than 12 minutes. A time management specialist selected six repair jobs and found their mean time to be 11.6 minutes. The standard deviation of the sample was 2.1 minutes. At $\alpha = 0.025$, is there enough evidence to support the shop's claim?

9–52.   Cushman and Wakefield reported that the average annual rent for office space in Tampa was $17.63 per square foot. A real estate agent selected a random sample of 15 rental properties (offices) and found the mean rent was $18.72 per square foot, and the standard deviation was $3.64. At $\alpha = 0.05$, can it be concluded that there is no difference in the rents?
Source: Snapshot, *USA TODAY*, September 13, 1995.

9–53.   A rental agent states that the average rent for a studio apartment in Shadyside is $750. A sample of 12 renters shows that the mean is $732 and the standard deviation is $17. Is there enough evidence to reject the agent's claim? Use $\alpha = 0.01$.

9–54.   A manufacturer claims that a new snowmaking machine can save money for ski resort owners. He states that in a sample of 10 tons, the cost of making a ton of snow was $5.75. The average of the old machine was $6.62. The standard deviation of the sample was $1.05. At $\alpha = 0.10$, does the new machine save money?

9–55.   An officer states that the average fine levied by the safety office against companies is at most $350. A company owner suspects it is higher. She samples 12 companies and finds that the average is $358. The standard deviation of the sample is $16. Test the claim that the average is higher than $350, at $\alpha = 0.05$.

9–56.   West Newton advertises that it is the cleanest town in New Jersey by stating that the average price of a car wash is $3.00. A sample of five car washes had an average price of $3.70 and a standard deviation of $0.30. At $\alpha = 0.05$, is the average price of a car wash really $3.00?

9–57.   A recent survey stated that the average single-person household received at least 37 telephone calls per month. To test the claim, a researcher surveyed 29 single-person households and found that the average number of calls was 34.9. The standard deviation of the sample was 6. At $\alpha = 0.05$, can the claim be substantiated?

9–58.   A student suspected the average cost of a Saturday night date was no longer $30.00. To test her hypothesis, she randomly selected 16 men from the dormitory and asked them how much they spent on a date last Saturday. She found that the average cost was $31.17. The standard deviation of the sample was $5.51. At $\alpha = 0.05$, is there enough evidence to support her claim?

*9–59.   From past experience, a teacher believes that the average score on a real estate exam is 75. A sample of 20 students' exam scores is as follows:

80, 68, 72, 73, 76, 81, 71, 71, 65, 50,
63, 71, 70, 70, 76, 75, 69, 70, 72, 74

Can the teacher conclude that the students' average is still 75? Use $\alpha = 0.01$.

*9–60.   A new laboratory technician read a report that the average number of students using the computer laboratory per hour was 16. To test this hypothesis, he selected a day at random and kept track of the number of students who used the lab over an eight-hour period. The results were as follows:

20, 24, 18, 16, 16, 19, 21, 23

At $\alpha = 0.05$, can the technician conclude that the average is actually 16?

9–5

Proportion Test

Objective 7. Test proportions using the z test.

Many hypothesis-testing situations involve proportions. Recall from Chapter 8 that a *proportion* is the same as a percentage of the population.

The following data were obtained from *The Book of Odds:*

- 59% of consumers purchase gifts for their father
- 85% of people over 21 said they have entered a sweepstakes
- 51% of Americans buy generic products
- 35% of Americans go out for dinner once a week

A hypothesis test involving a population proportion can be considered as a binomial experiment when there are only two outcomes and the probability of a success does not change from trial to trial. Recall from Section 6–4 in Chapter 6 that the mean is $\mu = np$ and the standard deviation is $\sigma = \sqrt{npq}$ for the binomial distribution.

Since the normal distribution can be used to approximate the binomial distribution when $np \geq 5$ and $nq \geq 5$, the standard normal distribution can be used to test hypotheses for proportions.

Formula for the z Test for Proportions

$$z = \frac{X - \mu}{\sigma} \quad \text{or} \quad z = \frac{X - np}{\sqrt{npq}}$$

where $\mu = np$ and $\sigma = \sqrt{npq}$.

The following examples illustrate the hypothesis-testing procedure for proportions. *Note that there is an additional step (Step 2) in this procedure for finding the mean and standard deviation.* (A detailed explanation of proportions can be found at the beginning of Section 8–4.)

Example 9–15

An educator estimates that the dropout rate for seniors at high schools in Ohio is 15%. Last year, 38 seniors from a random sample of 200 Ohio seniors withdrew. At $\alpha = 0.05$, is there enough evidence to reject the educator's claim?

Solution

STEP 1   State the hypotheses and identify the claim.

$$H_0: p = 0.15 \text{ (claim)} \quad \text{and} \quad H_1: p \neq 0.15$$

STEP 2   Find the mean and the standard deviation.

$$\mu = np = (200)(0.15) = 30$$
$$\sigma = \sqrt{npq} = \sqrt{(200)(0.15)(0.85)} = 5.05$$

STEP 3   Find the critical value(s). Since $\alpha = 0.05$ and the test is two-tailed, the critical values are ± 1.96.

STEP 4   Compute the test value.

$$z = \frac{X - \mu}{\sigma} = \frac{38 - 30}{5.05} = 1.58$$

STEP 5   Make the decision. Do not reject the null hypothesis, since the test value falls outside the critical region, as shown in Figure 9–22.

Figure 9–22

Critical and Test Values for Example 9–15

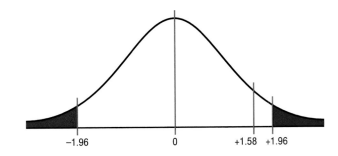

−1.96   0   +1.58 +1.96

STEP 6   Summarize the results. There is not enough evidence to reject the claim that the dropout rate for seniors in high schools in Ohio is 15%.

Example 9–16

A telephone company representative estimates that 40% of its customers have call-waiting service. To test this hypothesis, she selected a sample of 100 customers and found that 37% had call waiting. At $\alpha = 0.01$, is her estimate appropriate?

Solution

STEP 1   State the hypotheses and identify the claim.

$$H_0: p = 0.40 \text{ (claim)} \qquad \text{and} \qquad H_1: p \neq 0.40$$

STEP 2   Find the mean and the standard deviation.

$$\mu = np = (100)(0.40) = 40$$
$$\sigma = \sqrt{npq} = \sqrt{(100)(0.40)(0.60)} = 4.90$$

Note: $X = (100)(0.37) = 37$.

STEP 3   Find the critical value(s). Since $\alpha = 0.01$ and this test is two-tailed, the critical values are ± 2.58.

STEP 4   Compute the test value.

$$z = \frac{X - \mu}{\sigma} = \frac{37 - 40}{4.90} = -0.612$$

STEP 5   Make the decision. Do not reject the null hypothesis, since the test value falls in the noncritical region, as shown in Figure 9–23.

Figure 9–23

Critical and Test Values for Example 9–16

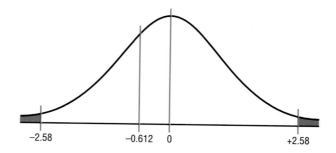

STEP 6   Summarize the results. There is not enough evidence to reject the claim that 40% of the telephone company's customers have call waiting.

Example 9–17

A statistician read that at least 77% of the population oppose replacing \$1 bills with \$1 coins. To see if this claim is valid, the statistician selected a sample of 80 people and found that 55 were opposed to replacing the \$1 bills. At $\alpha = 0.01$, can it be concluded that at least 77% of the population are opposed to the change?

Source: *USA Today*, June 14, 1995.

Interesting Fact

Many people have what is called "white-coat hyper-tension." This condition causes the blood pressure of people who have normal pressure to rise when they visit their physician. However, some people have the reverse effect. Their pressure is above normal in everyday life but actually drops to normal level at the doctor's office. Researchers suspect that for these people, a visit to a physician is a vacation from their fast-paced, high-pressure lives. Reprinted with permission from *Psychology Today* magazine Copyright © 1996 (Sussex Publishers, Inc.).

Solution

STEP 1   State the hypotheses and identify the claim.

$$H_0\colon p \geq 0.77 \text{ (claim)} \qquad \text{and} \qquad H_1\colon p < 0.77$$

STEP 2   Find the mean and the standard deviation.

$$\mu = np = (80)(0.77) = 61.6$$
$$\sigma = \sqrt{npq} = \sqrt{(80)(0.77)(0.23)} = 3.76$$

STEP 3   Find the critical value(s). Since $\alpha = 0.01$ and the test is left-tailed, the critical value is -2.33.

STEP 4   Compute the test value.

$$z = \frac{X - \mu}{\sigma} = \frac{55 - 61.6}{3.76} = -1.76$$

STEP 5   Do not reject the null hypothesis, since the test value does not fall in the critical region, as shown in Figure 9–24.

Figure 9–24

Critical and Test Values for Example 9–17

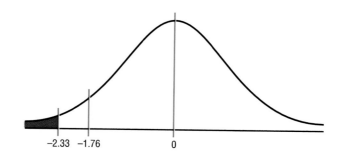

STEP 6   There is not enough evidence to reject the claim that at least 77% of the population oppose replacing \$1 bills with \$1 coins.

....

Example 9–18   An attorney claims that more than 25% of all lawyers advertise. A sample of 200 lawyers in a certain city showed that 63 had used some form of advertising. At $\alpha = 0.05$, is there enough evidence to support the attorney's claim?

Solution

STEP 1 $H_0: p \leq 0.25$ and $H_1: p > 0.25$ (claim).

STEP 2 $\mu = np = (200)(0.25) = 50$

$\sigma = \sqrt{npq} = \sqrt{(200)(025)(0.75)} = 6.12$

STEP 3   The critical value is $+1.65$.

STEP 4 $z = \dfrac{X - \mu}{\sigma} = \dfrac{63 - 50}{6.12} = 2.12$

STEP 5   Reject the null hypothesis, since the test value falls in the critical region, as shown in Figure 9–25.

Figure 9–25
....
Critical and Test Values for Example 9–18

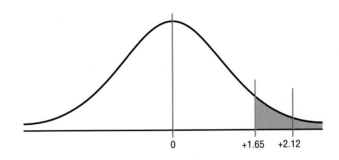

STEP 6   There is enough evidence to support the attorney's claim that more than 25% of the lawyers use some form of advertising.

....

Tests for proportions can also be conducted by using a formula equivalent to the one shown in the preceding examples. This formula is

$$z = \frac{\hat{p} - p}{\sqrt{pq/n}}$$

where

$\hat{p} = X/n$, the sample proportion
p = hypothesized population proportion
n = sample size

The denominator $\sqrt{pq/n}$ is the standard error of the proportions. *P*-values for proportions are found the same way as shown in Section 9–3.

Exercises

9–61. (W) Give three examples of proportions.

9–62. (W) Why is a proportion considered a binomial variable?

9–63. (W) When one is testing hypotheses using proportions, what are the necessary requirements?

9–64. (W) What are the mean and the standard deviation of a proportion?

For Exercises 9–65 through 9–75, perform each of the following steps.
a. State the hypotheses and identify the claim.
b. Find the mean and the standard deviation.
c. Find the critical value(s)
d. Compute the test value.
e. Make the decision.
f. Summarize the results.

9–65. A toy manufacturer claims that at least 23% of the 14-year-old residents of a certain city own a skateboard. A sample of forty 14-year-olds shows that seven own a skateboard. Is there enough evidence to support the manufacturer's claim at $\alpha = 0.05$?

9–66. From past records, a hospital found that 37% of all full-term babies born in the hospital weighed more than 7 pounds 2 ounces. This year a sample of 100 babies showed that 23 weighted over 7 pounds 2 ounces. At $\alpha = 0.01$, is there enough evidence to say the percentage has changed?

9–67. A study on crime suggested that at least 40% of all arsonists were under 21 years of age. Checking local crime statistics, a researcher found that 30 out of 80 arson suspects were under 21. At $\alpha = 0.10$, should the crime study's statement be believed?

9–68. A recent study found that, at most, 32% of people who have been in a plane crash have died. In a sample of 100 people who were in a plane crash, 38 died. Is the study's statement believable? Use $\alpha = 0.05$.

9–69. An item in *USA Today* reported that 63% of Americans owned an answering machine. A survey of 143 employees at a large school showed that 85 owned an answering machine. At $\alpha = 0.05$, can it be concluded that the percentage is the same as stated in *USA Today?*
Source: *USA Today,* May 22, 1995.

9–70. The *Statistical Abstract* reported that 17% of adults attended a musical play in the past year. To test this claim, a researcher surveyed 90 people and found that 22 had attended a musical play in the past year. At $\alpha = 0.05$, can it be concluded that this figure is correct?
Source: *Statistical Abstract of the United States 1994.*

9–71. A recent study claimed that at least 15% of all eighth-grade students are overweight. In a sample of 80 students, 9 were found to be overweight. At $\alpha = 0.05$, is there enough evidence to support the claim?

9–72. At a large university, a study found that no more than 25% of the students who commute travel more than 14 miles to campus. At $\alpha = 0.10$, are the findings supported if in a sample of 100 students, 30 drove more than 14 miles?

9–73. A telephone company wants to advertise that more than 30% of all its customers have at least two telephones. To support this ad, the company selects a sample of 200 customers and finds that 72 have more than two telephones. Does the evidence support the ad? Use $\alpha = 0.05$.

9–74. Experts claim that 10% of murders are committed by women. Is there enough evidence to reject the claim if in a sample of 67 murders, 10 were committed by women? Use $\alpha = 0.01$.

9–75. Researchers suspect that 18% of all high school students smoke at least one pack of cigarettes a day. At Wilson High School, with an enrollment of 300 students, a study found that 50 students smoked at least one pack of cigarettes a day. At $\alpha = 0.05$, can the study conclude that 18% of all high school students smoke at least one pack of cigarettes a day?

When *np* or *nq* is not 5 or more, the binomial table (Table B in Appendix C) must be used to find critical values in hypothesis tests involving proportions.

***9–76.** A coin is tossed nine times and three heads appear. Can one conclude that the coin is not balanced? Use $\alpha = 0.10$. (*Hint:* Use the binomial table and find $P(X = 3)$ with $p = 0.5$ and $n = 9$.)

***9–77.** Fashion buyers have suggested that 30% of all men purchase blue suits. If in a sample of 10 men who purchase suits, 6 select blue suits, should the null hypothesis be rejected at $\alpha = 0.05$?

***9–78.** In the past, 20% of all airline passengers flew first class. In a sample of 15 passengers, 5 flew first class. At $\alpha = 0.10$, can one conclude that the proportions have changed?

9–6

Variance or Standard Deviation Test

Objective 8. Test variances or standard deviation using the chi-square test.

In Chapter 8, the chi-square distribution was used to construct a confidence interval for a single variance or standard deviation. This distribution is also used to test a claim about a single variance or standard deviation.

In order to find the area under the chi-square distribution, use Table G in Appendix C. There are three cases to consider:

1. Finding the chi-square critical value for a specific α when the hypothesis test is right-tailed.
2. Finding the chi-square critical value for a specific α when the hypothesis test is left-tailed.
3. Finding the chi-square critical values for a specific α when the hypothesis test is two-tailed.

Example 9–19

Find the critical chi-square value for 15 degrees of freedom when $\alpha = 0.05$ and the test is right-tailed.

Solution

The distribution is shown in Figure 9–26.

Figure 9–26

Chi-Square Distribution for Example 9–19

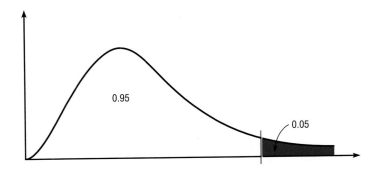

Find the α value at the top of Table G and find the corresponding degrees of freedom in the left column. The critical value is located where the two columns meet—in this case, 24.996. See Figure 9–27.

Figure 9–27

Locating the Critical Value in Table G for Example 9–19

Degrees of freedom	α									
	0.995	0.99	0.975	0.95	0.90	0.10	0.05	0.025	0.01	0.005
1										
2										
⋮										
15								24.996			
16										
⋮										

Example 9–20

Find the critical chi-square value for 10 degrees of freedom when $\alpha = 0.05$ and the test is left-tailed.

Solution

This distribution is shown in Figure 9–28.

Figure 9–28

Chi-Square Distribution for Example 9–20

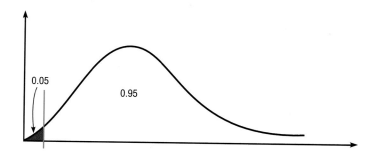

When the test is one-tailed left, the α value must be subtracted from 1: $1 - 0.05 = 0.95$. The left side of the table is used, because the chi-square table gives the area to the right of the critical value, and the chi-square statistic cannot be negative. The table is set up so that it gives the values for the area to the right of the critical value. In this case, 95% of the area will be to the right of the value.

For 0.95 and 10 degrees of freedom, the critical value is 3.940. See Figure 9–29.

Figure 9–29

Locating the Critical Value in Table G for Example 9–20

Degrees of freedom	α									
	0.995	0.99	0.975	0.95	0.90	0.10	0.05	0.025	0.01	0.005
1										
2										
⋮										
10				3.940						
⋮										

Example 9–21

Find the critical chi-square values for 22 degrees of freedom when $\alpha = 0.05$ and a two-tailed test is conducted.

Solution

When a two-tailed test is conducted, the area must be split, as shown in Figure 9–30. Note that the area to the right of the larger value is 0.025 (0.05/2), and the area to the right of the smaller value is 0.975 (1.00 − 0.05/2).

Remember that chi-square values cannot be negative. Hence, one must use α values in the table of 0.025 and 0.975. With 22 degrees of freedom, the critical values are 36.781 and 10.982, respectively.

Figure 9–30
Chi-Square Distribution
for Example 9–21

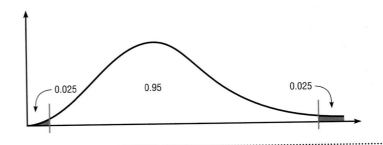

After the degrees of freedom reach 30, Table G gives values only for multiples of 10 (40, 50, 60, etc.). When the exact degrees of freedom one is seeking are not specified in the table, the closest smaller value should be used. For example, if the given degrees of freedom are 36, use the table value for 30 degrees of freedom. This guideline keeps the type I error equal to or below the α value.

When one is testing a claim about a single variance, there are three possible test situations: right-tailed test, left-tailed test, and two-tailed test.

If a researcher believes the variance of a population to be greater than some specific value, say 225, then the researcher states the hypotheses as

$$H_0: \sigma^2 \le 225$$
$$H_1: \sigma^2 > 225$$

and conducts a right-tailed test.

If the researcher believes the variance of a population to be less than 225, then the researcher states the hypotheses as

$$H_0: \sigma^2 \ge 225$$
$$H_1: \sigma^2 < 225$$

and conducts a left-tailed test.

Finally, if a researcher does not wish to specify a direction, he or she states the hypotheses as

$$H_0: \sigma^2 = 225$$
$$H_1: \sigma^2 \ne 225$$

and conducts a two-tailed test.

Formula for the Chi-Square Test for a Single Variance

$$\chi^2 = \frac{(n-1)s^2}{\sigma^2}$$

with degrees of freedom equal to $n - 1$ and where

n = sample size
s^2 = sample variance
σ^2 = population variance

One might ask, "Why is it important to test variances?" There are several reasons. First, in any situation where consistency is required, such as in manufacturing, one would like to have the smallest variation possible in the products. For example, when bolts are manufactured, the variation in diameters due to the process must be kept to a minimum, or the nuts will not fit them properly. In education, consistency is required on a test. That is, if the same students take the same test several times, they should get approximately the same grades, and the variance of each of the students' grades should be small. On the other hand, if the test is to be used to judge learning, the overall standard deviation of all of the grades should be large so that one can differentiate those who have learned the subject from those who have not learned it.

Three assumptions are made for the chi-square test, as outlined next.

Assumptions for the Chi-Square Test for a Single Variance

1. The sample must be randomly selected from the population.
2. The population must be normally distributed for the variable under study.
3. The observations must be independent of each other.

The hypothesis-testing procedure follows the same five steps listed for hypothesis tests in the preceding sections.

STEP 1   State the hypotheses and identify the claim.

STEP 2   Find the critical value(s).

STEP 3   Compute the test value.

STEP 4   Make the decision.

STEP 5   Summarize the results.

The next three examples illustrate the hypothesis-testing procedure for variances.

Example 9–22

An instructor wishes to see whether the variation in scores of the 23 students in her class is less than the variance of the population. The variance of the class is 198. Is there enough evidence to support the claim that the variation of the students is less than the population variance ($\sigma^2 = 225$) at $\alpha = 0.05$? Assume that the scores are normally distributed.

Solution

STEP 1   State the hypotheses and identify the claim.

$$H_0\colon \sigma^2 \geq 225 \quad \text{and} \quad H_1\colon \sigma^2 < 225 \text{ (claim)}$$

STEP 2   Find the critical value. Since this test is left-tailed and $\alpha = 0.05$, use the value $1 - 0.05 = 0.95$. The degrees of freedom are $n - 1 = 23 - 1 = 22$. Hence, the critical value is 12.338. Note that the critical region is on the left, as shown in Figure 9–31.

Figure 9–31

Critical Value for
Example 9–22

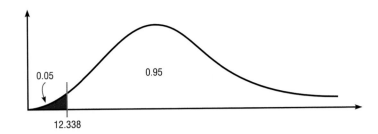

STEP 3   Compute the test value.

$$\chi^2 = \frac{(n-1)s^2}{\sigma^2} = \frac{(23-1)(198)}{225} = 19.36$$

STEP 4   Make the decision. Since the test value 19.36 falls in the noncritical
region, as shown in Figure 9–32, the decision is to not reject the null
hypothesis.

Figure 9–32

Critical and Test Values
for Example 9–22

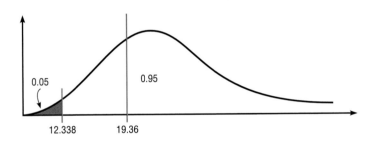

STEP 5   Summarize the results. There is not enough evidence to support the
claim that the variation in test scores of the instructor's students is less
than the variation in scores of the population.

Example 9–23

A medical researcher believes that the standard deviation of the temperatures of
newborn infants is greater than 0.6°. A sample of 15 infants was found to have a
standard deviation of 0.8°. At $\alpha = 0.10$, does the evidence support the researcher's
belief? Assume that the variable is normally distributed.

Solution

STEP 1   State the hypotheses and identify the claim.

$$H_0: \sigma^2 \leq 0.36 \qquad \text{and} \qquad H_1: \sigma^2 > 0.36 \text{ (claim)}$$

Note that since standard deviations are given in the problem, they must
be squared to get variances.

STEP 2   Find the critical value. Since this test is right-tailed with d.f. of 15 − 1
= 14 and $\alpha = 0.10$, the critical value is 21.064.

STEP 3   Compute the test value.

$$\chi^2 = \frac{(n-1)s^2}{\sigma^2} = \frac{(15-1)(0.64)}{0.36} = 24.89$$

STEP 4   Make the decision. The decision is to reject the null hypothesis, since the test value 24.89 is greater than the critical value 21.064 and falls in the critical region, as shown in Figure 9–33.

Figure 9–33

Critical and Test Values for Example 9–23

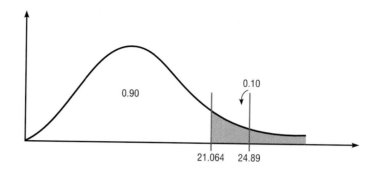

STEP 5   Summarize the results. There is enough evidence to support the claim that the standard deviation of the temperatures is greater than 0.6°.

Approximate *P*-values for the chi-square test can be found using Table G much the same way as *P*-values for the *t* test. In Example 9–23, one would look across the row with 14 degrees of freedom and find the two values that 24.89 falls between. They are 23.685 and 26.119. These values correspond to 0.05 and 0.025; hence, $0.025 < p < 0.05$. Thus, the null hypothesis can be rejected at $\alpha = 0.05$. Since $\alpha = 0.05$ is less than $\alpha = 0.10$ stated in Example 9–23, the decision is to reject the null hypothesis.

Example 9–24

A cigarette manufacturer wishes to test the claim that the variance of the nicotine content of its cigarettes is 0.644. Nicotine content is measured in milligrams, and assume that it is normally distributed. A sample of 20 cigarettes has a standard deviation of 1.00 milligram. At $\alpha = 0.05$, is there enough evidence to support the manufacturer's claim?

Solution

STEP 1   State the hypotheses and identify the claim.

$$H_0\!: \sigma^2 = 0.644 \text{ (claim)} \qquad \text{and} \qquad H_1\!: \sigma^2 \neq 0.644$$

STEP 2   Find the critical values. Since this test is a two-tailed test at $\alpha = 0.05$, the critical values for 0.025 and 0.975 must be found. The degrees of freedom are 19; hence, the critical values are 32.852 and 8.907. The critical or rejection regions are shown in Figure 9–34.

Figure 9–34

Critical Values for Example 9–24

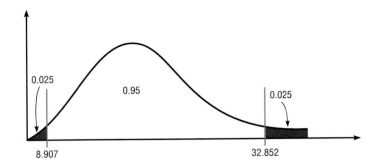

STEP 3 Compute the test value.

$$\chi^2 = \frac{(n-1)s^2}{\sigma^2} = \frac{(20-1)(1.0)^2}{0.644} = 29.5$$

Since the standard deviation s is given in the problem, it must be squared for the formula.

STEP 4 Make the decision. Do not reject the null hypothesis, since the test value falls between the critical values ($8.907 < 29.5 < 32.852$) and in the noncritical region, as shown in Figure 9–35.

Figure 9–35

Critical and Test Values for Example 9–24

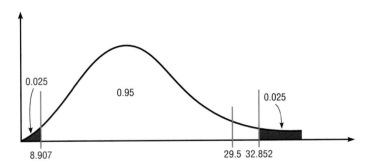

STEP 5 Summarize the results. There is not enough evidence to reject the manufacturer's claim that the variance of the nicotine content of the cigarettes is equal to 0.644.

Exercises

9–79. Using Table G, find the critical value(s) for each, show the critical and noncritical regions, and state the appropriate null and alternative hypotheses for $\sigma^2 = 225$.
a. $\alpha = 0.05, n = 18$, right-tailed
b. $\alpha = 0.10, n = 23$, left-tailed
c. $\alpha = 0.05, n = 15$, two-tailed
d. $\alpha = 0.10, n = 8$, two-tailed
e. $\alpha = 0.01, n = 17$, right-tailed

f. $\alpha = 0.025, n = 20$, left-tailed
g. $\alpha = 0.01, n = 13$, two-tailed
h. $\alpha = 0.025, n = 29$, left-tailed

For Exercises 9–80 through 9–87, assume that the variables are normally or approximately normally distributed.

9–80. A manufacturer of pacemakers wants the batteries used in the devices to have a standard deviation less than or equal to 1.5 months. A sample of 25 batteries has a standard deviation of 1.8 months. Should the manufacturer use the batteries? Select $\alpha = 0.05$.

9–81. An automobile mechanic knows from past studies that the standard deviation of the time it takes to inspect a car is 16.8 minutes. This year, the standard deviation of 24 cars was 12.5 minutes. At $\alpha = 0.10$, can the mechanic conclude that the standard deviation has changed?

9–82. Eighteen 5-gallon containers of fruit punch are selected, and each is tested to determine its weight in ounces. The variance of the sample is 6.5. Test the claim that the population variance is greater than 6.2, at $\alpha = 0.01$.

9–83. A company claims that the variance of the sugar content of its yogurt is less than or equal to 25. (The sugar content is measured in milligrams per ounce.) A sample of 20 servings is selected, and the sugar content is measured. The variance of the sample is found to be 36. At $\alpha = 0.10$, is there enough evidence to reject the claim?

9–84. A researcher suggests that the variances of the ages of freshmen at Oak Park College is greater than 1.6, as past studies have shown. A sample of 50 freshmen is selected, and the variance is found to be 2.3 years. At $\alpha = 0.05$, is the variation of the freshmen's ages greater than 1.6?

9–85. The manager of a large company claims that the standard deviation of the time (in minutes) that it takes a telephone call to be transferred to the correct office in her company is 1.2 minutes or less. A sample of 15 calls is selected and the calls timed. The standard deviation of the sample is 1.8 minutes. At $\alpha = 0.01$, is the standard deviation less than or equal to 1.2 minutes?

9–86. A machine fills 12-ounce bottles with soda. In order for the machine to function properly, the standard deviation of the sample must be less than or equal to 0.03 ounce. A sample of 8 bottles is selected, and the number of ounces of soda in each bottle is as given below. At $\alpha = 0.05$, is the machine functioning properly?

12.03, 12.1, 12.02, 11.98,
12.00, 12.05, 11.97, 11.99

9–87. A construction company purchases steel cables. A sample of 12 cables is selected, and the breaking strength (in pounds) of each is found. The data are shown below. The lot is rejected if the standard deviation of the sample is greater than 2 pounds. At $\alpha = 0.01$, should the lot be rejected?

2001, 1998, 2002, 2000, 1998, 1999,
1997, 2005, 2003, 2001, 1999, 2006

9–7

Confidence Intervals and Hypothesis Testing

Objective 9. Test hypotheses using confidence intervals.

In hypothesis testing, another concept that might be of interest to students in elementary statistics is the relationship between hypothesis testing and confidence intervals.

There is a relationship between confidence intervals and hypothesis testing. When the null hypothesis is rejected in a hypothesis-testing situation, the confidence interval for the mean using the same level of significance *will not* contain the hypothesized mean. Likewise, when the null hypothesis is not rejected, the confidence interval computed using the same level of significance *will* contain the hypothesized mean. The next two examples show this concept for two-tailed tests.

Example 9–25

Sugar is packed in 5-pound bags. An inspector suspects the bags may not contain 5 pounds. A sample of 50 bags produces a mean of 4.6 pounds and a standard deviation of 0.7 pound. Is there enough evidence to conclude that the bags do not contain 5 pounds as stated, at $\alpha = 0.05$? Also, find the 95% confidence interval of the true mean.

Solution

H_0: $\mu = 5$, and H_1: $\mu \neq 5$ (claim). The critical values are $+1.96$ and -1.96. The test value is

$$z = \frac{\bar{X} - \mu}{s/\sqrt{n}} = \frac{4.6 - 5.00}{0.7/\sqrt{50}} = \frac{-0.4}{0.099} = -4.04$$

Since $-4.04 < -1.96$, the null hypothesis is rejected. There is enough evidence to support the claim that the bags do not weigh 5 pounds.

The 95% confidence for the mean is given by

$$\bar{X} - z_{\alpha/2} \cdot \frac{s}{\sqrt{n}} < \mu < \bar{X} + z_{\alpha/2} \cdot \frac{s}{\sqrt{n}}$$

$$4.6 - (1.96)\left(\frac{0.7}{\sqrt{50}}\right) < \mu < 4.6 + (1.96)\left(\frac{0.7}{\sqrt{50}}\right)$$

$$4.406 < \mu < 4.794$$

Notice that the 95% confidence interval of μ does *not* contain the hypothesized value $\mu = 5$. Hence, there is agreement between the hypothesis test and the confidence interval.

Example 9–26

A researcher claims that adult hogs fed a special diet will have an average weight of 200 pounds. A sample of 10 hogs has an average weight of 198.2 pounds and a standard deviation of 3.3 pounds. At $\alpha = 0.05$, is the claim justified? Also, find the 95% confidence interval of the true mean.

Solution

H_0: $\mu = 200$ pounds (claim) and H_1: $\mu \neq 200$ pounds. The t test must be used since σ is unknown and $n < 30$. The critical values at $\alpha = 0.05$ with 9 degrees of freedom are $+2.262$ and -2.262. The test value is

$$t = \frac{\bar{X} - \mu}{s/\sqrt{n}} = \frac{198.2 - 200}{3.3/\sqrt{10}} = \frac{-1.8}{1.0436} = -1.72$$

Thus, the null hypothesis is not rejected. There is not enough evidence to reject the claim that the weight of the adult hogs is 200 pounds.

The 95% confidence interval of the mean is

$$\bar{X} - t_{\alpha/2} \cdot \frac{s}{\sqrt{n}} < \mu < \bar{X} + t_{\alpha/2} \cdot \frac{s}{\sqrt{n}}$$

$$198.2 - (2.262)\left(\frac{3.3}{\sqrt{10}}\right) < \mu < 198.2 + (2.262)\left(\frac{3.3}{\sqrt{10}}\right)$$

$$198.2 - 2.361 < \mu < 198.2 + 2.361$$

$$195.84 < \mu < 200.56$$

The 95% confidence interval does contain the hypothesized mean $\mu = 200$.

In summary, then, when the null hypothesis is rejected, the confidence interval computed at the same significance level will not contain the value of the mean that is stated in the null hypothesis. On the other hand, when the null hypothesis is not rejected, the confidence interval computed at the same significance level will contain the value of the mean stated in the null hypothesis. These results are true for other hypothesis-testing situations and are not limited to means tests.

The relationship between confidence intervals and hypothesis testing presented here is valid for two-tailed tests. The relationship between one-tailed hypothesis tests and one-sided or one-tailed confidence intervals is also valid; however, this technique is beyond the scope of this textbook.

Exercises

9–88. Explain how confidence intervals are related to hypothesis testing.

9–89. A ski-shop manager claims that the average of the sales for her shop is $1800 a day during the winter months. Ten winter days are selected at random, and the mean of the sales is $1830. The standard deviation of the population is $200. Can one reject the claim at $\alpha = 0.05$? Find the 95% confidence interval of the mean. Does the confidence interval interpretation agree with the hypothesis test results? Explain. Assume that the variable is normally distributed.

9–90. Charter bus records show that in past years, the buses carried an average of 42 people per trip to Niagara Falls. The standard deviation of the population in the past was found to be 8. This year, the average of 10 trips showed a mean of 48 people booked. Can one reject the claim, at $\alpha = 0.10$, that the average is still the same? Find the 90% confidence interval of the mean. Does the confidence interval interpretation agree with the hypothesis-testing results? Explain. Assume that the variable is normally distributed.

9–91. The sales manager of a rental agency claims that the monthly maintenance fee for a condominium in the Lakewood region is $86. Past surveys showed that the standard deviation of the population is $6. A sample of 15 owners shows that they pay an average of $84. Test the manager's claim at $\alpha = 0.01$. Find the 99% confidence interval of the mean. Does the confidence interval interpretation agree with the results of the hypothesis test? Explain. Assume that the variable is normally distributed.

9–92. The average time it takes a person in a one-person canoe to complete a certain river course is 47 minutes. Because of rapid currents in the spring, a group of 10 people traverse the course in 42 minutes. The standard deviation, known from previous trips, is 7 minutes. Test the claim that this group's time was different because of the strong currents. Use $\alpha = 0.10$. Find the 90% confidence level of the true mean. Does the confidence interval interpretation agree with the results of the hypothesis test? Explain. Assume that the variable is normally distributed.

9–93. From past studies the average time college freshmen spend studying is 22 hours per week. The standard deviation is 4 hours. This year, 60 students were surveyed, and the average time that they spent studying was 20.8 hours. Test the claim that the time students spend studying has changed. Use $\alpha = 0.01$. It is believed that the standard deviation is unchanged. Find the 99% confidence interval of the mean. Do the results agree? Explain.

9–94. A survey taken several years ago found that the average time a person spent reading the local daily newspaper was 10.8 minutes. The standard deviation of the population was 3 minutes. In order to see whether the average time had changed since the newspaper's format was revised, the newspaper editor surveyed 36 individuals. The average time that the 36 people spent reading the paper was 12.2 minutes. At $\alpha = 0.02$, is there a change in the average time an individual spends reading the newspaper? Find the 98% confidence interval of the mean. Do the results agree? Explain.

9–8

Summary

This chapter introduces the basic concepts of hypothesis testing. A statistical hypothesis is a conjecture about a population. There are two types of statistical hypotheses: the null and the alternative hypotheses. The null hypothesis states that there is no difference, and the alternative hypothesis specifies the difference. To test the null hypothesis, researchers use a statistical test. Many test values are computed by using

$$\text{test value} = \frac{(\text{observed value}) - (\text{expected value})}{\text{standard error}}$$

Two common statistical tests are the z test and the t test. The z test is used when the population standard deviation is known and $n \geq 30$. When σ is not known but the sample size is greater than or equal to 30, s can be used. The z test is also used to test proportions when $np \geq 5$ and $nq \geq 5$.

Researchers compute a test value from the sample data in order to decide whether the null hypothesis should or should not be rejected. Statistical tests can be one-tailed or two-tailed, depending on the hypotheses.

The null hypothesis is rejected when the difference between the population parameter and the sample statistic is said to be significant. The difference is significant when the test value falls in the critical region of the distribution. The critical region is determined by α, the level of significance of the test. The level is the probability of committing a type I error. This error occurs when the null hypothesis is rejected when it is true. Three generally agreed-upon significance levels are 0.10, 0.05, and 0.01.

A second kind of error, the type II error, can occur when the null hypothesis is not rejected when it is false. When the population standard deviation is not known, the sample standard deviation can be used, but a t test should be conducted in this situation when the sample size is less than 30.

Finally, one can test a single variance by using a chi-square test.

All hypothesis-testing situations should include the following steps:

1. State the null and alternative hypotheses and identify the claim.
2. State an alpha level and find the critical value(s).
3. Compute the test value.
4. Make the decision to reject or not reject the null hypothesis.
5. Summarize the results.

Important Terms

α (alpha) 328

Alternative hypothesis 324

β (beta) 328

Critical or rejection region 328

Critical value 328

Degrees of freedom 346

Hypothesis testing 324

Left-tailed test 329

Level of significance 328

Noncritical or nonrejection region 328

Null hypothesis 324

One-tailed test 329

P-value 340

Right-tailed test 329

Statistical hypothesis 324

Statistical test 327

Test value 327

t test 346

Two-tailed test 330

type I error 328

type II error 328

z test 335

Important Formulas

Formula for the z test for means:

$$z = \frac{\overline{X} - \mu}{\sigma/\sqrt{n}} \quad \text{for any value } n$$

$$z = \frac{\overline{X} - \mu}{s/\sqrt{n}} \quad \text{for} \quad n \geq 30$$

Formula for the t test for means:

$$t = \frac{\overline{X} - \mu}{s/\sqrt{n}} \quad \text{for} \quad n < 30$$

Formula for the z test for proportions:

$$z = \frac{X - \mu}{\sigma} \quad \text{or} \quad z = \frac{X - np}{\sqrt{npq}}$$

Formula for the chi-square test for variance or standard deviation:

$$\chi^2 = \frac{(n-1)s^2}{\sigma^2}$$

Review Exercises

For exercises 9–95 through 9–113, perform each of the following steps.

a. State the hypotheses and identify the claim.
b. Find the critical value(s).
c. Compute the test value.
d. Make the decision.
e. Summarize the results.

9–95. The Medical Rehabilitation Education Foundation found that the average cost of cardiac rehabilitation is $16,411. An administrator at Pine Valley Rehabilitation Center sampled 40 cardiac patients and found the mean cost was $14,706 and the standard deviation was $2,016. At $\alpha = 0.05$, can it be concluded that the average cost is different from $16,411?

Source: Snapshot, *USA Today*, September 18, 1995.

9–96. An automobile dealer believes that the average cost of accessories in new automobiles is $3000 over the base sticker price. He selects 50 new automobiles at random and finds that the average cost of the accessories is $3256. The standard deviation of the sample is $2300. Test his belief at $\alpha = 0.05$.

9–97. A recent study stated that if a person smoked, the average of the number of cigarettes he or she smoked was 14 per day. To test the claim, a researcher selected a random sample of 40 smokers and found that the mean number of cigarettes smoked per day was 18. The standard deviation of the sample was 6. At $\alpha = 0.05$, is the number of cigarettes a person smokes per day actually equal to 14?

9–98. A high school counselor wishes to test the theory that the average age of the dropouts in her school district is 16.3 years. She samples 32 recent dropouts and finds that their mean age is 16.9 years. At $\alpha = 0.01$, is the theory correct? The standard deviation of the population is 0.3.

9–99. In a certain city, a researcher wishes to determine whether the average age of its citizens is really 61.2 years, as the mayor claims. A sample of 22 residents has an average age of 59.8 years. The standard deviation of the sample is 1.5 years. At $\alpha = 0.01$, is the average age of the residents really 61.2 years? Assume that the variable is approximately normally distributed.

9–100. The average temperature during the summer months for the northeastern part of the United States is 67.0°. A sample of 10 cities had an average temperature of 69.6° for the summer of 1995. The standard deviation of the sample is 1.1°. At $\alpha = 0.10$, can it be concluded that the summer of 1995 was warmer than average?

Source: *USA Today*, September 14, 1995.

9–101. A recent study claimed that the average age of murder victims in a small city was less than or equal to 23.2 years. A sample of 18 recent victims had a mean of 22.6 years and a standard deviation of 2 years. At $\alpha = 0.05$, is the average age higher than originally believed? Assume that the variable is approximately normally distributed.

9–102. A magazine article stated that the average age of men who were getting divorced for the first time was less than 40 years. A researcher decided to test this theory at $\alpha = 0.025$. She selected a sample of 20 men who were recently divorced and found that the average age was 38.6 years. The standard deviation of the sample was 4 years. Should the null hypothesis be rejected on the basis of the sample? Assume that the variable is approximately normally distributed.

9–103.   The financial aid director of a college believes that at least 30% of the students are receiving some sort of financial aid. To see whether his belief is correct, the director selects a sample of 60 students and finds that 15 are receiving financial aid. At $\alpha = 0.05$, can the director conclude that at least 30% of the students are receiving financial aid?

9–104.   A dietitian read in a survey that at least 60% of adults eat eggs for breakfast at least four times a week. To test this claim, she selected a random sample of 100 adults and asked them how many days a week they ate eggs. In her sample, 54% responded that they ate eggs at least four times a week. At $\alpha = 0.10$, do her results support the survey?

9–105.   A contractor desires to build new homes with fireplaces. He read in a survey that 80% of all home buyers want a fireplace. To test this figure, he selected a sample of 30 home buyers and found that 20 wanted a fireplace. At $\alpha = 0.02$, should he arrive at the same conclusion as the survey?

9–106.   A radio manufacturer claims that 65% of teenagers 13–16 years old have their own portable radios. A researcher wishes to test the claim and selects a random sample of 80 teenagers. She finds that 57 have their own portable radios. At $\alpha = 0.05$, should the claim be rejected?

9–107.   A football coach claims that the average weight of all the opposing teams' members is 225 pounds. For a test of the claim, a sample of 50 players is taken from all the opposing teams. The mean is found to be 230 pounds and a standard deviation is 15 pounds. At $\alpha = 0.01$, test the coach's claim. Find the P-value and make the decision.

9–108.   An advertisement claims that Fasto Stomach Calm will provide relief from indigestion in less than 10 minutes. For a test of the claim, 35 individuals were given the product; the average time until relief was 9.25 minutes. From past studies, the standard deviation is known to be 2 minutes. Can one conclude that the claim is justified? Find the P-value and let $\alpha = 0.05$.

9–109.   A film editor feels that the standard deviation for the number of minutes in a video is 3.4 minutes. A sample of 24 videos has a standard deviation of 4.2 minutes. At $\alpha = 0.05$, is the sample standard deviation different from what the editor hypothesized?

9–110.   The standard deviation of the fuel consumption of a certain automobile is hypothesized to be greater than or equal to 4.3 miles per gallon. A sample of 20 automobiles produced a standard deviation of 2.6 miles per gallon. Is the standard deviation really less than previously thought? Use $\alpha = 0.05$.

9–111.   A manufacturer claims that the standard deviation of the drying time of a certain type of paint is 18 minutes. A sample of five test panels produced a standard deviation of 21 minutes. Can one conclude, at $\alpha = 0.05$, that the claim is correct?

9–112.   In order to see whether people are keeping their car tires inflated to the correct level, 35 pounds per square inch (psi), a tire company manager selects a sample of 36 tires and checks the pressure. The mean of the sample is 33.5 psi, and the standard deviation is 3. Are the tires properly inflated? Use $\alpha = 0.10$. Find the 90% confidence interval of the mean. Do the results agree? Explain.

9–113.   A biologist knows that the average length of a leaf of a certain full-grown plant is 4 inches. The standard deviation of the population is 0.6 inch. A sample of 20 leaves of that type of plant given a new type of plant food had an average length of 4.2 inches. Is there reason to believe that the new food is responsible for a change in the growth of the leaves? Use $\alpha = 0.01$. Find the 99% confidence interval of the mean. Do the results concur? Explain. Assume that the variable is approximately normally distributed.

Statistics Today

How Much Better Is Better?   Revisited

Now that you have learned the techniques of hypothesis testing presented in this chapter, you realize that the difference between the sample mean and the population mean must be *significant* before one can conclude that the students really scored above average. The superintendent should follow the steps in the hypothesis-testing procedure and be able to reject the null hypothesis before announcing that his students scored higher than average.

Data Analysis

The Data Bank is found in Appendix D.

1. From the Data Bank, select a random sample of at least 30 individuals, and test one or more of the following hypotheses by using the z test.
a. For serum cholesterol, H_0: $\mu = 220$ milligram percent (mg%).
b. For systolic pressure, H_0: $\mu = 120$ millimeters of mercury (mmHg).
c. For IQ, H_0: $\mu = 100$.
d. For sodium level, H_0: $\mu = 140$ milliequivalents per liter (mEq/l).

2. Select a random sample of 15 individuals and test one or more of the hypotheses in Exercise 1 by using the t test.

3. Select a random sample of at least 30 individuals, and using the z test for proportions, test one or more of the following hypotheses.
a. For educational level, H_0: $p = 0.50$ for level 2.
b. For smoking status, H_0: $p = 0.20$ for level 1.
c. For exercise level, H_0: $p = 0.10$ for level 1.
d. For gender, H_0: $p = 0.50$ for males.

4. Select a sample of 20 individuals and test the hypothesis H_0: $\sigma^2 = 225$ for IQ level.

Quiz

Determine whether each statement is true or false. If the statement is false, explain why.

1. No error is committed when the null hypothesis is rejected when it is false.

2. When one is conducting the t test, the population must be approximately normally distributed.

3. The test value separates the critical region from the noncritical region.

4. The values of a chi-square test cannot be negative.

5. The chi-square test for variances is always one-tailed.

Select the best answer.

6. When the value of α is increased, the probability of committing a type I error is
a. Decreased
b. Increased
c. The same
d. None of the above.

7. If one wishes to test the claim that the mean of the population is 100, the appropriate null hypothesis is
a. $\bar{X} = 100$
b. $\mu \geq 100$
c. $\mu \leq 100$
d. $\mu = 100$

8. The degrees of freedom for the chi-square test for variances or standard deviation are

a. One
b. n
c. $n - 1$
d. None of the above.

9. For the z test, if σ is unknown and $n \geq 30$, one can substitute for σ
a. n
b. s
c. χ^2
d. t

Complete the following statements with the best answer.

10. Rejecting the null hypothesis when it is true is called a _____ error.

11. The probability of a type II error is referred to as _____.

12. A conjecture about a population parameter is called a _____.

13. To test the hypothesis H_0: $\mu \leq 87$, one would use a _____ tailed test.

14. The degrees of freedom for the t test are _____.

For the following exercises where applicable:

a. State the hypotheses
b. Find the critical value(s)
c. Compute the test value
d. Make the decision
e. Summarize the results

15. A sociologist wishes to see if it is true that for a certain group of professional women, the average age at which they have their first baby is 28.6. The sociologist selects a sample of 36 women and finds their mean age for first giving birth was 29.2. The standard deviation of the population is 3.1 years. At $\alpha = 0.10$, does the evidence support the hypothesis?

16. A real estate agent believes that the average closing cost of purchasing a new home is $6500 over the purchase price. She selects 40 new home sales at random and finds that the average closing costs are $6600. The standard deviation of the population is $120. Test her belief at $\alpha = 0.05$.

17. A recent study stated that if a person chewed gum, the average number of sticks of gum he or she chewed was eight. To test the claim, a researcher selected a random sample of 36 gum chewers and found the mean number of sticks of gum chewed per day was nine. The standard deviation was 1. At $\alpha = 0.05$, is the number of sticks of gum a person chews per day actually greater than eight?

18. A high school counselor wishes to test the theory that the average number of dropouts in his school district per year is 21. He reviews the last 17 years and finds that the mean number of dropouts is 18.7. At $\alpha = 0.01$, is the theory correct? The standard deviation of the sample is 1.5.

19. In a New York modeling agency, a researcher wishes to see if the average height of female models is really 67 inches, as the chief claims. A sample of 20 models has an average height of 65.8 inches. The standard deviation of the sample is 1.7 inches. At $\alpha = 0.05$, is the average height of the models really less than 67 inches?

20. A taxi company claims that its drivers average at least 12.4 years of experience. In a study of 15 taxi drivers, the average experience was 11.2 years. The standard deviation was 2. At $\alpha = 0.10$, is the number of years' experience of the taxi drivers really less than the taxi company claimed?

21. A recent study in a small city stated that the average age of robbery victims was 63.5. A sample of 20 recent victims had a mean of 63.7 years and a standard deviation of 1.9. At $\alpha = 0.05$, is the average age higher than originally believed?

22. A magazine article stated that the average age of women who are getting married for the first time is 26 years. A researcher decided to test this hypothesis at $\alpha = 0.02$. She selected a sample of 25 women who were recently married for the first time and found the average was 25.1. The standard deviation was 3 years. Should the null hypothesis be rejected on the basis of the sample?

23. The president of a local college reported that at least 25% of the students have received scholarships. To see if this is true, the registrar selected a sample of 100 students and found that 22 received scholarships. At $\alpha = 0.05$, can it be concluded that at least 25% of the students have scholarships?

24. A dietitian read in a survey that at least 55% of adults do not eat breakfast at least three days a week. To verify this, she selected a random sample of 80 adults and asked them how many days a week they skipped breakfast. A total of 50% responded that they skipped breakfast at least three days a week. At $\alpha = 0.10$, do her results support the survey?

25. A contractor wanted to build new homes with two-car garages. He read in a survey that 70% of all home buyers want a two-car garage. To verify the figure, he selected a sample of 30 home buyers and found that 23 wanted a two-car garage. At $\alpha = 0.02$, should he arrive at the same conclusions as the survey?

26. A magazine claims that 75% of all teenage boys have their own radios. A researcher wished to test the claim and selected a random sample of 60 teenage boys. She found that 54 had their own radios. At $\alpha = 0.01$, should the claim be rejected?

27. Find the P-value for the z test in Exercise 15.

28. Find the P-value for the z test in Exercise 16.

29. A copyeditor thinks the standard deviation for the number of pages in a romance novel is greater than six. A sample of 25 novels has a standard deviation of nine pages. At $\alpha = 0.05$, is it higher as the editor hypothesized?

30. It has been hypothesized that the standard deviation of the germination time of radish seeds is eight days. A sample of 60 radish plants' germination time was six days. At $\alpha = 0.01$, can it be concluded that the hypothesis is correct?

31. The standard deviation of the pollution by-products released in the burning of one gallon of gas is 2.3 ounces. A sample of 20 automobiles tested produced a standard deviation of 1.9 ounces. Is the standard deviation really less than previously thought? Use $\alpha = 0.05$.

32. A manufacturer claims that the standard deviation of the strength of wrapping cord is nine pounds. A sample of 10 wrapping cords produced a standard deviation of 11 pounds. Can it be concluded that at $\alpha = 0.05$, the claim is correct?

33. Find the 90% confidence interval of the mean in Exercise 15.

34. Find the 95% confidence interval for the mean in Exercise 16.

Critical Thinking Challenge

Hypothesis Testing versus a Jury Trial   Hypothesis testing can be compared to a jury trial. For example, the hypotheses would be the following:

H_0: The defendant is innocent.
H_1: The defendant is not innocent.

The results of a trial can be shown as follows:

	H_0 True (innocent)	H_0 False (not innocent)
Reject H_0 (convict)	**Type I** error	Correct decision
Do not reject H_0 (acquit)	Correct decision	**Type II** error

Write a paper that explains the results and consequences of each block in the table (e.g., committing a type I error).

Data Projects

Use MINITAB, the TI-83, or a computer program of your choice to complete the following exercises.

1. Choose a variable such as the number of miles students live from the college or the number of daily admissions to an ice-skating rink for a one-month period. Before collecting the data, decide what a likely average might be, then complete the following:
a. Write a brief statement of the purpose of the study.
b. Define the population.
c. State the hypotheses for the study.
d. Select an α value.
e. State how the sample was selected.
f. Show the raw data.
g. Decide which statistical test is appropriate and compute the test statistic (z or t). Why is the test appropriate?
h. Find the critical value(s).
i. State the decision.
j. Summarize the results.

2. Decide on a question that could be answered with a "yes" or "no" or "not sure (undecided)." For example, "Are you satisfied with the variety of food the cafeteria serves?" Before collecting the data, hypothesize what you expect the proportion of students who will respond "yes" will be. Then complete the following:
a. Write a brief statement of the purpose of the study.
b. Define the population.
c. State the hypotheses for the study.
d. Select an α value.
e. State how the sample was selected.
f. Show the raw data.
g. Decide which statistical test is appropriate and compute the test statistic (z or t). Why is the test appropriate?
h. Find the critical value(s).
i. State the decision.
j. Summarize the results.

Do you think the project supported your initial hypothesis? Explain your answer.

TI-83 Calculator

Note: For most of the hypothesis tests and confidence intervals, the calculator will accept raw data in lists denoted by *Data* or summary statistics denoted by *Stats.*

Also in hypothesis testing, the calculator gives *P*-values instead of critical values. When the *P*-value is less than the α value, the decision is to reject the null hypothesis. If the *P*-value is greater than the α value, the decision is not to reject the null hypothesis. For example, if the *P*-value for a test is 0.032568971 and α is 0.05, the decision will be to reject the null hypothesis, since $0.032568971 < 0.05$. If $\alpha = 0.01$, the decision will be not to reject the null hypothesis, since $0.032568971 > 0.01$. For an in-depth explanation of *P*-values, see Section 9–3.

z Test: One Mean

A. One sample *z* test (Raw data)
1. Enter the data into L_1.
2. Press **STAT** and move the cursor to TESTS.
3. Press either **ENTER** or **1**.
4. Select Data and press **ENTER**, then press ▼.
5. Type in the value for μ_0.
6. Type in the value for σ. Make sure L_1 is selected and Freq. is 1.
7. Select the correct alternative hypothesis—either $\mu: \neq \mu_0$, $\mu < \mu_0$, or $\mu > \mu_0$, where μ_0 represents the hypothesized mean.
8. Select Calculate and press **ENTER**.
Example 12: Test the claim H_1: $\mu < 18$ when $\sigma = 6$ for the following data: 27, 16, 9, 14, 32, 15, 16, 18, 16, 13 Use $\alpha = 0.05$.

Input

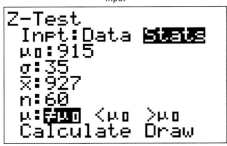

Output

The alternative hypothesis is $\mu < 18$. The test value is -0.2108185107, which is not significant at $\alpha = 0.05$. The sample mean is $\overline{X} = 17.6$ and the sample standard deviation is 6.818276094. The sample size is 10.

B. One sample *z* test (Stats)
1. Select **STAT**. Move the cursor to TESTS and press either **ENTER** or **1**.
2. Select Stats and press **ENTER**, then press ▼.
3. Type the values for μ_0, σ, $\overline{X}$, and n.
4. Select the correct alternative hypothesis.
5. Select Calculate and press **ENTER**.
Example 13: Test the claim H_1: $\mu \neq 915$ when $\sigma = 35$, $\overline{X} = 927$, and $n = 60$. Use $\alpha = 0.05$.

Input

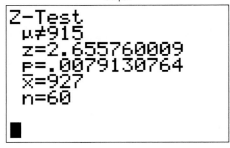

Output

The alternative hypothesis is H_1: $\mu \neq 915$. The *z* test value is 2.655760009. The *P*-value is 0.0079130764. The decision is to reject H_0, since $0.0079130764 < 0.05$.

t Test: One Mean

A. One sample *t* test (Data)
1. Enter the data into L_1.
2. Press **STAT** and move the cursor to TESTS. Then press **2**.
3. Select Data and press **ENTER**, then press ▼.
4. Type in the value for μ_0 and make sure L_1 is selected and Freq. is 1.
5. Select the correct alternative hypothesis.
6. Select Calculate and press **ENTER**.

Example 14: Test the claim H_1: $\mu \neq 2$ at $\alpha = 0.10$, using the data values 6, 8, 3, 2, 0, 0, 1, 5, 4, 3, 3, 2.

Input

```
T-Test
 Inpt:DATA Stats
 µ0:2
 List:L1
 Freq:1
 µ:≠µ0 <µ0 >µ0
 Calculate Draw
```

Output

```
T-Test
 µ≠2
 t=1.569156352
 P=.1449105032
 x̄=3.083333333
 Sx=2.391588796
 n=12
```

The alternative hypothesis is H_1: $\mu \neq 2$. The test value is 1.569156352. The P-value is 0.1449105032, which is not significant at $\alpha = 0.10$, since $0.1449105032 > 0.10$. The values for $\bar{X}$ and s and n are also given.

B. One sample t test (Stats)
1. Press **STAT** and move the cursor to TESTS. Then press **2**.
2. Move the cursor to Stats. Press **ENTER**, then press ▼.
3. Type the values for μ_0, $\bar{X}$, s, and n.
4. Select the correct alternative hypothesis.
5. Select Calculate and press **ENTER**.
Example 15: Test the claim H_1: $\mu < 16$ at $\alpha = 0.10$ when $\bar{X} = 15.6$, $s = 0.3$, and $n = 8$.

Input

```
T-Test
 Inpt:Data Stats
 µ0:16
 x̄:15.6
 Sx:.3
 n:8
 µ:≠µ0 <µ0 >µ0
 Calculate Draw
```

Output

```
T-Test
 µ<16
 t=-3.771236166
 P=.0034858275
 x̄=15.6
 Sx=.3
 n=8
```

The test value is -3.771236166, and the P-value is 0.0034858275. The decision is to reject H_0, since $0.0034858275 < 0.10$.

***z* Test: One Proportion**

1. Press **STAT** and move the cursor to TESTS.
2. Press **5**.
3. Enter the hypothesized proportion p_0, X, and n.
4. Select the correct alternative hypothesis.
5. Select Calculate and press **ENTER**.
Example 16: Test the claim H_1: $p \neq 0.15$ when $X = 38$ and $n = 200$ at $\alpha = 0.05$.

Input

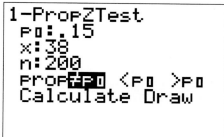

```
1-PropZTest
 P0:.15
 x:38
 n:200
 Prop≠P0 <P0 >P0
 Calculate Draw
```

Output

```
1-PropZTest
 Prop≠.15
 z=1.584236069
 P=.1131400131
 p̂=.19
 n=200
```

The test value is 1.584236069. The P-value is 0.1131400131. The decision is not to reject H_0. The values for $\hat{p}$ and n are also given.

chapter

10

Testing the Difference between Means, Variances, and Proportions

Objectives

After completing this chapter, you should be able to

1. Test the difference between two large sample means using the z test.

2. Test the difference between two variances or standard deviations.

3. Test the difference between two means for small independent samples.

4. Test the difference between two means for small dependent samples.

5. Test the difference between two proportions.

Statistics Today

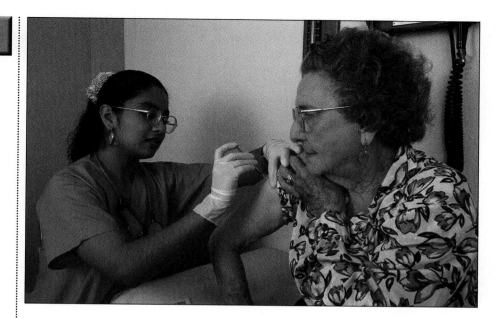

To Vaccinate or Not to Vaccinate? Small or Large?

Influenza is a serious disease among the elderly, especially those living in nursing homes. Those residents are more susceptible to influenza than elderly persons living in the community because they are usually older and more debilitated, and they live in a closed environment where they are exposed more so than community residents to the virus if it is introduced into the home. Three researchers decided to investigate the use of vaccine and its value in determining outbreaks of influenza in small nursing homes.

These researchers surveyed 83 licensed homes in seven counties in Michigan. Part of the study consisted of comparing the number of people being vaccinated in small nursing homes (100 or fewer beds) with the number in larger nursing homes (more than 100 beds). Unlike the statistical methods presented in the previous chapter, these researchers used the techniques explained in this chapter to compare two sample proportions to see if there was a significant difference in the vaccination rates of patients in small nursing homes as compared to those in larger nursing homes.

Source: Nancy Arden, Arnold S. Monto, and Suzanne E. Ohmit, "Vaccine Use and the Risk of Outbreaks in a Sample of Nursing Homes during an Influenza Epidemic," *American Journal of Public Health* 85, no. 3 (March 1995), pp. 399–401.

10–1

Introduction

The basic concepts of hypothesis testing were explained in Chapter 9. With the z and t tests, a sample mean, variance, or proportion can be compared with a specific population mean, variance, or proportion in order to determine whether the null hypothesis should be rejected.

There are, however, many instances when researchers wish to compare two sample means using experimental and control groups. For example, the average lifetimes of two different brands of bus tires might be compared to see whether there is any difference in tread wear. Two different brands of fertilizer might be tested to see whether one is better than the other for growing plants. Or two brands of cough syrup might be tested to see whether one brand is more effective than the other.

In the comparison of two means, the same basic steps for hypothesis testing shown in Chapter 9 are used, and the z and t tests are also used. When comparing two means using the t test, the researcher must decide if the two samples are *independent* or *dependent*. The concepts of independent samples and dependent samples will be explained in Sections 10–4 and 10–5.

Furthermore, when the samples are independent, there are two different formulas that can be used depending on whether or not the variances are equal. To determine if the variances are equal, use the F test. Finally, the z test can be used to compare two proportions, as shown in Section 10–6.

10–2

Testing the Difference between Two Means: Large Samples

Objective 1. Test the difference between two large sample means using the z test.

Suppose a researcher wishes to determine whether there is a difference in the average age of nursing students who enroll in a nursing program at a community college and those who enroll in a nursing program at a university. In this case, the researcher is not interested in the average age of all beginning nursing students; instead, he is interested in *comparing* the means of the two groups. His research question is: Does the mean age of nursing students who enroll at a community college differ from the mean age of nursing students who enroll at a university? Here, the hypotheses are:

$$H_0: \mu_1 = \mu_2$$
$$H_1: \mu_1 \neq \mu_2$$

where

μ_1 = mean age of all beginning nursing students at the community college

μ_2 = mean age of all beginning nursing students at the university

Another way of stating the hypotheses for this situation is

$$H_0: \mu_1 - \mu_2 = 0$$
$$H_1: \mu_1 - \mu_2 \neq 0$$

If there is no difference in population means, subtracting them will give a difference of zero. If they are different, subtracting will give a number other than zero. Both methods of stating hypotheses are correct; however, the first method will be used in this book.

Assumptions for the Test to Determine the Difference between Two Means

1. The samples must be independent of each other. That is, there can be no relationship between the subjects in each sample.
2. The populations from which the samples were obtained must be normally distributed, and the standard deviations of the variable must be known, or the samples must be greater than or equal to 30.

The theory behind testing the difference between two means is based on selecting pairs of samples and comparing the means of the pairs. The population means need not be known.

All possible pairs of samples are taken from populations. The means for each pair of samples are computed and then subtracted, and the differences are plotted. If both populations have the same mean, then most of the differences will be zero or close to zero. Occasionally, there will be a few large differences due to chance alone, some positive and some negative. If the differences are plotted, the curve will be shaped like the normal distribution and have a mean of zero, as shown in Figure 10–1.

Figure 10–1

Differences of Means of Pairs of Samples

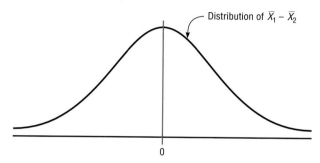

Distribution of $\bar{X}_1 - \bar{X}_2$

0

The variance of the difference $\bar{X}_1 - \bar{X}_2$ is equal to the sum of the individual variances of $\bar{X}_1$ and $\bar{X}_2$. That is,

$$\sigma^2_{\bar{X}_1 - \bar{X}_2} = \sigma^2_{\bar{X}_1} + \sigma^2_{\bar{X}_2}$$

where

$$\sigma^2_{\bar{X}_1} = \frac{\sigma^2_1}{n_1} \quad \text{and} \quad \sigma^2_{\bar{X}_2} = \frac{\sigma^2_2}{n_2}$$

So the standard deviation of $\bar{X}_1 - \bar{X}_2$ is

$$\sqrt{\frac{\sigma^2_1}{n_1} + \frac{\sigma^2_2}{n_2}}$$

Formula for the *z* Test for Comparing Two Means from Independent Populations

$$z = \frac{(\bar{X}_1 - \bar{X}_2) - (\mu_1 - \mu_2)}{\sqrt{\dfrac{\sigma^2_1}{n_1} + \dfrac{\sigma^2_2}{n_2}}}$$

This formula is based on the general format of

$$\text{test value} = \frac{(\text{observed value}) - (\text{expected value})}{\text{standard error}}$$

where $\bar{X}_1 - \bar{X}_2$ is the observed difference, and the expected difference $\mu_1 - \mu_2$ is zero when the null hypothesis is $\mu_1 = \mu_2$, since that is equivalent to $\mu_1 - \mu_2 = 0$. Finally, the standard error of difference is

$$\sqrt{\frac{\sigma^2_1}{n_1} + \frac{\sigma^2_2}{n_2}}$$

In the comparison of two sample means, the difference may be due to chance, in which case the null hypothesis will not be rejected, and the researcher can assume that the means of the populations are basically the same. The difference in this case is not significant. See Figure 10–2(a). On the other hand, if the difference is significant, the null hypothesis is rejected, and the researcher can conclude that the population means are different. See Figure 10–2(b).

Figure 10–2

Hypothesis-Testing Situations in the Comparison of Means

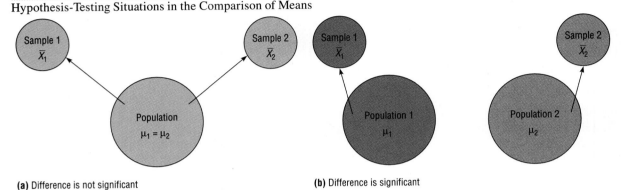

(a) Difference is not significant

Do not reject H_0: $\mu_1 = \mu_2$ since $\overline{X}_1 - \overline{X}_2$ is not significant.

(b) Difference is significant

Reject H_0: $\mu_1 = \mu_2$ since $\overline{X}_1 - \overline{X}_2$ is significant.

These tests can also be one-tailed, using the following hypothesis.

Right-Tailed		Left-Tailed	
H_0: $\mu_1 \leq \mu_2$	H_0: $\mu_1 - \mu_2 \leq 0$	H_0: $\mu_1 \geq \mu_2$	H_0: $\mu_1 - \mu_2 \geq 0$
or		or	
H_1: $\mu_1 > \mu_2$	H_1: $\mu_1 - \mu_2 > 0$	H_1: $\mu_1 < \mu_2$	H_1: $\mu_1 - \mu_2 < 0$

The same critical values used in Section 9–3 are used here. They can be obtained from Table E in Appendix C.

If σ_1^2 and σ_2^2 are not known, the researcher can use the variances obtained from each sample, s_1^2 and s_2^2, but both sample sizes must be 30 or more. The formula then is

$$z = \frac{(\overline{X}_1 - \overline{X}_2) - (\mu_1 - \mu_2)}{\sqrt{\dfrac{s_1^2}{n_1} + \dfrac{s_2^2}{n_2}}}$$

provided that $n_1 \geq 30$ and $n_2 \geq 30$.

When one or both sample sizes are less than 30 and σ_1 and σ_2 are unknown, the t test must be used, as shown in Section 10–4.

The basic format for hypothesis testing is reviewed here.

STEP 1   State the hypotheses and identify the claim.

STEP 2   Find the critical value(s).

STEP 3   Compute the test value.

STEP 4   Make the decision.

STEP 5   Summarize the results.

Example 10–1

A survey found that the average hotel room rate in New Orleans is $88.42 and the average room rate in Phoenix is $80.61. Assume that the data were obtained from two samples of 50 hotels each and that the standard deviations were $5.62 and $4.83 respectively. At $\alpha = 0.05$, can it be concluded that there is no significant difference in the rates?

Source: *USA Today*, January 10, 1995.

Solution

STEP 1   State the hypothesis and identify the claim.

$$H_0: \mu_1 = \mu_2 \text{(claim)} \quad \text{and} \quad H_1: \mu_1 \neq \mu_2$$

STEP 2   Find the critical values. Since $\alpha = 0.05$, the critical values are $+1.96$ and -1.96.

STEP 3   Compute the test value.

$$z = \frac{(\overline{X}_1 - \overline{X}_2) - (\mu_1 - \mu_2)}{\sqrt{\dfrac{\sigma_1^2}{n_1} + \dfrac{\sigma_2^2}{n_2}}} = \frac{(88.42 - 80.61) - 0}{\sqrt{\dfrac{5.62^2}{50} + \dfrac{4.83^2}{50}}} = 7.45$$

STEP 4   Make the decision. Reject the null hypothesis at $\alpha = 0.05$, since $7.45 > 1.96$. See Figure 10–3.

Figure 10–3

Critical and Test Values
for Example 10–1

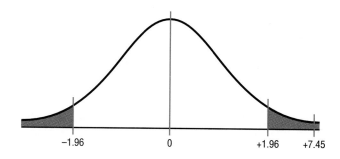

| −1.96 | 0 | +1.96 | +7.45 |

STEP 5   Summarize the results. There is enough evidence to reject the claim that the means are equal. Hence, there is a significant difference in the rates.

P-values for this test can be determined using the same procedure as shown in Section 9–3. For example, if the test value for a two-tailed test is 1.40, then the *P*-value obtained from Table E is 0.1616. This value is obtained by looking up the area for $z = 1.40$, which is 0.4192. Then 0.4192 is subtracted from 0.5000 to get 0.0808. Finally, this value is doubled to get 0.1616 since the test is two-tailed. If $\alpha = 0.05$, the decision would be do not reject the null hypothesis.

Sometimes, the researcher is interested in testing a specific difference in means other than zero. For example, he or she might hypothesize that the nursing students at a community college are, on the average, 3.2 years older than those at a university. In this case, the hypotheses are

$$H_0: \mu_1 - \mu_2 \leq 3.2 \quad \text{and} \quad H_1: \mu_1 - \mu_2 > 3.2$$

The formula for the z test is still

$$z = \frac{(\bar{X}_1 - \bar{X}_2) - (\mu_1 - \mu_2)}{\sqrt{\dfrac{\sigma_1^2}{n_1} + \dfrac{\sigma_2^2}{n_2}}}$$

where $\mu_1 - \mu_2$ is the hypothesized difference or expected value. In this case, $\mu_1 - \mu_2 = 3.2$.

Confidence intervals for the difference between two means can also be found. When hypothesizing a difference of 0, if the confidence interval contains 0, the null hypothesis is not rejected. If the confidence interval does not contain 0, the null hypothesis is rejected.

Confidence intervals for the difference between two means can be found by using the following formula:

Formula for Confidence Interval for Difference between Two Means: Large Samples

$$(\bar{X}_1 - \bar{X}_2) - (z_{\alpha/2})\sqrt{\frac{\sigma_1^2}{n_1} + \frac{\sigma_2^2}{n_2}} < \mu_1 - \mu_2$$

$$< (\bar{X}_1 - \bar{X}_2) + z_{\alpha/2}\sqrt{\frac{\sigma_1^2}{n_1} + \frac{\sigma_2^2}{n_2}}$$

When $n_1 \geq 30$ and $n_2 \geq 30$, then s_1^2 and s_2^2 can be used in place of σ_1^2 and σ_2^2.

Example 10–2

Find the 95% confidence interval for the difference between the means for the data in Example 10–1.

Solution

Substitute in the formula using $z_{\alpha/2} = 1.96$.

$$(\bar{X}_1 - \bar{X}_2) - z_{\alpha/2}\sqrt{\frac{\sigma_1^2}{n_1} + \frac{\sigma_2^2}{n_2}} < \mu_1 - \mu_2$$

$$< (\bar{X}_1 - \bar{X}_2) + z_{\alpha/2}\sqrt{\frac{\sigma_1^2}{n_1} + \frac{\sigma_2^2}{n_2}}$$

$$(88.42 - 80.61) - 1.96\sqrt{\frac{5.62^2}{50} + \frac{4.83^2}{50}} < \mu_1 - \mu_2$$

$$< (88.42 - 80.61) + 1.96\sqrt{\frac{5.62^2}{50} + \frac{4.83^2}{50}}$$

$$7.81 - 2.05 < \mu_1 - \mu_2 < 7.81 + 2.05$$

$$5.76 < \mu_1 - \mu_2 < 9.86$$

Since the confidence interval does not contain zero, the decision is to reject the null hypothesis, which agrees with the previous result.

Exercises

10–1. (W) Explain the difference between testing a single mean and testing the difference between two means.

10–2. (W) When a researcher selects all possible pairs of samples from a population in order to find the difference between the means of each pair, what will be the shape of the distribution of the differences when the original distributions are normally distributed? What will be the mean of the distribution? What will be the standard deviation of the distribution?

10–3. (W) What two assumptions must be met when one is using the z test to test differences between two means? When can the sample standard deviations, s_1 and s_2, be used in place of the population standard deviations σ_1 and σ_2?

10–4. (W) Show two different ways to state that the means of two populations are equal.

For Exercises 10–5 through 10–17, perform each of the following steps.

a. State the hypotheses and identify the claim.
b. Find the critical value(s).
c. Compute the test value.
d. Make the decision.
e. Summarize the results.

10–5. A publishing company wishes to test the claim that there is a difference between two overnight delivery companies in the speed with which their material is delivered. The average speed of material delivered over a 30-day period is shown here. At $\alpha = 0.01$, is there enough evidence to support the claim that there is a difference between the delivery times of the two companies?

Company 1	Company 2
$\overline{X}_1 = 16$ hours	$\overline{X}_2 = 18$ hours
$\sigma_1 = 3.2$	$\sigma_2 = 3$
$n_1 = 30$	$n_2 = 30$

10–6. A study was conducted to see if there was a difference between spouses and significant others in coping skills when living with or caring for a person with multiple sclerosis. These skills were measured by questionnaire responses. The results of the two groups are given below on one factor, ambivalence. At $\alpha = 0.10$, is there a difference in the means of the two groups?

Spouses	Significant others
$\overline{X}_1 = 2$	$\overline{X}_2 = 1.7$
$s_1 = 0.6$	$s_2 = 0.7$
$n_1 = 120$	$n_2 = 34$

Source: Elsie E. Gulick, "Coping among Spouses or Significant Others of Persons with Multiple Sclerosis," *Nursing Research* 44, no. 4 (July/August 1995), p. 224.

10–7. A medical researcher wishes to see whether the pulse rates of smokers are higher than the pulse rates of nonsmokers. Samples of 100 smokers and 100 nonsmokers are selected. The results are shown here. Can the researcher conclude, at $\alpha = 0.05$, that smokers have higher pulse rates than nonsmokers?

Smokers	Nonsmokers
$\overline{X}_1 = 90$	$\overline{X}_2 = 88$
$s_1 = 5$	$s_2 = 6$
$n_1 = 100$	$n_2 = 100$

10–8. A statistician claims that the average score on a standardized test of students who major in psychology is greater than that of students who major in mathematics. The results of the test, given to 50 students in each group, are shown here. Is there enough evidence to support the statistician's claim at $\alpha = 0.01$?

Psychology	Mathematics
$\overline{X}_1 = 118$	$\overline{X}_2 = 115$
$\sigma_1 = 15$	$\sigma_2 = 15$
$n_1 = 50$	$n_2 = 50$

10–9. Using data from the "Noise Level in an Urban Hospital" study cited in Exercise 8–19, test the claim that the noise level in the corridors is higher than that in the clinics. Use $\alpha = 0.02$. The data are shown here.

Corridors	Clinics
$\overline{X}_1 = 61.2$ dBA	$\overline{X}_2 = 59.4$ dBA
$s_1 = 7.9$	$s_2 = 7.9$
$n_1 = 84$	$n_2 = 34$

Source: M. Bayo, A. Garcia, and A. Garcia, "Noise Levels in an Urban Hospital and Workers' Subjective Responses," *Archives of Environmental Health* 50, no. 38 (May/June 1995).

10–10.   A real estate agent compares the selling prices of homes in two suburbs of Seattle to see whether there is a difference in price. The results of the study are shown here. Is there evidence to reject the claim that the average cost of a home in both locations is the same? Use $\alpha = 0.01$.

Suburb 1	Suburb 2
$\overline{X}_1 = \$63{,}255$	$\overline{X}_2 = \$59{,}102$
$s_1 = \$5602$	$s_2 = \$4731$
$n_1 = 35$	$n_2 = 40$

10–11.   In a study of women science majors, the following data were obtained on two groups, those who left their profession within a few months after graduation (leavers) and those who remained in their profession after they graduated (stayers). Test the claim that those who stayed had a higher science grade point average than those who left. Use $\alpha = 0.05$.

Leavers	Stayers
$\overline{X}_1 = 3.16$	$\overline{X}_2 = 3.28$
$s_1 = 0.52$	$s_2 = 0.46$
$n_1 = 103$	$n_2 = 225$

Source: Paula Rayman and Belle Brett, "Women Science Majors: What Makes a Difference in Persistence after Graduation?", *The Journal of Higher Education* 66, no. 4 (July/August 1995), pp. 388–414.

10–12.   A college admissions officer believes that students enrolling from Shawnee School District have higher SAT scores than those who enroll from West River School District. A sample of 30 students from each district is selected, and the results are as shown below. Is the belief of the admissions officer substantiated? Use $\alpha = 0.05$.

Shawnee	West River
$\overline{X}_1 = 656$	$\overline{X}_2 = 560$
$\sigma_1 = 100$	$\sigma_2 = 100$
$n_1 = 30$	$n_2 = 30$

10–13.   Is there a difference in the cranking powers of two brands of car batteries at 32°F? Use $\alpha = 0.10$.

Brand 1	Brand 2
$\overline{X}_1 = 320$ amperes	$\overline{X}_2 = 383$ amperes
$s_1 = 15$ amperes	$s_2 = 10$ amperes
$n_1 = 36$	$n_2 = 36$

10–14.   Is there a difference in average miles traveled for each of two taxi companies during a randomly selected week? The data are shown below. Use $\alpha = 0.05$. Assume that the populations are normally distributed.

Moonview Cab Company	Starlight Taxi Company
$\overline{X}_1 = 837$	$\overline{X}_2 = 753$
$\sigma_1 = 30$	$\sigma_2 = 40$
$n_1 = 35$	$n_2 = 40$

10–15.   In the study cited in Exercise 10–11, the researchers collected the data shown here on a self-esteem questionnaire. At $\alpha = 0.05$, can it be concluded that there is no difference in the self-esteem scores of the two groups?

Leavers	Stayers
$\overline{X}_1 = 3.05$	$\overline{X}_2 = 2.96$
$s_1 = 0.75$	$s_2 = 0.75$
$n_1 = 103$	$n_2 = 225$

Source: Paula Rayman and Belle Brett, "Women Science Majors: What Makes a Difference in Persistence after Graduation?", *The Journal of Higher Education* 66, no. 4 (July/August 1995), pp. 388–414.

10–16.   The dean of students wants to see whether there is a significant difference in ages of resident students and commuting students. She selects a sample of 50 students from each group. The ages are shown here. At $\alpha = 0.05$, decide if there is enough evidence to reject the claim of no difference in the ages of the two groups. Use the standard deviations from the samples.

Resident students							
22	25	27	23	26	28	26	24
25	20	26	24	27	26	18	19
18	30	26	18	18	19	32	23
19	19	18	29	19	22	18	22
26	19	19	21	23	18	20	18
22	21	19	21	21	22	18	20
19	23						

Commuter students							
18	20	19	18	22	25	24	35
23	18	23	22	28	25	20	24
26	30	22	22	22	21	18	20
19	26	35	19	19	18	19	32
29	23	21	19	36	27	27	20
20	21	18	19	23	20	19	19
20	25						

10–17.   A researcher claims that students in a private school have exam scores at most 8 points higher than those of students in public schools. Random samples of 60 students from each type of school are selected and given an exam. The results are shown below. At $\alpha = 0.05$, test the claim.

Private school	Public school
$\overline{X}_1 = 110$	$\overline{X}_2 = 104$
$s_1 = 15$	$s_2 = 15$
$n_1 = 60$	$n_2 = 60$

10–18.   A study of teenagers found the following information on the length of time (in minutes) each talked on the telephone. Find the 95% confidence level of the true differences in means.

Boys	Girls
$\overline{X}_1 = 21$	$\overline{X}_2 = 18$
$\sigma_1 = 2.1$	$\sigma_2 = 3.2$
$n_1 = 50$	$n_2 = 50$

10–19.   Two brands of cigarettes are selected and their nicotine content compared. The data are shown here. Find the 99% confidence interval of the true difference in the means.

Brand A	Brand B
$\overline{X}_1 = 28.6$ milligrams	$\overline{X}_2 = 32.9$ milligrams
$\sigma_1 = 5.1$ milligrams	$\sigma_2 = 4.4$ milligrams
$n_1 = 30$	$n_2 = 40$

10–20.   Two brands of batteries are tested and their voltage compared. The data follow. Find the 95% confidence interval of the true difference in the means. Assume that both variables are normally distributed.

Brand X	Brand Y
$\overline{X}_1 = 9.2$ volts	$\overline{X}_2 = 8.8$ volts
$\sigma_1 = 0.3$ volt	$\sigma_2 = 0.1$ volt
$n_1 = 27$	$n_2 = 30$

10–21.   Two groups of students are given a problem-solving test, and the results are compared. Find the 90% confidence interval of the true difference in means.

Mathematics majors	Computer science majors
$\overline{X}_1 = 83.6$	$\overline{X}_2 = 79.2$
$s_1 = 4.3$	$s_2 = 3.8$
$n_1 = 36$	$n_2 = 36$

10–3

Testing the Difference between Two Variances

Objective 2. Test the difference between two variances or standard deviations.

In addition to comparing two means, statisticians are also interested in comparing two variances or standard deviations. For example, is the variation in the temperatures for a certain month for two cities different?

In another situation, a researcher may be interested in comparing the variance of the cholesterol of men with the variance of the cholesterol of women. For the comparison of two variances or standard deviations, an **F test** is used. The F test should not be confused with the chi-square test, which compares a single sample variance to a specific population variance, as shown in Chapter 9.

If two independent samples are selected from two normally distributed populations in which the variances are equal ($\sigma_1^2 = \sigma_2^2$) and if the variances s_1^2 and s_2^2 are compared as $\dfrac{s_1^2}{s_2^2}$, the sampling distribution of the variances is called the **F distribution.**

Characteristics of the *F* Distribution

1. The values of *F* cannot be negative, because variances are always positive or zero.
2. The distribution is positively skewed.
3. The mean value of *F* is approximately equal to 1.
4. The *F* distribution is a family of curves based on the degrees of freedom of the variance of the numerator and the degrees of freedom of the variance of the denominator.

Figure 10–4 shows the shapes of several curves for the *F* distribution.

Figure 10–4

The *F* Family of Curves

Formula for the *F* Test

$$F = \frac{s_1^2}{s_2^2}$$

where s_1^2 is the larger of the two variances.

The *F* test has two terms for the degrees of freedoms: that of the numerator, $n_1 - 1$, and that of the denominator, $n_2 - 1$, where n_1 is the sample size from which the larger variance was obtained.

When one is finding the *F* test value, *the larger of the variances is placed in the numerator of the F formula;* this is not necessarily the variance of the larger of the two sample sizes.

Table H in Appendix C gives the *F* critical values for $\alpha = 0.005, 0.01, 0.025, 0.05,$ and 0.10 (each α value involves a separate table in Table H). These are one-tailed values; if a two-tailed test is being conducted, then the $\alpha/2$ value must be used. For example, if a two-tailed test with $\alpha = 0.05$ is being conducted, then the $0.05/2 = 0.025$ table of Table H should be used.

Example 10–3

Find the critical value for a right-tailed F test when $\alpha = 0.05$, the degrees of freedom for the numerator (abbreviated d.f.N.) are 15, and the degrees of freedom for the denominator (d.f.D.) are 21.

Solution

Since this test is right-tailed with $\alpha = 0.05$, use the 0.05 table. The d.f.N. is listed across the top, and the d.f.D. is listed in the left column. The critical value is found where the row and column intersect in the table. In this case, it is 2.18. See Figure 10–5.

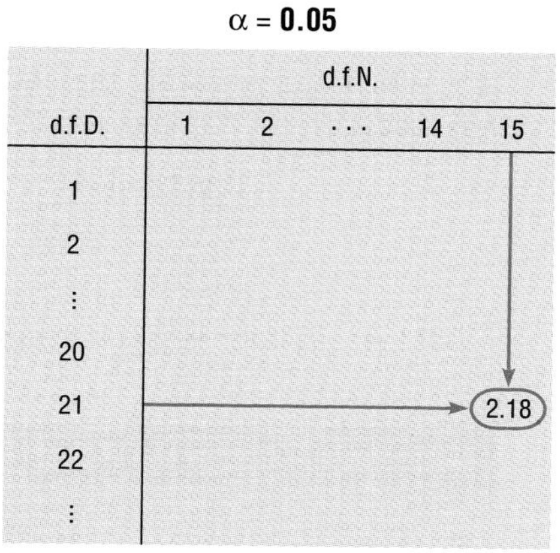

Figure 10–5

Finding the Critical Value in Table H for Example 10–3

$\alpha = 0.05$

As noted previously, when the F test is used, the larger variance is always placed in the numerator of the formula. Hence, when one is conducting a two-tailed test, α is split; and even though there are two values, only the right one is used. The reason is that the F test value is always greater than or equal to 1.

Example 10–4

Find the critical value for a two-tailed F test with $\alpha = 0.05$ when the sample size from which the variance for the numerator was obtained was 21 and the sample size from which the variance for the denominator was obtained was 12.

Solution

Since this is a two-tailed test with $\alpha = 0.05$, the $0.05/2 = 0.025$ table must be used. Here, d.f.N. $= 21 - 1 = 20$, and d.f.D. $= 12 - 1 = 11$; hence, the critical value is 3.23. See Figure 10–6.

Figure 10–6

Finding the Critical
Value in Table H for
Example 10–4

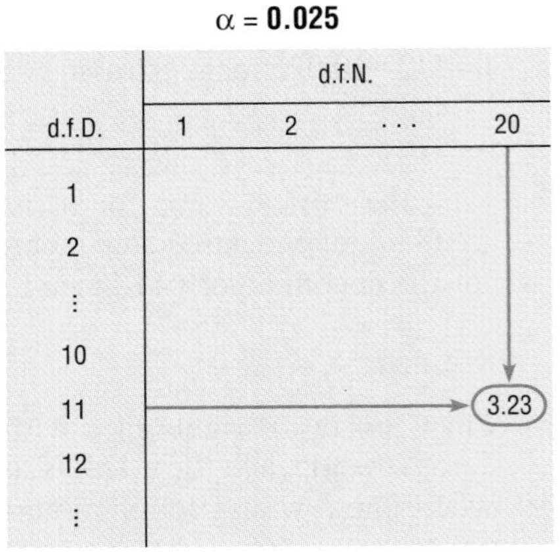

$\alpha = 0.025$

		d.f.N.		
d.f.D.	1	2	. . .	20
1				
2				
⋮				
10				
11				(3.23)
12				
⋮				

When the degrees of freedom values cannot be found in the table, the closest value on the smaller side should be used. For example, if d.f.N. = 14, this value is between the given table values of 12 and 15; therefore, 12 should be used to be on the safe side.

When one is testing the equality of two variances, the following hypotheses are used.

Right-tailed	Left-tailed	Two-tailed
$H_0: \sigma_1^2 \leq \sigma_2^2$	$H_0: \sigma_1^2 \geq \sigma_2^2$	$H_0: \sigma_1^2 = \sigma_2^2$
$H_1: \sigma_1^2 > \sigma_2^2$	$H_1: \sigma_1^2 < \sigma_2^2$	$H_1: \sigma_1^2 \neq \sigma_2^2$

There are four key points to keep in mind when using the F test.

Notes for the Use of the F Test

1. The larger variance should always be designated as s_1^2 and be placed in the numerator of the formula.

$$F = \frac{s_1^2}{s_2^2}$$

2. For a two-tailed test, the α value must be divided by 2 and the critical value be placed on the right side of the F curve.
3. If the standard deviations instead of the variances are given in the problem, they must be squared for the formula for the F test.
4. When the degrees of freedom cannot be found in Table H, the closest value on the smaller side should be used.

Assumptions for Testing the Difference between Two Variances

1. The populations from which the samples were obtained must be normally distributed. (*Note:* The test should not be used when the distributions depart from normality.)
2. The samples must be independent of each other.

Remember also that in tests of hypotheses, the following five steps should be taken.

STEP 1 State the hypotheses and identify the claim.

STEP 2 Find the critical value.

STEP 3 Compute the test value.

STEP 4 Make the decision.

STEP 5 Summarize the results.

Example 10–5

A medical researcher wishes to see whether the variances of the heart rates (in beats per minute) of smokers are different from the variances of heart rates of people who do not smoke. Two samples are selected, and the data are as shown. Using $\alpha = 0.05$, is there enough evidence to support the claim?

Smokers	Nonsmokers
$n_1 = 26$	$n_2 = 18$
$s_1^2 = 36$	$s_2^2 = 10$

Solution

STEP 1 State the hypotheses and identify the claim.

$$H_0: \sigma_1^2 = \sigma_2^2 \quad \text{and} \quad H_1: \sigma_1^2 \neq \sigma_1^2 \text{ (claim)}$$

STEP 2 Find the critical value. Use the 0.025 table in Table H since $\alpha = 0.05$ and this is a two-tailed test. Here, d.f.N. $= 26 - 1 = 25$, and d.f.D. $= 18 - 1 = 17$. The critical value is 2.56 (d.f.N. $= 24$ was used). See Figure 10–7.

Figure 10–7

Critical Value for Example 10–5

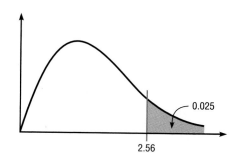

2.56

0.025

STEP 3 Compute the test value.

$$F = \frac{s_1^2}{s_2^2} = \frac{36}{10} = 3.6$$

STEP 4 Make the decision. Reject the null hypothesis, since $3.6 > 2.56$.

STEP 5 Summarize the results. There is enough evidence to support the claim that the variances are different.

Example 10–6

An instructor hypothesizes that the standard deviation of the final exam grades in her statistics class is larger for the male students than it is for the female students. The data from the final exam for the last semester are as shown. Is there enough evidence to support her claim, using $\alpha = 0.01$?

Males	Females
$n_1 = 16$	$n_2 = 18$
$s_1 = 4.2$	$s_2 = 2.3$

Solution

STEP 1 State the hypotheses and identify the claim.

$$H_0: \sigma_1^2 \leq \sigma_2^2 \quad \text{and} \quad H_1: \sigma_1^2 > \sigma_2^2 \text{ (claim)}$$

STEP 2 Find the critical value. Here, d.f.N. $= 16 - 1 = 15$, and d.f.D. $= 18 - 1 = 17$. From the 0.01 table, the critical value is 3.31.

STEP 3 Compute the test value.

$$F = \frac{s_1^2}{s_2^2} = \frac{(4.2)^2}{(2.3)^2} = 3.33$$

STEP 4 Make the decision. Reject the null hypothesis, since $3.33 > 3.31$.

STEP 5 Summarize the results. There is enough evidence to support the claim that the standard deviation of the final exam grades for the male students is larger than that for the female students.

Exercises

10–22. (**W**) What is the formula for the F test?

10–23. (**W**) When one is computing the F test value, what condition is placed on the variance that is in the numerator?

10–24. (**W**) Why is the critical region always on the right side in the use of the F test?

10–25. (**W**) What is one application of the F test?

10–26. (**W**) What are the two different degrees of freedoms associated with the F distribution?

10–27. (**W**) What are the characteristics of the F distribution?

10–28. (**W**) Using Table H, find the critical value for each.

a. sample 1: $\sigma_1^2 = 128, n_1 = 23$
    sample 2: $\sigma_2^2 = 162, n_2 = 16$
    two-tailed, $\alpha = 0.01$

b. sample 1: $\sigma_1^2 = 37, n_1 = 14$
    sample 2: $\sigma_2^2 = 89, n_2 = 25$
    right-tailed, $\alpha = 0.01$

c. sample 1: $\sigma_1^2 = 232, n_1 = 30$
    sample 2: $\sigma_2^2 = 387, n_2 = 46$
    two-tailed, $\alpha = 0.05$

d. sample 1: $\sigma_1^2 = 164, n_1 = 21$
    sample 2: $\sigma_2^2 = 53, n_2 = 17$
    two-tailed, $\alpha = 0.10$

e.   sample 1: $\sigma_1^2 = 92.8$, $n_1 = 11$
sample 2: $\sigma_2^2 = 43.6$, $n_2 = 11$
right-tailed, $\alpha = 0.05$

For Exercises 10–29 through 10–42, perform the following steps. Assume that all variables are normally distributed.

a.   State the hypotheses and identify the claim.
b.   Find the critical value.
c.   Compute the test value.
d.   Make the decision.
e.   Summarize the results.

10–29. An instructor claims that when a composition course is taught in conjunction with a word-processing course, the variance in the final grades will be larger than when the composition course is taught without the word-processing component. Two groups are randomly selected. The variance of the exams of the group that also had word-processing instruction is 103, and the variance of the exams of the students who did not have the word-processing component is 73. Each sample consists of 20 students. At $\alpha = 0.05$, can the instructor's claim be supported?

10–30. A consumer advocate claims that there is no difference in the variance of the number of hours that two companies' batteries will last. A sample of 10 batteries is selected from company X, and the variance of hours is 24. A sample of 10 batteries from company Y has a variance of 40. At $\alpha = 0.10$, can the consumer advocate conclude that there is no difference in the variance of the life of the batteries?

10–31. A researcher claims that the variance of the IQ scores of women who major in psychology is larger than the variance of the IQ scores of men who major in psychology. A sample of IQ scores of 22 women had a variance of 192, and a sample of IQ scores of 18 men had a variance of 84. At $\alpha = 0.05$, can the researcher conclude that the hypothesis is correct?

10–32. In the hospital study cited in Exercise 8–19, it was found that the standard deviation of the sound levels from 20 areas designated as "casualty doors" was 4.1 dBA and the standard deviation of 24 areas designated as operating theaters was 7.5 dBA. At $\alpha = 0.05$, can one substantiate the claim that there is no difference in the standard deviations?

Source: M. Bayo, A. Garcia, and A. Garcia, "Noise Levels in an Urban Hospital and Workers' Subjective Responses," *Archives of Environmental Health* 50, no. 3 (May/June 1995), p. 249.

10–33. A nurse claims that the variations of the lengths of newborn males is different from the variations of the lengths of newborn females. A sample of 15 newborn males is selected, and the standard deviation is 1.3 inches. The standard deviation of a sample of 15 newborn females is 0.9 inch. At $\alpha = 0.10$, can the nurse conclude that the variation of the lengths if different?

10–34. Two brands of refrigerators are selected and tested to determine whether there is a difference in the variation in temperatures. Each is set at 36°. After three hours, the standard deviation of the temperature of a sample of five brand A refrigerators is 1.6°, and the standard deviation of the temperatures of nine brand B refrigerators is 0.7°. At $\alpha = 0.05$, can one substantiate the claim that the standard deviations of the temperatures of both brands are equal?

10–35. A researcher claims that the variation of blood pressure of overweight individuals is greater than the variation of blood pressure of normal-weight individuals. The standard deviation of the pressures of 28 overweight people was found to be 6.2 mmHg, and the standard deviation of the pressures of 25 normal-weight people was 2.7 mmHg. At $\alpha = 0.01$, can the researcher conclude that the blood pressures of overweight individuals are more variable than those of individuals who are of normal weight?

10–36. An educator claims that the variation in the number of years of teaching experience of senior high school teachers is greater than the variation in the number of years of teaching experience of elementary school teachers. Two groups are randomly selected. The variance of the number of years of teaching experience of 18 elementary teachers is 1.9, and the variance of the years of experience of 26 senior high school teachers is 2.8. At $\alpha = 0.10$, can the educator conclude that the variation in the years of teaching experience of senior high school teachers is greater than the variation of the elementary school teachers?

10–37. A researcher wishes to test the variation in the number of pounds lost by women who follow two popular liquid diets. Ten women follow diet A for four months, and the standard deviation of the weight loss is 6.3 pounds. Twelve women follow diet B for four months, and the standard deviation of the weight loss is 4.8 pounds. At $\alpha = 0.05$, can the researcher substantiate the claim that the variation in pounds lost following diet A is greater than the variation in pounds lost following diet B?

10–38. The variations of stopping distances of two brands of automobile tires are tested. A sample of 25 Eagle Claw tires produces a standard deviation of stopping distance of 10.6 feet. A sample of 18 Mega Tread tires has a standard deviation of 14.2 feet. At $\alpha = 0.10$, can one substantiate the claim that there is no difference in the standard deviations of the stopping distances of the two brands of tires?

10–39. Two fast-food restaurants with drive-through windows are selected, and the standard deviations of the time it takes to fill the orders at the drive-through windows are compared. A sample of 10 orders selected from the first restaurant has a standard deviation of 3.5 minutes, and a sample of 12 orders from the second restaurant has a standard deviation of 4.8 minutes. At $\alpha = 0.01$, can one substantiate the claim that there is a difference in the standard deviations?

10–40. A researcher claims that the variation in the salaries of elementary school teachers is greater than the variation in the salaries of secondary school teachers. A sample of the salaries of 30 elementary school teachers has a variance of $8324, and a sample of the salaries of 30 secondary school teachers has a variance of $2862. At $\alpha = 0.05$, can the researcher conclude that the variation in the elementary school teachers' salaries is greater than the variation in the secondary teachers' salaries?

10–41. The weights in ounces of a sample of running shoes for men and women are shown below. Calculate the variances for each sample and test the claim that the variances are equal at $\alpha = 0.05$.

Men			Women		
11.9	10.4	12.6	10.6	10.2	8.8
12.3	11.1	14.7	9.6	9.5	9.5
9.2	10.8	12.9	10.1	11.2	9.3
11.2	11.7	13.3	9.4	10.3	9.5
13.8	12.8	14.5	9.8	10.3	11.0

Source: *Consumer Reports* May 1995, p. 316.

*10–42. Upright vacuum cleaners have either a hard body type or a soft body type. Shown below are the weights in pounds of a sample of each type. At $\alpha = 0.05$, can the claim that there is a difference in the variances of the weights of the two types be substantiated?

Hard Body Types				Soft Body Types			
21	17	17	20	24	13	11	13
16	17	15	20	12	15		
23	16	17	17				
13	15	16	18				
18							

Source: *Consumer Reports* January 1995, p. 46.

10–4

Testing the Difference between Two Means: Small Independent Samples

Objective 3. Test the difference between two means for small independent samples.

In Section 10–2, the z test was used to test the difference between two means when the population standard deviations were known and the variables were normally or approximately normally distributed, or when both sample sizes were greater than or equal to 30. In many situations, however, these conditions cannot be met—that is, the population standard deviations are not known, and one or both sample sizes are less than 30. In these cases, a t test is used to test the difference between means when the two samples are independent and when the samples are taken from two normally or approximately normally distributed populations. Samples are **independent** when they are not related.

There are actually two different options for the use of t tests. *One option is used when the variances of the populations are not equal, and the other option is used when the variances are equal.* To determine whether two sample variances are equal, the researcher can use an F test, as shown in the previous section.

Note, however, that not all statisticians are in agreement about using the F test before using the t test. Some believe that conducting the F and t tests at the same level of significance will change the overall level of significance of the t test. Their reasons are beyond the scope of this textbook.

Formulas for the *t* Tests—For Testing the Difference between Two Means—Small Samples

Variances are assumed to be unequal:

$$t = \frac{(\overline{X}_1 - \overline{X}_2) - (\mu_1 - \mu_2)}{\sqrt{\dfrac{s_1^2}{n_1} + \dfrac{s_2^2}{n_2}}}$$

where the degrees of freedom are equal to the smaller of $n_1 - 1$ or $n_2 - 1$.
Variances are assumed to be equal:

$$t = \frac{(\overline{X}_1 - \overline{X}_2) - (\mu_1 - \mu_2)}{\sqrt{\dfrac{(n_1 - 1)s_1^2 + (n_2 - 1)s_2^2}{n_1 + n_2 - 2}}\sqrt{\dfrac{1}{n_1} + \dfrac{1}{n_2}}}$$

where the degrees of freedom are equal to $n_1 + n_2 - 2$.

When the variances are unequal, the first formula

$$t = \frac{(\overline{X}_1 - \overline{X}_2) - (\mu_1 - \mu_2)}{\sqrt{\dfrac{s_1^2}{n_1} + \dfrac{s_2^2}{n_2}}}$$

follows the format of

$$\text{test value} = \frac{(\text{observed value}) - (\text{expected value})}{\text{standard error}}$$

where $\overline{X}_1 - \overline{X}_2$ is the observed difference between sample means and where the expected value $\mu_1 - \mu_2$ is equal to zero when no difference between population means is hypothesized. The denominator $\sqrt{(s_1^2/n_1) + (s_2^2/n_2)}$ is the standard error of the difference between two means. Since mathematical derivation of the standard error is somewhat complicated, it will be omitted here.

When the variances are assumed to be equal, the second formula

$$t = \frac{(\overline{X}_1 - \overline{X}_2) - (\mu_1 - \mu_2)}{\sqrt{\dfrac{(n_1 - 1)s_1^2 + (n_2 - 1)s_2^2}{n_1 + n_2 - 2}}\sqrt{\dfrac{1}{n_1} + \dfrac{1}{n_2}}}$$

also follows the format of

$$\text{test value} = \frac{(\text{observed value}) - (\text{expected value})}{\text{standard error}}$$

For the numerator, the terms are the same as in the first formula. However, a note of explanation is needed for the denominator of the second test statistic. Since both populations are assumed to have the same variance, the standard error is computed with what is called a pooled estimate of the variance. A **pooled estimate of the variance** is a weighted average of the variance using the two sample variances and the *degrees of freedom* of each variance as the weights. Again, since the algebraic derivation of the standard error is somewhat complicated, it is omitted.

Example 10–7

The average size of a farm in Greene County, Pennsylvania, is 199 acres, and the average size of a farm in Indiana County, Pennsylvania, is 191 acres. Assume the data were obtained from two samples with standard deviations of 12 acres and 38 acres, respectively, and the sample sizes are 10 farms from Greene County and 8 farms in Indiana County. Can it be concluded at $\alpha = 0.05$ that the average size of the farms in the two counties is different? Assume the populations are normally distributed.

Source: *Pittsburgh Tribune-Review,* August 28, 1994.

Solution

The approach here will be to use the F test to determine whether or not the variances are equal. The null hypothesis is that the variances are equal.

The critical value for the F test found in Table H (Appendix C) for $\alpha = 0.05$ is 4.20, since there are 7 and 9 degrees of freedom. (Note: Use the 0.025 table.) The test value is

$$F = \frac{s_1^2}{s_2^2} = \frac{38^2}{12^2} = 10.03$$

Since $10.03 > 4.20$, the decision is to reject the null hypothesis and conclude that the variances are not equal. Hence the first formula is used to test the difference between the means, as shown next.

STEP 1   State the hypotheses and identify the claim for the means.

$$H_0: \mu_1 = \mu_2 \quad \text{and} \quad H_1: \mu_1 \neq \mu_2 \text{ (claim)}$$

STEP 2   Find the critical values. Since the test is two-tailed, since $\alpha = 0.05$, and since the variances are unequal, the degrees of freedom are the smaller of $n_1 - 1$ or $n_2 - 1$. In this case, the degrees of freedom are $8 - 1 = 7$. Hence, from Table F, the critical values are $+2.365$ and -2.365.

STEP 3   Compute the test value. Since the variances are unequal, use the first formula.

$$t = \frac{(\bar{X}_1 - \bar{X}_2) - (\mu_1 - \mu_2)}{\sqrt{\dfrac{s_1^2}{n_1} + \dfrac{s_2^2}{n_2}}} = \frac{(199 - 191) - 0}{\sqrt{\dfrac{12^2}{10} + \dfrac{38^2}{8}}} = 0.57$$

STEP 4   Make the decision. Do not reject the null hypothesis, since $0.57 < 2.365$. See Figure 10–8.

Figure 10–8

Critical and Test Values for Example 10–7

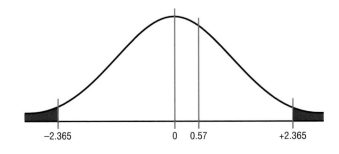

−2.365	0 0.57	+2.365

STEP 5   Summarize the results. There is not enough evidence to support the claim that the average size of the farms is different.

....

Example 10–8

A researcher wishes to determine whether the salaries of professional nurses employed by private hospitals are higher than those of nurses employed by government-owned hospitals. She selects a sample of nurses from each type of hospital and calculates the means and standard deviations of their salaries. At $\alpha = 0.01$, can she conclude that the private hospitals pay more than the government hospitals? Assume that the populations are approximately normally distributed.

Private	Government
$\overline{X}_1 = \$26{,}800$	$\overline{X}_2 = \$25{,}400$
$s_1 = \$600$	$s_2 = \$450$
$n_1 = 10$	$n_2 = 8$

Solution

The F test will be used to determine whether or not the variances are equal. The null hypothesis is that the variances are equal.

The critical value obtained from Table H for $\alpha = 0.01$ is 8.51, using 9 and 7 degrees of freedom.

The test value is

$$F = \frac{s_1^2}{s_2^2} = \frac{600^2}{450^2} = 1.78$$

Since $1.78 < 8.51$, the decision is do not reject the null hypothesis and conclude that the variances are equal. Hence the second formula is used to test the difference between the two means, as shown next.

STEP 1   State the hypotheses and identify the claim.

$$H_0\text{: } \mu_1 \leq \mu_2 \quad \text{and} \quad H_1\text{: } \mu_1 > \mu_2 \text{ (claim)}$$

STEP 2   Find the critical value. Since $\alpha = 0.01$ and the test is a one-tailed right test with equal variances, the degrees of freedom are $n_1 + n_2 - 2$, which is $10 + 8 - 2 = 16$. The critical value is $+2.583$.

STEP 3   Compute the test value. Use the second formula, since the variances are assumed to be equal.

$$t = \frac{(\overline{X}_1 - \overline{X}_2) - (\mu_1 - \mu_2)}{\sqrt{\dfrac{(n_1 - 1)s_1^2 + (n_2 - 1)s_2^2}{n_1 + n_2 - 2}}\sqrt{\dfrac{1}{n_1} + \dfrac{1}{n_2}}}$$

$$= \frac{(26{,}800 - 25{,}400) - 0}{\sqrt{\dfrac{(10 - 1)(600)^2 + (8 - 1)(450)^2}{10 + 8 - 2}}\sqrt{\dfrac{1}{10} + \dfrac{1}{8}}}$$

$$= 5.47$$

STEP 4   Make the decision. Reject the null hypothesis, since $5.47 > 2.583$. See Figure 10–9.

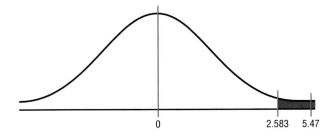

STEP 5   Summarize the results. There is enough evidence to support the claim that the salaries paid to nurses employed by private hospitals are higher than those paid to nurses employed by government-owned hospitals.

....

P-values are found from Table F the same way as shown in Section 9–4. Using d.f. $= 16$, the test value 5.47 falls beyond 0.005; hence, the null hypothesis is rejected.

Confidence intervals can also be found for the difference between two means with the following formulas:

Confidence Intervals for the Difference of Two Means: Small Independent Samples

Variances unequal

$$(\bar{X}_1 - \bar{X}_2) - t_{\alpha/2}\sqrt{\frac{s_1^2}{n_1} + \frac{s_2^2}{n_2}} < \mu_1 - \mu_2$$

$$< (\bar{X}_1 - \bar{X}_2) + t_{\alpha/2}\sqrt{\frac{s_1^2}{n_1} + \frac{s_2^2}{n_2}}$$

d.f. = the smaller value of $n_1 - 1$ or $n_2 - 1$

Variances equal

$$(\bar{X}_1 - \bar{X}_2) - t_{\alpha/2}\sqrt{\frac{(n_1 - 1)s_1^2 + (n_2 - 1)s_2^2}{n_1 + n_2 - 2}} \cdot \sqrt{\frac{1}{n_1} + \frac{1}{n_2}}$$

$$< \mu_1 - \mu_2 < (\bar{X}_1 - \bar{X}_2) + t_{\alpha/2}\sqrt{\frac{(n_1 - 1)s_1^2 + (n_2 - 1)s_2^2}{n_1 + n_2 - 2}} \cdot \sqrt{\frac{1}{n_1} + \frac{1}{n_2}}$$

d.f. $= n_1 + n_2 - 2$

Remember that when one is testing the difference between two independent means, two different statistical test formulas can be used. One formula is used when the variances are equal, the other when the variances are not equal. As shown in Section 10–3, some statisticians use an F test to determine whether two variances are equal.

Example 10–9

Find the 95% confidence interval for the data in Example 10–7.

Solution

Substitute in the formula

$$(\overline{X}_1 - \overline{X}_2) - t_{\alpha/2} \sqrt{\frac{s_1^2}{n_1} + \frac{s_2^2}{n_2}} < \mu_1 - \mu_2$$

$$< (\overline{X}_1 - \overline{X}_2) + t_{\alpha/2} \sqrt{\frac{s_1^2}{n_1} + \frac{s_2^2}{n_2}}$$

$$(199 - 191) - 2.365 \sqrt{\frac{12^2}{10} + \frac{38^2}{8}} < \mu_1 - \mu_2$$

$$< (199 - 191) + 2.365 \sqrt{\frac{12^2}{10} + \frac{38^2}{8}}$$

$$-25.02 < \mu_1 - \mu_2 < 41.02$$

Since 0 is contained in the interval, the decision is not to reject the null hypothesis, H_0: $\mu_1 = \mu_2$.

Computer Application for t tests for the Difference of Two Means: Independent Samples

MINITAB   Two sample independent t tests:

1. Enter the data into C1 and C2.
2. Click Stat > Basic Statistics > 2-Sample t.
3. Click Samples in different columns and press Tab.
4. Highlight C1 in the dialog box; hit Select to enter it into the first text box. Press Tab.
5. Highlight C2 in the dialog box; hit Select to enter it into the second text box. Press Tab.
6. Select the appropriate Alternative hypothesis. Press Tab.
7. Click Assume equal variances if necessary.
8. Click OK.

Example: Test the claim H_0: $\mu_1 = \mu_2$ and assume equal variances using the following data. Use $\alpha = 0.05$.

Data:

C1	12	15	18	13	10	9	6	3	11	5
C2	18	20	13	7	9	4				

MINITAB printout for this example:

Two Sample T-Test and Confidence Interval

```
Two sample T for C1 vs C2
        N      Mean     StDev    SE Mean
C1     10     10.20      4.64        1.5
C2      6     11.83      6.31        2.6
```

```
95% CI for mu C1 - mu C2: ( -7.5,  4.2)
T-Test mu C1 = mu C2 (vs not =): T = -0.60 P = 0.56 DF = 14
Both use Pooled StDev = 5.29
```

Summary: In this case, the test value is -0.60, and the *P*-value is 0.56. The decision is not to reject the null hypothesis at $\alpha = 0.05$.

Exercises

For Exercises 10–43 through 10–53, perform each of the following steps. Assume that all variables are normally or approximately normally distributed. Be sure to test for equality of variance first.

a. State the hypotheses and identify the claim.
b. Find the critical value(s).
c. Compute the test value.
d. Make the decision.
e. Summarize the results.

10–43. A real estate agent wishes to determine whether tax assessors and real estate appraisers agree on the values of homes. A random sample of the two groups appraised 10 homes. The data are shown here. Is there a significant difference in the values of the homes for each group? Let $\alpha = 0.05$. Find the 95% confidence interval for the difference of the means.

Real estate appraisers	Tax assessors
$\overline{X}_1 = \$83{,}256$	$\overline{X}_2 = \$88{,}354$
$s_1 = \$3256$	$s_2 = \$2341$
$n_1 = 10$	$n_2 = 10$

10–44. A researcher suggests that male nurses earn more than female nurses. A survey of 16 male nurses and 20 female nurses reports the following data. Is there enough evidence to support the claim that male nurses earn more than female nurses? Use $\alpha = 0.05$.

Male	Female
$\overline{X}_1 = \$23{,}800$	$\overline{X}_2 = \$23{,}750$
$s_1 = \$300$	$s_2 = \$250$
$n_1 = 16$	$n_2 = 20$

10–45. An instructor thinks that math majors can write and debug computer programs faster than business majors. A sample of 12 math majors took an average of 36 minutes to write a specific program and debug it; a sample of 18 business majors took an average of 39 minutes. The standard deviations were 4 minutes and 9 minutes, respectively. At $\alpha = 0.10$, is there evidence to support the claim?

10–46. A researcher estimates that high school girls miss more days of school than high school boys. A sample of 16 girls showed that they missed an average of 3.9 days of school per school year; a sample of 22 boys showed that they missed an average of 3.6 days of school per year. The standard deviations are 0.6 and 0.8 respectively. At $\alpha = 0.01$, is there enough evidence to support the researcher's claim?

10–47. A leading manufacturer claims that his brand of television (brand A) will cost less to repair over a five-year period than any other brand. A researcher selects a sample of 10 sets of brand A and 10 sets of another leading brand (brand B) and finds the average repair costs over a five-year period. The data are shown here. At $\alpha = 0.01$, is there enough evidence to support the manufacturer's claim?

Brand A	Brand B
$\overline{X}_1 = \$896$	$\overline{X}_2 = \$902$
$s_1 = \$14$	$s_2 = \$16$
$n_1 = 10$	$n_2 = 10$

10–48. An automotive engineer wishes to see whether a new automobile will get more miles per gallon with high-octane gasoline than with regular octane. The engineer selects two samples of cars and uses regular gasoline for one and high-octane for the other. At $\alpha = 0.05$, do the cars using the high-octane gasoline get better mileage?

High-octane	Regular
$\overline{X}_1 = 26.7$	$\overline{X}_2 = 23$
$s_1 = 2.1$	$s_2 = 1.6$
$n_1 = 5$	$n_2 = 12$

10–49. The local branch of the Internal Revenue Service spent an average of 21 minutes helping each of 10 people prepare their tax returns. The standard deviation was 5.6 minutes. A volunteer tax preparer spent an average of 27 minutes helping 14 people prepare their taxes. The standard deviation was

4.3 minutes. At $\alpha = 0.02$, is there a difference in the average time spent by the two services? Find the 98% confidence interval for the two means.

10–50.   The average price of seven ABC dishwashers was $815, and the average price of nine XYZ dishwashers was $845. The standard deviations were $19 and $9, respectively. At $\alpha = 0.05$, can one conclude that the XYZ dishwashers cost more?

10–51.   The average monthly premium paid by 12 administrators for hospitalization insurance is $56. The standard deviation is $3. The average monthly premium paid by 27 nurses is $63. The standard deviation is $5.75. At $\alpha = 0.05$, do the nurses pay more for hospitalization insurance?

10–52.   Health Care Knowledge Systems reported that an insured woman spends on average 2.3 days in the hospital for a routine childbirth, while an uninsured woman spends on average 1.9 days. Assume two samples of 16 women each were used and the standard deviations are both equal to 0.6 days. At $\alpha = 0.01$, test the claim that the means are equal. Find the 99% confidence interval for the differences of the means.

Source: Michael D. Shook and Robert L. Shook, *The Book of Odds* (New York: Plume, 1991).

*10–53.   The times (in minutes) it took six white mice to learn to run a simple maze and the times it took six brown mice to learn to run the same maze are given here. At $\alpha = 0.05$, does the color of the mice make a difference in their learning rate? Find the 95% confidence interval for the difference of the means.

White Mice	18	24	20	13	15	12
Brown Mice	25	16	19	14	16	10

10–5

Testing the Difference between Two Means: Small Dependent Samples

Objective 4. Test the difference between two means for small dependent samples.

In the previous section, the *t* test was used to compare two sample means when the samples were independent. In this section, a different version of the *t* test is discussed. This version is used when the samples are dependent. Samples are considered to be **dependent samples** when the subjects are paired or matched in some way.

For example, suppose a medical researcher wants to see whether a drug will affect the reaction time of its users. To test this hypothesis, the researcher must pretest the subjects in the sample first. That is, they are given a test to ascertain their normal reaction times. Then after taking the drug, the subjects are tested again, using a posttest. Finally, the means of the two tests are compared to see whether there is a difference. Since the same subjects are used in both cases, the samples are *related;* subjects scoring high on the pretest will generally score high on the posttest, even after consuming the drug. Likewise, those scoring lower on the pretest will tend to score lower on the posttest. In order to take this effect into account, the researcher employs a *t* test using the differences between the pretest values and the posttest values. This way makes sure only the gain or loss in values is compared.

Here are some other examples of dependent samples. A researcher may want to design an SAT preparation course to help students raise their test scores the second time they take the SAT exam. Hence, the differences between the two exams are compared. A medical specialist may want to see whether a new counseling program will help subjects lose weight. Therefore, the preweights of the subjects will be compared with the postweights.

Besides samples in which the same subjects are used in a prepost situation, there are other cases where the samples are considered dependent. For example, students might be matched or paired according to some variable that is pertinent to the study; then one student is assigned to one group and the other student is assigned to a second group. For instance, in a study involving learning, students can be selected and paired according to their IQs. That is, two students with the same IQ will be paired. Then one will be assigned to one sample group (which

On the basis of the conclusions in this study, state several hypotheses that may have been used to support these conclusions. Comment on how you think these hypotheses were tested. For example, how would one determine whether standardized tests are biased against girls?

Public School Teachers Show Bias for Boys

By Katy Kelly
USA TODAY

Public schools favor boys, says a report out today.

"There is an illusion that schools are treating boys and girls equally," says Alice McKee, American Association of University Women, which commissioned the report.

Based on a review of recent research, it suggests:

• Teachers ask academic questions of boys 80% more often than of girls.
• School curricula generally ignore or stereotype females.
• Most standardized tests are biased against girls.

• Preschool boys get more individual attention, even hugs, from teachers than girls do.

Bias is not intentional: "When teachers are made aware they're doing this . . . they are willing and eager to change," McKee says.

The report lists 40 recommendations, including requiring course work on gender issues for teacher certification.

New classroom materials are needed too, McKee says: "Boys and girls should be able to study women Nobel Prize winners in addition to Betsy Ross sewing the flag."

might receive instruction by computers), and the other student will be assigned to another sample group (which might receive instruction by the lecture-discussion method). These assignments will be done randomly. Since a student's IQ is important to learning, it is a variable that should be controlled. By matching subjects on IQ, the researcher can eliminate the variable's influence, for the most part. Matching, then, helps to reduce type II error by eliminating extraneous variables.

Two notes of caution should be mentioned. First, when subjects are matched according to one variable, the matching process does not eliminate the influence of other variables. Matching students according to IQ does not account for their mathematical ability or their familiarity with computers. Since all variables influencing a study cannot be controlled, it is up to the researcher to determine which variables should be used in matching. Second, when the same subjects are used for a prepost study, sometimes the knowledge that they are participating in a study can influence the results. For example, if people are placed in a special program, they may be more highly motivated to succeed simply because they have been selected to participate; the program itself may have little effect on their success.

When the samples are dependent, a special *t* test for dependent means is used. This test employs the difference in values of the matched pairs. The hypotheses are as follows, where μ_D is the symbol for the expected mean of the differences of the matched pairs.

Two-tailed	Left-tailed	Right-tailed
$H_0: \mu_D = 0$	$H_0: \mu_D \geq 0$	$H_0: \mu_D \leq 0$
$H_1: \mu_D \neq 0$	$H_1: \mu_D < 0$	$H_1: \mu_D > 0$

The general procedure for finding the test value involves several steps. First, find the differences of the values of the pairs of data.

$$D = X_1 - X_2$$

Second, find the mean $(\overline{D})$ of the differences, using the formula

$$\overline{D} = \frac{\Sigma D}{n}$$

where n = the number of data pairs.

Third, find the standard deviation (s_D) of the differences, using the formula

$$s_D = \sqrt{\frac{\Sigma D^2 - \frac{(\Sigma D)^2}{n}}{n - 1}}$$

Fourth, find the estimated standard error $(s_{\overline{D}})$ of the differences, which is

$$s_{\overline{D}} = \frac{s_D}{\sqrt{n}}$$

Finally, find the test value, using the formula

$$t = \frac{\overline{D} - \mu_D}{\frac{s_D}{\sqrt{n}}} \text{ with d.f.} = n - 1$$

The formula in the final step follows the basic format of

$$\text{test value} = \frac{(\text{observed value}) - (\text{expected value})}{\text{standard error}}$$

where the observed value is the mean of the differences. The expected value μ_D is zero if the hypothesis is $\mu_D = 0$. The standard error of the difference is the standard deviation of the difference divided by the square root of the sample size. Both populations must be normally or approximately normally distributed. Example 10–10 illustrates the hypothesis-testing procedure in detail.

Example 10–10

A physical education director claims by taking a special vitamin, a weight lifter can increase his strength. Eight athletes are selected and given a test of strength, using the standard bench press. After two weeks of regular training, supplemented with the vitamin, they are tested again. Test the effectiveness of the vitamin regimen at $\alpha = 0.05$. Each value in the data that follow represents the maximum number of pounds the athlete can bench press. Assume that the variable is approximately normally distributed.

Athlete	1	2	3	4	5	6	7	8
Before (X_1)	210	230	182	205	262	253	219	216
After (X_2)	219	236	179	204	270	250	222	216

Solution

STEP 1   State the hypothesis and identify the claim. In order for the vitamin to be effective, the "before" weights must be significantly less than the "after" weights; hence, the mean of the differences must be less than zero.

$$H_0\text{: }\mu_D \geq 0 \qquad \text{and} \qquad H_1\text{: }\mu_D < 0 \text{ (claim)}$$

STEP 2   Find the critical value. The degrees of freedom are $n - 1$. In this case, d.f. $= 8 - 1 = 7$. The critical value for a left-tailed test with $\alpha = 0.05$ is -1.895.

STEP 3   Compute the test value.

   a.   Make a table.

Before (X_1)	After (X_2)	A $D = (X_1 - X_2)$	B $D^2 = (X_1 - X_2)^2$
210	219		
230	236		
182	179		
205	204		
262	270		
253	250		
219	222		
216	216		

   b.   Find the differences and place the results in column A.

$$
\begin{aligned}
210 - 219 &= -9 \\
230 - 236 &= -6 \\
182 - 179 &= +3 \\
205 - 204 &= +1 \\
262 - 270 &= -8 \\
253 - 250 &= +3 \\
219 - 222 &= -3 \\
216 - 216 &= \underline{0} \\
\Sigma D &= -19
\end{aligned}
$$

   c.   Find the mean of the differences.

$$\overline{D} = \frac{\Sigma D}{n} = \frac{-19}{8} = -2.375$$

   d.   Square the differences and place the results in column B.

$$
\begin{aligned}
(-9)^2 &= 81 \\
(-6)^2 &= 36 \\
(+3)^2 &= 9 \\
(+1)^2 &= 1 \\
(-8)^2 &= 64 \\
(+3)^2 &= 9 \\
(-3)^2 &= 9 \\
0^2 &= \underline{0} \\
\Sigma D^2 &= 209
\end{aligned}
$$

The completed table is shown next.

Before (X_1)	After (X_2)	A $D = (X_1 - X_2)$	B $D^2 = (X_1 - X_2)^2$
210	219	−9	81
230	236	−6	36
182	179	+3	9
205	204	+1	1
262	270	−8	64
253	250	+3	9
219	222	−3	9
216	216	0	0
		$\Sigma D = -19$	$\Sigma D^2 = 209$

e. Find the standard deviation of the differences.

$$s_D = \sqrt{\dfrac{\Sigma D^2 - \dfrac{(\Sigma D)^2}{n}}{n-1}} = \sqrt{\dfrac{209 - \dfrac{(-19)^2}{8}}{8-1}} = 4.84$$

f. Find the test value.

$$t = \dfrac{\overline{D} - \mu_D}{\dfrac{s_D}{\sqrt{n}}} = \dfrac{-2.375 - 0}{\dfrac{4.84}{\sqrt{8}}} = -1.388$$

STEP 4   Make the decision. The decision is not to reject the null hypothesis at $\alpha = 0.05$, since $-1.388 > -1.895$, as shown in Figure 10–10.

Figure 10–10

Critical and Test Values for Example 10–10

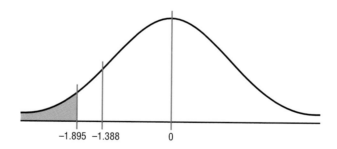

−1.895 −1.388   0

STEP 5   Summarize the results. There is not enough evidence to support the claim that the vitamin increases the strength of weight lifters.

The formulas for this t test are summarized next.

Formulas for the t Test for Dependent Samples

$$t = \dfrac{\overline{D} - \mu_D}{\dfrac{s_D}{\sqrt{n}}}$$

with d.f. $= n - 1$ and where

$$\overline{D} = \dfrac{\Sigma D}{n} \quad \text{and} \quad s_D = \sqrt{\dfrac{\Sigma D^2 - \dfrac{(\Sigma D)^2}{n}}{n-1}}$$

....

| Example 10–11 |

A dietitian wishes to see if a person's cholesterol level will change if the diet is supplemented by a certain mineral. Six subjects were pretested and then took the mineral supplement for a six-week period. The results are shown in the table. (Cholesterol level is measured in milligrams per deciliter.) Can it be concluded that the cholesterol level has been changed at $\alpha = 0.10$? Assume the variable is approximately normally distributed.

Subject	1	2	3	4	5	6
Before (X_1)	210	235	208	190	172	244
After (X_2)	190	170	210	188	173	228

Solution

STEP 1 State the hypotheses and identify the claim. If the diet is effective, the "before" cholesterol levels should be different from the "after" levels.

$H_0: \mu_D = 0$ and $H_1: \mu_D \neq 0$ (claim)

STEP 2 Find the critical value. The degrees of freedom are 5. At $\alpha = 0.10$, the critical values are ± 2.015.

STEP 3 Compute the test value.

a. Make a table.

Before (X_1)	After (X_2)	**A** $D = (X_1 - X_2)$	**B** $D^2 = (X_1 - X_2)^2$
210	190		
235	170		
208	210		
190	188		
172	173		
244	228		

b. Find the differences and place the results in column A.

$$210 - 190 = 20$$
$$235 - 170 = 65$$
$$208 - 210 = -2$$
$$190 - 188 = 2$$
$$172 - 173 = -1$$
$$244 - 228 = \underline{16}$$
$$\Sigma D = 100$$

c. Find the mean of the difference.

$$\overline{D} = \frac{\Sigma D}{n} = \frac{100}{6} = 16.7$$

d. Square the differences and place the results in column B.

$$(20)^2 = 400$$
$$(65)^2 = 4225$$
$$(-2)^2 = 4$$
$$(2)^2 = 4$$
$$(-1)^2 = 1$$
$$(16)^2 = 256$$
$$\Sigma D^2 = 4890$$

Then complete the table as shown.

Before (X_1)	After (X_2)	A $D = (X_1 - X_2)$	B $D^2 = (X_1 - X_2)^2$
210	190	20	400
235	170	65	4225
208	210	−2	4
190	188	2	4
172	173	−1	1
244	228	16	256
		$\Sigma D = 100$	$\Sigma D^2 = 4890$

e. Find the standard deviation of the differences.

$$s_D = \sqrt{\frac{\Sigma D^2 - \dfrac{(\Sigma D)^2}{n}}{n - 1}} = \sqrt{\frac{4890 - \dfrac{(100)^2}{6}}{5}} = 25.39$$

f. Find the test value.

$$t = \frac{\overline{D} - \mu_D}{\dfrac{s_D}{\sqrt{n}}} = \frac{16.7 - 0}{\dfrac{25.39}{\sqrt{6}}} = 1.611$$

STEP 4 Make the decision. The decision is not to reject the null hypothesis, since the test value 1.611 is in the noncritical region, as shown in Figure 10–11.

Figure 10–11

Critical and Test Values for Example 10–11

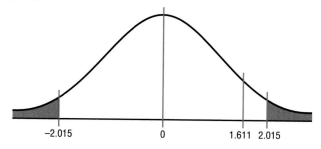

−2.015   0   1.611 2.015

STEP 5 Summarize the results. There is not enough evidence to support the claim that the mineral changes a person's cholesterol level.

The steps for this t test are summarized in Procedure Table 8.

Procedure Table 8

Testing the Difference between Means for Dependent Samples

STEP 1: State the hypotheses and identify the claim.

STEP 2: Find the critical value(s).

STEP 3: Compute the test value.

  a. Make a table, as shown.

		A	B
X_1	X_2	$D = (X_1 - X_2)$	$D^2 = (X_1 - X_2)^2$
$\vdots$	$\vdots$		
		$\Sigma D = $ _____	$\Sigma D^2 = $ _____

  b. Find the differences and place the results in column A.

$$D = (X_1 - X_2)$$

  c. Find the mean of the differences.

$$\overline{D} = \frac{\Sigma D}{n}$$

  d. Square the differences and place the results in column B. Complete the table.

$$D^2 = (X_1 - X_2)^2$$

  e. Find the standard deviation of the differences.

$$s_D = \sqrt{\frac{\Sigma D^2 - \dfrac{(\Sigma D)^2}{n}}{n - 1}}$$

  f. Find the test value.

$$t = \frac{\overline{D} - \mu_D}{\dfrac{s_D}{\sqrt{n}}}$$

STEP 4: Make the decision.

STEP 5: Summarize the results.

If a specific difference is hypothesized, the following formula should be used:

$$t = \frac{\overline{D} - \mu_D}{s_D/\sqrt{n}}$$

where μ_D is the hypothesized difference.

For example, if a dietitian claims that people on a specific diet will lose an average of 3 pounds in a week, the hypotheses are

$$H_0: \mu_D = 3 \quad \text{and} \quad H_1: \mu_D \neq 3$$

The value 3 will be substituted in the test statistic for μ_D.

P-values for the *t* test are found in Table F. For a right-tailed test with d.f. = 5 and $t = 1.611$, the *P*-value is found between 1.476 and 2.015; hence, $0.05 < p < 0.10$. Thus, the null hypothesis can be rejected at $\alpha = 0.10$.

Confidence intervals can be found for the mean differences with the following formula.

Confidence Interval for the Mean Difference

$$\overline{D} - t_{\alpha/2}\frac{s_D}{\sqrt{n}} < \mu_D < \overline{D} + t_{\alpha/2}\frac{s_D}{\sqrt{n}}$$

$$\text{d.f.} = n - 1$$

Example 10–12

Find the 90% confidence interval for the data in Example 10–11.

Solution

Substitute in the formula

$$\overline{D} - t_{\alpha/2}\frac{s_D}{\sqrt{n}} < \mu_D < \overline{D} + t_{\alpha/2}\frac{s_D}{\sqrt{n}}$$

$$16.7 - 2.015 \cdot \frac{25.4}{\sqrt{6}} < \mu_D < 16.7 + 2.015 \cdot \frac{25.4}{\sqrt{6}}$$

$$16.7 - 20.89 < \mu_D < 16.7 + 20.89$$

$$-4.19 < \mu_D < 37.59$$

Since 0 is contained in the interval, the decision is not to reject the null hypothesis, $H_0: \mu_D = 0$.

Computer Applications for *t* Test for Dependent Samples

MINITAB   Two sample dependent *t* tests:

1. Type the data into C1 and C2.
2. Click on Calc > Calculator.
3. Type C3 in the Store result in Variable box and press Tab.
4. Type C1 − C2 in the Expression box.
5. Click on OK.
6. Choose Stat > Basic Statistics > 1-Sample *t*.
7. Highlight C3 to enter it into Variables text box and click on Select.
8. Click on Test mean. Press Tab.
9. Type 0 for the value of the hypothesized mean. Press Tab.
10. Select appropriate alternative hypothesis and Click on OK.

In this study, the same subjects were used to determine if music helped in exercise workouts. What type of test may have been used to compare the groups? Also, the study found that men increased their time by 30% and women by 25%. How might these groups be compared? Based on the information in the article, do you think people should work out to music? Explain your answer.

Wired for a Workout: Music May Help You Last Longer

We don't know whether it has charms to soothe a savage breast or not, but there is evidence that music may boost your workout.

Twenty-six people pedaled longer on stationary bikes when working out to music than when they were cycling to just the rhythms in their heads. Men moving to music increased their time on the bikes by almost 30 percent, and women were able to go about 25 percent longer before feeling exhausted (*Physical Therapy,* May 1994).

All subjects exercised to the sounds of their own choosing—mostly rock-and-roll for these college-age subjects. Study co-author Esther Haskvitz, Ph.D., P.T., assistant professor of physical therapy at Springfield College, Springfield, Massachusetts, suspects that there's nothing magic about rock—the key is choosing something you enjoy hearing.

These researchers didn't turn up any new music-to-muscle connection that prolongs endurance. "The effect may have been mostly due to the distraction of the music," says Dr. Haskvitz. Although this study didn't test it, she says that listening to the radio or informational tapes may help a workout as well. For some people, knowing that they'll get the news may not only boost their time to exhaustion but it may also motivate them to get moving in the first place.

Example: Test the claim H_0: $\mu_D = 0$ when

A	33	35	28	29	32	34	30	34
B	27	29	36	34	30	29	28	24

Use $\alpha = 0.05$.

MINITAB printout for this example:

```
T-Test of the Mean

Test of mu = 0.00 vs mu not = 0.00

Variable    N    Mean    StDev    SE Mean      T        P
C3          8    2.25    6.02       2.13     1.06     0.33
```

Summary: Since the test value is 1.06, and the *P*-value is 0.33, the decision is not to reject the null hypothesis.

Exercises

10–54. (**W**) Explain the difference between independent and dependent samples.

10–55. Classify each as independent or dependent samples.
a. Heights of identical twins
b. Test scores of the same students in English and psychology
c. The effectiveness of two different brands of aspirin
d. Effects of a drug on reaction time, measured by a "before" and an "after" test
e. The effectiveness of two different diets on two different groups of individuals

For Exercises 10–56 through 10–64, perform each of the following steps. Assume that all variables are normally or approximately normally distributed.

a. State the hypotheses and identify the claim.
b. Find the critical value(s).
c. Compute the test value.
d. Make the decision.
e. Summarize the results.

10–56. A program for reducing the number of days missed by food handlers in a certain restaurant chain was conducted. The owners hypothesized that after the program, the workers would miss fewer days of work due to illness. The table shows the number of days 10 workers missed per month before and after completing the program. Is there enough evidence to support the claim, at $\alpha = 0.05$, that the food handlers missed fewer days after the program?

Before	2	3	6	7	4	5	3	1	0	0
After	1	4	3	8	3	3	1	0	1	0

10–57. As an aid for improving students' study habits, nine students were randomly selected to attend a seminar on the importance of education in life. The table shows the number of hours each student studied per week before and after the seminar. At $\alpha = 0.10$, did attending the seminar increase the number of hours the students studied per week?

Before	9	12	6	15	3	18	10	13	7
After	9	17	9	20	2	21	15	22	6

10–58. A doctor is interested in determining whether a film about exercise will change 10 people's attitudes about exercise. The results of his questionnaire are shown below. A higher numerical value shows a more favorable attitude toward exercise. Is there enough evidence to support the claim, at $\alpha = 0.05$, that there was a change in attitude? Find the 95% confidence interval for the difference of the two means.

Before	12	11	14	9	8	6	8	5	4	7
After	13	12	10	9	8	8	7	6	5	5

10–59. Suppose a researcher wishes to test the effects of a new diet designed to reduce the blood sodium level of 10 patients. The patients' sodium levels are checked before they are placed on the diet and after two weeks on the diet. At $\alpha = 0.05$, is there enough evidence to conclude that the diet is effective in lowering the sodium content of their blood?

Patient	Before	After
1	146	135
2	138	133
3	152	147
4	163	156
5	136	138
6	147	141
7	148	139
8	141	132
9	143	138
10	142	131

10–60. An office manager wishes to see whether the typing speed of 10 secretaries can be increased by changing over to computers. The number of words typed per minute is given here. At $\alpha = 0.10$, is there enough evidence to conclude that by using computers, the secretaries can type more words per minute?

Secretary	Typewriter	Computer
1	63	68
2	72	80
3	85	95
4	97	93
5	82	80
6	101	106
7	73	82
8	62	78
9	58	65
10	75	83

10–61. A composition teacher wishes to see whether a new grammar program will reduce the number of grammatical errors her students make when writing a two-page essay. The data are shown here. At $\alpha = 0.025$, can it be concluded that the number of errors has been reduced?

Student	1	2	3	4	5	6
Errors before	12	9	0	5	4	3
Errors after	9	6	1	3	2	3

10–62.   A sports-shoe manufacturer claims that joggers who wear its brand of shoe will jog faster than those who don't. A sample of eight joggers is taken, and they agree to test the claim on a 1-mile track. The rates (in minutes) of the joggers while wearing the manufacturer's shoe and while wearing any other brand of shoe are shown here. Test the claim at $\sigma = 0.025$.

Runner	1	2	3	4	5	6	7	8
Manufacturer's brand	8.2	6.3	9.2	8.6	6.8	8.7	8.0	6.9
Other brand	7.1	6.8	9.8	8.0	5.8	8.0	7.4	8.0

10–63.   A researcher wanted to compare the pulse rates of identical twins to see whether there was any difference. Eight sets of twins were selected. The rates are given in the table as number of beats per minute. At

$\alpha = 0.01$, is there a significant difference in the average pulse rates of twins? Find the 99% confidence interval for the difference of the two.

Twin A	87	92	78	83	88	90	84	93
Twin B	83	95	79	83	86	93	80	86

10–64.   A physician claims that a person's diastolic blood pressure can be lowered if the person listens to a special relaxation tape each evening. Ten patients are pretested and then given the tape, which is to be played each evening for one week. Then their blood pressure readings are taken again. The data are shown in the table (in millimeters of mercury). At $\alpha = 0.025$, can the physician conclude that using the tape may lower a person's blood pressure?

Patient	1	2	3	4	5	6	7	8	9	10
Before	86	92	95	84	80	78	98	95	94	96
After	84	83	81	78	82	74	86	83	80	82

10–65.   Instead of finding the mean of the differences between X_1 and X_2 by subtracting $X_1 - X_2$, one can find it by finding the means of X_1 and X_2 and then subtracting the means. Show that these two procedures will yield the same results.

10–6

Testing the Difference between Proportions

Objective 5. Test the difference between two proportions.

The z test with some modifications can be used to test the equality of two proportions. For example, a researcher might ask: Is the proportion of men who exercise regularly less than the proportion of women who exercise regularly? Is there a difference in the percentage of students who own a personal computer and the percentage of nonstudents who own one? Is there a difference in the proportion of college graduates who pay cash for purchases and the proportion of noncollege graduates who pay cash?

Recall from Chapter 8 that the symbol $\hat{p}$ ("p hat") is the sample proportion used to estimate the population proportion, denoted by p. For example, if in a sample of 30 college students, 9 are on probation, then the sample proportion is $\hat{p} = \frac{9}{30}$, or 0.3. The population proportion p is the number of all students who are on probation divided by the number of students who attend the college. The formula for $\hat{p}$ is

$$\hat{p} = \frac{X}{n}$$

where

X = number of units that possess the characteristic of interest
n = sample size

When one is testing the difference between two population proportions, p_1 and p_2, the hypotheses can be stated as follows, if no difference between the proportions is hypothesized.

$$H_0: p_1 = p_2 \quad \text{or} \quad H_0: p_1 - p_2 = 0$$
$$H_1: p_1 \neq p_2 \qquad\quad H_1: p_1 - p_2 \neq 0$$

Speaking of | **STATISTICS**

In the study below, 50 sets of twins were used. Why do you think twins were used? What type of statistics might have been used to determine the nature of the results? What are some risks associated with the drugs used in the study?

A Pill to Remember: Do Better Brain Cells Come in a Bottle?

If you wonder why you can remember next week's engagements but your sister needs a reminder, scientists say it may have to do with your anti-inflammatory drugs. According to a recent study, these may help prevent gaps in the mental database that signals Alzheimer's disease.

This study of 50 sets of twins took stock of their drug-taking history and compared that with when one or both of the twins had been overtaken by Alzheimer's. Habitual use of anti-inflammatories—prednisone, cortisone shots, ACTH, aspirin or non-steroidal anti-inflammatories—was associated with the delay or prevention Alzheimer's disease (*Neurology*, February 1994).

For a disease that has notoriously eluded understanding, "this finding has two broad implications," says study leader John C. Breitner, M.D., M.P.H., chairperson of the division of geriatric psychiatry, Duke University Medical Center, Durham, North Car-

olina. "If more research confirms this link, this could turn out to be an important therapeutic advance. It could also offer a very important clue to the biology of the disease." If Alzheimer's turns out to have to do with inflammation, this could up-end current theories about how the disease develops and lead to new therapies.

The strongest association between these drugs and diminished risk of the disease was in people who took the strongest anti-inflammatories for diseases like arthritis. "These drugs can have side effects, such as gastrointestinal hemorrhage, so they're not to be taken lightly," says Dr. Breitner.

The study is potentially good news for people already on these drugs, but it shouldn't be interpreted as a green light to start popping aspirin or other inflammation-resolving pills on your own. "If you have any concerns about Alzheimer's, consult your physician," he says. "He or she will be able to assess the risks and benefits of any

Reprinted by permission of *Prevention* Magazine. Copyright 1995 Rodale Press, Inc. All rights reserved.
Note: This excerpt is for reading comprehension only. For the most up-to-date information about health, please consult a health professional.

Similar statements using $\geq$ and $<$ or $\leq$ and $>$ can be formed for one-tailed tests.

For two proportions, $\hat{p}_1 = X_1/n_1$ is used to estimate p_1 and $\hat{p}_2 = X_2/n_2$ is used to estimate p_2. The standard error of difference is

$$\sigma_{(\hat{p}_1 - \hat{p}_2)} = \sqrt{\sigma_{p_1}^2 + \sigma_{p_2}^2} = \sqrt{\frac{p_1 q_1}{n_1} + \frac{p_2 q_2}{n_2}}$$

where $\sigma_{p_1}^2$ and $\sigma_{p_2}^2$ are the variances of the proportions, $q_1 = 1 - p_1$, $q_2 = 1 - p_2$, and n_1 and n_2 are the respective sample sizes.

Since p_1 and p_2 are unknown, a weighted estimate of p can be computed by using the formula

$$\bar{p} = \frac{n_1 \hat{p}_1 + n_2 \hat{p}_2}{n_1 + n_2}$$

and $\bar{q} = 1 - \bar{p}$. This weighted estimate is based on the hypothesis that $p_1 = p_2$. Hence, $\bar{p}$ is a better estimate than either $\hat{p}_1$ or $\hat{p}_2$, since it is a combined average using both $\hat{p}_1$ and $\hat{p}_2$.

Since $\hat{p}_1 = X_1/n_1$ and $\hat{p}_2 = X_2/n_2$, $\bar{p}$ can be simplified to

$$\bar{p} = \frac{X_1 + X_2}{n_1 + n_2}$$

Finally, the standard error of difference in terms of the weighted estimate is

$$\sigma_{(\hat{p}_1 - \hat{p}_2)} = \sqrt{\bar{p}\,\bar{q}\left(\frac{1}{n_1} + \frac{1}{n_2}\right)}$$

The formula for the test value is shown next.

Formula for the z Test for Comparing Two Proportions

$$z = \frac{(\hat{p}_1 - \hat{p}_2) - (p_1 - p_2)}{\sqrt{\bar{p}\,\bar{q}\left(\dfrac{1}{n_1} + \dfrac{1}{n_2}\right)}}$$

where
$$\bar{p} = \frac{X_1 + X_2}{n_1 + n_2} \qquad \hat{p}_1 = \frac{X_1}{n_1}$$

$$\bar{q} = 1 - \bar{p} \qquad \hat{p}_2 = \frac{X_2}{n_2}$$

This formula follows the format

$$\text{test value} = \frac{(\text{observed value}) - (\text{expected value})}{\text{standard error}}$$

There are two requirements for use of the z test: (1) The samples must be independent of each other, and (2) $n_1 p_1$ and $n_1 q_1$ must be 5 or more and $n_2 p_2$ and $n_2 q_2$ must be 5 or more.

....

Example 10–13

In the nursing home study mentioned in the chapter-opening "Statistics Today," the researchers found that 12 out of 34 small nursing homes had a resident vaccination rate of less than 80%, while 17 out of 24 large nursing homes had a vaccination rate of less than 80%. At $\alpha = 0.05$, can it be concluded that there is no difference in the proportions of the small and large nursing homes with a resident vaccination rate of less than 80%?

Source: Nancy Arden, Arnold S. Monto, and Suzanne E. Ohmit, "Vaccine Use and the Risk of Outbreaks in a Sample of Nursing Homes during an Influenza Epidemic," *American Journal of Public Health* 85, no. 3 (March 1995), pp. 399–401.

Solution

Let $\hat{p}_1$ be the proportion of the small nursing homes with a vaccination rate of less than 80% and $\hat{p}_2$ be the proportion of the large nursing homes with a vaccination rate of less than 80%. Then

$$\hat{p}_1 = \frac{X_1}{n_1} = \frac{12}{34} = 0.35 \quad \text{and} \quad \hat{p}_2 = \frac{X_2}{n_2} = \frac{17}{24} = 0.71$$

$$\bar{p} = \frac{X_1 + X_2}{n_1 + n_2} = \frac{12 + 17}{34 + 24} = \frac{29}{58} = 0.5$$

$$\bar{q} = 1 - \bar{p} = 1 - 0.5 = 0.5$$

Now, follow the steps in hypothesis testing.

STEP 1   State the hypotheses and identify the claim.

$$H_0: p_1 = p_2 \text{ (claim)} \quad \text{and} \quad H_1: p_1 \neq p_2$$

STEP 2   Find the critical values. Since $\alpha = 0.05$, the critical values are $+1.96$ and -1.96.

STEP 3   Compute the test value.

$$z = \frac{(\hat{p}_1 - \hat{p}_2) - (p_1 - p_2)}{\sqrt{\bar{p}\,\bar{q}\left(\dfrac{1}{n_1} + \dfrac{1}{n_2}\right)}}$$

$$= \frac{(0.35 - 0.71) - 0}{\sqrt{(0.5)(0.5)\left(\dfrac{1}{34} + \dfrac{1}{24}\right)}} = \frac{-0.36}{0.1333} = -2.7$$

STEP 4   Make the decision. Reject the null hypothesis, since $-2.7 < -1.96$. See Figure 10-12.

Figure 10-12

Critical and Test Values
for Example 10-13

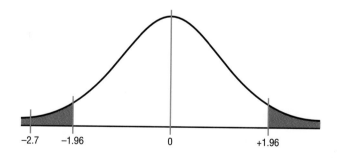

-2.7   -1.96   0   +1.96

STEP 5   Summarize the results. There is enough evidence to reject the claim that there is no difference in the proportions.

Example 10-14

A sample of 50 randomly selected men with high triglyceride levels consumed 2 tablespoons of oat bran daily for six weeks. After six weeks, 60% of the men had lowered their triglyceride level. A sample of 80 men consumed 2 tablespoons of wheat bran for six weeks. After six weeks, 25% had lower triclyceride levels. Is there a significant difference in the two proportions, at the 0.01 significance level?

Solution

Since the statistics are given in percentages, $\hat{p}_1 = 60\%$, or 0.60, and $\hat{p}_2 = 25\%$, or 0.25. In order to compute $\bar{p}$, one must find X_1 and X_2.

$$X_1 = (0.60)(50) = 30 \qquad X_2 = (0.25)(80) = 20$$

$$\bar{p} = \frac{X_1 + X_2}{n_1 + n_2} = \frac{30 + 20}{50 + 80} = \frac{50}{130} = 0.385$$

$$\bar{q} = 1 - \bar{p} = 1 - 0.385 = 0.615$$

STEP 1 State the hypotheses and identify the claim.

$$H_0: p_1 = p_2 \qquad \text{and} \qquad H_1: p_1 \neq p_2 \text{ (claim)}$$

STEP 2 Find the critical values. Since $\alpha = 0.01$, the critical values are $+2.58$ and -2.58.

STEP 3 Compute the test value.

$$z = \frac{(\hat{p}_1 - \hat{p}_2) - (p_1 - p_2)}{\sqrt{\bar{p}\,\bar{q}\left(\dfrac{1}{n_1} + \dfrac{1}{n_2}\right)}}$$

$$= \frac{(0.60 - 0.25) - 0}{\sqrt{(0.385)(0.615)\left(\dfrac{1}{50} + \dfrac{1}{80}\right)}} = 3.99$$

STEP 4 Make the decision. Reject the null hypothesis, since $3.99 > 2.58$. See Figure 10–13.

Figure 10–13

Critical and Test Values for Example 10–14

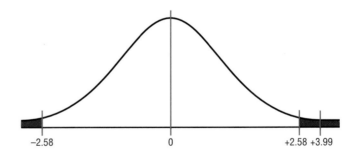

$-2.58 \qquad 0 \qquad +2.58 \; +3.99$

STEP 5 Summarize the results. There is enough evidence to support the claim that there is a difference in proportions.

The P-value for the difference of proportions can be found from Table E, as shown in Section 9–3. For Example 10–14, 3.99 is beyond 3.09; hence, the null hypothesis can be rejected since the P-value is less than 0.001.

The formula for the confidence interval for the difference between two proportions is shown next.

Confidence Interval for the Difference between Two Proportions

$$(\hat{p}_1 - \hat{p}_2) - z_{\alpha/2}\sqrt{\frac{\hat{p}_1\hat{q}_1}{n_1} + \frac{\hat{p}_2\hat{q}_2}{n_2}} < (p_1 - p_2) < (\hat{p}_1 - \hat{p}_2) + z_{\alpha/2}\sqrt{\frac{\hat{p}_1\hat{q}_1}{n_1} + \frac{\hat{p}_2\hat{q}_2}{n_2}}$$

Example 10–15

Find the 95% confidence interval for the difference of proportions for the data in Example 10–13.

Solution

$$\hat{p}_1 = \frac{12}{34} = 0.35 \quad \hat{q}_1 = 0.65$$

$$\hat{p}_2 = \frac{17}{24} = 0.71 \quad \hat{q}_2 = 0.29$$

Substitute in the formula.

$$(\hat{p}_1 - \hat{p}_2) - z_{\alpha/2}\sqrt{\frac{\hat{p}_1\hat{q}_1}{n_1} + \frac{\hat{p}_2\hat{q}_2}{n_2}} < (p_1 - p_2) <$$

$$(\hat{p}_1 - \hat{p}_2) + z_{\alpha/2}\sqrt{\frac{\hat{p}_1\hat{q}_1}{n_1} + \frac{\hat{p}_2\hat{q}_2}{n_2}}$$

$$(0.35 - 0.71) - 1.96\sqrt{\frac{(0.35)(0.65)}{34} + \frac{(0.71)(0.29)}{24}}$$

$$< (p_1 - p_2) < (0.35 - 0.71) + 1.96\sqrt{\frac{(0.35)(0.65)}{34} + \frac{(0.71)(0.29)}{24}}$$

$$-0.36 - 0.242 < (p_1 - p_2) < -0.36 + 0.242$$

$$-0.602 < (p_1 - p_2) < -0.118$$

Since 0 is not contained in the interval, the decision is to reject the null hypothesis, H_0: $p_1 = p_2$.

Exercises

10–66. Find the proportions $\hat{p}$ and $\hat{q}$ for each.
a. $n = 48$, $X = 34$
b. $n = 75$, $X = 28$
c. $n = 100$, $X = 50$
d. $n = 24$, $X = 6$
e. $n = 144$, $X = 12$

10–67. Find each X given $\hat{p}$.
a. $\hat{p} = 0.16$, $n = 100$
b. $\hat{p} = 0.08$, $n = 50$
c. $\hat{p} = 6\%$, $n = 80$
d. $\hat{p} = 52\%$, $n = 200$
e. $\hat{p} = 20\%$, $n = 150$

10–68. Find $\bar{p}$ and $\bar{q}$ for each.
a. $X_1 = 60$, $n_1 = 100$, $X_2 = 40$, $n_2 = 100$
b. $X_1 = 22$, $n_1 = 50$, $X_2 = 18$, $n_2 = 30$
c. $X_1 = 18$, $n_1 = 60$, $X_2 = 20$, $n_2 = 80$
d. $X_1 = 5$, $n_1 = 32$, $X_2 = 12$, $n_2 = 48$
e. $X_1 = 12$, $n_1 = 75$, $X_2 = 15$, $n_2 = 50$

For Exercises 10–69 through 10–80, perform the following steps.

a. State the hypotheses and identify the claim.
b. Find the critical value(s).
c. Compute the test value.
d. Make the decision.
e. Summarize the results.

10–69. A sample of 150 people from a certain industrial community showed that 80 people suffered from a lung disease. A sample of 100 people from a rural community showed that 30 suffered from the same lung disease. At $\alpha = 0.05$, is there a difference between the proportion of people who suffer from the disease in the two communities?

10–70. A recent survey showed that in a sample of 80 surgeons, 15 smoked; in a sample of 50 general practitioners, 5 smoked. At $\alpha = 0.05$, is the proportion of surgeons who smoke higher than the proportion of general practitioners who smoke?

10–71. In a sample of 100 store customers, 43 used a MasterCard. In another sample of 100, 58 used a Visa card. At $\alpha = 0.05$, is there a difference in the proportion of people who use each type of credit card?

10–72. In Cleveland, a sample of 73 mail carriers showed that 10 had been bitten by an animal during one week. In Philadelphia, in a sample of 80 mail carriers, 16 had received animal bites. Is there a significant difference in the proportions? Use $\alpha = 0.05$. Find the 95% confidence interval for the difference of the two proportions.

10–73. A survey found that 83% of the men questioned preferred computer-assisted instruction to lecture, and 75% of the women preferred computer-assisted instruction to lecture. There were 100 individuals in each sample. At $\alpha = 0.05$, test the claim that there is no difference in the proportion of men and the proportion of women who favor computer-assisted instruction over lecture. Find the 95% confidence interval for the difference of the two proportions.

10–74. In a sample of 200 surgeons, 15% thought the government should control health care. In a sample of 200 general practitioners, 21% felt this way. At $\alpha = 0.10$, is there a difference in the proportions? Find the 90% confidence interval for the difference of the two proportions.

10–75. In a sample of 80 Americans, 55% wished that they were rich. In a sample of 90 Europeans, 45% wished that they were rich. At $\alpha = 0.01$, is there a difference in the proportions? Find the 99% confidence interval for the difference of the two proportions.

10–76. In a sample of 200 men, 130 said they used seat belts. In a sample of 300 women, 63 said they used seat belts. Test the claim that men are more safety-conscious than women, at $\alpha = 0.01$.

10–77. A survey of 80 homes in a Washington, D.C., suburb showed that 45 were air conditioned. A sample of 120 homes in a Pittsburgh suburb showed that 63 had air conditioning. At $\alpha = 0.05$, is there a difference

in the two proportions? Find the 95% confidence interval for the difference of the two proportions.

10–78. A recent study showed that in a sample of 100 people, 30% had visited Disneyland. In another sample of 100 people, 24% had visited Disney World. Are the proportions of people who visited each park different? Use $\alpha = 0.02$.

10–79. A sample of 200 teenagers shows that 50 believe that war is inevitable, and a sample of 300 people over 60 shows that 93 believe war is inevitable. Is the proportion of teenagers who believe war is inevitable different from the proportion of people over 60 who do? Use $\alpha = 0.01$. Find the 99% confidence interval for the difference of the two proportions.

10–80. In a sample of 50 high school seniors, eight had their own cars. In a sample of 75 college freshmen, 20 had their own cars. At $\alpha = 0.05$, can it be concluded that a higher proportion of college freshmen have their own cars?

*10–81. If there is a significant difference between p_1 and p_2 and between p_2 and p_3, can one conclude that there is a significant difference between p_1 and p_3?

10–82. Find the 95% confidence interval for the true difference in proportions for the data of a study in which 40% of the 200 males surveyed opposed the death penalty and 56% of the 100 females surveyed opposed the death penalty.

10–83. Find the 99% confidence interval for the difference in the population proportions for the data of a study in which 80% of the 150 Republicans surveyed favored the bill for a salary increase and 60% of the 200 Democrats surveyed favored the bill for a salary increase.

10–7

Summary

Many times, researchers are interested in comparing two population parameters, such as means or proportions. This comparison can be accomplished using special z and t tests. If the samples are independent and the variances are known, the z test is used. The z test is also used when the variances are unknown but both sample sizes are 30 or more. If the variances are not known and one or both sample sizes are less than 30, the t test must be used. For independent samples, a further requirement is that one must determine whether the variances of the populations are equal.

The F-test is used to determine whether or not the variances are equal. Different formulas are used in each case. If the samples are dependent, the t test for dependent samples is used. Finally, a z test is used to compare two proportions.

Speaking of   STATISTICS

In this study, two proportions are compared. Do you think there is a significant difference between the proportions? What other information would be necessary to test the significance of the difference? State a possible null and a possible alternative hypothesis for the study.

Clear Landings: How to Tame Airplane Ear Pain

Preparation for landing should begin long before you bring your seat backs forward and return your trays to their full upright and locked positions. New research suggests you get ready at least 30 minutes before your flight leaves the ground, if you suffer from ear pain. That's when you should pop a decongestant that contains pseudoephedrine.

In a study of 190 fliers with recurrent ear pain, only 32 percent of the people who received pseudoephedrine before takeoff had ear pain, compared to 62 percent in the placebo group. This research confirms anecdotal reports of the effectiveness of this pain-prevention strategy (*Annals of Emergency Medicine,* June 1994).

Decongestants dilate the eustachian tubes and make it easier to clear the ears of the pressure built up inside. The medicine also helps decrease secretions that might block the tubes, says study co-author Jeffrey Jones, M.D., director of the department of emergency medicine at Butterworth Hospital (Grand Rapids, Michigan). Prevention is essential since time is the only thing that can take away that full-ear feeling of unequal pressure once it occurs. One dose should last all day—so you don't have to keep popping decongestants if your flight is delayed. The only side effect seen was drowsiness.

Researchers warn against this decongestant if you have thyroid disease, heart disease, high blood pressure, diabetes or an enlarged prostate. In these cases Dr. Jones recommends other ear-clearing techniques: Gum chewing or yawning are perennial favorites. But the Valsalva maneuver—holding your nose, closing your mouth and trying to exhale—is a no-no. "You're forcing air through tubes that may already be blocked, and you can actually injure your ears," says Dr. Jones. A better way would be the Frenzel maneuver: Pinch your nose and push your tongue against the back part of the roof of your mouth. That works a little air through the eustachian tubes without damage, he says.

Now that your ears are clear, try this with your baby: Wait until the plane is starting to descend to give him a bottle. (The action of sucking on the bottle can help clear his ears.) It helps the ears even more to keep the child upright while he's feeding. For older kids, gum, candy or lozenges keep them doing ear-clearing swallows. Or if they've complained about ear pain before, decongestants for kids could be in order with your doctor's approval.

Important Terms

Dependent samples 399

F distribution 385

F test 385

Independent
samples 392

Pooled estimate of the
variance 393

Important Formulas

Formula for the z test for comparing two means from independent populations:

$$z = \frac{(\overline{X}_1 - \overline{X}_2) - (\mu_1 - \mu_2)}{\sqrt{\dfrac{\sigma_1^2}{n_1} + \dfrac{\sigma_2^2}{n_2}}}$$

Formula for the confidence interval (large samples):

$$(\overline{X}_1 - \overline{X}_2) - (z_{\alpha/2})\sqrt{\frac{\sigma_1^2}{n_1} + \frac{\sigma_2^2}{n_2}} < \mu_1 - \mu_2 <$$

$$(\overline{X}_1 - \overline{X}_2) + (z_{\alpha/2})\sqrt{\frac{\sigma_1^2}{n_1} + \frac{\sigma_2^2}{n_2}}$$

Formula for the F test for comparing two variances:

$$F = \frac{s_1^2}{s_2^2}$$

Formula for the t test for comparing two means (independent samples, variances not equal):

$$t = \frac{(\overline{X}_1 - \overline{X}_2) - (\mu_1 - \mu_2)}{\sqrt{\dfrac{s_1^2}{n_1} + \dfrac{s_2^2}{n_2}}}$$

and d.f. = the smaller of $n_1 - 1$ or $n_2 - 1$.

Formula for the t test for comparing two means (independent samples, variances equal):

$$t = \frac{(\overline{X}_1 - \overline{X}_2) - (\mu_1 - \mu_2)}{\sqrt{\dfrac{(n_1 - 1)s_1^2 + (n_2 - 1)s_2^2}{(n_1 + n_2 - 2)}} \sqrt{\dfrac{1}{n_1} + \dfrac{1}{n_2}}}$$

and d.f. = $n_1 + n_2 - 2$.

Formula for the confidence interval for difference of two means (small independent samples, variances unequal):

$$(\overline{X}_1 - \overline{X}_2) - t_{\alpha/2} \sqrt{\frac{s_1^2}{n_1} + \frac{s_2^2}{n_2}} < \mu_1 - \mu_2 <$$

$$(\overline{X}_1 - \overline{X}_2) + t_{\alpha/2} \sqrt{\frac{s_1^2}{n_1} + \frac{s_2^2}{n_2}}$$

and d.f. = smaller of $n_1 - 1$ and $n_2 - 1$.

Formula for the confidence interval for difference of two means (small independent samples, variances equal):

$$(\overline{X}_1 - \overline{X}_2) - t_{\alpha/2} \sqrt{\frac{(n_1 - 1)s_1^2 + (n_2 - 1)s_2^2}{n_1 + n_2 - 2}} \cdot \sqrt{\frac{1}{n_1} + \frac{1}{n_2}}$$

$$< \mu_1 - \mu_2 <$$

$$(\overline{X}_1 - \overline{X}_2) + t_{\alpha/2} \sqrt{\frac{(n_1 - 1)s_1^2 + (n_2 - 1)s_2^2}{n_1 + n_2 - 2}} \cdot \sqrt{\frac{1}{n_1} + \frac{1}{n_2}}$$

and d.f. = $n_1 + n_2 - 2$.

Formula for the t test for comparing two means from dependent samples:

$$t = \frac{\overline{D} - \mu_D}{\dfrac{s_D}{\sqrt{n}}}$$

where $\overline{D}$ is the mean of the differences,

$$\overline{D} = \frac{\Sigma D}{n}$$

and s_D is the standard deviation of the differences,

$$s_D = \sqrt{\frac{\Sigma D^2 - \dfrac{(\Sigma D)^2}{n}}{n - 1}}$$

Formula for confidence interval for the mean of the difference for dependent samples:

$$\overline{D} - t_{\alpha/2} \frac{s_D}{\sqrt{n}} < \mu_D < \overline{D} + t_{\alpha/2} \frac{s_D}{\sqrt{n}}$$

and d.f. = $n - 1$.

Formula for the z test for comparing two proportions:

$$z = \frac{(\hat{p}_1 - \hat{p}_2) - (p_1 - p_2)}{\sqrt{\overline{p}\,\overline{q}\left(\dfrac{1}{n_1} + \dfrac{1}{n_2}\right)}}$$

where $\overline{p} = \dfrac{X_1 + X_2}{n_1 + n_2} \qquad \hat{p}_1 = \dfrac{X_1}{n_1}$

$\overline{q} = 1 - \overline{p} \qquad \hat{p}_2 = \dfrac{X_2}{n_2}$

Formula for confidence interval for the difference of two proportions:

$$(\hat{p}_1 - \hat{p}_2) - z_{\alpha/2} \sqrt{\frac{\hat{p}_1 \hat{q}_1}{n_1} + \frac{\hat{p}_2 \hat{q}_2}{n_2}} < (p_1 - p_2) <$$

$$(\hat{p}_1 - \hat{p}_2) + z_{\alpha/2} \sqrt{\frac{\hat{p}_1 \hat{q}_1}{n_1} + \frac{\hat{p}_2 \hat{q}_2}{n_2}}$$

Review Exercises

For each problem, perform the following steps. Assume that all variables are normally or approximately normally distributed.

a. State the hypotheses and identify the claim.
b. Find the critical value(s).
c. Compute the test value.
d. Make the decision.
e. Summarize the results.

10–84.   The average annual cost of automobile insurance in 1992 for residents of North Carolina was $541.07, while for residents of Indiana it was $584.17. Test the claim at $\alpha = 0.10$ that there is no difference in the means for both states. Assume samples of 100 residents were used and the standard deviation was $81 for both samples. Find the 90% confidence interval for the difference in the means.

Source: *In Sync* (Erie Insurance, Erie, PA), Fall 1995.

10–85. Two groups of people are surveyed. In a sample of 50 drivers who are single people, the drivers average 106 miles per week for pleasure trips. In a sample of 65 married people, the drivers average 68 miles per week for pleasure trips. The sample standard deviations are 15 and 9 miles, respectively. At $\alpha = 0.01$, can it be concluded that single people do more driving for pleasure trips than married people?

10–86. An educator wishes to compare the variances of the amount of money spent per pupil in two states. The data are given below. At $\alpha = 0.05$, is there a significant difference in the variances of the amounts the states spend per pupil?

State 1	State 2
$s_1^2 = \$585$	$s_2^2 = \$261$
$n_1 = 18$	$n_2 = 16$

10–87. In the hospital study cited in Exercise 8–19, the standard deviation of the noise levels of the 11 intensive care units was 4.1 dBA and the standard deviation of the noise levels of 24 nonmedical care areas, such as kitchens and machine rooms, was 13.2. At $\alpha = 0.10$, is there a significant difference between the standard deviations of these two areas?

Source: M. Bayo, A. Garcia, and A. Garcia, "Noise Levels in an Urban Hospital and Workers' Subjective Responses," *Archives of Environmental Health* 50, no. 3 (May/June 1995), p. 249.

10–88. A researcher wants to compare the variances of the heights (in inches) of major league baseball players with those of players in the minor leagues. A sample of 25 players from each league is selected, and the variances of the heights for each league are 2.25 and 4.85, respectively. At $\alpha = 0.10$, is there a significant difference between the variances of the heights for the two leagues?

10–89. A traffic safety commissioner believes the variation in the number of speeding tickets given on Route 19 is larger than the variation in the number of speeding tickets given on Route 22. Ten weeks are randomly selected; the standard deviation of the number of tickets issued for Route 19 is 6.3, and the standard deviation of the number of tickets issued for Route 22 is 2.8. At $\alpha = 0.05$, can the commissioner conclude that the variance of speeding tickets issued on Route 19 is greater than the variance of speeding tickets issued on Route 22?

10–90. The variations in the number of absentees per day in two schools are being compared. A sample of 30 days is selected; the standard deviation of the number of absentees in school A is 4.9, and for school B it is 2.5. At $\alpha = 0.01$, can one conclude that there is a difference in the two standard deviations?

10–91. A researcher claims that the variation in the number of days factory workers miss per year due to illness is greater than the variation in the number of days hospital workers miss per year. A sample of 42 workers from a large hospital has a standard deviation of 2.1 days, and a sample of 65 workers from a large factory has a standard deviation of 3.2 days. Test the claim, at $\alpha = 0.10$.

10–92. The average price of 15 cans of tomato soup from different stores is $0.73, and the standard deviation is $0.05. The average price of 24 cans of chicken noodle soup is $0.91, and the standard deviation is $0.03. At $\alpha = 0.01$, is there a significant difference in price?

10–93. The average temperature over a 25-day period in June for Birmingham, Alabama, is 78°, and the standard deviation is 3°. The average temperature over the same 25 days for Chicago, Illinois, is 75°, with a standard deviation of 5°. At $\alpha = 0.01$, can one conclude that it is warmer in Birmingham?

10–94. A sample of 15 teachers from Rhode Island has an average salary of $35,270, with a standard deviation of $3256. A sample of 30 teachers from New York has an average salary of $29,512, with a standard deviation of $1432. Is there a significant difference in teachers' salaries between the two states? Use $\alpha = 0.02$. Find the 99% confidence interval for the difference of the two means.

10–95. The average income of 16 families who reside in a large metropolitan city is $54,356, and the standard deviation is $8256. The average income of 12 families who reside in a suburb of the same city is $46,512, with a standard deviation of $1311. At $\alpha = 0.05$, can one conclude that the income of the families who reside within the city is greater than that of those who reside in the suburb?

10–96. In an effort to improve the vocabulary of 10 students, a teacher provides a weekly one-hour tutoring session for them. A pretest is given before the sessions and a posttest is given afterward. The results are shown in the table. At $\alpha = 0.01$, can the teacher conclude that the tutoring sessions helped to improve the students' vocabulary?

Before	1	2	3	4	5	6	7	8	9	10
Pretest	83	76	92	64	82	68	70	71	72	63
Posttest	88	82	100	72	81	75	79	68	81	70

10–97. In an effort to increase production of an automobile part, the factory manager decides to play music in the manufacturing area. Eight workers are selected, and the number of items each produced for a specific day is recorded. After one week of music, the same workers are monitored again. The data are given in the following table. At $\alpha = 0.05$, can the manager conclude that the music has increased production?

Worker	1	2	3	4	5	6	7	8
Before	6	8	10	9	5	12	9	7
After	10	12	9	12	8	13	8	10

10–98. St. Petersburg, Russia, has 207 foggy days out of 365 days while Stockholm, Sweden, has 166 foggy days out of 365. At $\alpha = 0.02$, can it be concluded that the proportions of foggy days for the two cities are different? Find the 98% confidence interval for the difference of the two proportions.

Source: Jack Williams, *USA Today, 1995: The Weather Almanac* (New York: Vantage Books, 1994), p. 355.

10–99. In a recent survey of 50 apartment residents, 32 had microwave ovens. In a survey of 60 homeowners, 24 had microwave ovens. At $\alpha = 0.05$, can one conclude that the proportions are equal? Find the 95% confidence interval for the difference of the two proportions.

Statistics Today

To Vaccinate or Not to Vaccinate? Small or Large?   Revisited

Using a *t* test to compare two proportions, the researchers found that the proportion of residents in smaller nursing homes who were vaccinated (80.8%) was statistically greater than that of residents in large nursing homes who were vaccinated (68.7%). Using statistical methods presented in later chapters, they also found that the larger size of the nursing home and the lower frequency of vaccination were significant predictions of influenza outbreaks in nursing homes.

Data Analysis

The Data Bank is found in Appendix D.

1. From the Data Bank, select a variable and compare the mean of the variable for a random sample of at least 30 men with the mean of the variable for the random sample of at least 30 women. Use a *z* test.

2. Repeat the experiment in Exercise 1 using a different variable and two samples of size 15. Compare the means by using a *t* test. Assume that the variances are equal.

3. Compare the proportion of men who are smokers with the proportion of women who are smokers. Use the data in the Data Bank. Choose random samples of size 30 or more. Use the *z* test for proportions.

Quiz

Determine whether each statement is true or false. If the statement is false, explain why.

1. When one is testing the difference between two means for small samples, it is not important to distinguish whether or not the samples are independent of each other.

2. If the same diet is given to two groups of randomly selected individuals, the samples are considered to be dependent.

3. When computing the *F* test value, one always places the larger variance in the numerator of the fraction.

4. Tests for variances are always two-tailed.

Select the best answer.

5. To test the equality of two variances, one would use a(n) _____ test.

a. z
b. t
c. chi-square
d. F

6. To test the equality of two proportions, one would use a(n) _____ test.
a. z   *c.* chi-square
b. t   *d.* F

7. The mean value of the F is approximately equal to
a. 0
b. 0.5
c. 1
d. It cannot be determined.

8. What test can be used to test the difference between two small sample means?
a. z   *c.* chi-square
b. t   *d.* F

Complete the following statements with the best answer.

9. If one hypothesizes that there is no difference between means, this is represented as H_0: _____.

10. When one is testing the difference between two means, a _____ estimate of the variances is used when the variances are equal.

11. When the t test is used for testing the equality of two means, the populations must be _____.

12. The values of F cannot be _____.

13. The formula for the F test for variances is _____.

For each of the following problems, perform the following steps.
a. State the hypotheses.
b. Find the critical value(s).
c. Compute the test value.
d. Make the decision.
e. Summarize the results.

14. A researcher wishes to see if there is a difference in the cholesterol levels of two groups of men. A random sample of 30 men between the ages of 25 and 40 is selected and tested. The average level is 223. A second sample of 25 men between the ages of 41 and 56 is selected and tested. The average of this group is 229. The population standard deviation for both groups is 6. At $\alpha = 0.01$, is there a difference in the cholesterol levels between the two groups? Find the 99% confidence interval for the difference of the two means.

15. Two groups of people are surveyed. It is found that in a sample of 40 drivers, single men drive an average of 85 work-related miles per week, while a sample of 45 married men shows that they travel an average of 80 work-related miles per week. The standard deviations are 12 and 8 miles, respectively. At $\alpha = 0.01$, test the claim that single men do more work-related driving than married men.

16. A politician wishes to compare the variances of the amount of money spent for road repair in two different counties. The data are given here. At $\alpha = 0.05$, is there a significant difference in the variances of the amounts spent in the two counties?

County A	County B
$s_1 = \$11,596$	$s_2 = \$14,837$
$n_1 = 15$	$n_2 = 18$

17. A researcher wants to compare the variances of the heights (in inches) of four-year college basketball players with those of players in junior colleges. A sample of 30 players from each type of school is selected, and the variances of the heights for each type are 2.43 and 3.15, respectively. At $\alpha = 0.10$, is there a significant difference between the variances of the heights in the two types of schools?

18. A traffic safety commissioner claims that the variation in the number of drunk driving citations given on Route 30 is not different from the variation in the number of drunk driving citations given on Route 65. Six weeks are randomly selected; the standard deviation of the number of citations issued for Route 30 is 5.8, and the standard deviation of the number of citations for Route 65 is 3.9. At $\alpha = 0.05$, can it be concluded that there is no difference in the variances of the number of citations issued on the two routes?

19. The variances of the amount of fat in two different types of ground beef are compared. Eight samples of the first type, Super Lean, have a variance of 18.2 grams; 12 of the second type, Ultimate Lean, have a variance of 9.4 grams. At $\alpha = 0.10$, can it be concluded that there is a difference in the variances of the two types of ground beef?

20. It is hypothesized that the variations of the number of days high school teachers miss per year due to illness are greater than the variations of the number of days nurses miss per year. A sample of 56 high school teachers has a standard deviation of 3.4 days, while a sample of 70 nurses has a standard deviation of 2.8. Test the hypothesis at $\alpha = 0.10$.

21. The variations in the number of retail thefts per day in two shopping malls are being compared. A sample of 21 days is selected. The standard deviation of the number of retail thefts in mall A is 6.8, and for mall B, it is 5.3. At $\alpha = 0.05$, can it be concluded that there is a difference in the two standard deviations?

22. The average price of a sample of 12 bottles of diet salad dressing taken from different stores is $1.43. The standard deviation is $0.09. The average price of a sample of 16 low-calorie frozen desserts is $1.03. The standard deviation is $0.10. At $\alpha = 0.01$, is there a significant difference in price? Find the 99% confidence interval of the difference in the means.

23. The average temperature for a 20-day period in May for Pasadena, California, was 82°, and the standard deviation was 4°. The average temperature for the same 20 days for Greenville, South Carolina, was 78°, with a standard deviation of 5°. At $\alpha = 0.01$, can it be concluded that it is warmer in Pasadena?

24. A sample of 12 chemists from Washington state shows an average salary of $39,420 with a standard deviation of $1659, while a sample of 26 chemists from New Mexico has an average salary of $30,215 with a standard deviation of $4116. Is there a significant difference between the two states in chemists' salaries at $\alpha = 0.02$? Find the 98% confidence interval of the difference in the means.

25. The average income of 15 families who reside in a large metropolitan East Coast city is $62,456. The standard deviation is $9652. The average income of 11 families who reside in a rural area of the Midwest is $60,213, with a standard deviation of $2009. At $\alpha = 0.05$, can it be concluded that the families who live in the cities have a higher income than those who live in the rural areas?

26. In an effort to improve the mathematical skills of 10 students, a teacher provides a weekly one-hour tutoring session for the students. A pretest is given before the sessions, and a posttest is given after. The results are shown here. At $\alpha = 0.01$, can it be concluded that the sessions help to improve the students' mathematical skills?

Student	1	2	3	4	5	6	7	8	9	10
Pretest	82	76	91	62	81	67	71	69	80	85
Posttest	88	80	98	80	80	73	74	78	85	93

27. In order to increase egg production, a farmer decided to increase the amount of time the lights in his hen house were on. Ten hens were selected, and the number of eggs each produced was recorded. After one week of lengthened light time, the same hens were monitored again. The data are given here. At $\alpha = 0.05$, can it be concluded that the increased light time increased egg production?

Hen	1	2	3	4	5	6	7	8	9	10
Before	4	3	8	7	6	4	9	7	6	5
After	6	5	9	7	4	5	10	6	9	6

28. In a sample of 80 workers from a factory in city A, it was found that 5% were unable to read, while in a sample of 50 workers in city B, 8% were unable to read. Can it be concluded that there is a difference in the proportions of nonreaders in the two cities? Use $\alpha = 0.10$. Find the 90% confidence interval for the difference of the two proportions.

29. In a recent survey of 45 apartment residents, 28 had phone answering machines. In a survey of 55 homeowners, 20 had phone answering machines. At $\alpha = 0.05$, can it be concluded that the proportions are equal? Find the 95% confidence interval for the difference of the two proportions.

Critical Thinking Challenges

1. Reducing Flight Fatigue In the article at the top of the next page, researchers for Japan Airlines are trying to reduce flight fatigue by masking cabin noise. No data or statistics are given for the results of the study. Design a statistical study to see if the noise-canceling system reduced flight fatigue in airline passengers by answering the following questions:
a. How could airline fatigue be measured?
b. How could a population be defined?
c. How could a sample be selected?
d. Suggest other features that might influence flight fatigue (duration of the flights, time of day, etc.). How might these be controlled?

e. What statistical tests might be used to analyze the data?
f. Find some information on jet lag in books and periodicals in the library and write a brief summary of these findings.

2. Biceps and Brainpower In the article at the bottom of the next page, researchers concluded that physical exercise can keep the brain sharp into old age. After reading the study, answer the following questions:
a. Do you think the conclusions derived from studying rats would be valid for humans?

A Dull Roar

—Charles N. Barnard

One culprit causing flight fatigue is cabin noise—which comes not only from jet engines but also from the rush of air over the airplane fuselage. On the theory that you can't escape this racket but maybe you can disguise it, Japan Airlines offers a **noise-canceling system** through special battery-powered headphones produced by Sony. The system generates a 250 Hz noise of its own, which masks and flattens out other sounds between 60 and 2,000 Hz. Passengers can use the headphones in the usual way for movies and audio channels, or to lull themselves to sleep with "white noise."

Does this help with jet lag? Well, a good long sleep always speeds up *my* lag!

Source: "A Dull Roar," *Modern Maturity* 38, no. 1 (January/February 1995), p. 20. Used with permission.

Building Biceps Could Boost Brainpower, Too

By Ellen Hale
Gannett News Service

Exercise can keep the brain sharp into old age and might help prevent Alzheimer's disease and other mental disorders that accompany aging, says a new study that provides some of the first direct evidence linking physical activity and mental ability.

The study, reported in the journal *Nature*, is the first to show that growth factors in the brain—compounds responsible for the brain's health—can be controlled by exercise.

Combined with previous research that shows exercisers live longer and score higher on tests of mental function, the new findings add hard proof of the importance of physical activity in the aging process.

"Here's another argument for getting active and staying active," says Dr. Carl Cotman of the University of California at Irvine.

Cotman's research was on rodents, but the effects of exercise are nearly identical in humans and rats, and rats have "surprisingly similar" exercise habits, Cotman says.

In his study, which promises to be controversial, rats were permitted to choose how much they wanted to exercise, and each had its own activity habits—just like humans. Some were "couch" rats, Cotman says, rarely getting on the treadmill; others were "runaholics," with one obsessively logging five miles every night on the wheel. "Those little feet must have been paddling away like crazy," Cotman says.

The rats that exercised had much higher levels of BDNF (brain-derived neurotrophic factor), the most widely distributed growth factor in the brain and one reported to decline with the onset of Alzheimer's.

Cotman predicts there is a minimum level of exercise that provides the maximum benefit. The rat that ran five miles nightly, for example, did not raise its level of growth factor much more than those that ran a mile or two.

Source: Ellen Hale, "Building Biceps Could Boost Brainpower, Too," *USA Today,* January 12, 1995. Copyright 1995. *USA TODAY.* Reprinted with permission.

b. What could be a possible hypothesis for a study such as this?

c. What statistical test could be used to test the hypothesis?

d. Cite several reasons why the study might be controversial.

e. What factors other than exercise might influence the results of the study?

Data Projects

Where appropriate, use MINITAB, the TI-83, or a computer program of your choice to complete the following exercises.

1.   Choose a variable for which you would like to determine if there is a difference in the averages for two groups. Make sure that the samples are independent. For example, you may wish to see if men see more movies or spend more money on lunch than women. Select a sample of data values (10–50) and complete the following:

a.   Write a brief statement as to the purpose of the study.
b.   Define the population.
c.   State the hypotheses for the study.
d.   Select an α value.
e.   State how the sample was selected.
f.   Show the raw data.
g.   Decide which statistical test is appropriate and compute the test statistic (z or t). Why is the test appropriate?
h.   Find the critical value(s).
i.   State the decision.
j.   Summarize the results.

2.   Choose a variable that will permit using dependent samples. For example, you might wish to see if a person's weight has changed after a diet. Select a sample of data (10 to 50) value pairs (e.g., before and after), and then complete the following:

a.   Write a brief statement as to the purpose of the study.
b.   Define the population.
c.   State the hypotheses for the study.
d.   Select an α value.
e.   State how the sample was selected.
f.   Show the raw data.
g.   Decide which statistical test is appropriate and compute the test statistic (z or t). Why is the test appropriate?
h.   Find the critical value(s).
i.   State the decision.
j.   Summarize the results.

3.   Choose a variable that will enable you to compare proportions of two groups. For example, you might want to see if the proportion of freshmen who buy used books is lower than (or higher than or the same as) the proportion of sophomores who buy used books. After collecting 30 or more responses from the two groups, complete the following:

a.   Write a brief statement as to the purpose of the study.
b.   Define the population.
c.   State the hypotheses for the study.
d.   Select an α value.
e.   State how the sample was selected.
f.   Show the raw data.
g.   Decide which statistical test is appropriate and compute the test statistic (z or t). Why is the test appropriate?
h.   Find the critical value(s).
i.   State the decision.
j.   Summarize the results.

TI-83 Calculator

z Test for Two Means

A.   z test for two means (Data)
1.   Enter the data into L_1 and L_2.
2.   Press **STAT** and move the cursor to TESTS.
3.   Press **3**.
4.   Select Data and press **ENTER**, then press ▼.
5.   Enter the values for σ_1 and σ_2. Make sure L1 and L2 are selected and Freq 1 and Freq 2 = 1.
6.   Select the correct alternative hypothesis.
7.   Select Calculate and press **ENTER**.

Example 17: Test the claim $H_1: \mu_1 > \mu_2$ when $\sigma_1 = 2.60$ and $\sigma_2 = 2.71$ at $\alpha = 0.10$.

A					B				
10	12	15	18	13	5	8	10	9	9
15	16	14	18	12	11	12	16	8	8
15	16	14	18	16	9	10	11	7	6

Input

```
2-SampZTest
 Inpt:Data Stats
 σ1:2.6
 σ2:2.71
 List1:L₁
 List2:L₂
 Freq1:1
↓Freq2:1
```

```
μ1:≠μ2 <μ2 >μ2
Calculate Draw
```

Output

```
2-SampZTest
 μ₁>μ₂
 z=5.706367974
 P=5.7873457E-9
 x̄₁=14.8
 x̄₂=9.266666667
↓Sx₁=2.36643191
```

```
 Sx₂=2.65832027
 n₁=15
 n₂=15
```

Input

```
2-SampZTest
 Inpt:Data Stats
 σ1:.02
 σ2:.01
 x̄1:.38
 n1:10
 x̄2:.37
↓n2:9
```

```
μ1:≠μ2 <μ2 >μ2
Calculate Draw
```

Output

```
2-SampZTest
 μ₁≠μ₂
 z=1.398757212
 P=.1618859056
 x̄₁=.38
 x̄₂=.37
↓n₁=10
```

```
 n₂=9
```

The test value is 5.706367974, and the P-value is 5.7873457E-9. The decision is to reject the null hypothesis at $\alpha = 0.10$, since 5.7873457E-9 < 0.10. The calculator also gives the values of the sample means and standard deviations.

B. z test for two means (Stats)
1. Press **STAT** and move the cursor to TESTS.
2. Press **3**.
3. Select Stats and press **ENTER**, then press ▼.
4. Enter the values for σ_1, σ_2, $\overline{X}_1$, n_1, $\overline{X}_2$, and n_2.
5. Select the correct alternative hypothesis.
6. Select Calculate and press **ENTER**.

Example 18: Test the claim $H_1: \mu_1 \neq \mu_2$ at $\alpha = 0.05$ when

A	B
$\overline{X}_1 =$	$\overline{X}_2 =$
0.38	0.37
$\sigma_1 = 0.02$	$\sigma_2 = 0.01$
$n_1 = 10$	$n_2 = 9$

The test value is 1.398757212, and the P-value is 0.1618859056. The decision is not to reject the null hypothesis, since 2(0.1618859056) > 0.05.

z Confidence Interval: Two Means

A. z confidence interval for two means (Data)
1. Enter the data into L_1 and L_2.
2. Press **STAT** and move the cursor to TESTS.
3. Press **9**.
4. Select Data and press **ENTER**, then press ▼.
5. Enter the values for σ_1 and σ_2. Make sure L1 and L2 are selected and Freq 1 and Freq 2 = 1.
6. Enter the correct confidence level.
7. Select Calculate and press **ENTER**.

Example 19: Find the 95% confidence for the difference between two means for the data shown. Assume $\sigma_1 = 2.6$ and $\sigma_2 = 2.71$.

A					B				
10	12	15	18	13	5	8	10	9	9
15	16	14	18	12	11	12	16	8	8
15	16	14	18	16	9	10	11	7	6

Input

```
2-SampZInt
 Inpt:DATA Stats
 σ1:2.6
 σ2:2.71
 List1:L₁
 List2:L₂
 Freq1:1
↓Freq2:1
```

```
 C-Level:.95
 Calculate
```

Output

```
2-SampZInt
 (3.6328,7.4339)
 x̄1=14.8
 x̄2=9.266666667
 Sx1=2.36643191
 Sx2=2.65832027
↓n1=15
```

```
 n2=15
█
```

The 95% confidence interval is $3.6328 < \mu_1 - \mu_2 < 7.4339$. The calculator also gives the values for $\overline{X}_1$, $\overline{X}_2$, s_1, and s_2.

B. z confidence interval for two means (Stats)
1. Press **STAT** and move the cursor to TESTS.
2. Press **9**.
3. Select Stats and press **ENTER,** then press ▼.
4. Enter the values for σ_1, σ_2, $\overline{X}_1$, n_1, $\overline{X}_2$, and n_2.
5. Select the correct confidence level.
6. Select Calculate and press **ENTER.**

Example 20: Find the 95% confidence interval for the difference between two means when

A	B
$\overline{X}_1 =$	$\overline{X}_2 =$
0.38	0.37
$\sigma_1 = 0.02$	$\sigma_2 = 0.01$
$n_1 = 10$	$n_2 = 9$

Input

```
2-SampZInt
 Inpt:Data Stats
 σ1:.02
 σ2:.01
 x̄1:.38
 n1:10
 x̄2:.37
↓n2:9
```

```
 C-Level:.95
 Calculate
```

Output

```
2-SampZInt
 (-.004,.02401)
 x̄1=.38
 x̄2=.37
 n1=10
 n2=9
```

The 95% confidence interval for $\mu_1 - \mu_2$ is $-0.004 < \mu_1 - \mu_2 < 0.02401$. The calculator also gives the values for $\overline{X}_1$ and $\overline{X}_2$.

F Test: Two Variances

A. F test for two variances (Data)
1. Enter the data into L_1 and L_2.
2. Press **STAT** and move the cursor to TESTS.
3. Press **D (ALPHA X⁻¹).**
4. Select Data; press **ENTER,** then press ▼. Make sure L1 and L2 are selected and Freq 1 and Freq 2 = 1.
5. Select the correct alternative hypothesis.
6. Select Calculate and press **ENTER.**

Example 21: Test the claim H_1: $\sigma_1 \neq \sigma_2$ at $\alpha = 0.05$ for the following data:

A			B		
63	73	80	86	93	64
60	86	83	82	81	75
70	72	82	88	63	63

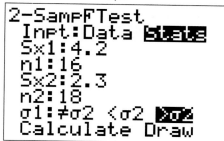

Input

```
2-SampFTest
 Inpt:Data Stats
List1:L₁
List2:L₂
Freq1:1
Freq2:1
σ1:≠σ2 <σ2 >σ2
Calculate Draw
```

Output

```
2-SampFTest
 σ1≠σ2
 F=.6224404513
 p=.5176283012
 Sx1=9.0967027
 Sx2=11.5301537
↓x̄1=74.33333333
```

```
 x̄2=77.22222222
 n1=9
 n2=9
```

Input

```
2-SampFTest
 Inpt:Data Stats
Sx1:4.2
n1:16
Sx2:2.3
n2:18
σ1:≠σ2 <σ2 >σ2
Calculate Draw
```

Output

```
2-SampFTest
 σ1>σ2
 F=3.334593573
 p=.0096648303
 Sx1=4.2
 Sx2=2.3
↓n1=16
```

```
 n2=18
```

The test value is 0.6224404513, and the P-value is 0.5176283012. The decision is not to reject the null hypothesis, since $0.5176283012 > 0.05$. The calculator also gives various other statistics.

B. F test for two variances (Stats)
1. Press **STAT** and move the cursor to TESTS.
2. Press **D (ALPHA X⁻¹)**.
3. Select Stats and **ENTER,** then press ▼.
4. Enter the sample 1 standard deviation in Sx1 and sample size in n1.
5. Enter the sample 2 standard deviation in Sx2 and sample size in n2.
6. Select the correct alternative hypothesis.
7. Select Calculate and press **ENTER.**

Example 22: Test the claim H_1 $\sigma_1 > \sigma_2$ at $\alpha = 0.01$, when

A	B
$s_1 = 4.2$	$s_2 = 2.3$
$n_1 = 16$	$n_2 = 18$

The test value is 3.334593573, and the P-value is 0.0096648303. The decision is to reject the null hypothesis, since $0.0096648303 < 0.01$.

t Test: Two Means, Independent Samples

A. t test for two means (Data)
1. Enter the data into L_1 and L_2.
2. Press **STAT** and move the cursor to TESTS.
3. Press **4**.
4. Select Data; press **ENTER,** then press ▼. Make sure L1 and L2 are selected and Freq 1 and Freq 2 = 1.
5. Select the correct alternative hypothesis.
6. Select either No or Yes for Pooled. (*Note:* If the variances are assumed equal, select Yes. If the variances are assumed unequal, select No.)
7. Select Calculate and press **ENTER.**

Example 23: Test the claim H_1: $\mu_1 = \mu_2$ at $\alpha = 0.05$. Assume the variances are equal.

A					B				
32	38	37	36	36	30	36	35	36	31
34	39	36	37	42	34	37	33	32	

Input

```
2-SampTTest
 Inpt:DATA Stats
 List1:L₁
 List2:L₂
 Freq1:1
 Freq2:1
 μ1:≠μ2 <μ2 >μ2
↓Pooled:No Yes
─────────────────
 Calculate Draw
```

Output

```
2-SampTTest
 μ₁≠μ₂
 t=2.459726496
 P=.0249148008
 df=17
 x̄₁=36.7
↓x̄₂=33.77777778
```

```
↑Sx₁=2.71006355
 Sx₂=2.43812314
 SxP=2.58565677
 n₁=10
 n₂=9
```

The test value is 2.459726496, and the P-value is 0.0249148008. The decision is to reject the null hypothesis. The calculator also gives the values for d.f., $\overline{X}_1$, $\overline{X}_2$, s_1, s_2, n_1, and n_2.

B.   t test for two means (Stats)
1.   Press **STAT** and move the cursor to TESTS.
2.   Press **4.**
3.   Select STATS and press **ENTER,** then press ▼.
4.   Enter the values for $\overline{X}_1$, s_1, n_1, $\overline{X}_2$, s_2, and n_2.
5.   Select the correct alternative hypothesis.
6.   Select Yes if variances are assumed equal; otherwise, select No.
7.   Select Calculate and press **ENTER.**

Example 24: Test the claim H_1: $\mu_1 \neq \mu_2$ at $\alpha = 0.05$ when

A	B
$\overline{X}_1 = 262$	$\overline{X}_2 = 271$
$s_1 = 6$	$s_2 = 14$
$n_1 = 10$	$n_2 = 8$

(Assume the variances are not equal.)

Input

```
2-SampTTest
 Inpt:Data Stats
 x̄1:262
 Sx1:6
 n1:10
 x̄2:271
 Sx2:14
↓n2:8
```

```
 μ1:≠μ2 <μ2 >μ2
 Pooled:No Yes
 Calculate Draw
```

Output

```
2-SampTTest
 μ₁≠μ₂
 t=-1.697811025
 P=.123561132
 df=9.056199105
 x̄₁=262
↓x̄₂=271
```

```
 Sx₁=6
 Sx₂=14
 n₁=10
 n₂=8
```

The test value is -1.697811025, and the P-value is 0.123561132. The decision is not to reject the null hypothesis, since $2(0.123561132) > 0.05$. (*Note:* the degrees of freedom are calculated differently from in the textbook.)

t **Confidence Interval: Two Means**

A.   t confidence interval for two means (Data)
1.   Enter the data into L_1 and L_2.
2.   Press **STAT** and move the cursor to TESTS.
3.   Press **0.**
4.   Select Data and press **ENTER,** then press ▼. Make sure L1 and L2 are selected and Freq 1 and Freq 2 = 1.
5.   Select the correct confidence level.
6.   Select Yes or No for pooled standard deviation.
7.   Select Calculate and press **ENTER.**

Example 25: Find the 95% confidence interval for the difference between the two means. Assume the variances are equal.

	A					B				
32	38	37	36	36	30	36	35	36	31	
34	39	36	37	42	34	37	33	32		

Input

```
2-SampTInt
 Inpt:DATA Stats
 List1:L1
 List2:L2
 Freq1:1
 Freq2:1
 C-Level:.95
↓Pooled:No Yes

 Calculate
```

Output

```
2-SampTInt
 (.4157,5.4287)
 df=17
 x̄1=36.7
 x̄2=33.77777778
 Sx1=2.71006355
↓Sx2=2.43812314

 SxP=2.58565677
 n1=10
 n2=9
```

The 95% confidence interval is $0.4157 < \mu_1 - \mu_2 < 5.4287$. The calculator also gives various other statistics.

B. *t* confidence interval for two means (Stats)
1. Press **STAT** and move the cursor to TESTS.
2. Press **0**.
3. Select Stats and press **ENTER,** then press ▼.
4. Enter the values for $\bar{X}_1$, s_1, n_1, $\bar{X}_2$, s_2, and n_2.
5. Select the correct confidence level.
6. Select No or Yes for Pooled.
7. Select Calculate and press **ENTER.**

Example 26: Find the 95% confidence interval for the difference between the two means when

	A	B
	$\bar{X}_1 = 262$	$\bar{X}_2 = 271$
	$s_1 = 6$	$s_2 = 14$
	$n_1 = 10$	$n_2 = 8$

(Assume the variances are not equal.)

Input

```
2-SampTInt
 Inpt:Data Stats
 x̄1:262
 Sx1:6
 n1:10
 x̄2:271
 Sx2:14
↓n2:8

 C-Level:.95
 Pooled:No Yes
 Calculate
```

Output

```
2-SampTInt
 (-20.98,2.9802)
 df=9.056199105
 x̄1=262
 x̄2=271
 Sx1=6
↓Sx2=14

 n1=10
 n2=8
```

The 95% confidence interval for $\mu_1 - \mu_2$ is $-20.98 < \mu_1 - \mu_2, < 2.9802$.

t Test: Two Means, Dependent Samples

Note: When this type of problem occurs, use the one sample *t* test, enter the differences in L_1 and select.

1. Enter the data into L_1.
2. Press **STAT** and move the cursor to TESTS.
3. Press **2**.
4. Select **Data** and press **ENTER,** then press ▼.
Make sure L1 is selected and Freq = 1.
5. Enter **0** for the value of μ_0.
6. Select the correct alternative hypothesis.
7. Select Calculate and press **ENTER.**

Example 27: Test the claim $H_1: \mu_D > 0$ at $\alpha = 0.10$ when

Before X_1	210	235	208	190	172	244
After X_2	190	170	210	188	173	228

Note: Subtract $X_1 - X_2$ and enter these values in L_1:
20, 65, −2, 2, −1, 16

Input

```
T-Test
 Inpt:DATA Stats
 μ0:0
 List:L1
 Freq:1
 μ:≠μ0 <μ0 >μ0
 Calculate Draw
```

Output

```
T-Test
 μ>0
 t=1.607891603
 P=.0843852916
 x̄=16.66666667
 Sx=25.39028686
 n=6
```

The test value is 1.607891603, and the P-value is 0.0843852916. The decision is to reject the null hypothesis at = 0.10, since 0.0843852916 < 0.10.

z Test: Two Proportions

1. Press **STAT** and move the cursor to TESTS.
2. Press **6.**
3. Enter the values for X_1, n_1, X_2, and n_2.
4. Select the correct alternative hypothesis.
5. Select Calculate and press **ENTER.**

Example 28: Test the claim H_1: $p_1 \neq p_2$ at $\alpha = 0.05$ when

$$X_1 = 43 \qquad X_2 = 90$$
$$n_1 = 100 \qquad n_2 = 200$$

Input

```
2-PropZTest
 x1:43
 n1:100
 x2:90
 n2:200
 p1:≠p2 <p2 >p2
 Calculate Draw
```

Output

```
2-PropZTest
 p1≠p2
 z=-.3287165459
 P=.7423700749
 p̂1=.43
 p̂2=.45
↓p̂=.4433333333

 n1=100
 n2=200
```

The test value is −0.3287165459, and the P-value is 0.7423700749. The decision is not to reject the null hypothesis, since 2(0.7423700749) > 0.05.

z Confidence: Two Proportions

1. Press **STATS** and move the cursor to TESTS.
2. Press **B (ALPHA MATRIX).**
3. Enter the values for X_1, n_1, X_2, and n_2.
4. Type in the correct confidence level.
5. Select Calculate and press **ENTER.**

Example 29: Find the 99% confidence for $p_1 - p_2$ when

$$X_1 = 120 \qquad X_2 = 120$$
$$n_1 = 150 \qquad n_2 = 200$$

Input

```
2-PropZInt
 x1:120
 n1:150
 x2:120
 n2:200
 C-Level:.99
 Calculate
```

Output

```
2-PropZInt
 (.07737,.32263)
 p̂1=.8
 p̂2=.6
 n1=150
 n2=200
```

The 99% confidence interval is $0.07737 < p_1 - p_2 < 0.32263$.

Hypothesis-Testing Summary 1

1. Comparison of a sample mean with a specific population mean.

Example: H_0: $\mu = 100$

a. Use the z test when σ is known:

$$z = \frac{\bar{X} - \mu}{\dfrac{\sigma}{\sqrt{n}}}$$

b. Use the t test when σ is unknown:

$$t = \frac{\bar{X} - \mu}{\dfrac{s}{\sqrt{n}}} \quad \text{with} \quad \text{d.f.} = n - 1$$

2. Comparison of a sample variance or standard deviation with a specific population variance or standard deviation.

Example: H_0: $\sigma^2 = 225$

Use the chi-square test:

$$\chi^2 = \frac{(n - 1)s^2}{\sigma^2} \quad \text{with} \quad \text{d.f.} = n - 1$$

3. Comparison of two sample means.

Example: H_0: $\mu_1 = \mu_2$

a. Use the z test when the population variances are known:

$$z = \frac{(\bar{X}_1 - \bar{X}_2) - (\mu_1 - \mu_2)}{\sqrt{\dfrac{\sigma_1^2}{n_1} + \dfrac{\sigma_2^2}{n_2}}}$$

b. Use the t test for independent samples when the population variances are unknown and the sample variances are unequal:

$$t = \frac{(\bar{X}_1 - \bar{X}_2) - (\mu_1 - \mu_2)}{\sqrt{\dfrac{s_1^2}{n_1} + \dfrac{s_2^2}{n_2}}}$$

with d.f. = the smaller of $n_1 - 1$ or $n_2 - 1$

c. Use the t test for independent samples when the population variances are unknown and the sample variances are equal:

$$t = \frac{(\bar{X}_1 - \bar{X}_2) - (\mu_1 - \mu_2)}{\sqrt{\dfrac{(n_1 - 1)s_1^2 + (n_2 - 1)s_2^2}{n_1 + n_2 - 2}}\sqrt{\dfrac{1}{n_1} + \dfrac{1}{n_2}}}$$

with d.f. = $n_1 + n_2 - 2$

d. Use the t test for dependent samples when the means are related.

Example: H_0: $\mu_D = 0$

$$t = \frac{\bar{D} - \mu_D}{\dfrac{s_D}{\sqrt{n}}} \quad \text{with} \quad \text{d.f.} = n - 1$$

where n = number of pairs.

4. Comparison of a sample proportion with a specific population proportion.

Example: H_0: $p = 0.32$

Use the z test:

$$z = \frac{X - np}{\sqrt{npq}} \quad \text{or} \quad z = \frac{X - \mu}{\sigma}$$

5. Comparison of two sample proportions.

Example: H_0: $p_1 = p_2$

Use the z test:

$$z = \frac{(\hat{p}_1 - \hat{p}_2) - (p_1 - p_2)}{\sqrt{\bar{p}\,\bar{q}\left(\dfrac{1}{n_1} + \dfrac{1}{n_2}\right)}}$$

where

$$\bar{p} = \frac{X_1 + X_2}{n_1 + n_2} \qquad \hat{p}_1 = \frac{X_1}{n_1}$$

$$\bar{q} = 1 - \bar{p} \qquad \hat{p}_2 = \frac{X_2}{n_2}$$

6. Comparison of two sample variances or standard deviations.

Example: H_0: $\sigma_1^2 = \sigma_2^2$

Use the F test:

$$F = \frac{s_1^2}{s_2^2}$$

where

$$s_1^2 = \text{larger variance} \quad \text{d.f.N.} = n_1 - 1$$
$$s_2^2 = \text{smaller variance} \quad \text{d.f.D.} = n_2 - 1$$

chapter

11

Correlation and Regression

Outline

Objectives

After completing this chapter, you should be able to

1. Draw a scatter plot for a set of ordered pairs.

2. Find the correlation coefficient.

3. Test the hypothesis H_0: $p = 0$.

4. Find the equation of the regression line.

5. Find the coefficient of determination.

6. Find the standard error of estimate.

7. Find a prediction interval.

Statistics Today

Do Dust Storms Affect Respiratory Health?

Southeast Washington state has a long history of seasonal dust storms. Several researchers decided to see what effect, if any, these storms had on the respiratory health of the people living in the area. They undertook (among other things) to see if there was a relationship between the amount of dust and sand particles in the air when the storms occur and the number of hospital emergency room visits for respiratory disorders at three community hospitals in southeast Washington. Using methods of correlation and regression, which are explained in this chapter, they were able to determine the effect of these dust storms on local residents.

Source: B. Hefflin, B. Jalaludin, N. Cobb, C. Johnson, L. Jecha, and R. Etzel, "Surveillance for Dust Storms and Respiratory Diseases in Washington State, 1991," *Archives of Environmental Health* 49, no. 3 (May/June 1994), pp. 170–74. Reprinted with permission of the Helen Dwight Reid Educational Foundation. Published by Heldref Publications, 1319 18th St. N.W. Washington, DC. 20036-1802. Copyright 1994.

11–1

Introduction

In the previous chapters, two areas of inferential statistics, hypothesis testing and confidence intervals, were explained. Another area of inferential statistics involves determining whether a relationship between two or more numerical or quantitative variables exists. For example, a businessperson may want to know whether the volume of sales for a given month is related to the amount of advertising the firm does that month. Educators are interested in determining whether the number of hours a student studies is related to the student's score on a particular exam. Medical researchers are interested in questions such as "Is caffeine related to heart damage?" or "Is there a relationship between a person's age and his or her blood pressure?" A zoologist may want to know whether the birth weight of a certain animal is related to its life span. These are only a few of the many questions that can be answered by using the techniques of correlation and regression analysis. **Correlation** is a statistical method used to determine whether a relationship between variables exists. **Regression** is a statistical method used to describe the nature of the relationship between variables—that is, positive or negative, linear or nonlinear.

The purpose of this chapter is to answer the following questions statistically:

1. Are two or more variables related?
2. If so, what is the strength of the relationship?
3. What type of relationship exists?
4. What kind of predictions can be made from the relationship?

To answer the first two questions, statisticians use a measure to determine whether two or more variables are related and also to determine the strength of the relationship between or among the variables. This measure is called a *correlation coefficient*. For example, there are many variables that contribute to heart disease, among them lack of exercise, smoking, heredity, age, stress, and diet. Of these variables, some are more important than others; therefore, a physician who wants to help a patient must know which factors are most important.

Another consideration in determining relationships is to ascertain what type of relationship exists. There are two types of relationships: simple and multiple. In a **simple relationship,** there are only two variables under study. For example, a manager may wish to see whether the number of years the salespeople have been working for the company has anything to do with the amount of sales they make. This type of study involves a simple relationship, since there are only two variables, years of experience and amount of sales.

In **multiple relationships,** many variables are under study. For example, an educator may wish to investigate the relationship between a student's success in college and factors such as the number of hours devoted to studying, the student's GPA, and the student's high school background. This type of study involves several variables. This topic is beyond the scope of this book.

Simple relationships can also be positive or negative. A **positive relationship** exists when both variables increase or decrease at the same time. For instance, a person's height and weight are related; and the relationship is positive, since the taller a person is, generally, the more the person weighs. In a **negative relationship,** as one variable increases, the other variable decreases, and vice versa. For example, if one measures the strength of people over 60 years of age, one will find that as age increases, strength generally decreases. The word *generally* is used here because there are exceptions.

Finally, the fourth question asks what type of predictions can be made. Predictions are made in all areas and on a daily basis. Examples include weather forecasting, stock market analyses, sales predictions, crop predictions, gasoline predictions, and sports predictions. Some predictions are more accurate than others, due to the strength of the relationship. That is, the stronger the relationship is between variables, the more accurate the prediction is.

<table>
<tr><td>

11–2

Scatter Plots

Objective 1. Draw a scatter plot for a set of ordered pairs.

</td><td>

In simple correlation and regression studies, the researcher collects data on two numerical or quantitative variables to see whether a relationship exists between the variables. For example, if a researcher wishes to see whether there is a relationship between number of hours studied and test scores on an exam, he must select a random sample of students, determine the hours each studied, and obtain their grades on the exam. A table can be made for the data, as shown here.

</td></tr>
</table>

Student	Hours studied, x	Grade, y (%)
A	6	82
B	2	63
C	1	57
D	5	88
E	2	68
F	3	75

The two variables for this study are called the independent variable and the dependent variable. The **independent variable** is the variable in regression that can be controlled or manipulated. In this case, "number of hours studied" is the independent variable and is designated as the x variable. The **dependent variable** is the variable in regression that cannot be controlled or manipulated. The grade the student received on the exam is the dependent variable, designated as the y variable. The reason for this distinction between the variables is that one assumes that the grade the student earns *depends* on the number of hours the student studied. Also, one assumes that, to some extent, the student can regulate or *control* the number of hours he or she studies for the exam.

The determination of the x and y variables is not always clear-cut and sometimes is an arbitrary decision. For example, if a researcher studies the effects of age on a person's blood pressure, the researcher can generally assume that age affects blood pressure. Hence, the variable "age" can be called the *independent variable* and the variable "blood pressure" can be called the *dependent variable.* On the other hand, if a researcher is studying the attitudes of husbands on a certain issue and the attitudes of their wives on the same issue, it is difficult to say which variable is the independent variable and which is the dependent variable. In this study, the researcher can arbitrarily designate the variables as independent and dependent.

The independent and dependent variables can be plotted on a graph called a *scatter plot.* The independent variable, x, is plotted on the horizontal axis and the dependent variable, y, is plotted on the vertical axis.

A **scatter plot** is a graph of the ordered pairs (x, y) of numbers consisting of the independent variable, x, and the dependent variable, y.

The scatter plot is a visual way to describe the nature of the relationship between the independent and dependent variables. The scales of the variables can be different and the coordinates of the axes are determined by the smallest and largest data values of the variables.

The procedure for drawing a scatter plot is shown in the next three examples.

Example 11–1

Construct a scatter plot for the data obtained in a study of age and systolic blood pressure of six randomly selected subjects. The data are shown in the following table.

Subject	Age, x	Pressure, y
A	43	128
B	48	120
C	56	135
D	61	143
E	67	141
F	70	152

Solution

STEP 1   Drawn and label the x and y axes.

STEP 2   Plot each point on the graph, as shown in Figure 11–1.

Figure 11–1

Scatter Plot for
Example 11–1

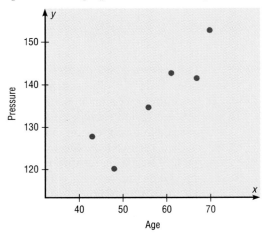

Example 11–2

Construct a scatter plot for the data obtained in a study on the number of absences and the final grades of seven randomly selected students from a statistics class. The data are shown here.

Student	Number of absences, x	Final grade, $y(\%)$
A	6	82
B	2	86
C	15	43
D	9	74
E	12	58
F	5	90
G	8	78

Solution

STEP 1   Draw and label the x and y axes.

STEP 2   Plot each point on the graph, as shown in Figure 11–2.

Figure 11–2

Scatter Plot for
Example 11–2

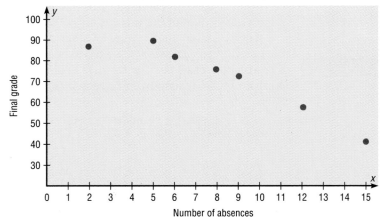

Example 11–3

Construct a scatter plot for the data obtained in a study on the number of hours nine people exercise each week and the amount of milk (in ounces) each person consumes per week. The data follow.

Subject	Hours, x	Amount, y
A	3	48
B	0	8
C	2	32
D	5	64
E	8	10
F	5	32
G	10	56
H	2	72
I	1	48

Solution

STEP 1   Draw and label the x and y axes.

STEP 2   Plot each point on the graph, as shown in Figure 11–3.

Figure 11–3

Scatter Plot for
Example 11–3

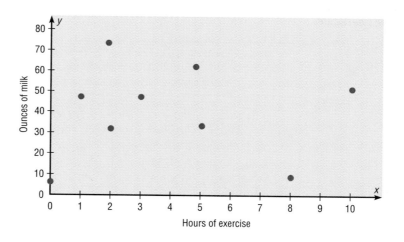

After the plot is drawn, it should be analyzed to determine which type of relationship, if any, exists. For example, the plot shown in Figure 11–1 suggests a positive relationship, since as a person's age increases, blood pressure tends to increase also. The plot of the data shown in Figure 11–2 suggests a negative relationship, since as the number of absences increases, the final grade decreases. Finally, the plot of the data shown in Figure 11–3 shows no specific type of relationship, since no pattern is discernible.

Note that the data shown in Figures 11–1 and 11–2 also suggest a linear relationship, since the points seem to fit a straight line, although not perfectly. Sometimes a scatter plot, such as the one in Figure 11–4, will show a curvilinear relationship between the data. In this situation, the methods shown in this section and the next cannot be used. Methods for curvilinear relationships are beyond the scope of this book.

Figure 11–4

Scatter Plot Suggesting a Curvilinear Relationship

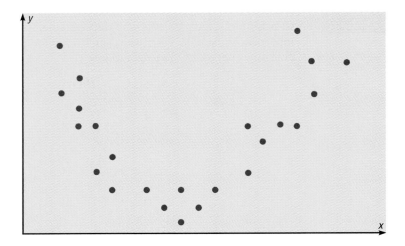

11–3

Correlation

Correlation Coefficient

Objective 2. Find the correlation coefficient.

As stated in Section 11–1, statisticians use a measure called the *correlation coefficient* to determine the strength of the relationship between two variables. There are several types of correlation coefficients. The one explained in this section is called the **Pearson product moment correlation coefficient** (PPMC), named after statistician Karl Pearson, who pioneered the research in this area.

The **correlation coefficient** computed from the sample data measures the strength and direction of a relationship between two variables. The symbol for the sample correlation coefficient is **r**. The symbol for the population correlation coefficient is ρ (Greek letter rho).

The *range of the correlation coefficient* is from -1 to $+1$. If there is a *strong positive linear relationship* between the variables, the value of r will be close to $+1$. If there is a *strong negative linear relationship* between the variables, the value of r will be close to -1. When there is no linear relationship between the variables or only a weak relationship, the value of r will be close to 0. See Figure 11–5.

Figure 11–5

Range of Values for the Correlation Coefficient

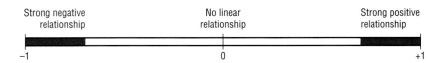

There are several ways to compute the value of the correlation coefficient. One method is to use the formula shown next.

Formula for the Correlation Coefficient r

$$r = \frac{n(\sum xy) - (\sum x)(\sum y)}{\sqrt{[n(\sum x^2) - (\sum x)^2][n(\sum y^2) - (\sum y)^2]}}$$

where n is the number of data pairs.

Round the value of r to three decimal places.

The formula looks somewhat complicated, but using a table to compute the values, as shown in the next example, makes it somewhat easier to determine the value of r.

Example 11–4

Compute the value of the correlation coefficient for the data obtained in the study of age and blood pressure given in Example 11–1.

Solution

STEP 1   Make a table, as shown here.

Subject	Age, x	Pressure, y	xy	x^2	y^2
A	43	128			
B	48	120			
C	56	135			
D	61	143			
E	67	141			
F	70	152			

STEP 2   Find the values of xy, x^2, and y^2 and place these values in the corresponding columns of the table.

The completed table is shown next.

Subject	Age, x	Pressure, y	xy	x^2	y^2
A	43	128	5,504	1,849	16,384
B	48	120	5,760	2,304	14,400
C	56	135	7,560	3,136	18,225
D	61	143	8,723	3,721	20,449
E	67	141	9,447	4,489	19,881
F	70	152	10,640	4,900	23,104
	$\Sigma x = 345$	$\Sigma y = 819$	$\Sigma xy = 47,634$	$\Sigma x^2 = 20,399$	$\Sigma y^2 = 112,443$

STEP 3   Substitute in the formula and solve for r.

$$r = \frac{n(\Sigma xy) - (\Sigma x)(\Sigma y)}{\sqrt{[n(\Sigma x^2) - (\Sigma x)^2][n(\Sigma y^2) - (\Sigma y)^2]}}$$

$$= \frac{(6)(47,634) - (345)(819)}{\sqrt{[(6)(20,399) - (345)^2][(6)(112,443) - (819)^2]}} = 0.897$$

The correlation coefficient suggests a strong positive relationship between age and blood pressure.

Example 11–5

Compute the value of the correlation coefficient for the data obtained in the study of the number of absences and the final grade of the seven students in the statistics class given in Example 11–2.

Solution

STEP 1   Make a table.

STEP 2   Find the values of xy, x^2, and y^2 and place these values in the corresponding columns of the table.

Student	Number of absences, x	Final grade, y (%)	xy	x^2	y^2
A	6	82	492	36	6,724
B	2	86	172	4	7,396
C	15	43	645	225	1,849
D	9	74	666	81	5,476
E	12	58	696	144	3,364
F	5	90	450	25	8,100
G	8	78	624	64	6,084
	$\sum x = 57$	$\sum y = 511$	$\sum xy = 3745$	$\sum x^2 = 579$	$\sum y^2 = 38{,}993$

STEP 3 Substitute in the formula and solve for r.

$$r = \frac{n(\sum xy) - (\sum x)(\sum y)}{\sqrt{[n(\sum x^2) - (\sum x)^2][n(\sum y^2) - (\sum y)^2]}}$$

$$= \frac{(7)(3745) - (57)(511)}{\sqrt{[(7)(579) - (57)^2][(7)(38{,}993) - (511)^2]}} = -0.944$$

The value of r suggests a strong negative relationship between a student's final grade and the number of absences a student has.

Example 11–6

Compute the value of the correlation coefficient for the data given in Example 11–3 for the number of hours a person exercises and the amount of milk a person consumes per week.

Solution

STEP 1 Make a table.

STEP 2 Find the values of xy, x^2, and y^2 and place these values in the corresponding columns of the table.

Subject	Hours, x	Amount, y	xy	x^2	y^2
A	3	48	144	9	2,304
B	0	8	0	0	64
C	2	32	64	4	1,024
D	5	64	320	25	4,096
E	8	10	80	64	100
F	5	32	160	25	1,024
G	10	56	560	100	3,136
H	2	72	144	4	5,184
I	1	48	48	1	2,304
	$\sum x = 36$	$\sum y = 370$	$\sum xy = 1520$	$\sum x^2 = 232$	$\sum y^2 = 19{,}236$

STEP 3 Substitute the formula and solve for r.

$$r = \frac{n(\sum xy) - (\sum x)(\sum y)}{\sqrt{[n(\sum x^2) - (\sum x)^2][n(\sum y^2) - (\sum y)^2]}}$$

$$= \frac{(9)(1520) - (36)(370)}{\sqrt{[(9)(232) - (36)^2][9(19{,}236) - (370)^2]}} = 0.067$$

The value of r indicates a very weak positive relationship between the variables.

In Example 11–4, the value of r was high (close to 1.00); in Example 11–6, the value of r was much lower (close to 0). This question then arises: "When is the value of r due to chance and when does it suggest a significant relationship between the variables?" This question will be answered next.

The Significance of the Correlation Coefficient

Objective 3. Test the hypothesis H_0: $\rho = 0$.

As stated before, the range of the correlation coefficient is between -1 and $+1$. When the value of r is near $+1$ or -1, there is a strong relationship. When the value of r is near 0, the relationship is weak or nonexistent. Since the value of r is computed from data obtained from samples, there are two possibilities: either the value of r is high enough to conclude that there is a significant relationship between the variables, or the value of r is due to chance.

In order to make this determination, one uses a hypothesis-testing procedure. This procedure is similar to the one used in the previous chapters.

STEP 1 State the hypotheses.

STEP 2 Find the critical values.

STEP 3 Compute the test value.

STEP 4 Make the decision.

STEP 5 Summarize the results.

The population correlation coefficient is computed from taking all possible (x, y) pairs; it is designated by the Greek letter ρ (rho). The sample correlation coefficient can then be used as an estimator of ρ if the following assumptions are valid.

1. The variables x and y are *linearly* related.
2. The variables are *random* variables.
3. The two variables have a *bivariate normal distribution.*

This means for any given value of x, the y variable is normally distributed.

Formally defined, the **population correlation coefficient**, p, is the correlation computed by using all possible pairs of data values (x, y) taken from a population.

In hypothesis testing, one of the following is true:

H_0: $\rho = 0$ This null hypothesis means that there is no correlation between the x and y variables in the population.

H_1: $\rho \neq 0$ This alternative hypothesis means that there is a significant correlation between the variables in the population.

When the null hypothesis is rejected at a specific level, it means that there is a significant difference between the value of r and 0. When the null hypothesis is not rejected, it means that the value of r is not significantly different from 0 (zero) and is probably due to chance.

There are several methods that can be used to test the significance of the correlation coefficient. Two methods will be shown in this section. The first uses the t test.

Historical Note

A mathematician named Karl Pearson (1857–1936) became interested in Francis Galton's work and saw that the correlation and regression theory could be applied to other areas besides heredity. Pearson developed the correlation coefficient that bears his name.

Formula for the *t* test for the Correlation Coefficient

$$t = r \sqrt{\frac{n-2}{1-r^2}}$$

with degrees of freedom equal to $n - 2$.

Although hypothesis tests can be one-tailed, most hypotheses involving the correlation coefficient are two-tailed. Recall that p represents the population correlation coefficient. Also, if there is no relationship, the value of the correlation coefficient will be 0. Hence, the hypotheses will be

$$H_0: \rho = 0 \quad \text{and} \quad H_1: \rho \neq 0$$

One does not have to identify the claim here, since the question will always be whether or not there is a significant relationship between the variables.

The two-tailed critical values are used. These values are found in Table F in Appendix C. Also, when one is testing the significance of a correlation coefficient, both variables, x and y, must come from normally distributed populations.

Example 11–7

Test the significance of the correlation coefficient found in Example 11–4. Use $\alpha = 0.05$ and $r = 0.897$.

Solution

STEP 1 State the hypotheses.

$$H_0: \rho = 0 \quad \text{and} \quad H_1: \rho \neq 0$$

STEP 2 Find the critical values. Since $\alpha = 0.05$ and there are $6 - 2 = 4$ degrees of freedom, the critical values obtained from Table F are ± 2.776, as shown in Figure 11–6.

Figure 11–6

Critical Values for Example 11–7

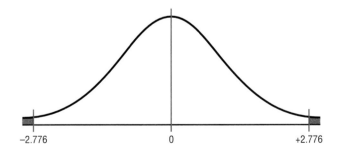

−2.776   0   +2.776

STEP 3 Compute the test value.

$$t = r \sqrt{\frac{n-2}{1-r^2}} = (0.897) \sqrt{\frac{6-2}{1-(0.897)^2}} = 4.059$$

STEP 4 Make the decision. Reject the null hypothesis, since the test value falls in the critical region, as shown in Figure 11–7.

Figure 11–7

Test Value for
Example 11–7

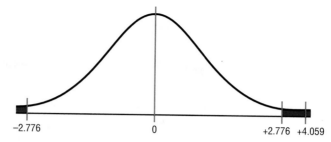

-2.776   0   +2.776 +4.059

STEP 5 Summarize the results. There is a significant relationship between the variables of age and blood pressure.

The second method of testing the significance of r is to use Table I in Appendix C. This table shows the values of the correlation coefficient that are significant for a specific α level and a specific number of degrees of freedom. For example, for 7 degrees of freedom and $\alpha = 0.05$, the table gives a critical value of 0.666. Any value of r greater than $+0.666$ or less than -0.666 will be significant, and the null hypothesis will be rejected. See Figure 11–8. When Table I is used, one need not compute the t test value. Table I is for two-tailed tests only.

Figure 11–8

Finding the Critical
Value from Table I

d.f.	$\alpha = 0.05$	$\alpha = 0.01$
1		
2		
3		
4		
5		
6		
7	.666	

Example 11–8

Using Table I, test the significance of the correlation coefficient, $r = 0.067$, obtained in Example 11–6, at $\alpha = 0.01$.

Solution

Since the sample size is 9, there are 7 degrees of freedom. When $\alpha = 0.01$ and with 7 degrees of freedom, the value obtained from Table I is 0.798. For a significant relationship, a value of r greater than $+0.798$ or less than -0.798 is needed. Since $r = 0.067$, the null hypothesis is not rejected. Hence, there is not enough evidence to say that there is a significant relationship between the variables. See Figure 11–9.

Figure 11–9

Rejection and
Nonrejection Regions
for Example 11–8

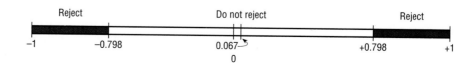

Reject   Do not reject   Reject

-1   -0.798   0.067   +0.798   +1
                          0

Correlation and Causation

Researchers must understand the nature of the relationship between the independent variable x and the dependent variable y. When a hypothesis test indicates that a significant relationship exists between the variables, researchers must consider the possibilities outlined next.

Possible Relationships between Variables

When the null hypothesis has been rejected for a specific α value, any of the following five possibilities can exist.

1. *There is a direct cause-and-effect relationship between the variables.* That is, x causes y. For example, water causes plants to grow, poison causes death, and heat causes ice to melt.

2. *There is a reverse cause-and-effect relationship between the variables.* That is, y causes x. For example, suppose a researcher believes excessive coffee consumption causes nervousness, but the researcher fails to consider that the reverse situation may occur. That is, it may be that an extremely nervous person craves coffee to calm his or her nerves.

3. *The relationship between the variables may be caused by a third variable.* For example, if a statistician correlated the number of deaths due to drowning and the number of cans of soft drink consumed during the summer, he or she would probably find a significant relationship. However, the soft drink is not necessarily responsible for the deaths, since both variables may be related to heat and humidity.

4. *There may be a complexity of interrelationships among many variables.* For example, a researcher may find a significant relationship between students' high school grades and college grades. But there probably are many other variables involved, such as IQ, hours of study, influence of parents, motivation, age, and instructors.

5. *The relationship may be coincidental.* For example, a researcher may be able to find a significant relationship between the increase in the number of people who are exercising and the increase in the number of people who are committing crimes. But common sense dictates that any relationship between these two values must be due to coincidence.

In correlation and regression studies, it is difficult to control all variables. This study shows some of the consequences when researchers overlook certain aspects in studies. Suggest ways that the extraneous variables might be controlled in future studies.

Coffee Not Disease Culprit, Study Says

NEW YORK (AP)—Two new studies suggest that coffee drinking, even up to 5 1/2 cups per day, does not increase the risk of heart disease, and other studies that claim to have found increased risks might have missed the true culprits, a researcher says.

"It might not be the coffee cup in one hand, it might be the cigarette or coffee roll in the other," said Dr. Peter W.F. Wilson, the author of one of the new studies.

He noted in a telephone interview Thursday that many coffee drinkers, particularly heavy coffee drinkers, are smokers. And one of the new studies found that coffee drinkers had excess fat in their diets.

The findings of the new studies conflict sharply with a study reported in November 1985 by Johns Hopkins University scientists in Baltimore.

The Hopkins scientists found that coffee drinkers who consumed five or more cups of coffee per day had three times the heart-disease risk of non-coffee drinkers.

The reason for the discrepancy appears to be that many of the coffee drinkers in the Hopkins study also smoked—and it was the smoking that increased their heart-disease risk, said Wilson.

Wilson, director of laboratories for the Framingham Heart Study in Framingham, Mass., said Thursday at a conference sponsored by the American Heart Association in Charleston, S.C., that he had examined the coffee intake of 3,937 participants in the Framingham study during 1956–66 and an additional 2,277 during the years 1972–1982.

In contrast to the subjects in the Hopkins study, most of these coffee drinkers consumed two or three cups per day, Wilson said. Only 10 percent drank six or more cups per day.

He then looked at blood cholesterol levels and heart and blood vessel disease in the two groups. "We ran these analyses for coronary heart disease, heart attack, sudden death and stroke and in absolutely every analysis, we found no link with coffee," Wilson said.

He found that coffee consumption was linked to a significant decrease in total blood cholesterol in men, and to a moderate increase in total cholesterol in women.

Source: Associated Press. Reprinted with permission.

Thus, when the null hypothesis is rejected, the researcher must consider all possibilities and select the appropriate one as determined by the study. Remember, correlation does not necessarily imply causation.

Exercises

11–1. (**W**) What is meant by the statement that two variables are related?

11–2. (**W**) How is a relationship between two variables measured in statistics? Explain.

11–3. (**W**) What is the symbol for the sample correlation coefficient? The population correlation coefficient?

11–4. (**W**) What is the range of values for the correlation coefficient?

11–5. (**W**) What is meant when the relationship between the two variables is positive? Negative?

11–6. (**W**) Give examples of two variables that are positively correlated and two that are negatively correlated.

11–7. (**W**) Give an example of a correlation study and identify the independent and dependent variables.

11–8. (**W**) What is the diagram of the independent and dependent variables called? Why is drawing this diagram important?

11–9. (**W**) What is the name of the correlation coefficient used in this section?

11–10. (**W**) What statistical test is used to test the significance of the correlation coefficient?

11–11. (**W**) When two variables are correlated, can the researcher be sure that one variable causes the other? Why or why not?

For Exercises 11–12 through 11–27, perform the following steps.
a. Draw the scatter plot for the variables.
b. Compute the value of the correlation coefficient.
c. State the hypotheses.
d. Test the significance of the correlation coefficient at $\alpha = 0.05$, using Table I.
e. Give a brief explanation of the type of relationship.

11–12. A manager wishes to find out whether there is a relationship between the number of radio ads aired per week and the amount of sales (in thousands of dollars) of a product. The data for the sample follow. (The information in this exercise will be used for Exercises 11–40 and 11–64.)

No. of ads, x	2	5	8	8	10	12
Sales, y	$2	$4	$7	$6	$9	$10

11–13. A researcher wishes to determine if a person's age is related to the number of hours he or she exercises per week. The data for the sample are shown here. (The information in this exercise will be used for Exercises 11–41, 11–64, and 11–81.)

Age, x	18	26	32	38	52	59
Hours, y	10	5	2	3	1.5	1

11–14. A study was conducted to determine the relationship between a person's monthly income and the number of meals that person eats away from home per month. The data from the sample are shown here. (The information in this exercise will be used for Exercises 11–42, 11–64, and 11–82.)

Income, x	$500	$1200	$1500	$945	$850	$400	$540
Meals, y	8	12	16	10	9	3	7

11–15. The director of an alumni association for a small college wants to determine whether there is any type of relationship between the amount of an alumnus's contribution and the years the alumnus has been out of school. The data follow. (The information in this exercise will be used for Exercises 11–43, 11–65, and 11–83.)

Years, x	1	5	3	10	7	6
Contribution, y	$500	$100	$300	$50	$75	$80

11–16. A store manager wishes to find out whether there is a relationship between the age of her employees and the number of sick days they take each year. The data for the sample follow. (The information in this exercise will be used for Exercises 11–44, 11–65, and 11–84.)

Age	18	26	39	48	53	58
Days	16	12	9	5	6	2

11–17. An educator wants to see how strong the relationship is between a student's score on a test and his or her grade point average. The data obtained from the sample follow. (The information in this exercise will be used for Exercises 11–45 and 11–65.)

Test score, x	98	105	100	100	106	95	116	112
GPA, y	2.1	2.4	3.2	2.7	2.2	2.3	3.8	3.4

11–18. An English instructor is interested in finding the strength of a relationship between the final exam grades of students enrolled in Composition I and Composition II classes. The data are given below in percentages. (The information in this exercise will be used for Exercises 11–46 and 11–66.)

Comp I, x	83	97	80	95	73	78	91	86
Comp II, y	78	95	83	97	78	72	90	80

11–19. An insurance company wants to determine the strength of the relationship between the number of hours a person works per week and the number of injuries or accidents that person has. The data follow. (The information in this exercise will be used for Exercises 11–47 and 11–66.)

Hours worked, x	40	32	36	44	41
No. of accidents, y	1	0	3	8	5

11–20. An emergency service wishes to see whether a relationship exists between the outside temperature and the number of emergency calls it receives. The data follow. (The information in this exercise will be used for Exercises 11–48 and 11–66.)

Temperature, x	68	74	82	88	93	99	101
No. of calls, y	7	4	8	10	11	9	13

11–21. A researcher wishes to see if there is a relationship between the number of calories a sandwich has and the sodium content of the sandwich. Several different types of sandwiches are used. The data follow. (The information in this exercise will be used in Exercise 11–49.)

Calories, x	419	419	386	240	354	231	174
Sodium (mg), y	495	645	1202	581	990	1159	787

Source: *Consumer Reports*, September 1995, p. 588.

11–22. A medical researcher wishes to determine how the dosage (in milligrams) of a drug affects the heart rate of the patient. The data for seven patients are given here. (The information in this exercise will be used for Exercise 11–50.)

Drug dosage, x	0.125	0.20	0.25	0.30	0.35	0.40	0.50
Heart rate, y	95	90	93	92	88	80	82

11–23. A researcher wishes to determine if there is a relationship between the energy efficiency rating (EER) of an air conditioner and its cost. The EER is determined by the Appliance Manufacturers' Association. The data follow. (The information in this exercise will be used for Exercise 11–51.)

EER, x	9.5	10	10	10	10.3	9.1	8.6
Cost, y	$370	360	400	400	420	350	310

Source: *Consumer Reports*, June 1995, p. 406.

11–24. In the air conditioner study cited in the previous exercise, the researcher also wishes to see if there is a relationship between the air conditioner's moisture removal capacity (in pints per hour) and its weight (in pounds). The data are shown here. (The information in this exercise will be used for Exercise 11–52.)

Moisture removal capacity, x	1.9	1.6	2.0	1.6	2.0	2.0	2.1
Weight, y	57	60	72	55	69	55	48

11–25. A psychologist selects six families with two children each, a boy and a girl. She is interested in comparing the IQs of each gender to determine whether there is a relationship between them. The data are given here. (The information in this exercise will be used for Exercise 11–53.)

IQ, females, x	107	95	116	109	101	98
IQ, males, y	107	102	112	104	105	103

11–26. A researcher wishes to determine whether there is a relationship between the age of a copy machine and the monthly maintenance cost. The data follow. (The information in this exercise will be used for Exercise 11–54.)

Age, x	3	5	2	1	2	4	3
Cost, y	$80	$100	$75	$60	$80	$93	$84

11–27. A study was conducted at a large hospital to determine whether there was a relationship between the number of years nurses were employed and the number of nurses who voluntarily resign. The results are shown below. (The information in this exercise will be used for Exercise 11–28 and 11–55.)

Years employed, x	5	6	3	7	8	2
Resignations, y	7	4	7	1	2	8

*11–28. One of the formulas for computing r is

$$r = \frac{\Sigma(x - \bar{x})(y - \bar{y})}{(n - 1)(s_x)(s_y)}$$

Using the data in Exercise 11–27, compute r with this formula. Compare the results.

*11–29. (W) Compute r for the following data set. Explain the reason for this value of r. Now, interchange the values of x and y and compute r again. Compare this value with the previous one. Explain the results of the comparison.

x	1	2	3	4	5
y	3	5	7	9	11

*11–30. (W) Compute r for the following data and test the hypothesis $H_0: \rho = 0$. Draw the scatter plot; then explain the results.

x	−3	−2	−1	0	1	2	3
y	9	4	1	0	1	4	9

11–4

Regression

Objective 4. Find the equation of the regression line.

In studying relationships between two variables, collect the data and then construct a scatter plot. The purpose of the scatter plot, as indicated previously, is to determine the nature of the relationship. The possibilities include a positive linear relationship, a negative linear relationship, a curvilinear relationship, or no discernible relationship. After the scatter plot is drawn, the next steps are to compute the value of the correlation coefficient and to test the significance of the relationship. If the value of the correlation coefficient is significant, the next step is to determine the equation of the **regression line,** which is the data's line of best fit. (*Note:* Determining the regression line when *r* is not significant and then making predictions using the regression line is meaningless.) The purpose of the regression line is to enable the researcher to see the trend and make predictions on the basis of the data.

Line of Best Fit

Figure 11–10 shows a scatter plot for the data of two variables. It also shows that several lines can be drawn on the graph near the points. Given a scatter plot, one must be able to draw the *line of best fit. Best fit* means that the sum of the squares of the vertical distances from each point to the line are at a minimum. The reason one needs a line of best fit is that the values of *y* will be predicted from the values of *x*; hence, the closer the points are to the line, the better the fit and the prediction will be. See Figure 11–11.

Figure 11–10

Scatter Plot with Three Lines Fit to the Data

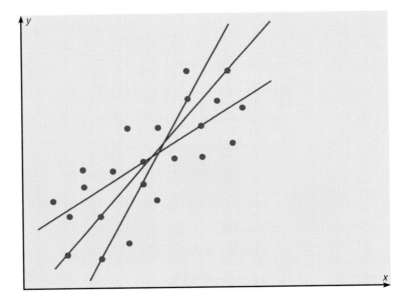

Historical Note

Francis Galton drew the line of best fit visually. An assistant of Karl Pearson's named G. Yule devised the mathematical solution using the least-squares method, employing a mathematical technique developed by Adrien Marie Legendre about 100 years earlier.

Figure 11–12 shows the relationship between the values of the correlation coefficient and the variability of the scores about the regression line. The closer the points fit the regression line, the higher the absolute value of *r* is and the closer it will be to +1 or −1. Figure 11–12(a) shows *r* as positive and approximately equal to 0.50; Figure 11–12(b) shows *r* approximately equal to 0.90; and Figure 11–12(c) shows *r* equal to +1. For positive values of *r*, the line slopes upward from left to right.

Figure 11–11

Line of Best Fit for a
Set of Data Points

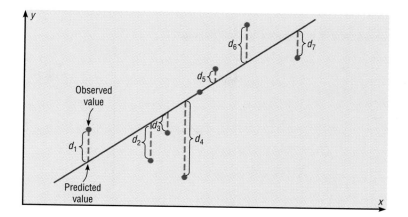

Figure 11–12

Relationship between
the Correlation
Coefficient and the Line
of Best Fit

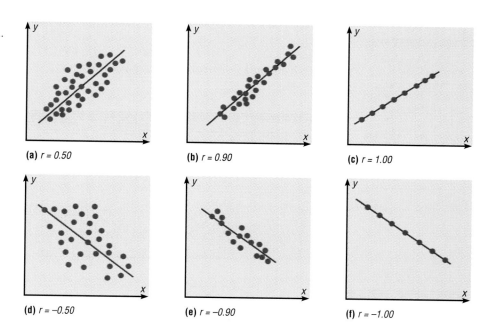

(a) r = 0.50

(b) r = 0.90

(c) r = 1.00

(d) r = −0.50

(e) r = −0.90

(f) r = −1.00

Figure 11–12(d) shows r as negative and approximately equal to -0.50; Figure 11–12(e) shows r approximately equal to -0.90; and Figure 11–12(f) shows r equal to -1. For negative values of r, the line slopes downward from left to right.

**Determination of the
Regression Line
Equation**

In algebra, the equation of a line is usually given as $y = mx + b$, where m is the slope of the line and b is the y intercept. (Students who need an algebraic review of the properties of a line should refer to Appendix A Section A–3 before studying this section.) In statistics, the equation of the regression line is written as $y' = a + bx$, where a is the y' intercept and b is the slope of the line. See Figure 11–13.

Figure 11–13

A Line as Represented in Algebra and in Statistics

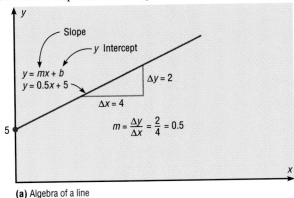

(a) Algebra of a line

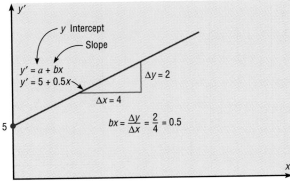

(b) Statistical notation for a regression line

Historical Note

In 1795, Adrien Marie Legendre (1752–1833) measured the meridian arc on the earth's surface from Barcelona, Spain, to Dunkirk, England. This measure was used as the basis for the measure of the meter. Legendre developed the least-squares method around the year 1805.

There are several methods for finding the equation of the regression line. Two formulas are given here. *These formulas use the same values that are used in computing the value of the correlation coefficient.* The mathematical development of these formulas is beyond the scope of this book.

Formulas for the Regression Line $y' = a + bx$

$$a = \frac{(\sum y)(\sum x^2) - (\sum x)(\sum xy)}{n(\sum x^2) - (\sum x)^2}$$

$$b = \frac{n(\sum xy) - (\sum x)(\sum y)}{n(\sum x^2) - (\sum x)^2}$$

where a is the y' intercept and b is the slope of the line.
Note: Round the values of a and b to three decimal places.

Example 11–9

Find the equation of the regression line for the data in Example 11–4, and graph the line on the scatter plot of the data.

Solution

The values needed for the equation are $n = 6$, $\sum x = 345$, $\sum y = 819$, $\sum xy = 47,634$, and $\sum x^2 = 20,399$. Substituting in the formulas, one gets

$$a = \frac{(\sum y)(\sum x^2) - (\sum x)(\sum xy)}{n(\sum x^2) - (\sum x)^2} = \frac{(819)(20,399) - (345)(47,634)}{(6)(20,399) - (345)^2} = 81.048$$

$$b = \frac{n(\sum xy) - (\sum x)(\sum y)}{n(\sum x^2) - (\sum x)^2} = \frac{(6)(47,634) - (345)(819)}{(6)(20,399) - (345)^2} = 0.964$$

Hence, the equation of the regression line, $y' = a + bx$, is

$$y' = 81.048 + 0.964x$$

The graph of the line is shown in Figure 11–14.

Figure 11–14

Regression Line for
Example 11–9

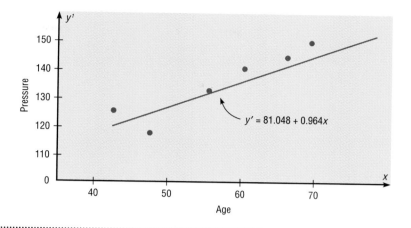

Note: When drawing the scatter plot and the regression line, it is sometimes desirable to *truncate* the graph (see Chapter 2). The reason is to show the line drawn in the range of the independent and dependent variables. For example, the regression line in Figure 11–14 is drawn between the x values of approximately 43 and 82 and the y values of approximately 120 and 152. The range of the x values in the original data shown in Example 11–4 is $70 - 43 = 27$ and the range of the y values is $152 - 120 = 32$. Notice that the x axis has been truncated; the distance between 0 and 40 is not shown in the proper scale compared to the distance between 40 and 50, 50 and 60, etc. The y axis has been similarly truncated.

The important thing to remember is that when the x axis and sometimes the y axis have been truncated, do not use the y-intercept value, a, to graph the line. To be on the safe side when graphing the regression line, use a value for x selected from the range of x values.

Example 11–10

Find the equation of the regression line for the data in Example 11–5, and graph the line on the scatter plot.

Solution

The values needed for the equation are $n = 7$, $\sum x = 57$, $\sum y = 511$, $\sum xy = 3745$, and $\sum x^2 = 579$. Substituting in the formulas, one gets

$$a = \frac{(\sum y)(\sum x^2) - (\sum x)(\sum xy)}{n(\sum x^2) - (\sum x)^2} = \frac{(511)(579) - (57)(3745)}{(7)(579) - (57)^2} = 102.493$$

$$b = \frac{n(\sum xy) - (\sum x)(\sum y)}{n(\sum x^2) - (\sum x)^2} = \frac{(7)(3745) - (57)(511)}{(7)(579) - (57)^2} = -3.622$$

Hence, the equation of the regression line, $y' = a + bx$, is

$$y' = 102.493 - 3.622x$$

The graph of the line is shown in Figure 11–15.

Figure 11–15

Regression Line for Example 11–10

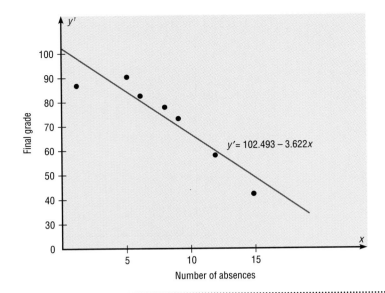

$y' = 102.493 - 3.622x$

The sign of the correlation coefficient and the sign of the slope of the regression line will always be the same. That is, if r is positive, then b will be positive; if r is negative, then b will be negative. The reason is that the numerators of the formulas are the same and determine the signs of r and b, and the denominators are always positive.

The regression line can be used to make predictions for the dependent variable. The method for making predictions is shown in the next example.

Example 11–11

Using the equation of the regression line found in Example 11–9, predict the blood pressure for a person who is 50 years old.

Solution

Substituting 50 for x in the regression line $y' = 81.048 + 0.964x$ gives

$$y' = 81.048 + (0.964)(50) = 129.248 \text{ (rounded to 129)}$$

The value obtained in Example 11–11 is a point prediction, and with point predictions, no degree of accuracy or confidence can be determined. More information on prediction is given in Section 11–5, including prediction intervals.

When r is not significantly different from 0, the best predictor of y is the mean of the data values of y. For valid predictions, the value of the correlation coefficient must be significant. Also, two other assumptions must be met.

Assumptions for Valid Predictions in Regression

1. For any specific value of the independent variable x, the value of the dependent variable y must be normally distributed about the regression line. See Figure 11–16(a).
2. The standard deviation of each of the dependent variables must be the same for each value of the independent variable. See Figure 11–16(b).

Speaking of STATISTICS

Here is a relationship study. Identify the two variables used in the study. Which one is the independent variable? Which is the dependent variable? Do you think the aging variable should be ruled out completely? Explain your answer.

Clear Thinking: Do Clogged Vessels Muddle Thought?

If you want your brain to stay in the fast lane forever, take a hike today. Research now offers more evidence that what you do to keep your body fit may also tone your mind. A look inside the arteries of nearly 5,000 people in the Netherlands (ages 55 to 94) plus some written tests taken by their owners gives credence to the suspicion that often it's not age but arteries that are at the root of mental decline. These physical and mental tests showed that people with artery disease (including built-up plaque or previous heart attacks and strokes) had lower mental performance than did people whose arteries were clean, regardless of age (*British Medical Journal*, June 18, 1994).

This suggests that a mushy mind isn't an inevitable part of aging. Sure, there's more than one cause of mental decline, but "this study refocuses attention on something that can be treated or prevented, namely, atherosclerosis," says Vladimir Hachinski, M.D., Richard and Beryl Ivey Professor, and chairman, department of clinical neurological science, University of Western Ontario.

Thinking trouble may start when buildups in the arteries break off and block small vessels in the brain. For a while, the brain can create new pathways around the damaged parts. But when the new paths get blocked, the damage is dramatic because you've used up the brain's capacity to repair itself. "Lesions in the brain don't just add up, they multiply," Dr. Hachinski explains.

All the more reason for prevention. "Certainly, one would want to recommend exercise and clean living [read: eating the right things]," he says. "But that's not the whole story. People who already are at high risk can get antiplatelet agents like aspirin."

A good checkup should be enough to determine if you're in a mindblowing state and need artery-cleaning treatment. Otherwise, it's smart to start now with "clean living" and regular checkups.—**by Marty Munson and Teresa Yeykal**

Reprinted by permission of *Prevention* Magazine. Copyright 1994 Rodale Press, Inc. All rights reserved.
Note: This excerpt is for reading comprehension only. For the most up-to-date information about health, please consult a health professional.

Figure 11–16

Assumptions for Predictions

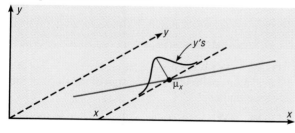

(a) Dependent variable *y* normally distributed

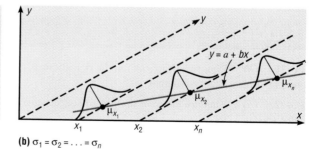

(b) $\sigma_1 = \sigma_2 = \ldots = \sigma_n$

When predictions are made beyond the bounds of the data, they must be interpreted cautiously. For example, in 1979, some experts predicted that the United States would run out of oil by the year 2003. This prediction was based on the current consumption and on known oil reserves at that time. However, since then, the automobile industry has produced many new fuel-efficient vehicles. Also, there are many as yet undiscovered oil fields. Finally, science may someday discover a way to run a car on something as unlikely but as common as peanut oil. In addition, the price of a gallon of gasoline was predicted to reach $10 a few years later. Fortunately this has not come to pass. *Remember that when predictions are made, they are based on present conditions or on the premise that present trends will continue.* This assumption may or may not prove true in the future.

The procedures for finding the value of the correlation coefficient and the regression line equation are summarized in Procedure Table 9.

Procedure Table 9

Finding the Correlation Coefficient and the Regression Line Equation

STEP 1: Make a table, as shown in Step 2.

STEP 2: Find the values of xy, x^2, and y^2. Place them in the appropriate columns and sum each column.

x	y	xy	x^2	y^2
.	.	.	.	.
.	.	.	.	.
.	.	.	.	.
$\sum x =$	$\sum y =$	$\sum xy =$	$\sum x^2 =$	$\sum y^2 =$

STEP 3: Substitute in the formula to find the value of r.

$$r = \frac{n(\sum xy) - (\sum x)(\sum y)}{\sqrt{[n(\sum x^2) - (\sum x)^2][n(\sum y^2) - (\sum y)^2]}}$$

STEP 4: When r is significant, substitute in the formulas to find the values of a and b for the regression line equation $y' = a + bx$.

$$a = \frac{(\sum y)(\sum x^2) - (\sum x)(\sum xy)}{n(\sum x^2) - (\sum x)^2} \qquad b = \frac{n(\sum xy) - (\sum x)(\sum y)}{n(\sum x^2) - (\sum x)^2}$$

Exercises

11–31. (W) What two things should be done before one performs a regression analysis?

11–32. (W) What are the assumptions for regression analysis?

11–33. (W) What is the general form for the regression line used in statistics?

11–34. (W) What is the symbol for the slope? For the y intercept?

11–35. (W) What is meant by the *line of best fit?*

11–36. (W) When all the points fall on the regression line, what is the value of the correlation coefficient?

11–37. (W) What is the relationship of the sign of the correlation coefficient and the sign of the slope of the regression line?

11–38. (W) As the value of the correlation coefficient increases from 0 to 1, or decreases from 0 to −1, how do the points of the scatter plot fit the regression line?

11–39. (W) How is the value of the correlation coefficient related to the accuracy of the predicted value for a specific value of x?

11–40. For the data in Exercise 11–12, find the equation of the regression line, and predict y' when $x = 7$ ads.

11–41. For the data in Exercise 11–13, find the equation of the regression line, and predict y' when $x = 35$ years.

11–42. For the data in Exercise 11–14, find the equation of the regression line, and predict y' when $x = \$1100$.

11–43. For the data in Exercise 11–15, find the equation of the regression line, and predict y' when $x = 4$ years.

11–44. For the data in Exercise 11–16, find the equation of the regression line, and predict y' when $x = 47$ years.

11–45. For the data in Exercise 11–17, find the equation of the regression line, and predict y' when $x = 104$.

11–46. For the data in Exercise 11–18, find the equation of the regression line, and predict y' when $x = 88$.

11–47. For the data in Exercise 11–19, find the equation of the regression line, and predict y' when $x = 35$ hours.

11–48. For the data in Exercise 11–20, find the equation of the regression line, and predict y' when $x = 80°$.

11–49. For the data in Exercise 11–21, find the equation of the regression line, and predict y' when $x = 400$ calories.

11–50. For the data in Exercise 11–22, find the equation of the regression line, and predict y' when $x = 0.27$ milligram.

11–51. For the data in Exercise 11–23, find the equation of the regression line, and predict y' when $x = 9$ EER.

11–52. For the data in Exercise 11–24, find the equation of the regression line, and predict y' when $x = 1.8$ pints.

11–53. For the data in Exercise 11–25, find the equation of the regression line, and predict y' when $x = 104$.

11–54. For the data in Exercise 11–26, find the equation of the regression line, and predict y' when $x = 4$ years.

11–55. For the data in Exercise 11–27, find the equation of the regression line, and predict y' when $x = 4$ years.

LAFF-A-DAY

"Explain that to me."

Source: Reprinted with special permission of King Features Syndicate, Inc.

For Exercises 11–56 through 11–63, do a complete regression analysis by performing the following steps.

a. Draw a scatter plot.
b. Compute the correlation coefficient.
c. State the hypotheses.
d. Test the hypotheses at $\alpha = 0.05$.
e. Determine the regression line equation.
f. Plot the regression line on the scatter plot.
g. Summarize the results.

11–56. The following data were obtained for the amount of rainfall (in inches) and the yield of bushels of oats per acre for a selected sample. Predict the yield when there are 5 inches of rain.

Rainfall, x	2.5	3	5.7	8	9.3	9.7	10.2
Bushels, y	38	42	41	45	46	44	48

11–57. The following data were obtained from a survey of the number of years people smoked and the percentage of lung damage. Predict the percentage of lung damage for a person who has smoked for 30 years.

Years, x	22	14	31	36	9	41	19
Damage, y	20	14	54	63	17	71	23

11–58. A researcher wishes to determine whether a person's age is related to the number of hours he or she jogs per week. The data for the sample follow.

Age, x	34	22	48	56	62
Hours, y	5.5	7	3.5	3	1

11–59. A study was conducted to determine the relationship between monthly income and the amount a person spends per month for recreation. The data from the sample follow.

Income, x	$800	$1200	$1000	$900	$850	$907	$1100
Amount, y	$60	$200	$160	$135	$45	$90	$150

11–60. A statistics instructor is interested in finding the strength of a relationship between the final exam grades of students enrolled in Statistics I and in Statistics II. The data are given here in percentages.

Statistics I, x	87	92	68	72	95	78	83	98
Statistics II, y	83	88	70	74	90	74	83	99

11–61. An educator wants to see how the number of absences a student in her class has affects the student's final grade. The data obtained from a sample follow.

No. of absences, x	10	12	2	0	8	5
Final grade, y	70	65	96	94	75	82

11–62. A physician wishes to know whether there is a relationship between a father's weight (in pounds) and his newborn son's weight (in pounds). The data are given here.

Father's weight, x	176	160	187	210	196	142	205	215
Son's weight, y	6.6	8.2	9.2	7.1	8.8	9.3	7.4	8.6

11–63. A sales manager wishes to determine the relationship between the number of years of experience a salesperson has and the amount of his or her monthly sales. The data are given below.

Years of experience, x	2	10	4	6	2
Sales, y	$1000	$5000	$3000	$5000	$3000

*11–64. For Exercises 11–12, 11–13, and 11–14, find the mean of the x and y variables, then substitute the mean of the x variable into the corresponding regression line equations found in Exercises 11–40, 11–41, and 11–42 and find y'. Compare the value of y' with $\bar{y}$ for each exercise. Generalize the results.

*11–65. The y intercept value, a, can also be found by using the following equation:

$$a = \bar{y} - b\bar{x}$$

Verify this result by using the data in Exercises 11–15 and 11–43, 11–16 and 11–44, and 11–17 and 11–45.

*11–66. The value of the correlation coefficient can also be found by using the formula

$$r = \frac{bs_x}{s_y}$$

Verify this result for Exercises 11–18, 11–19, and 11–20.

11–5

Coefficient of Determination and Standard Error of Estimate

The previous sections stated that if the correlation coefficient is significant, the equation of the regression line can be determined. Also, for various values of the independent variable x, the corresponding values of the dependent variable y can be predicted. Several other measures are associated with the correlation and regression techniques. They include the coefficient of determination, the standard error of estimate, and the prediction interval. But before these concepts can be explained, the different types of variation associated with the regression model must be defined.

Types of Variation for the Regression Model

Consider the following hypothetical regression model.

x	1	2	3	4	5
y	10	8	12	16	20

The equation of the regression line is $y' = 4.8 + 2.8x$, and $r = 0.919$. The sample y values are 10, 8, 12, 16, and 20. The predicted values, designated by y', for each x can be found by substituting each x value into the regression equation and finding y'. For example, when $x = 1$,

$$y' = 4.8 + 2.8x = 4.8 + (2.8)(1) = 7.6$$

Now, for each x, there are an observed y value and a predicted y' value. For example, when $x = 1$, $y = 10$ and $y' = 7.6$. Recall that the closer the observed values are to the predicted values, the better the fit is and the closer r is to $+1$ or -1.

The total variation $\sum (y - \bar{y})^2$ is the sum of the squares of the vertical distances each point is from the mean. The total variation can be divided into two parts: that which is attributed to the relationship of x and y and that which is due to chance. The variation obtained from the relationship (i.e., from the predicted y' values) is $\sum (y' - \bar{y})^2$ and is called the *explained variation*. Most of the variations can be explained by the relationship. The closer the value r is to $+1$ or -1, the better the points fit the line, and the closer $\sum (y' - \bar{y})^2$ is to $\sum (y - \bar{y})^2$. In fact, if all points fall on the regression line, $\sum (y' - \bar{y})^2$ will equal $\sum (y - \bar{y})^2$, since y' would be equal to y in each case.

On the other hand, the variation due to chance, found by $\sum (y - y')^2$, is called the *unexplained variation*. This variation cannot be attributed to the relationships. When the unexplained variation is small, the value of r is close to $+1$ or -1. If all points fall on the regression line, the unexplained variation, $\sum (y - y')^2$, will be 0. Hence, the total variation is equal to the sum of the explained variation and the unexplained variation. That is,

$$\sum (y - \bar{y})^2 = \sum (y' - \bar{y})^2 + \sum (y - y')^2$$

These values are shown in Figure 11–17. For a single point, the differences are called *deviations*. For the hypothetical regression model given earlier, for $x = 1$ and $y = 10$, one gets $y' = 7.6$ and $\bar{y} = 13.2$.

Figure 11–17

Deviations for the Regression Equation

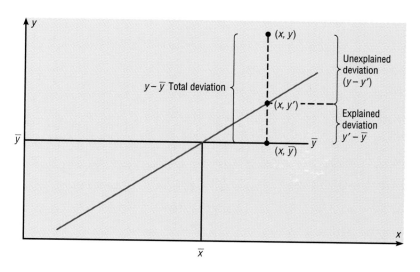

The procedure for finding the three types of variation is illustrated next.

STEP 1   Find the predicted y' values.

$$
\begin{array}{ll}
\text{for } x = 1 & y' = 4.8 + 2.8x = 4.8 + (2.8)(1) = 7.6 \\
\text{for } x = 2 & y' = 4.8 + (2.8)(2) = 10.4 \\
\text{for } x = 3 & y' = 4.8 + (2.8)(3) = 13.2 \\
\text{for } x = 4 & y' = 4.8 + (2.8)(4) = 16 \\
\text{for } x = 5 & y' = 4.8 + (2.8)(5) = 18.8
\end{array}
$$

Hence, the values for this example are as follows:

x	y	y'
1	10	7.6
2	8	10.4
3	12	13.2
4	16	16
5	20	18.8

STEP 2   Find the mean of the y values.

$$\bar{y} = \frac{10 + 8 + 12 + 16 + 20}{5} = 13.2$$

STEP 3   Find the total variation, $\sum (y - \bar{y})^2$.

$$(10 - 13.2)^2 = 10.24$$
$$(8 - 13.2)^2 = 27.04$$
$$(12 - 13.2)^2 = 1.44$$
$$(16 - 13.2)^2 = 7.84$$
$$(20 - 13.2)^2 = \underline{46.24}$$
$$\sum (y - \bar{y})^2 = 92.8$$

STEP 4   Find the explained variation, $\sum (y' - \bar{y})^2$.

$$(7.6 - 13.2)^2 = 31.36$$
$$(10.4 - 13.2)^2 = 7.84$$
$$(13.2 - 13.2)^2 = 0.00$$
$$(16 - 13.2)^2 = 7.84$$
$$(18.8 - 13.2)^2 = \underline{31.36}$$
$$\sum (y' - \bar{y})^2 = 78.4$$

STEP 5   Find the unexplained variation, $\sum (y - y')^2$.

$$(10 - 7.6)^2 = 5.76$$
$$(8 - 10.4)^2 = 5.76$$
$$(12 - 13.2)^2 = 1.44$$
$$(16 - 16)^2 = 0.00$$
$$(20 - 18.8)^2 = \underline{1.44}$$
$$\sum (y - y')^2 = 14.4$$

Notice that

$$\text{total variation} = \text{explained variation} + \text{unexplained variation}$$
$$92.8 \quad = \quad 78.4 \quad + \quad 14.4$$

Coefficient of Determination

Objective 5. Find the coefficient of determination.

The *coefficient of determination* is the ratio of the explained variation to the total variation and is denoted by r^2. That is,

$$r^2 = \frac{\text{explained variation}}{\text{total variation}}$$

Historical Note

Karl Pearson recommended in 1897 that the French government close all its casinos and turn the gambling devices over to the academic community to use in the study of probability.

For the example, $r^2 = 78.4/92.8 = 0.845$. The term r^2 is usually expressed as a percentage. So in this case, 84.5% of the total variation is explained by the regression line using the independent variable.

Another way to arrive at the value for r^2 is to square the correlation coefficient. In this case, $r = 0.919$, and $r^2 = 0.845$, which is the same value found by using the variation ratio.

The **coefficient of determination** is a measure of the variation of the dependent variable that is explained by the regression line and the independent variable. The symbol for the coefficient of determination is r^2.

Of course, it is usually easier to find the coefficient of determination by squaring r and converting it to a percentage. Therefore, if $r = 0.90$, then $r^2 = 0.81$, which is equivalent to 81%. This result means that 81% of the variation in the dependent variable is accounted for by the variations in the independent variable. The rest of the variation, 0.19 or 19%, is unexplained. This value is called the *coefficient of nondetermination* and is found by subtracting the coefficient of determination from 1. As the value of r approaches 0, r^2 decreases more rapidly. For example, if $r = 0.6$, then $r^2 = 0.36$, which means that only 36% of the variation in the dependent variable can be attributed to the variation in the independent variable.

Formula for the Coefficient of Nondetermination

$$1.00 - r^2$$

Standard Error of Estimate

Objective 6. Find the standard error of estimate.

When a y' value is predicted for a specific x value, the prediction is a point prediction. However, a prediction interval about the y' value can be constructed, just as a confidence interval was constructed for an estimate of the population mean. The prediction interval uses a statistic called the *standard error of estimate*.

The **standard error of estimate**, denoted by s_{est} is the standard deviation of the observed y values about the predicted y' values. The formula for the standard error of estimate is

$$s_{est} = \sqrt{\frac{\Sigma (y - y')^2}{n - 2}}$$

The standard error of estimate is similar to the standard deviation, but the mean is not used. As can be seen from the formula, the standard error of estimate is the square root of the unexplained variation—i.e., the variation due to the difference of the observed values and the expected values—divided by $n - 2$. So the closer the observed values are to the predicted values, the smaller the standard error of estimate will be.

The next example shows how to compute the standard error of estimate.

Example 11–12

A researcher collects the following data and determines that there is a significant relationship between the age of a copy machine and its monthly maintenance cost. The regression equation is $y' = 55.57 + 8.13x$. Find the standard error of estimate.

Machine	Age, x (Years)	Monthly cost, y
A	1	$ 62
B	2	78
C	3	70
D	4	90
E	4	93
F	6	103

Solution

STEP 1 Make a table, as shown.

x	y	y'	$y - y'$	$(y - y')^2$
1	62			
2	78			
3	70			
4	90			
4	93			
6	103			

STEP 2 Using the regression line equation, $y' = 55.57 + 8.13x$, compute the predicted values y' for each x and place the results in the column labeled y'.

$$x = 1 \quad y' = 55.57 + (8.13)(1) = 63.7$$
$$x = 2 \quad y' = 55.57 + (8.13)(2) = 71.83$$
$$x = 3 \quad y' = 55.57 + (8.13)(3) = 79.96$$
$$x = 4 \quad y' = 55.57 + (8.13)(4) = 88.09$$
$$x = 6 \quad y' = 55.57 + (8.13)(6) = 104.35$$

STEP 3 For each y, subtract y' and place the answer in the column labeled $y - y'$.

$$62 - 63.7 = -1.7 \qquad 90 - (88.09) = 1.91$$
$$78 - 71.83 = 6.17 \qquad 93 - (88.09) = 4.91$$
$$70 - 79.96 = -9.96 \qquad 103 - (104.35) = -1.35$$

STEP 4 Square the numbers found in Step 3 and place the squares in the column labeled $(y - y')^2$.

STEP 5 Find the sum of the numbers in the last column. The completed table follows.

x	y	y'	$y - y'$	$(y - y')^2$
1	62	63.7	-1.7	2.89
2	78	71.83	6.17	38.0689
3	70	79.96	-9.96	99.2016
4	90	88.09	1.91	3.6481
4	93	88.09	4.91	24.1081
6	103	104.35	-1.35	1.8225
				169.7392

STEP 6 Substitute in the formula and find s_{est}.

$$s_{est} = \sqrt{\frac{\Sigma (y - y')^2}{n - 2}} = \sqrt{\frac{169.7392}{6 - 2}} = 6.51$$

In this case, the standard deviation of observed values about the predicted values is 6.51.

The standard error of estimate can also be found by using the formula

$$s_{est} = \sqrt{\frac{\Sigma y^2 - a \Sigma y - b \Sigma xy}{n - 2}}$$

Example 11–13

Find the standard error of estimate for the data for Example 11–12 by using the preceding formula. The equation of the regression line is $y' = 55.57 + 8.13x$.

Solution

STEP 1 Make a table.

STEP 2 Find the product of x and y values and place the results in the third column.

STEP 3 Square the y values and place the results in the fourth column.

STEP 4 Find the sums of the second, third, and fourth columns. The completed table is shown here.

x	y	xy	y^2
1	62	62	3,844
2	78	156	6,084
3	70	210	4,900
4	90	360	8,100
4	93	372	8,649
6	103	618	10,609
	$\Sigma y = 496$	$\Sigma xy = 1778$	$\Sigma y^2 = 42,186$

STEP 5 From the regression equation, $y' = 55.57 + 8.13x$, $a = 55.57$, and $b = 8.13$.

STEP 6 Substitute in the formula and solve for s_{est}.

$$s_{est} = \sqrt{\frac{\Sigma y^2 - a \Sigma y - b \Sigma xy}{n - 2}}$$

$$= \sqrt{\frac{42,186 - (55.57)(496) - (8.13)(1778)}{6 - 2}} = 6.48$$

This value is close to the value found in Example 11–12. The difference is due to rounding.

Prediction Interval

The standard error of estimate can be used for constructing a **prediction interval** (similar to a confidence interval) about a y' value.

Objective 7. Find a prediction interval.

When a specific value x is substituted into the regression equation, one gets y' which is a point estimate for y. For example, if the regression line equation for the age of a machine and the monthly maintenance cost is $y' = 55.57 + 8.13x$ (Example 11–12), then the predicted maintenance cost for a 3-year old machine would be $y' = 55.57 + 8.13(3)$ or \$79.96. Since this is a point estimate, one has no idea how accurate it is. But one can construct a prediction interval about that estimate. By selecting an α value, one can achieve a $1 - \alpha$ confidence that the interval contains the actual mean of the y values that correspond to the given value of x.

The reason is that there are possible sources of prediction errors in finding the regression line equation. One source occurs when finding the standard error of estimate, s_{est}. Two others are errors made in estimating the slope and the y intercept, since the equation of the regression line will change somewhat if different random samples are used when calculating the equation.

Formula for the Prediction Interval about a Value y'

$$y' - t_{\alpha/2}s_{est} \sqrt{1 + \frac{1}{n} + \frac{n(x - \overline{X})^2}{n\Sigma x^2 - (\Sigma x)^2}} < y < y'$$

$$+ t_{\alpha/2}s_{est} \sqrt{1 + \frac{1}{n} + \frac{n(x - \overline{X})^2}{n\Sigma x^2 - (\Sigma x)^2}}$$

with d.f. $= n - 2$

Example 11–14

For the data in Example 11–12, find the 95% prediction interval for the monthly maintenance cost of a machine that is 3 years old.

Solution

STEP 1 Find Σx, Σx^2 and $\overline{X}$.

$$\Sigma x = 20 \quad \Sigma x^2 = 82 \quad \overline{X} = \frac{20}{6} = 3.3$$

STEP 2 Find y' for $x = 3$

$$y' = 55.57 + 8.13x$$
$$y' = 55.57 + 8.13(3) = 79.96$$

STEP 3 Find s_{est}

$$s_{est} = 6.48, \text{ as shown in Example 11–13.}$$

STEP 4 Substitute in the formula and solve. $t_{\alpha/2} = 2.776$, d.f. $= 6 - 2 = 4$ for 95%

$$y' - t_{\alpha/2}s_{est} \sqrt{1 + \frac{1}{n} + \frac{n(x - \overline{X})^2}{n\Sigma x^2 - (\Sigma x)^2}} < y < y'$$

$$+ t_{\alpha/2}s_{est} \sqrt{1 + \frac{1}{n} + \frac{n(x - \overline{X})^2}{n\Sigma x^2 - (\Sigma x)^2}}$$

$$79.96 - (2.776)(6.48)\sqrt{1 + \frac{1}{6} + \frac{6(3 - 3.53)^2}{6(82) - (20)^2}} < y < 79.96$$

$$+ (2.776)(6.48)\sqrt{1 + \frac{1}{6} + \frac{6(3 - 3.5)^2}{6(82) - (20)^2}}$$

$$79.96 - (2.776)(6.48)(1.08) < y < 79.96 + (2.776)(6.48)(1.08)$$

$$79.96 - 19.43 < y < 79.96 + 19.43$$

$$60.53 < y < 99.39$$

Hence, one can be 95% confident that the interval $60.53 < y < 99.39$ contains the actual value of y.

Computer Applications for Correlation and Regression

MINITAB   Correlation, regression, and scatter plot:

1. Type the x values into C1.
2. Type the y values into C2.
3. Click Graph > Plot.
4. Highlight C2 and click Select to enter the y values into the first row and column of the Graph variables box (under Y).
5. Highlight C1 and click Select to enter the x values into the first row, second column of the Graph variables box (under X).
6. Click on OK.
   The computer will print a scatter plot.
7. Click Stat > Basic Statistics > Correlation.
8. Highlight C1 and C2 and click Select.
9. Click on OK.
   The computer will print a correlation matrix.
10. Click Stat > Regression > Regression.
11. Highlight C2 and click Select to enter the y values in the Response Box.
12. Highlight C1 and click Select to enter the x values in the Predictors Box.
13. Click on OK.

Example: Do a complete regression analysis using the following data.

Data:

C1	65	68	73	76	80	84	88	92	95	98
C2	15	22	20	24	27	27	32	38	42	40

MINITAB printout for this example:

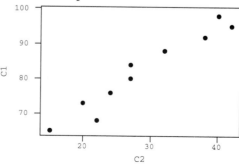

```
Correlations (Pearson)

Correlation of C1 and C2 = 0.968, P-Value = 0.000

Regression Analysis

The regression equation is
C2 = - 34.2 + 0.768 C1

Predictor        Coef      StDev         T        P
Constant      -34.231      5.838     -5.86    0.000
C1            0.76838    0.07067     10.87    0.000

S = 2.418       R-Sq = 93.7%     R-Sq(adj) = 92.9%
```

Summary: The scatter plot shows a somewhat positive linear relationship. The equation of the regression line is $y' = -34.2 + 0.768x_1$, and $r = 0.968$. Since the P-value is 0.000 and the t value is 10.87, the decision is to reject the null hypothesis at $\alpha = 0.05$.

Exercises

11–67. (W) What is meant by the *explained variation?* How is it computed?

11–68. (W) What is meant by the *unexplained variation?* How is it computed?

11–69. (W) What is meant by the *total variation?* How is it computed?

11–70. (W) Define *coefficient of determination.*

11–71. (W) How is the coefficient of determination found?

11–72. (W) Define *coefficient of nondetermination.*

11–73. (W) How is the coefficient of nondetermination found?

For Exercises 11–74 through 11–79, find the coefficients of determination and nondetermination and explain the meaning of each.

11–74. $r = 0.92$

11–75. $r = 0.75$

11–76. $r = 0.50$

11–77. $r = 0.37$

11–78. $r = 0.15$

11–79. $r = 0.05$

11–80. (W) Define *standard error of estimate* for regression. When can the standard error of estimate be used to construct a prediction interval about a value y'?

11–81. Compute the standard error of estimate for Exercise 11–13. The regression line equation was found in Exercise 11–41.

11–82. Compute the standard error of estimate for Exercise 11–14. The regression line equation was found in Exercise 11–42.

11–83. Compute the standard error of estimate for Exercise 11–15. The regression line equation was found in Exercise 11–43.

11–84. Compute the standard error of estimate for Exercise 11–16. The regression line equation was found in Exercise 11–44.

11–85. For the data in Exercises 11–13, 11–41, and 11–81, find the 90% prediction interval when $x = 20$ years.

11–86. For the data in Exercises 11–14, 11–42, and 11–82, find the 95% prediction interval when $x = \$1100$.

11–87. For the data in Exercises 11–15, 11–43, and 11–83, find the 90% prediction interval when $x = 4$ years.

11–88. For the data in Exercises 11–16, 11–44, and 11–84, find the 98% prediction interval when $x = 47$ years.

11–6

Summary

Many relationships among variables exist in the real world. One way to determine whether a relationship exists is to use the statistical techniques known as correlation and regression. The strength and direction of the relationship is measured by the value of the correlation coefficient. It can assume values between and including -1 and $+1$. The closer the value of the correlation coefficient is to $+1$ or -1, the stronger the relationship is between the variables. A value of $+1$ or -1 indicates a perfect relationship. A positive relationship between two variables means that for small values of the independent variable, the values of the dependent variable will be small, and that for large values of the independent variable, the values of the dependent variable will be large. A negative relationship between two variables means that for small values of the independent variable, the values of the dependent variable will be large, and that for large values of the independent variable, the values of the dependent variable will be small.

Relationships can be linear or curvilinear. To determine the shape, one draws a scatter plot of the variables. If the relationship is linear, the data can be approximated by a straight line, called the *regression line* or the *line of best fit*. The closer the value of r is to $+1$ or -1, the closer the points will fit the line.

The coefficient of determination is a better indicator of the strength of a relationship than the correlation coefficient. It is better because it identifies the percentage of variation of the dependent variable that is directly attributable to the variation of the independent variable. The coefficient of determination is obtained by squaring the correlation coefficient and converting the result to a percentage.

Another statistic used in correlation and regression is the standard error of estimate, which is an estimate of the standard deviation of the y values about the

"At this point in my report, I'll ask all of you to follow me to the conference room directly below us!"

Source: Cartoon by Bradford Veley, Marquette, Michigan. Reprinted with permission.

predicted y' values. The standard error of estimate can be used to construct a prediction interval about a specific value point estimate y' of the mean of the y values for a given value of x.

Finally, remember that a significant relationship between two variables does not necessarily mean that one variable is a direct cause of the other variable. In some cases this is true, but other possibilities that should be considered include a complex relationship involving other (perhaps unknown) variables, a third variable interacting with both variables, or a relationship due solely to chance.

Important Terms

Coefficient of determination 459

Correlation 433

Correlation coefficient 438

Dependent variable 435

Independent variable 435

Multiple relationship 434

Negative relationship 434

Pearson product moment correlation coefficient 438

Population correlation coefficient 441

Positive relationship 434

Prediction interval 461

Regression 434

Regression line 448

Scatter plot 435

Simple relationship 434

Standard error of estimate 459

Important Formulas

Formula for the correlation coefficient:

$$r = \frac{n(\sum xy) - (\sum x)(\sum y)}{\sqrt{[n(\sum x^2) - (\sum x)^2][n(\sum y^2) - (\sum y)^2]}}$$

Formula for the t test for the correlation coefficient:

$$t = r\sqrt{\frac{n-2}{1-r^2}} \qquad \text{d.f.} = n - 2$$

The regression line equation: $y' = a + bx$, where

$$a = \frac{(\sum y)(\sum x^2) - (\sum x)(\sum xy)}{n(\sum x^2) - (\sum x)^2}$$

$$b = \frac{n(\sum xy) - (\sum x)(\sum y)}{n(\sum x^2) - (\sum x)^2}$$

Formula for the standard error of estimate:

$$s_{est} = \sqrt{\frac{\sum(y - y')^2}{n-2}}$$

or

$$s_{est} = \sqrt{\frac{\sum y^2 - a\sum y - b\sum xy}{n-2}}$$

Formula for the prediction interval for a value y'

$$y' - t_{\alpha/2}s_{est}\sqrt{1 + \frac{1}{n} + \frac{n(x-\overline{X})^2}{n\sum x^2 - (\sum x)^2}} < y <$$

$$y' + t_{\alpha/2}s_{est}\sqrt{1 + \frac{1}{n} + \frac{n(x-\overline{X})^2}{n\sum x^2 - (\sum x)^2}}$$

$$\text{d.f.} = n - 2$$

Review Exercises

For Exercises 11–89 through 11–100, do a complete regression analysis by performing the following steps.
a. Draw the scatter plot.
b. Compute the value of the correlation coefficient.

c. Test the significance of the correlation coefficient at $\alpha = 0.01$, using Table I.
d. Determine the regression line equation.
e. Plot the regression line on the scatter plot.
f. Predict y' for a specific value of x.

11–89. A study is done to see whether there is a relationship between a student's grade point average and the number of hours the student studies per week. The data are shown here. If there is a significant relationship, predict the GPA of a student who studies 10 hours a week.

Hours, x	3	12	9	15	5	7	16
GPA, y	2.1	3.5	3.0	4.0	1.7	3.2	3.7

11–90. A study is conducted to determine the relationship between the number of calls a sales rep makes and the number of sales over a one-week period. The data are shown here. If there is a significant relationship, predict the number of sales when the rep makes 19 calls.

Calls, x	23	16	9	32	6	12
Sales, y	15	9	7	22	3	8

11–91. A study is done to see whether there is a relationship between a mother's age and the number of children she has. The data are shown here. If there is a significant relationship, predict the number of children of a mother whose age is 34.

Mother's age, x	18	22	29	20	27	32	33	36
No. of children, y	2	1	3	1	2	4	3	5

11–92. A study is conducted to determine the relationship between a driver's age and the number of accidents he or she has over a one-year period. The data are shown here. If there is a significant relationship, predict the number of accidents of a driver who is 28.

Driver's age, x	16	24	18	17	23	27	32
No. of accidents, y	3	2	5	2	0	1	1

11–93. A researcher desires to know whether the typing speed of a secretary (in words per minute) is related to the time (in hours) that it takes the secretary to learn to use a new word-processing program. The data follow.

Speed, x	Time, y
48	7
74	4
52	8
79	3.5
83	2
56	6
85	2.3
63	5
88	2.1
74	4.5
90	1.9
92	1.5

If there is a significant relationship, predict the time it will take the average secretary who has a typing speed of 72 words per minute to learn the word-processing program. (This information will be used for Exercise 11–97.)

11–94. A study was conducted with vegetarians to see whether the number of grams of protein each ate per day was related to diastolic blood pressure. The data are given here. (This information will be used for Exercise 11–98.) If there is a significant relationship, predict the diastolic pressure of a vegetarian who consumes 8 grams of protein per day.

Grams, x	4	6.5	5	5.5	8	10	9	8.2	10.5
Pressure, y	73	79	83	82	84	92	88	86	95

11–95. A study was conducted to determine whether there is a relationship between strength and speed. A sample of 20-year-old men was selected. Each was asked to do push-ups and to run a specific course. The number of push-ups and time it took to run the course (in seconds) are given in the table. If there is a significant relationship, predict the running time for a person who can do 18 push-ups.

Push-ups, x	5	8	10	10	11	13	15	18	23
Time, y	61	65	43	56	62	73	48	49	50

11–96. For Exercise 11–92, find the standard error of estimate.

11–97. For Exercise 11–93, find the standard error of estimate.

11–98. For Exercise 11–94, find the standard error of estimate.

11–99. For Exercise 11–93, find the 90% prediction interval for time when the speed is 72 words per minute.

11–100. For Exercise 11–94, find the 95% prediction interval for pressure when the number of grams is 8.

Do Dust Storms Affect Respiratory Health?   Revisited

The researchers correlated the dust pollutant levels in the atmosphere and the number of daily emergency room visits for several respiratory disorders, such as bronchitis, sinusitis, asthma, and pneumonia. Using the Pearson correlation coefficient, they found overall a significant but low correlation, $r = 0.13$, for bronchitis visits only. However, they found a much higher correlation value for sinusitis, P-value $= 0.08$, when pollutant levels exceeded maximums set by the Environmental Protection Agency (EPA). In addition, they found statistically significant correlation coefficients, $r = 0.94$, for sinusitis visits and $r = 0.74$ for upper-respiratory-tract infection visits two days after the dust pollutants exceeded the maximum levels set by the EPA.

Data Analysis

The Data Bank is found in Appendix D.

From the Data Bank, choose two variables that might be related: e.g., IQ and educational level; age and cholesterol level; exercise and weight; or weight and systolic pressure. Do a complete correlation and regression analysis by performing the following steps. Select a random sample of at least 10 subjects.

1. Draw a scatter plot.
2. Compute the correlation coefficient.
3. Test the hypothesis H_0: $\rho = 0$.
4. Find the regression line equation.
5. Summarize the results.

Quiz

Determine whether each statement is true or false. If the statement is false, explain why.

1. A negative relationship between two variables means that for the most part, as the x variable increases, the y variable increases.

2. A correlation coefficient of -1 implies a perfect linear relationship between the variables.

3. Even if the correlation coefficient is high or low, it may not be significant.

4. When the correlation coefficient is significant, one can assume x causes y.

5. It is not possible to have a significant correlation by chance alone.

6. In a multiple relationship, there are more than two variables under study.

Select the best answer.

7. The strength of the relationship between two variables is determined by the value of
   a. r
   b. a

   c. x
   d. s_{est}

8. To test the significance of r, a(n)_____ test is used.
   a. t
   b. F
   c. x^2
   d. None of the above

9. The test of significance for r has _____ degrees of freedom.
   a. 1
   b. n
   c. $n - 1$
   d. $n - 2$

10. The equation of the regression line used in statistics is
   a. $x = a + by$
   b. $y = bx + a$
   c. $y' = a + bx$
   d. $x = ay + b$

11. The coefficient of determination is _____.
   a. r
   b. r^2

c. *a*
d. *b*

Complete the following statements with the best answer.

12. A statistical graph of two variables is called a _____.

13. The *x* variable is called the _____ variable.

14. The range of *r* is from _____ to _____.

15. The sign of *r* and _____ will always be the same.

16. The regression line is called the _____.

17. If all the points fall on a straight line, the value of *r* will be _____ or _____.

For Problems 18–21, do a complete regression analysis.
a. Draw the scatter plot.
b. Compute the value of the correlation coefficient.
c. Test the significance of the correlation coefficient at $\alpha = 0.05$.
d. Determine the regression line equation.
e. Plot the regression line on the scatter plot.
f. Predict y' for a specific value of *x*.

18. The relationship between a father's age and the number of children he has is studied. The data are shown here. If there is a significant relationship, predict the number of children of a father whose age is 35.

Father's age, *x*	19	21	27	20	25	31	32	38
No. of children, *y*	1	2	3	0	1	4	3	4

19. A study is conducted to determine the relationship between a driver's age and the number of accidents he or she has over a one-year period. The data are shown here. If there is a significant relationship, predict the number of accidents of a driver who is 64.

Driver's age, *x*	63	65	60	62	66	67	59
No. of accidents, *y*	2	3	1	0	3	1	4

20. A researcher desires to know if the age of a child is related to the number of cavities he or she has. The data are shown here. If there is a significant relationship, predict the number of cavities for a child of 11.

Age of child, *x*	6	8	9	10	12	14
No. of cavities, *y*	2	1	3	4	6	5

21. A study is conducted with a group of dieters to see if the number of grams of fat each consumes per day is related to cholesterol level. The data are shown here. If there is a significant relationship, predict the cholesterol level of a dieter who consumes 8.5 grams of fat per day.

Fat grams, *x*	6.8	5.5	8.2	10	8.6
Cholesterol level, *y*	183	201	193	283	222
	9.1	8.6	10.4		
	250	190	218		

22. For Problem 20, find the standard error of the estimate.

23. For Problem 21, find the standard error of the estimate.

24. For Problem 20, find the 90% confidence interval of the number of cavities for a 7-year-old.

25. For Problem 21, find the 95% confidence interval of the cholesterol level of a person who consumes 10 grams of fat.

Critical Thinking Challenge

Curvilinear Regression   When the points in a scatter plot show a curvilinear trend rather than a linear trend, statisticians have methods of fitting curves rather than straight lines to the data, thus obtaining a better fit and a better prediction model. One type of curve that can be used is the logarithmic regression curve. The data shown are the number of items of a new product sold over a period of 15 months at a certain store. Notice that sales rise during the beginning months and then level off later on.

Month, *x*	1	3	6	8	10	12	15
No. of items sold, *y*	10	12	15	19	20	21	21

1. Draw the scatter plot for the data.

2. Find the equation of the regression line.

3. Describe how the line fits the data.

4. Using the log key on your calculator, transform the *x* values into log *x* values.

5. Using the log x values instead of the x values, find the equation of the a and b for the regression line.

6. Next, plot the curve $y = a + b \log x$ on the graph.

7. Compare the line $y = a + bx$ with the curve $y = a + b \log x$ and decide which one fits the data better.

8. Compute r using the x and y values and then compute r using the log x and y values. Which is higher?

9. In your opinion, which (the line or the logarithmic curve) would be a better predictor for the data? Why?

Data Projects

Where appropriate, use MINITAB, the TI-83, or a computer program of your choice to complete the following exercises.

1. Select two variables that might be related, such as the age of a person and the number of cigarettes the person smokes, or the number of credits a student has and the number of hours the student watches television. Sample at least 10 people.
a. Write a brief statement as to the purpose of the study.
b. Define the population.
c. State how the sample was selected.
d. Show the raw data.
e. Draw a scatter plot for the data values.
f. Write a statement analyzing the scatter plot.
g. Compute the value of the correlation coefficient.
h. Test the significance of r. (State the hypotheses, select α, find the critical values, make the decision, and analyze the results.)
i. Find the equation of the regression line and draw it on the scatter plot. (*Note:* Even if r is not significant, complete this step.)
j. Summarize the overall results.

2. Use MINITAB to answer the following.
a. Does a linear correlation exist between x and y?
b. If so, find the regression equation.
c. Explain how good a model the regression equation is by finding the coefficient of determination and coefficient of correlation and interpreting the strength of these values.
d. Find the prediction interval for y.

3. A study was conducted to determine if there is a relationship between the number of hours of television children between the ages of 6 and 12 watched per day and their grade point average. The data are shown here. If there is a significant relationship, predict the GPA of a child who watches 4.5 hours of television daily.

Hours of TV, x	5	3	2	4	1	1.5	2.3	4.8
GPA, y	2.3	3.0	2.9	3.2	3.8	3.5	3.1	2.2

TI-83 Calculator

Correlation and Regression

A. To graph a scatter plot

1. Enter the x values in L_1 and the y values in L_2.

2. Make sure the Window values are appropriate. Select an X_{min} slightly less than the smallest x data value and an X_{max} slightly larger than the largest x data value. Do the same for Y_{min} and Y_{max}. Also, you may need to change the X_{sc1} and Y_{sc1} values, depending on the data.

3. Press **2ND [STAT PLOT] ENTER**. The other y functions should be turned off.

4. Move the cursor to On and press **ENTER** on the stat plot 1 menu.

5. Move the cursor to the scatter plot Type, and press **ENTER**. Make sure the X list is L_1, and the Y list is L_2.

6. Press **GRAPH**.

Example 30: Draw a scatter plot for

x	65	68	73	76	80	84	88	92	95	98
y	15	22	20	24	27	27	32	38	42	40

Input

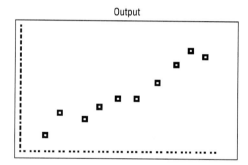

WINDOW
  Xmin=60
  Xmax=100
  Xscl=1
  Ymin=10
  Ymax=50
  Yscl=1
  Xres=1

Input

Plot1 Plot2 Plot3
On Off
Type:
Xlist:L₁
Ylist:L₂
Mark: □ + ·

Output

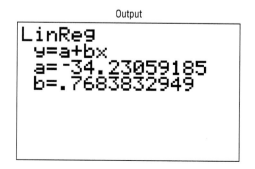

B. To find the equation of the regression line

1. Press **STAT** and move the cursor to Calc.

2. Press **8** then **ENTER**. The values for a and b will be displayed.

Example 31: Find the equation of the regression line for the data in the previous example.

Output

LinReg
  y=a+bx
  a=-34.23059185
  b=.7683832949

The equation of the regression line is
$y' = -34.23059185 + 0.7683832949x$.

C. To plot the regression line on the scatter plot

1. Press **Y =** and then **CLEAR** to clear any previous equations.

2. Press **VARS** and then **5**.

3. Move the cursor to EQ and press **ENTER**. The line will be in Y = screen.

4. Press **GRAPH**.

Example 32: Draw the regression line found in the previous example on the scatter plot.

Output

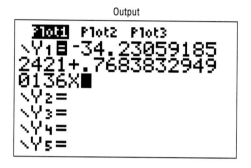

Plot1 Plot2 Plot3
\Y₁⊟-34.23059185
2421+.7683832949
0136X█
\Y₂=
\Y₃=
\Y₄=
\Y₅=

Output

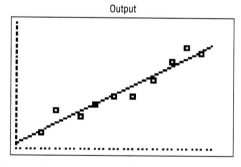

D. To test the significance of b and ρ.

1. Press **STAT** and move the cursor to TESTS.

2. Press **E (ALPHA SIN)**. Make sure the X List is L_1, the Y List is L_2, and the Freq. is 1.

3. Select the appropriate alternative hypothesis.

4. Move the cursor to Calculate and press **ENTER**.

Example 33: Test the hypothesis H_0: $\rho = 0$ for the data in Example 30. Use $\alpha = 0.05$.

Input

```
LinRegTTest
 Xlist:L₁
 Ylist:L₂
 Freq:1
 β & ρ:≠0 <0 >0
 RegEQ:
 Calculate
```

Output

```
LinRegTTest
 y=a+bx
 β≠0 and ρ≠0
 t=10.87243809
 p=4.5306836E⁻6
 df=8
↓a=-34.23059185
```

```
↑b=.7683832949
 s=2.418303889
 r²=.9366135353
 r=.9677879599
```

In this case, the t test value is 10.87243809. The P-value is 4.5306836 E-6, which is 0.000004, which is significant. The decision is to reject the null hypothesis at $\alpha = 0.05$, since $0.000004 < 0.05$, $r = 0.9677879599$, and $r^2 = 0.9366135353$.

chapter

12

Chi-Square and Analysis of Variance (ANOVA)

Outline

Objectives

After completing this chapter, you should be able to

1. Test a distribution for goodness of fit using chi-square.

2. Test two variables for independence using chi-square.

3. Test proportions for homogeneity using chi-square.

4. Use the ANOVA technique to determine if there is a significant difference among three or more means.

Statistics and Heredity

An Austrian monk, Gregor Mendel (1822–84), studied genetics, and his principles are the foundation for modern genetics. Mendel used his spare time to grow a variety of peas at the monastery. One of his many experiments involved crossbreeding peas that had smooth yellow seeds with peas that had wrinkled green seeds. He noticed that the results occurred with regularity. That is, some of the offspring had smooth yellow seeds, some had smooth green seeds, some had wrinkled yellow seeds, and some had wrinkled green seeds. Furthermore, after several experiments, the percentages of each type seemed to remain approximately the same. Mendel formulated his theory based on the assumption of dominant and recessive traits and tried to predict the results. He then crossbred his peas and examined 556 seeds over the next generation.

Finally, he compared the actual results with the theoretical results to see if his theory was correct. In order to do this, he used a "simple" chi-square test, which is explained in this chapter.

Source: J. Hodges, Jr., D. Krech, and R. Crutchfield, *Stat Lab, An Empirical Introduction to Statistics* (New York: McGraw-Hill, 1975), pp. 228–29. Used with permission.

12–1

Introduction

The chi-square distribution was used in Chapters 8 and 9 to find a confidence interval for a *variance* or standard deviation and to test a hypothesis about a single variance or standard deviation.

It can also be used for tests concerning *frequency distributions,* such as "If a sample of buyers is given a choice of automobile colors, will each color be selected with the same frequency?" The chi-square distribution can be used to test the *independence* of two variables. For example, "Are senators' opinions on gun control independent of party affiliations?" That is, do the Republicans feel one way and the Democrats feel differently, or do they have the same opinion?

The chi-square distribution can be used to test the *homogeneity of proportions.* For example, is the proportion of high school seniors who attend college immediately after graduating the same for the northern, southern, eastern, and western parts of the United States?

The chi-square test can also be used to test the normality of a variable.

In addition to the chi-square test, this chapter also explains another use for the *F* test. The *F* test, used to compare two variances shown in Chapter 10, can also be used to compare three or more means. This technique is called *analysis of variance,* or *ANOVA.* It is used to test claims involving three or more means. (*Note:* The *F* test can also be used to test the equality of two means. But since it is equivalent to the *t* test in this case, the *t* test is usually used instead of the *F* test when there are only two means.) For example, suppose a researcher wishes to see whether the means of the time it takes three groups of students to solve a computer problem using FORTRAN, BASIC, and PASCAL are different. The researcher will use the ANOVA technique for this test. The *z* and *t* tests should not be used when three or more means are compared, for reasons given later in this chapter.

12–2

Test for Goodness of Fit

Objective 1. Test a distribution for goodness of fit using chi-square.

In addition to being used to test a single variance, the chi-square statistic can be used to see whether a frequency distribution fits a specific pattern. For example, in order to meet customer demands, a manufacturer of running shoes may wish to see whether buyers show a preference for a specific style. A traffic engineer may wish to see whether accidents occur more often on some days than on others, so that he can increase police patrols accordingly. An emergency service may want to see whether it receives more calls at certain times of the day than others, so that it can provide adequate staffing.

When one is testing to see whether a frequency distribution fits a specific pattern, the chi-square **goodness-of-fit test** is used. For example, suppose a market analyst wished to see whether consumers have any preference among five flavors of a new fruit soda. A sample of 100 people provided the following data:

Cherry	Strawberry	Orange	Lime	Grape
32	28	16	14	10

If there were no preference, one would expect that each flavor would be selected with equal frequency. In this case, the equal frequency is 100/5 = 20. That is, *approximately* 20 people would select each flavor.

Since the frequencies for each flavor were obtained from a sample, these actual frequencies are called the **observed frequencies.** The frequencies obtained by calculation (as if there were no preference) are called the **expected frequencies.** A completed table for the test follows:

Frequency	Cherry	Strawberry	Orange	Lime	Grape
Observed	32	28	16	14	10
Expected	20	20	20	20	20

The observed frequencies will almost always differ from the expected frequencies due to sampling error; i.e., the values differ from sample to sample. But the question is: Are these differences significant (a preference exists), or are they due to chance? The chi-square goodness-of-fit test will enable the researcher to determine the answer.

Before computing the test value, one must state the hypotheses. The null hypothesis should be a statement indicating that there is no difference or no change. For this example, the hypotheses are as follows:

H_0: Consumers show no preference for flavors of the fruit soda.
H_1: Consumers show a preference.

In the goodness-of-fit test, the degrees of freedom are equal to the number of categories minus 1. For this example, there are five categories (cherry, strawberry, orange, lime, and grape); hence, the degrees of freedom are $5 - 1 = 4$. This is because the number of subjects in each of the first four categories is free to vary. But in order for the sum to be 100—the total number of subjects—the number of subjects in the last category is fixed.

Formula for the Chi-Square Goodness-of-Fit Test

$$\chi^2 = \Sigma \frac{(O - E)^2}{E}$$

with degrees of freedom equal to the number of categories minus 1, and where

O = observed frequency
E = expected frequency

Two assumptions are needed for the goodness-of-fit test. These assumptions are given next.

Assumptions for the Chi-Square Goodness-of-Fit Test

1. The data are obtained from a random sample.
2. The expected frequency for each category must be 5 or more.

The chi-square distribution shown in Chapter 8 comes from a set of p independent normally distributed variables with a mean of 0 and standard deviation of 1. These variables are designated as $z_1, z_2, \ldots, z_p$. Then the chi-square distribution with q degrees of freedom is defined as $z^2 = z_1^2 + z_2^2 + \ldots z_p^2$, where $q = p$.

This process is quite complicated, and a statistician rearranged the formula $\chi^2 = \Sigma \frac{(O - E)^2}{E}$ to make the computations easier.

In the formula, $\Sigma(O - E)$ will always be zero; hence, it is necessary to square the $(O - E)$ values. Then dividing by E yields a relative size value. The formula $\Sigma \frac{(O - E)^2}{E}$ is a discrete distribution that approximates the continuous chi-square distribution. When there is close agreement between the O and E values, the chi-square value is small and the null hypothesis is not rejected. When there are large differences between O and E, the chi-square value is large and the null hypothesis is rejected. This is shown graphically after the next example.

This test is a right-tailed test, since when the $(O - E)$ values are squared, the answer will be positive or zero. This formula is explained in the next example.

Example 12–1

Is there enough evidence to reject the claim that there is no preference in the selection of fruit soda flavors, using the data shown previously? Let $\alpha = 0.05$.

Solution

STEP 1   State the hypotheses and identify the claim.

H_0: Consumers show no preference for flavors (claim).
H_1: Consumers show a preference.

STEP 2   Find the critical value. The degrees of freedom are $5 - 1 = 4$, and $\alpha = 0.05$. Hence, the critical value from Table G is 9.488.

STEP 3   Compute the test value by subtracting the expected value from the corresponding observed value, squaring the result and dividing by the expected value, and finding the sum. The expected value for each category is 20, as shown previously.

$$\chi^2 = \Sigma \frac{(O - E)^2}{E}$$

$$= \frac{(32 - 20)^2}{20} + \frac{(28 - 20)^2}{20} + \frac{(16 - 20)^2}{20} + \frac{(14 - 20)^2}{20} + \frac{(10 - 20)^2}{20}$$

$$= 18.0$$

STEP 4   Make the decision. The decision is to reject the null hypothesis, since $18.0 > 9.488$, as shown in Figure 12–1.

Figure 12–1

Critical and Test Values for Example 12–1

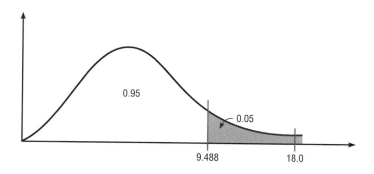

STEP 5   Summarize the results. There is enough evidence to reject the claim that consumers show no preference for the flavors.

To get some idea of why this test is called the goodness-of-fit test, examine graphs of the observed values and expected values. See Figure 12–2. From the graphs, one can see whether the observed values and expected values are close together or far apart.

Figure 12–2

Graphs of the Observed and Expected Values for Soda Flavors

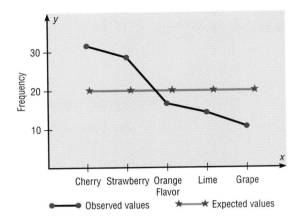

When the observed values and expected values are close together, the chi-square test value will be small. Then, the decision will be not to reject the null hypothesis—hence, there is "a good fit." See Figure 12–3(a). When the observed values and the expected values are far apart, the chi-square test value will be large. Then, the null hypothesis will be rejected—hence, there is "not a good fit." See Figure 12–3(b).

The procedure for the chi-square goodness-of-fit test is summarized in Procedure Table 10.

Figure 12–3

Results of the Goodness-of-Fit Test

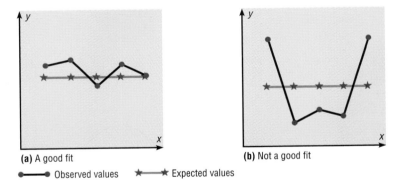

(a) A good fit   (b) Not a good fit

● Observed values   ★ Expected values

Procedure Table 10

The Chi-Square Goodness-of-Fit Test

STEP 1 State the hypotheses and identify the claim.

STEP 2 Find the critical value. The test is always right-tailed.

STEP 3 Compute the test value.

Find the sum of the $\dfrac{(O - E)^2}{E}$ values.

STEP 4 Make the decision.

STEP 5 Summarize the results.

When there is perfect agreement between the observed and the expected values, $\chi^2 = 0$. Also, χ^2 can never be negative. Finally, the test is right-tailed because "H_0: Good fit" and "H_1: Not a good fit" can be translated into H_0: χ^2 small and H_1: χ^2 large.

Example 12–2

The Russel Reynold Association surveyed retired senior executives who had returned to work. They gave the following reasons for returning to work: 38% were employed by another organization, 32% were *self-employed,* 23% were either *freelancing or consulting,* and 7% had formed their own companies. In order to see if these percentages are consistent with those of Allegheny County residents, a local researcher surveyed 300 retired executives who had returned to work and found that 122 were working for another company, 85 were self-employed, 76 were either freelancing or consulting, and 17 had formed their own companies. At $\alpha = 0.10$, can it be concluded that the percentages are the same for those people in Allegheny County?

Source: Michael L. Shook and Robert D. Shook, *The Book of Odds* (New York: Penguin Putnam Inc. 1991), p. 13.

Solution

STEP 1   State the hypotheses and identify the claim.

H_0: The reasons retired executives returned to work are distributed as follows: 38% are employed by another organization, 32% are self-employed, 23% are either freelancing or consulting, and 7% have formed their own companies (claim).

H_1: The distribution is not the same as stated in the null hypothesis.

STEP 2   Find the critical value. Since $\alpha = 0.10$ and the degrees of freedom are $4 - 1 = 3$, the critical value is 6.251.

STEP 3   Compute the test value. The expected values are computed as follows:

$$0.38 \times 300 = 114 \qquad 0.23 \times 300 = 69$$
$$0.32 \times 300 = 96 \qquad 0.07 \times 300 = 21$$

$$\chi^2 = \Sigma \frac{(O - E)^2}{E}$$

$$= \frac{(122 - 114)^2}{114} + \frac{(85 - 96)^2}{96} + \frac{(76 - 69)^2}{69} + \frac{(17 - 21)^2}{21}$$

$$= 3.2939$$

STEP 4   Make the decision. Since $3.2939 < 6.251$, the decision is not to reject the null hypothesis. See Figure 12–4.

Figure 12–4

Critical and Test Values
for Example 12–2

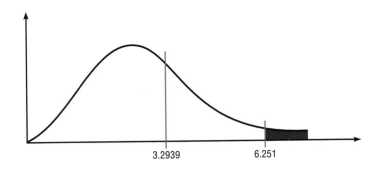

3.2939   6.251

STEP 5   Summarize the results. There is not enough evidence to reject the claim.
It can be concluded that the percentages are not significantly different
than those given in the null hypothesis.

Example 12–3

The advisor of an ecology club at a large college believes that the group consists
of 10% freshmen, 20% sophomores, 40% juniors, and 30% seniors. The membership for the club this year consisted of 14 freshmen, 19 sophomores, 51 juniors,
and 16 seniors. At $\alpha = 0.10$, test the advisor's conjecture.

Solution

STEP 1   State the hypotheses and identify the claim.

H_0: The club consists of 10% freshmen, 20% sophomores, 40% juniors,
and 30% seniors (claim).
H_1: The distribution is not the same as stated in the null hypothesis.

STEP 2   Find the critical value. Since $\alpha = 0.10$ and the degrees of freedom are
$4 - 1 = 3$, the critical value is 6.251.

STEP 3   Compute the test value. The expected values are computed as follows:

$$0.10 \times 100 = 10 \qquad 0.40 \times 100 = 40$$
$$0.20 \times 100 = 20 \qquad 0.30 \times 100 = 30$$

$$\chi^2 = \Sigma \frac{(O - E)^2}{E}$$

$$= \frac{(14 - 10)^2}{10} + \frac{(19 - 20)^2}{20} + \frac{(51 - 40)^2}{40} + \frac{(16 - 30)^2}{30}$$

$$= 11.208$$

STEP 4   Reject the null hypothesis, since $11.208 > 6.251$, as shown in Figure
12–5.

Figure 12–5

Critical and Test Values
for Example 12–3

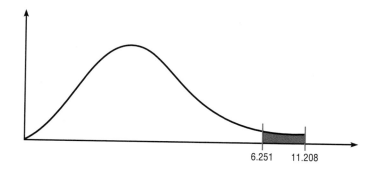

6.251   11.208

STEP 5   Summarize the results. There is enough evidence to reject the advisor's claim.

For use of the chi-square goodness-of-fit test, statisticians have determined that the expected frequencies should be at least 5, as stated in the assumptions. The reasoning is as follows: The chi-square distribution is continuous, whereas the goodness-of-fit test is discrete. However, the continuous distribution is a good approximation and can be used when the expected value for each class is at least 5. If an expected frequency of a class is less than 5, then that class can be combined with another class so that the expected frequency is 5 or more.

Test of Normality

The chi-square goodness-of-fit test can be used to test a variable to see if it is normally distributed. The null hypotheses are

H_0: The variable is normally distributed.
H_1: The variable is not normally distributed.

The next example shows how to do the calculations.

Example 12–4

Use chi-square to determine if the variable shown in the frequency distribution is normally distributed.

Unusual Stat

Drinking milk may lower your risk of stroke. A 22-year study of men over 55 found that only 4% of men who drank 16 ounces of milk every day suffered a stroke, compared with 8% of the nonmilk drinkers. (*Prime*, Fall 1996, p. 25)

Boundaries	Frequency
89.5–104.5	24
104.5–119.5	62
119.5–134.5	72
134.5–149.5	26
149.5–164.5	12
164.5–179.5	4
	200

Solution

First, find the mean and standard deviation of the variable. Then find the area under the standard normal distribution using z values and Table E for each class. Find the expected frequencies for each class by multiplying the area by 200. Finally, find the chi-square test value by using the formula $\chi^2 = \sum \dfrac{(O - E)^2}{E}$.

Boundaries	f	X_m	$f \cdot X_m$	$f \cdot X_m^2$
89.5–104.5	24	97	2328	225,816
104.5–119.5	62	112	6944	777,728
119.5–134.5	72	127	9144	1,161,288
134.5–149.5	26	142	3692	524,264
149.5–164.5	12	157	1884	295,788
164.5–179.5	4	172	688	118,336
	200		24,680	3,103,220

$$\overline{X} = \frac{24,680}{200} = 123.4$$

$$s = \sqrt{\frac{3,103,220 - 24,680^2/200}{199}} = \sqrt{290} = 17.03$$

The area below $z = 104.5$ is found as follows:

$$z = \frac{104.5 - 123.4}{17.03} = -1.11$$

The area for $z < -1.11$ is $0.500 - 0.3665 = 0.1335$.
The area between 104.5 and 119.5 is found as follows:

$$z = \frac{119.5 - 123.4}{17.03} = -0.23$$

The area for $-1.11 < z < -0.23$ is $0.3665 - 0.0910 = 0.2755$.
The area between 119.5 and 134.5 is found as follows:

$$z = \frac{134.5 - 123.4}{17.03} = 0.65$$

The area for $-0.23 < z < 0.65$ is $0.2422 + 0.0910 = 0.3332$.
The area between 134.5 and 149.5 is found as follows:

$$z = \frac{149.5 - 123.4}{17.03} = 1.53$$

The area for $0.65 < z < 1.53$ is $0.4370 - 0.2422 = 0.1948$.
The area between 149.5 and 164.5 is found as follows:

$$z = \frac{164.5 - 123.4}{17.03} = 2.41$$

The area for $1.53 < z < 2.41$ is $0.4920 - 0.4370 = 0.055$.
The area above 164.5 is found by

$$z = \frac{164.5 - 123.4}{17.03} = 2.41$$

$0.5000 - 0.4920 = 0.008$.

The frequencies are found by

$$0.1335 \cdot 200 = 26.7$$
$$0.2755 \cdot 200 = 55.1$$
$$0.3332 \cdot 200 = 66.64$$
$$0.1948 \cdot 200 = 38.96$$
$$0.055 \cdot 200 = 11.0$$
$$0.008 \cdot 200 = 1.6$$

Note: Since the expected frequency for the last category is less than five, it can be combined with the previous category.

The χ^2 is found by

O	24	62	72	26	16
E	26.7	55.1	66.64	38.96	12.6

$$\chi^2 = \frac{(24 - 26.7)^2}{26.7} + \frac{(62 - 55.1)^2}{55.1} + \frac{(72 - 66.64)^2}{66.64} + \frac{(26 - 38.96)^2}{38.96}$$
$$+ \frac{16 - 12.6}{12.6}$$
$$= 6.797$$

The C.V. with d.f. = 4 and $\alpha = 0.05$ is 9.488, so the null hypothesis is not rejected. Hence, the distribution can be considered approximately normal. $\alpha = 0.05$ is used.

Computer Applications for the Goodness-of-Fit Test

MINITAB   MINITAB does not have a special program to perform the chi-square goodness-of-fit test. The test value can be calculated using the following.

1. Enter the observed value in C1.
2. Enter the expected values in C2.
3. Click on Calc > Calculator.
4. Type C3 in the Store results in Variable box; Press Tab.
5. Type Sum ((C1 − C2)**2/C2) in the Expression box.
6. Click on OK.

The chi-square test statistic will be displayed in the first row of C3. To access this click on Worksheet.

Exercises

12–1. (W)   How does the goodness-of-fit test differ from the chi-square variance test?

12–2. (W)   How are the degrees of freedom computed for the goodness-of-fit test?

12–3. (W)   How are the expected values computed for the goodness-of-fit test?

12–4. (W)   When the expected frequencies are less than 5 for a specific class, what should be done so that one can use the goodness-of-fit test?

For Exercises 12–5 through 12–19, perform the following steps

a. State the hypotheses and identify the claim.
b. Find the critical value.
c. Compute the test value.
d. Make the decision.
e. Summarize the results.

12–5.   A staff member of an emergency medical service wishes to determine whether the number of accidents is equally distributed during the week.

A week was selected at random, and the following data were obtained. Is there evidence to reject the hypothesis that the number of accidents is equally distributed throughout the week, at $\alpha = 0.05$?

Day	Mon.	Tues.	Wed.	Thurs.	Fri.	Sat.	Sun.
No. of accidents	28	32	15	14	38	43	19

12–6.   A clothes manufacturer wants to know whether customers prefer any specific color over other colors in shirts. She selects a random sample of 100 shirts sold and notes the colors. The data are shown here. At $\alpha = 0.10$, is there a color preference for the shirts?

Color	White	Blue	Black	Red	Yellow	Green
No. sold	43	22	16	10	5	4

12–7.   The concessions manager at Twin Rivers Stadium wishes to see whether there is any preference in the flavors of popcorn that are sold during sporting events. A random sample of sales is selected, and the data are shown here. At $\alpha = 0.01$, are the flavors selected with equal frequency?

Flavor	Plain	Barbecue	Butter	Cheddar cheese
No. sold	25	18	32	45

12–8.   A bank manager wishes to see whether there is any preference in the times that customers use the bank. Six hours are selected, and the numbers of customers visiting the bank during each hour are as shown here. At $\alpha = 0.05$, do the customers show a preference for specific times?

Time	10:00	11:00	12:00	1:00	2:00	3:00
No. of customers	26	33	42	36	24	19

12–9.   The American Red Cross reports that 42% of Americans have type O blood, 44% have type A blood, 10% have type B blood, and 4% have type AB blood. A county medical examiner hypothesizes the distribution of blood types is the same in his county as it is nationally. A random sample of 200 people is selected, and the following data are tallied. At $\alpha = 0.10$, can the examiner conclude that his hypothesis is correct?

Type	A	O	B	AB
Frequency	58	65	55	22

Source: Robert D. Shook and Michael L. Shook, *The Book of Odds* (New York: Penguin Putnam Inc., 1991), p. 161.

12–10.   The chair of the history department of a college hypothesizes that the final grades are distributed as 40% A's, 30% B's, 20% C's, 5% D's, and 5% F's. At the end of the semester, the following numbers of grades were earned. For $\alpha = 0.05$, is the grade distribution for the department different from that expected?

Grade	A	B	C	D	F
Number	45	52	39	8	6

12–11.   *USA Today* reported that 21% of loans granted by credit unions were for home mortgages, 39% were for automobile purchases, 20% were for credit card and other unsecured loans, 12% were for real estate, and 8% were for other miscellaneous needs. In order to see if her credit union customers had similar needs, a manager surveyed a random sample of 100 loans and found that 25 were for home mortgages, 44 for automobile purchases, 19 for credit card and unsecured loans, 8 for real estate other than home loans, and 4 for miscellaneous needs. At $\alpha = 0.05$, is the distribution the same as reported in the newspaper?

Source: *USA Today*, July 21, 1995.

12–12.   In a check to see whether a die is fair, each number on the face should appear approximately $\frac{1}{6}$ of the time. A suspected die is selected and rolled 180 times, and the results are tallied as shown. At $\alpha = 0.10$, can one conclude that the die is loaded?

Number	1	2	3	4	5	6
Frequency	34	38	25	21	30	32

12–13.   A *USA Today* Snapshot states that 53% of adult shoppers prefer to pay cash for purchases, 30% use checks, 16% use credit cards, and 1% have no preference. The owner of a large store randomly selected 800 shoppers and asked their payment preferences. The results were that 400 paid cash, 210 paid by check, 170 paid with a credit card, and 20 had no preference. At $\alpha = 0.01$, can the owner conclude that her customers have the same preferences as those surveyed?

Source: *USA Today*, July 19, 1995.

12–14.   The owner of a sporting-goods store wishes to see whether his customers show any preference for the month in which they purchase hunting rifles. The sales of rifles for the end of last year are shown here. At $\alpha = 0.05$, test the claim that there is no preference for the month in which the customers purchase guns.

Month	Sept.	Oct.	Nov.	Dec.
No. sold	18	23	28	15

12–15. The dean of students of a college wishes to test the claim that the distribution of students is as follows: 40% business (BU), 25% computer science (CS), 15% science (SC), 10% social science (SS), 5% liberal arts (LA), and 5% general studies (GS). Last semester, the program enrollment was distributed as shown here. At $\alpha = 0.10$, is the distribution of students the same as hypothesized?

Major	BU	CS	SC	SS	LA	GS
Number	72	53	32	20	16	7

12–16. A quality control engineer for a manufacturing plant wishes to determine whether the number of defective items manufactured during the week is approximately the same on each day. A week is selected at random, and the number of defective items produced each day is shown here. At $\alpha = 0.05$, can the engineer conclude that the defective items are produced with the same frequency each day?

Day	Mon.	Tues.	Wed.	Thurs.	Fri.
Number	32	16	23	19	40

12–17. A software department manager of a computer store believes that 50% of her customers purchase word-processing programs, 25% purchase spreadsheet programs, and 25% purchase database programming. A sample of purchases shows the following distribution. At $\alpha = 0.05$, is her assumption correct?

Program	Word processing	Spreadsheet	Database
No. of purchases	38	23	19

*12–18. Three coins are tossed 72 times and the number of heads is as shown. At $\alpha = 0.05$, test the null hypothesis that the coins are balanced and randomly tossed. (*Hint:* Use the binomial distribution.)

No. of heads	0	1	2	3
Frequency	3	10	17	42

*12–19. Select a three-digit state lottery number over a period of 50 days. Count the number of times each digit, 0 through 9, occurs. Test the claim, at $\alpha = 0.05$, that the digits occur at random.

12–3

Tests Using Contingency Tables

When data can be tabulated in table form in terms of frequencies, several types of hypotheses can be tested using the chi-square test.

Two such tests are the independence of variables test and the homogeneity of proportions test. The test of independence of variables is used to determine whether two variables are independent of or related to each other when a single sample is selected. The test of homogeneity of proportions is used to determine whether the proportions for a variable are equal when several samples are selected from different populations. Both tests use the chi-square distribution and a contingency table, and the test value is found the same way. The independence test will be explained first.

Test for Independence

Objective 2. Test two variables for independence using chi-square.

The chi-square **independence test** can be used to test the independence of two variables. For example, suppose a new postoperative procedure is administered to a number of patients in a large hospital. One can ask the question, "Do the doctors feel differently about this procedure from the nurses, or do they feel basically the same way?" Note that the question is not whether or not they prefer the procedure but whether there is a difference of opinion between the two groups.

To answer this question, a researcher selects a sample of nurses and doctors and tabulates the data in table form, as shown.

Group	Prefer new procedure	Prefer old procedure	No preference
Nurses	100	80	20
Doctors	50	120	30

As the survey indicates, 100 nurses prefer the new procedure, 80 prefer the old procedure and 20 have no preference; 50 doctors prefer the new procedure, 120 like the old procedure, and 30 have no preference. Since the main question is whether there is a difference in opinion, the null hypothesis is stated as follows:

H_0: The opinion about the procedure is *independent* of the profession.

The alternative hypothesis is stated as follows:

H_1: The opinion about the procedure is *dependent* on the profession.

If the null hypothesis is not rejected, the test means that both professions feel basically the same way about the procedure, and the differences are due to chance. If the null hypothesis is rejected, the test means that one group feels differently about the procedure from the other. Remember that rejection does *not* mean that one group favors the procedure and the other does not. Perhaps both groups favor it or both dislike it, but in different proportions.

In order to test the null hypothesis using the chi-square independence test, one must compute the expected frequencies, assuming that the null hypothesis is true. These frequencies are computed by using the observed frequencies given in the table.

The Contingency Table

When data are arranged in table form for the chi-square independence test, the table is called a **contingency table.** The table is made up of R rows and C columns. The table here has two rows and three columns.

Group	Prefer new procedure	Prefer old procedure	No preference
Nurses	100	80	20
Doctors	50	120	30

Note that row and column headings do not count in determining the number of rows and columns.

A contingency table is designated as an $R \times C$ (rows times columns) table. In this case, $R = 2$ and $C = 3$; hence, this table is a 2×3 contingency table. Each block in the table is called a *cell* and is designated by its row and column position. For example, the cell with a frequency of 80 is designated as $C_{1,2}$, or row 1, column 2. The cells are shown below.

	Column 1	Column 2	Column 3
Row 1	$C_{1,1}$	$C_{1,2}$	$C_{1,3}$
Row 2	$C_{2,1}$	$C_{2,2}$	$C_{2,3}$

The degrees of freedom for any contingency table are (rows $-$ 1) times (columns $-$ 1); that is, d.f. $= (R - 1)(C - 1)$. In this case, $(2 - 1)(3 - 1) = (1)(2) = 2$. The reason for this formula for d.f. is that all the expected values except one are free to vary in each row and in each column.

Computation of Expected Frequencies

Using the previous table, one can compute the expected frequencies for each block (or cell) as shown next.

a.   Find the sum of each row and each column, and find the grand total, as shown.

Group	Prefer new procedure	Prefer old procedure	No preference	
				Row 1 sum
Nurses	100	80	20	200
				Row 2 sum
Doctors	+ 50	+ 120	+ 30	200
	150	200	50	400
	Column 1 sum	Column 2 sum	Column 3 sum	Grand total

b. For each cell, multiply the corresponding row sum by the column sum and divide by the grand total, to get the expected value:

$$\text{expected value} = \frac{\text{row sum} \times \text{column sum}}{\text{grand total}}$$

For example, for $C_{1,2}$, the expected value, denoted by $E_{1,2}$, is (refer to the previous tables)

$$E_{1,2} = \frac{(200)(200)}{400} = 100$$

For each cell, the expected values are computed as follows:

$$E_{1,1} = \frac{(200)(150)}{400} = 75 \qquad E_{1,2} = \frac{(200)(200)}{400} = 100 \qquad E_{1,3} = \frac{(200)(50)}{400} = 25$$

$$E_{2,1} = \frac{(200)(150)}{400} = 75 \qquad E_{2,2} = \frac{(200)(200)}{400} = 100 \qquad E_{2,3} = \frac{(200)(50)}{400} = 25$$

The expected values can now be placed in the corresponding cells along with the observed values, as shown.

Group	Prefer new procedure	Prefer old procedure	No preference	
Nurses	100 (75)	80 (100)	20 (25)	200
Doctors	50 (75)	120 (100)	30 (25)	200
	150	200	50	400

The rationale for the computation of the expected frequencies for a contingency table uses proportions. For $C_{1,1}$ a total of 150 out of 400 people prefer the new procedure. And since there are 200 nurses, one would expect, if the null hypothesis were true, (150/400)(200), or 75, of the nurses to be in favor of the new procedure.

Statistical Test

The formula for the test value for the independence test is the same as the one used for the goodness-of-fit test. It is

$$\chi^2 = \sum \frac{(O - E)^2}{E}$$

For the previous example, compute the $(O - E)^2/E$ values for each cell, and then find the sum.

$$\chi^2 = \sum \frac{(O - E)^2}{E}$$

$$= \frac{(100 - 75)^2}{75} + \frac{(80 - 100)^2}{100} + \frac{(20 - 25)^2}{25} + \frac{(50 - 75)^2}{75}$$

$$+ \frac{(120 - 100)^2}{100} + \frac{(30 - 25)^2}{25}$$

$$= 26.67$$

The final steps are to make the decision and summarize the results. This test is always a right-tailed test, and the degrees of freedom are $(R - 1)(C - 1) = (2 - 1)(3 - 1) = 2$. If $\alpha = 0.05$, the critical value is 5.991. Hence, the decision is to reject the null hypothesis, since $26.67 > 5.991$. See Figure 12–6.

Figure 12–6

Critical and Test Values for the Postoperative Procedures Example

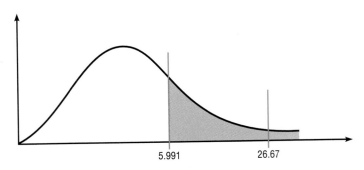

5.991   26.67

The conclusion is that there is enough evidence to support the claim that opinion is related to (dependent on) profession—i.e., that the doctors and nurses differ in their opinions about the procedure.

Two more examples illustrate the procedure for the chi-square test of independence.

Example 12–5

A sociologist wishes to see whether the number of years of college a person has completed is related to his or her place of residence. A sample of 88 people is selected and classified as shown.

Location	No college	Four-year degree	Advanced degree	Total
Urban	15	12	8	35
Suburban	8	15	9	32
Rural	6	8	7	21
Total	29	35	24	88

At $\alpha = 0.05$, can the sociologist conclude that the years of college education are dependent on the person's location?

STEP 1   State the hypotheses and identify the claim.

H_0: A person's place of residence is independent of the number of years of college completed.

H_1: A person's place of residence is dependent on the number of years of college completed (claim).

STEP 2 Find the critical value. The critical value is 9.488, since the degrees of freedom are $(3-1)(3-1) = (2)(2) = 4$.

STEP 3 Compute the test value. To compute the test value, one must first compute the expected values.

$$E_{1,1} = \frac{(35)(29)}{88} = 11.53 \qquad E_{1,2} = \frac{(35)(35)}{88} = 13.92 \qquad E_{1,3} = \frac{(35)(24)}{88} = 9.55$$

$$E_{2,1} = \frac{(32)(29)}{88} = 10.55 \qquad E_{2,2} = \frac{(32)(35)}{88} = 12.73 \qquad E_{2,3} = \frac{(32)(24)}{88} = 8.73$$

$$E_{3,1} = \frac{(21)(29)}{88} = 6.92 \qquad E_{3,2} = \frac{(21)(35)}{88} = 8.35 \qquad E_{3,3} = \frac{(21)(24)}{88} = 5.73$$

The completed table is as shown.

Location	No college	Four-year degree	Advanced degree	Total
Urban	15 (11.53)	12 (13.92)	8 (9.55)	35
Suburban	8 (10.55)	15 (12.73)	9 (8.73)	32
Rural	6 (6.92)	8 (8.35)	7 (5.73)	21
	29	35	24	88

Then, the chi-square test value is

$$\chi^2 = \sum \frac{(O-E)^2}{E}$$

$$= \frac{(15-11.53)^2}{11.53} + \frac{(12-13.92)^2}{13.92} + \frac{(8-9.55)^2}{9.55}$$

$$+ \frac{(8-10.55)^2}{10.55} + \frac{(15-12.73)^2}{12.73} + \frac{(9-8.73)^2}{8.73}$$

$$+ \frac{(6-6.92)^2}{6.92} + \frac{(8-8.35)^2}{8.35} + \frac{(7-5.73)^2}{5.73}$$

$$= 3.01$$

STEP 4 Make the decision. The decision is not to reject the null hypothesis, since $3.03 < 9.488$. See Figure 12–7.

Figure 12–7

Critical and Test Values for Example 12–5

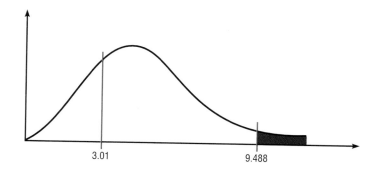

3.01   9.488

STEP 5   Summarize the results. There is not enough evidence to support the claim that a person's place of residence is dependent on the number of years of college completed.

Example 12–6

A researcher wishes to determine whether there is a relationship between the gender of an individual and the amount of alcohol consumed. A sample of 68 people is selected, and the following data are obtained.

Gender	Alcohol consumption			Total
	Low	Moderate	High	
Male	10	9	8	27
Female	13	16	12	41
Total	23	25	20	68

At $\alpha = 0.10$, can the researcher conclude that alcohol consumption is related to gender?

Solution

STEP 1   State the hypotheses and identify the claim.

H_0: The amount of alcohol that a person consumes is independent of the individual's gender.

H_1: The amount of alcohol that a person consumes is dependent on the individual's gender (claim).

STEP 2   Find the critical value. The critical value is 4.605, since the degrees of freedom are $(2 - 1)(3 - 1) = 2$.

STEP 3   Compute the test value. First, compute the expected values.

$$E_{1,1} = \frac{(27)(23)}{68} = 9.13 \qquad E_{1,2} = \frac{(27)(25)}{68} = 9.93 \qquad E_{1,3} = \frac{(27)(20)}{68} = 7.94$$

$$E_{2,1} = \frac{(41)(23)}{68} = 13.87 \qquad E_{2,2} = \frac{(41)(25)}{68} = 15.07 \qquad E_{2,3} = \frac{(41)(20)}{68} = 12.06$$

The completed table is shown next.

Gender	Alcohol consumption			Total
	Low	Moderate	High	
Male	10 (9.13)	9 (9.93)	8 (7.94)	27
Female	13 (13.87)	16 (15.07)	12 (12.06)	41
	23	25	20	68

Then, the test value is

$$\chi^2 = \sum \frac{(O - E)^2}{E}$$

$$= \frac{(10 - 9.13)^2}{9.13} + \frac{(9 - 9.93)^2}{9.93} + \frac{(8 - 7.94)^2}{7.94}$$

$$+ \frac{(13 - 13.87)^2}{13.87} + \frac{(16 - 15.07)^2}{15.07} + \frac{(12 - 12.06)^2}{12.06}$$

$$= 0.283$$

STEP 4   Make the decision. The decision is not to reject the null hypothesis, since $0.283 < 4.605$. See Figure 12–8.

Figure 12–8

Critical and Test Values
for Example 12–6

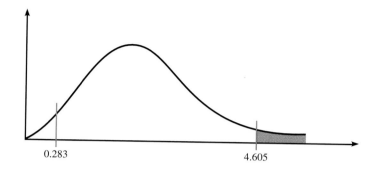

0.283   4.605

STEP 5   Summarize the results. There is not enough evidence to support the claim that the amount of alcohol a person consumes is dependent on the individual's gender.

Test for Homogeneity of Proportions

Objective 3. Test proportions for homogeneity using chi-square.

The second chi-square test that uses a contingency table is called the **homogeneity of proportions test.** In this situation, samples are selected from several different populations and the researcher is interested in determining whether the proportions of elements that have a common characteristic are the same for each population. The sample sizes are specified in advance, making either the row totals or column totals in the contingency table known before the samples are selected. For example, a researcher may select a sample of 50 freshmen, 50 sophomores, 50 juniors, and 50 seniors, and then find the proportion of students who are smokers in each level. The researcher will then compare the proportions for each group to see if they are equal. The hypotheses in this case would be

H_0: $p_1 = p_2 = p_3 = p_4$
H_1: At least one proportion is different from the others.

If one does not reject the null hypothesis, it can be assumed that the proportions are equal and the differences in them are due to chance. Hence, the proportion of students who smoke is the same for grade levels freshmen through senior. When the null hypothesis is rejected, it can be assumed that the proportions are not all equal. The computational procedure is the same as that for the test of independence, as shown in the next example.

Example 12–7

A researcher selected a sample of 150 seniors from each of three area high schools and asked each senior, "Do you drive to school in a car owned by either you or your parents?" The data are shown in the table. At $\alpha = 0.05$, can it be concluded that the proportion of students who drive their own or their parents' cars is the same at all three schools?

	School 1	School 2	School 3	Total
Yes	18	22	16	56
No	32	28	34	94
	50	50	50	150

Solution

STEP 1   State the hypotheses.

$$H_0: p_1 = p_2 = p_3$$
$$H_1: \text{At least one proportion is different from the others.}$$

STEP 2   Find the critical value. The formula for the degrees of freedom is the same as before: (rows $-$ 1) (columns $-$ 1) = (2 $-$ 1) (3 $-$ 1) = 1(2) = 2. The critical value is 5.991.

STEP 3   Compute the test value. First, compute the expected values.

$$E_{1,1} = \frac{(56)(50)}{150} = 18.67 \qquad E_{2,1} = \frac{(94)(50)}{150} = 31.33$$

$$E_{1,2} = \frac{(56)(50)}{150} = 18.67 \qquad E_{2,2} = \frac{(94)(50)}{150} = 31.33$$

$$E_{1,3} = \frac{(56)(50)}{150} = 18.67 \qquad E_{2,3} = \frac{(94)(50)}{150} = 31.33$$

The completed table is shown here.

	School 1	School 2	School 3	Total
Yes	18 (18.67)	22 (18.67)	16 (18.67)	56
No	32 (31.33)	28 (31.33)	34 (31.33)	94
	50	50	50	150

The test value is

$$\chi^2 = \sum \frac{(O - E)^2}{E}$$

$$= \frac{(18 - 18.67)^2}{18.67} + \frac{(22 - 18.67)^2}{18.67} + \frac{(16 - 18.67)^2}{18.67}$$

$$+ \frac{(32 - 31.33)^2}{31.33} + \frac{(28 - 31.33)^2}{31.33} + \frac{(34 - 31.33)^2}{31.33}$$

$$= 1.596$$

STEP 4   Make the decision. The decision is not to reject the null hypothesis, since $1.596 < 5.991$.

STEP 5   Summarize the results. Since there is not enough evidence to reject the null hypothesis, it can be concluded that the proportions of high school students who drive their own or their parents' cars to school are equal for each school.

Interesting Fact

Water is the most critical nutrient in your body. It is needed for just about everything that happens. Water is lost fast: two cups daily is lost just exhaling, 10 cups through normal waste and body cooling, and one to two quarts per hour running, biking, or working out. (*Fitness*, Special Edition, p. 84)

When the degrees of freedom for a contingency table are equal to 1—i.e., the table is a 2 $\times$ 2 table—some statisticians suggest using the *Yates correction for continuity*. The formula for the test is then

$$\chi^2 = \sum \frac{(|O - E| - 0.5)^2}{E}$$

Since the chi-square test is already conservative, most statisticians agree that the Yates correction is not necessary. (See Exercise 12–53.)

The procedure for the chi-square independence and homogeneity tests is summarized in Procedure Table 11.

Procedure Table 11

The Chi-Square Independence and Homogeneity Tests

STEP 1   State the hypotheses and identify the claim.

STEP 2   Find the critical value in the right tail.

STEP 3   Compute the test value. To compute the test value, first find the expected values. For each cell of the contingency table, use the formula

$$E = \frac{(\text{row sum})(\text{column sum})}{\text{grand total}}$$

to get the expected value. To find the test value, use the formula

$$\chi^2 = \sum \frac{(O - E)^2}{E}$$

STEP 4   Make the decision.

STEP 5   Summarize the results.

The assumptions for the two chi-square tests are given next.

Assumptions for the Chi-Square Independence and Homogeneity Tests

1.   The data are obtained from a random sample.
2.   The expected value in each cell must be 5 or more.

If the expected values are not 5 or more, combine categories.

Computer Applications for the Independence and Homogeneity of Proportions Tests

MINITAB   Chi-square contingency table test:

1.   Enter the data as they appear in the contingency table. The columns will be designated as C1, C2, C3, etc.
2.   Choose Stat > Tables > Chi-Square Test.
3.   Highlight all the columns in the dialog box (C1, C2, C3, etc.).
4.   Click Select.
5.   Click OK.

Example: Perform a chi-square independence test using the following data. Use $\alpha = 0.05$.

75	25
54	46
60	40

Data:

C1	75	54	60
C2	25	46	40

This study involves three groups: smokers, ex-smokers, and nonsmokers. Suggest how a chi-square independence test could be used to arrive at the conclusions. What hypothesis could be used in this study? Do you agree with the results of the study? Explain your answer.

At Least They're Checking for Radon

Smokers have a tough time kicking the habit—and it's not just cigarettes. They are also reluctant to give up other unhealthy acts.

University of Rhode Island researchers studied the readiness of 19,000 smokers, ex-smokers, and non-smokers to begin practicing 10 healthful behaviors. Smokers proved the least willing to wear seat belts, cut fat intake, exercise, eat more fiber, watch their weight, stay out of the sun, and use sunscreen.

People who had never smoked were most receptive to change, with ex-smokers usually falling in between. Only three behaviors—getting a Pap smear, going for a mammogram, and checking home radon levels—were unaffected by smoking status.

Health psychologists have been searching for a "gateway behavior"—a health-promoting practice that, once adopted, would lead folks to begin other healthy habits. "A lot of people think exercise might be that behavior," notes Joseph S. Rossi, Ph.D., research director at Rhode Island's Cancer Prevention Research Center. But giving up smoking may be an even better candidate, he told the Society of Behavioral Medicine.

Why does tobacco use coincide with so many other unhealthy practices? Perhaps cigarettes are so harmful that smokers consider it pointless to take up jogging or eat more broccoli, says Rossi. On the other hand he admits that it's hard to see why they wouldn't want to buckle up.

MINITAB printout for this example:

Chi-Square Test

Expected counts are printed below observed counts

	C1	C2	Total
1	75	25	100
	63.00	37.00	
2	54	46	100
	63.00	37.00	
3	60	40	100
	63.00	37.00	
Total	189	111	300

Chi-Sq = 2.286 + 3.892 +
         1.286 + 2.189 +
         0.143 + 0.243 = 10.039

DF = 2, P-Value = 0.007

Summary: The expected values are shown below the observed values in the table. The chi-square is 10.039. In this case, it is necessary to find the critical value from Table G. It is 5.991. The decision is to reject the null hypothesis.

Exercises

12–20. (W) How is the chi-square independence test similar to the goodness-of-fit test? How is it different?

12–21. (W) How are the degrees of freedom computed for the independence test?

12–22. (W) When the observed frequencies are close to the expected frequencies, what is the value of chi-square?

12–23. (W) Generally, how would the null and alternative hypotheses be stated for the chi-square independence test?

12–24. (W) What is the name of the table used in the independence test?

12–25. (W) How are the expected values computed for each cell in the table?

12–26. (W) Explain how the chi-square independence test differs from the chi-square homogeneity of proportions test.

12–27. (W) How are the null and alternative hypotheses stated for the test of homogeneity of proportions?

For Exercises 12–28 through 12–51, perform the following steps.

a. State the hypotheses and identify the claim.
b. Find the critical value.
c. Compute the test value.
d. Make the decision.
e. Summarize the results.

12–28. A study is being conducted to determine whether there is a relationship between jogging and blood pressure. A random sample of 210 subjects is selected, and they are classified as shown in the table. At $\alpha = 0.05$, test the claim that jogging and blood pressure are not related.

| | Blood pressure | | |
Jogging status	Low	Moderate	High
Joggers	34	57	21
Nonjoggers	15	63	20

12–29. A researcher wishes to see whether the age of an individual is related to milk consumption. A sample of 152 people is selected, and they are classified as shown in the table. At $\alpha = 0.10$, is there a relationship between milk consumption and age?

| | Milk consumption | | |
Age	Low	Moderate	High
21–30	12	16	18
31–40	15	27	9
41–50	12	10	5
51 and over	6	9	13

12–30. A survey of the 164 state representatives is conducted to see whether their opinions on a bill are related to their party affiliation. The following data are obtained. At $\alpha = 0.01$, can the researcher conclude that opinions are related to party affiliations?

| | Opinion | | |
Party	Approve	Disapprove	No opinion
Republican	27	15	13
Democrat	43	18	12
Independent	9	15	12

12–31. An automobile insurance company wishes to determine whether the age of the insured is related to the amount of liability he or she carries. A sample of 222 drivers shows the following data. At $\alpha = 0.05$, is the amount of coverage independent of the age of the driver?

| | Amount | | |
Age	Under $50,000	$50,000–$100,000	Over $100,000
21–30	16	25	3
31–40	23	44	15
41–50	15	31	18
51 and over	9	12	11

12–32. A researcher wishes to determine if on-line service or Internet use is independent of the type of user. A sample of 300 computer users shows the following data. At $\alpha = 0.10$, can the researcher conclude that usage is independent of the user?

| | Service usage | | |
	Increase	Same	Decrease
Business	79	21	0
Consumer	122	63	15

Source: *USA Today* Snapshot, August 31, 1995.

12–33. A greenhouse owner wishes to determine whether the number of plants she sells is related to the location of her portable plant stands. A sample of locations is selected, and the number of sales at each area is tallied over a three-day period. The results are shown here. At $\alpha = 0.01$, is the number of plants sold related to the location of the stand?

Location of plant stands	Number of plants sold		
	Mon.	**Tues.**	**Wed.**
Mall	17	12	4
Grocery	12	23	47
Roadside	23	15	52

12–34. An instructor wishes to see if the way people obtain information is independent of their educational background. A survey of 400 high school and college graduates yielded the following information. At $\alpha = 0.05$, can one conclude that the way people obtain information is independent of their educational background?

	Television	Newspapers	Other sources
High school	159	90	51
College	27	42	31

Source: *USA Today* Snapshot, 1993.

12–35. A university official wishes to determine whether the instructor's degree is related to the students' opinion of the quality of instruction received. A sample of students' evaluations of various instructors is selected; the data are shown here. At $\alpha = 0.10$, can the official conclude that the degree of the instructor is related to students' opinions about that instructor's effectiveness in the classroom?

	Degree		
Rating	**Bachelor's**	**Master's**	**Doctorate**
Excellent	14	9	4
Average	16	5	7
Poor	3	12	16

12–36. A researcher wishes to determine whether the marital status of a student is related to his or her grade in a statistics course. The data below were obtained from a random sample of 142 students. At $\alpha = 0.05$, is the marital status independent of the grade received in the course?

Marital status	Grade			
	A	**B**	**C**	**D or F**
Single	27	32	16	10
Married	14	19	16	8

12–37. A study is being conducted to determine whether the age of the customer is related to the type of movie he or she rents. A sample of renters gives the

data shown here. At $\alpha = 0.10$, is the type of movie selected related to the customer's age?

Age	Type of movie		
	Documentary	**Comedy**	**Mystery**
12–20	14	9	8
21–29	15	14	9
30–38	9	21	39
39–47	7	22	17
48 and over	6	38	12

12–38. A study was conducted to determine whether the preference for a two-wheel drive or a four-wheel drive vehicle is related to the gender of the purchaser. A sample of 90 buyers was selected, and the data are shown here. At $\alpha = 0.05$, can the researcher conclude that vehicle preference is independent of gender?

Gender	Two-wheel drive	Four-wheel drive
Male	23	43
Female	18	6

12–39. A survey at a ballpark shows the following selection of condiments made for the hot dogs purchased. At $\alpha = 0.10$, is the condiment chosen independent of the gender of the consumer?

Gender	Condiment		
	Catsup	**Mustard**	**Relish**
Men	15	18	10
Women	25	14	8

12–40. To test the effectiveness of a new drug, a researcher gives one group of individuals the new drug and another group a placebo. The results of the study are shown here. At $\alpha = 0.10$, can the researcher conclude that the drug is effective?

Medication	Effective	Not effective
Drug	32	9
Placebo	12	18

12–41. An educator wishes to determine whether there is a difference in the favorite subjects selected by males and females in college. A random sample provides the data given here. At $\alpha = 0.05$, can the educator conclude that subject selection is independent of the gender of the individual?

Gender	Subject		
	French	**History**	**English**
Male	324	223	191
Female	152	104	312

12–42. According to a recent survey, 32% of Americans say they are "very likely" to become organ donors. A researcher surveys 50 drivers in each of three neighborhoods to determine the percentage of those willing to donate their organs. The results are shown here. At $\alpha = 0.01$, can it be concluded that the proportions of those who will donate their organs are equal in all three neighborhoods?

	Neighborhood A	Neighborhood B	Neighborhood C
Will donate	28	14	21
Will not donate	22	36	29
Total	50	50	50

Source: Lewis H. Lapham, et al., *The Harper's Index Book* (New York: Henry Holt & Co., 1987), p. 27.

12–43. According to a recent survey, 64% of Americans between the ages of 6 and 17 cannot pass a basic fitness test. A physical education instructor wishes to determine if the percentages of such students in different schools in his school district are the same. He administers a basic fitness test to 120 students in each of four schools. The results are shown here. At $\alpha = 0.05$, test the claim that the proportions who pass the test are equal.

	Southside	West End	East Hills	Jefferson
Passed	49	38	46	34
Failed	71	82	74	86
Total	120	120	120	120

Source: Lewis H. Lapham, et al., *The Harper's Index Book* (New York: Henry Holt & Co., 1987), p. 57.

12–44. An advertising firm has decided to ask 92 customers at each of three local shopping malls if they are willing to take part in a market research survey. According to previous studies, 38% of Americans refuse to take part in such surveys. The results are shown here. At $\alpha = 0.01$, can it be concluded that the proportions of those who are willing to participate are equal?

	Mall A	Mall B	Mall C
Will participate	52	45	36
Will not participate	40	47	56
Total	92	92	92

Source: Lewis H. Lapham, et al., *The Harper's Index Book* (New York: Henry Holt & Co., 1987), p. 41.

12–45. An insurance firm wished to see if the proportion of drivers who admit to driving after drinking varies according to the age of the driver. The firm surveyed 86 drivers in each of four age groups to see if they admitted to driving after drinking. The results are shown here. At $\alpha = 0.05$, can it be concluded that the proportions of those who said yes are equal for the age groups?

	Ages 21–29	30–39	40–49	50 and over
Yes	32	28	26	21
No	54	58	60	65
Total	86	86	86	86

12–46. According to a recent survey, 59% of Americans aged 8 to 17 would prefer that their mother work outside the home, regardless of what she does now. A school district psychologist decided to select three samples of 60 students each in elementary, middle, and high school to see how the students in her district felt about the issue. At $\alpha = 0.10$, test the claim that the proportions of the students who prefer that their mother have a job are equal.

	Elementary	Middle	High
Prefers mother work	29	38	51
Prefers mother not work	31	22	9
Total	60	60	60

Source: Daniel Weiss, *100% American* (New York: Poseidon Press, 1988), p. 59.

12–47. A researcher surveyed 100 randomly selected lawyers in each of four areas of the country and asked them if they had performed *pro bono* work for 25 or fewer hours in the last year. The results are shown here. At $\alpha = 0.10$, is there enough evidence to reject the claim that the proportions of those who accepted *pro bono* work for 25 hours or less are the same in each area?

	North	South	East	West
Yes	43	39	22	28
No	57	61	78	72
Total	100	100	100	100

Source: Daniel Weiss, *100% American* (New York: Poseidon Press, 1988), p. 59.

12–48. On average, 79% of American fathers are in the delivery room when their children are born. A physician's assistant surveyed 300 first-time fathers to determine if they had been in the delivery room when their children were born. The results are shown here. At $\alpha = 0.05$, is there enough evidence to reject the claim that the proportions of those who were in the delivery room at the time of birth are the same?

	Hospital A	Hospital B	Hospital C	Hospital D
Present	66	60	57	56
Not present	9	15	18	19
Total	75	75	75	75

Source: Daniel Weiss, *100% American* (New York: Poseidon Press, 1988), p. 79.

12–49. A children's playground equipment manufacturer read in a survey that 55% of all American playground injuries occur on the monkey bars. The manufacturer wishes to investigate playground injuries in four different parts of the country to determine if the proportions of accidents on the monkey bars are equal. The results are shown here. At $\alpha = 0.05$, test the claim that the proportions are equal.

Accidents	North	South	East	West
On monkey bars	15	18	13	16
Not on monkey bars	15	12	17	14
Total	30	30	30	30

Source: Michael D. Shook and Robert L. Shook, *The Book of Odds* (New York: Penguin Putnam Inc., 1991), p. 96.

12–50. According to the American Automobile Association, 31 million Americans travel over the Thanksgiving holiday. To determine whether to stay open or not, a national restaurant chain surveyed 125 customers at each of four locations to see if they would be traveling over the holiday. The results are shown here. At $\alpha = 0.10$, can it be concluded that the proportions of Americans who will travel over the Thanksgiving holiday are equal?

	Location A	Location B	Location C	Location D
Will travel	37	52	46	49
Will not travel	88	73	79	76
Total	125	125	125	125

Source: Michael D. Shook and Robert L. Shook, *The Book of Odds* (New York: Penguin Putnam Inc., 1991), p. 67.

12–51. The vice president of a large supermarket chain wished to determine if his customers made a list before going grocery shopping. He surveyed 288 customers in three stores. The results are shown here. At $\alpha = 0.10$, test the claim that the proportions of the customers in the three stores who made a list before going shopping are equal.

	Store A	Store B	Store C
Made list	77	74	68
No list	19	22	28
Total	96	96	96

Source: Daniel Weiss, *100% American* (New York: Poseidon Press, 1988), p. 82.

*12–52. For a 2 × 2 table, a, b, c, and d are the observed values for each cell, as shown.

a	b
c	d

The chi-square test value can be computed as

$$\chi^2 = \frac{n(ad - bc)^2}{(a + b)(a + c)(c + d)(b + d)}$$

where $n = a + b + c + d$. Compute the χ^2 test value by using the above formula and the formula $\Sigma \, (O - E)^2/E$, and compare the results for the following table.

12	15
9	23

*12–53. For the contingency table shown in Exercise 12–52, compute the chi-square test value by using Yates's correction for continuity.

*12–54. When the chi-square test value is significant, and there is a relationship between the variables, the strength of this relationship can be measured by using the *contingency coefficient*. The formula for the contingency coefficient is

$$C = \sqrt{\frac{\chi^2}{\chi^2 + n}}$$

where χ^2 is the test value and n is the sum of frequencies of the cells. The contingency coefficient will always be less than 1. Compute the contingency coefficient for Exercises 12–28 and 12–40.

12–4

Analysis of Variance (ANOVA)

When an *F* test is used to test a hypothesis concerning the means of three or more populations, the technique is called **analysis of variance** (commonly abbreviated as ANOVA). At first glance, one might think that to compare the means of three or more samples, the *t* test can be used, comparing two means at a time. But there are several reasons why the *t* test should not be done.

Objective 4. Use the ANOVA technique to determine if there is a significant difference among three or more means.

First, when one is comparing two means at a time, the rest of the means under study are ignored. With the F test, all the means are compared simultaneously. Second, when one is comparing two means at a time and making all pairwise comparisons, the probability of rejecting the null hypothesis when it is true is increased, since the more t tests that are conducted, the greater is the likelihood of getting significant differences by chance alone. Third, the more means there are to compare, the more t tests are needed. For example, for the comparison of 3 means two at a time, 3 t tests are required. For the comparison of 5 means two at a time, 10 tests are required. And for the comparison of 10 means two at a time, 45 tests are required.

Assumptions for the F Test for Comparing Three or More Means

1. The populations from which the samples were obtained must be normally or approximately normally distributed.
2. The samples must be independent of each other.
3. The variances of the populations must be equal.

Even though one is comparing three or more means in this use of the F test, *variances* are used in the test instead of means.

With the F test, two different estimates of the population variance are made. The first estimate is called the **between-group variance,** and it involves computing the variance by using the means of the groups or between the groups. The second estimate, the **within-group variance** is made by computing the variance using all the data and is not affected by differences in the means. If there is no difference in the means, the between-group variance estimate will be approximately equal to the within-group variance estimate, and the F test value will be approximately equal to 1. However, when the means differ significantly, the between-group variance will be much larger than the within-group variance; the F test value will be significantly greater than 1; and the null hypothesis will be rejected. Since variances are compared, this procedure is called *analysis of variance* (ANOVA).

For a test of the difference among three or more means, the following hypotheses should be used:

H_0: $\mu_1 = \mu_2 = \cdots = \mu_n$.
H_1: At least one mean is different from the others.

As stated previously, a significant test value means that there is a high probability that this difference in means is not due to chance, but it does not indicate where the difference lies.

The degrees of freedom for this F test are d.f.N. $= k - 1$ where k is the number of groups, and d.f.D. $= N - k$ where N is the sum of the sample sizes of the groups, $N = n_1 + n_2 + \cdots + n_k$. The sample sizes need not be equal. The F test to compare means is always right-tailed.

The next two examples illustrate the computational procedure for the ANOVA technique for comparing three or more means, and the steps are summarized in Procedure Table 12, shown after the examples.

Example 12–8

A researcher wishes to try three different techniques to lower the blood pressure of individuals diagnosed with high blood pressure. The subjects are randomly assigned to three groups; the first group takes medication, the second group exercises, and the third group diets. After four weeks, the reduction in each person's blood pressure is recorded. At $\alpha = 0.05$, test the claim that there is no difference among the means. The data follow.

Medication	Exercise	Diet
10	6	5
12	8	9
9	3	12
15	0	8
13	2	4
$\overline{X}_1 = 11.8$	$\overline{X}_2 = 3.8$	$\overline{X}_3 = 7.6$
$s_1^2 = 5.7$	$s_2^2 = 10.2$	$s_3^2 = 10.3$

Solution

STEP 1   State the hypotheses and identify the claim.

H_0: $\mu_1 = \mu_2 = \mu_3$ (claim).
H_1: At least one mean is different from the others.

STEP 2   Find the critical value. Since $k = 3$ and $N = 15$,

$$\text{d.f.N.} = k - 1 = 3 - 1 = 2$$
$$\text{d.f.D.} = N - k = 15 - 3 = 12$$

The critical value is 3.89, obtained from Table H with $\alpha = 0.05$.

STEP 3   Compute the test value, using the procedure outlined here.

    a.   Find the mean and variance of each sample (these values are shown below the data).

    b.   Find the grand mean. The *grand mean,* denoted by X_{GM}, is the mean of all values in the samples.

$$\overline{X}_{GM} = \frac{\Sigma X}{N} = \frac{10 + 12 + 9 + \cdots + 4}{15} = \frac{116}{15} = 7.73$$

When samples are equal in size, find $\overline{X}_{GM}$ by summing the $\overline{X}$'s and dividing by k = the number of groups.

    c.   Find the between-group variance, denoted by s_B^2.

$$s_B^2 = \frac{\Sigma n_i (\overline{X}_i - \overline{X}_{GM})^2}{k - 1}$$

$$= \frac{5(11.8 - 7.73)^2 + 5(3.8 - 7.73)^2 + 5(7.6 - 7.73)^2}{3 - 1}$$

$$= \frac{160.13}{2} = 80.07$$

Note: This formula finds the variance between the means using the sample sizes as weights and considers the differences in the means.

   d.   Find the within-group variance, denoted by s_W^2.

$$s_W^2 = \frac{\Sigma(n_i - 1)s_i^2}{\Sigma(n_i - 1)}$$

$$= \frac{(5 - 1)(5.7) + (5 - 1)(10.2) + (5 - 1)(10.3)}{(5 - 1) + (5 - 1) + (5 - 1)}$$

$$= \frac{104.80}{12} = 8.73$$

Note: This formula finds the variance within the group means, again using the sample sizes as weights, but it does not involve the differences in the means.

   e.   Find the *F* test value.

$$F = \frac{s_B^2}{s_W^2} = \frac{80.07}{8.73} = 9.17$$

STEP 4   Make the decision. The decision is to reject the null hypothesis, since $9.17 > 3.89$.

STEP 5   Summarize the results. There is enough evidence to reject the claim and conclude that at least one mean is different from the others.

The numerator of the fraction obtained in Step 3, part *c*, of the computational procedure is called the **sum of squares between groups,** denoted by SS_B. The numerator of the fraction obtained in Step 3, part *d*, of the computational procedure is called the **sum of squares within groups,** denoted by SS_W. This statistic is also called the *sum of squares for the error*. SS_B is divided by d.f.N. to obtain the between-group variance. These two variances are sometimes called **mean squares,** denoted by MS_B and MS_W. These terms are used to summarize the analysis of variance and are placed in a summary table, as shown in Table 12–1.

Table 12–1   Analysis of Variance Summary Table

Source	Sum of squares	d.f.	Mean square	F
Between	SS_B	$k - 1$	MS_B	
Within (error)	SS_W	$N - k$	MS_W	
Total				

In the table,

SS_B = sum of squares between groups

SS_W = sum of squares within groups

   k = number of groups

   $N = n_1 + n_2 + \cdots + n_k$ = sum of the sample sizes for the groups

$$MS_B = \frac{SS_B}{k-1}$$

$$MS_W = \frac{SS_W}{N-k}$$

$$F = \frac{MS_B}{MS_W}$$

The totals are obtained by adding the corresponding columns. For Example 12–8, the ANOVA summary table is shown in Table 12–2.

Table 12–2	Analysis of Variance Summary Table for Example 12–8				
Source	**Sum of squares**	**d.f.**	**Mean square**	**F**
Between	160.13	2	80.07	9.17
Within (error)	104.80	12	8.73	
Total	264.93	14		

Most computer programs will print out an ANOVA summary table.

Example 12–9

A marketing specialist wishes to see whether there is a difference in the average time a customer has to wait in a checkout line in three large self-service department stores. The times (in minutes) are shown next.

Store A	Store B	Store C
3	5	1
2	8	3
5	9	4
6	6	2
3	2	7
1	5	3
$\overline{X}_1 = 3.33$	$\overline{X}_2 = 5.83$	$\overline{X}_3 = 3.33$
$s_1^2 = 3.47$	$s_2^2 = 6.17$	$s_3^2 = 4.27$

At $\alpha = 0.05$, is there a significant difference in the mean waiting times of customers for each store?

Solution

STEP 1 State the hypotheses and identify the claim.

H_0: $\mu_1 = \mu_2 = \mu_3$.
H_1: At least one mean is different from the others (claim).

STEP 2 Find the critical value. Since $k = 3$, $N = 18$, and $\alpha = 0.05$,

d.f.N. $= k - 1 = 3 - 1 = 2$
d.f.D. $= N - k = 18 - 3 = 15$

The critical value is 3.68.

STEP 3 Compute the test value.

    a. Find the mean and variance of each sample (these values are shown below the data).

    b. Find the grand mean.

$$\overline{X}_{GM} = \frac{\Sigma X}{N} = \frac{3 + 2 + \cdots + 3}{18} = \frac{75}{18} = 4.17$$

    c. Find the between-group variance.

$$s_B^2 = \frac{\Sigma n_i (\overline{X}_i - \overline{X}_{GM})^2}{k - 1}$$

$$= \frac{6(3.33 - 4.17)^2 + 6(5.83 - 4.17)^2 + 6(3.33 - 4.17)^2}{3 - 1}$$

$$= \frac{25}{2} = 12.5$$

    d. Find the within-group variance.

$$s_W^2 = \frac{\Sigma(n_i - 1)s_i^2}{\Sigma(n_i - 1)}$$

$$= \frac{(6 - 1)(3.47) + (6 - 1)(6.17) + (6 - 1)(4.27)}{(6 - 1) + (6 - 1) + (6 - 1)}$$

$$= \frac{69.6}{15} = 4.64$$

    e. Find the *F* test value.

$$F = \frac{s_B^2}{s_W^2} = \frac{12.5}{4.64} = 2.69$$

STEP 4 Make the decision. Since $2.69 < 3.68$, the decision is not to reject the null hypothesis.

STEP 5 Summarize the results. There is not enough evidence to support the claim that there is a difference among the means. The ANOVA summary table for this example is shown in Table 12–3.

Table 12–3 Analysis of Variance Summary Table for Example 12–9

Source	Sum of squares	d.f.	Mean square	F
Between	25	2	12.5	2.69
Within	69.6	15	4.64	
Total	94.6	17		

The procedure for computing the F test value for the ANOVA is summarized in Procedure Table 12.

Procedure Table 12

Finding the F Test Value for the Analysis of Variance

STEP 1 Find the mean and variance of each sample:

$$(\overline{X}_1, s_1^2), (\overline{X}_2, s_2^2), \cdots, (\overline{X}_k, s_k^2).$$

STEP 2 Find the grand mean

$$\overline{X}_{\mathrm{GM}} = \frac{\Sigma X}{N}$$

STEP 3 Find the between-group variance.

$$s_B^2 = \frac{\Sigma n_i (\overline{X}_i - \overline{X}_{\mathrm{GM}})^2}{k - 1}$$

STEP 4 Find the within-group variance.

$$s_W^2 = \frac{\Sigma (n_i - 1) s_i^2}{\Sigma (n_i - 1)}$$

STEP 5 Find the F test value.

$$F = \frac{s_B^2}{s_W^2}$$

The degrees of freedom are

$$\mathrm{d.f.N.} = k - 1$$

where k is the number of groups,

$$\text{and } \mathrm{d.f.D.} = N - k$$

where N is the sum of the sample sizes of the groups,

$$N = n_1 + n_2 + \cdots + n_k$$

The P-values for the ANOVA are shown on computer printouts such as those in the next computer application. They give the probabilities for the F value. If the P-value is less than the α value, the null hypothesis should be rejected.

Computer Applications for One-Way ANOVA

MINITAB One-way ANOVA:

1. Enter the data into C1, C2, C3, etc.
2. Click Stat > ANOVA > One-way (Unstacked).
3. Highlight C1, C2, C3, etc. and click Select.
4. Click OK.

Example: Perform a one-way ANOVA using the following data. (Use $\alpha = 0.05$.)

Data:

C1	9	6	15	4	3
C2	8	7	12	3	5
C3	12	15	18	9	10

MINITAB printout for this example:

One-way Analysis of Variance

Analysis of Variance

Source	DF	SS	MS	F	P
Factor	2	104.9	52.5	3.25	0.075
Error	12	194.0	16.2		
Total	14	298.9			

Summary: The computer will print the ANOVA table as shown. The test value is 3.25, and the P-value is 0.075. The decision is not to reject the null hypothesis.

Exercises

12–55. (**W**) What test is used to compare three or more means?

12–56. (**W**) State three reasons why multiple t tests cannot be used to compare three or more means.

12–57. (**W**) What are the assumptions for ANOVA?

12–58. (**W**) Define *between-group variance* and *within-group variance.*

12–59. (**W**) What is the formula for comparing three or more means?

12–60. (**W**) State the hypotheses used in the ANOVA test.

12–61. (**W**) When there is no significant difference between three or more means the value of F will be close to what number?

For Exercises 12–62 through 12–73, assume that all variables are normally distributed, that the samples are independent, and that the population variances are equal. Also, for each exercise, perform the following steps.

a. State the hypotheses and identify the claim.
b. Find the critical value.
c. Compute the test value.
d. Make the decision.
e. Summarize the results.

12–62. A researcher wishes to see whether there is a difference in the average age of nurses, doctors, and X-ray technicians at a local hospital. Employees are randomly selected, and their ages are recorded as shown in the table. At $\alpha = 0.05$, can the researcher conclude that the average ages of the three groups differ?

Nurses	Doctors	X-ray technicians
23	60	33
25	36	28
26	29	35
35	56	29
42	32	23
22	54	41
	58	

12–63. Three types of computer disks are selected, and the number of defects in each is recorded below. At $\alpha = 0.05$, can one conclude that there is a difference in the means of the number of defects for the three groups?

Type A	Type B	Type C
0	2	1
1	0	0
0	3	1
2	5	1
3	3	0
2	4	2
0	6	0
1	0	0
1	2	1
0	5	2

12–64. The grade point averages of students participating in college sports programs are compared. The data are shown here. At $\alpha = 0.10$, can one conclude that there is a difference in the mean GPA of the three groups?

Football	Basketball	Hockey
3.2	3.8	2.6
2.6	3.1	1.9
2.4	2.6	1.7
2.4	3.9	2.5
1.8	3.3	1.9

12–65. Three different relaxation techniques are given to randomly selected patients in an effort to reduce their stress levels. A special instrument has been designed to measure the percentage of stress reduction in each person. The data are shown in the table. At $\alpha = 0.05$, can one conclude that there is a difference in the means of the percentages?

Technique I	Technique II	Technique III
3	12	15
10	12	14
5	17	18
1	13	14
13	18	20
3	9	22
4	14	16

12–66. A researcher wishes to see whether there is any difference in the weight gains of athletes following one of three special diets. Athletes are randomly assigned to three groups and placed on the diet for six weeks. The weight gains (in pounds) are shown here. At $\alpha = 0.05$, can the researcher conclude that there is a difference in the diets?

Diet A	Diet B	Diet C
3	10	8
6	12	3
7	11	2
4	14	5
	8	
	6	

12–67. A consumer magazine rated dishwashers as excellent, very good, and good. A researcher wishes to see if the average prices for the three groups differ. At $\alpha = 0.10$, is there a difference in the average prices of the machines rated?

Excellent	Very good	Good
$565	$330	$350
400	840	379
369	510	280
550	470	320
460	380	
400	375	
400	450	
	290	
	319	

Source: *Consumer Reports,* August 1995, p. 536.

12–68. Workers are randomly assigned to four machines on an assembly line. The number of defective parts produced by each worker for one day is recorded. The data are shown here. At $\alpha = 0.05$, can one conclude that the mean number of defective parts produced by the workers is the same?

Machine 1	Machine 2	Machine 3	Machine 4
3	8	10	9
2	6	9	15
0	2	8	3
6	0	11	0
4	1	12	2
3	9	15	0
5	7	17	1

12–69. A researcher wishes to see if there is a difference in the weights (in pounds) of four types of lawnmowers. At $\alpha = 0.10$, can one conclude that the weights differ?

Gas (self-propelled)	Gas (push; rear bag)	Electric	Manual
95	73	55	37
101	69	52	24
108	72	51	25
107	71	37	29
97	67	57	22
101	62	54	17
	68	34	17
	71	45	22
		41	20
		53	18
			21

Source: *Consumer Reports,* June 1995, pp. 400, 402.

12–70. A researcher tests the lifetimes (in hours) of three cassette tapes. The data are shown here. At $\alpha = 0.10$, is there a difference in the means?

Tape 1	Tape 2	Tape 3
196	98	94
183	91	106
112	101	85
107	99	102
189	84	101

12–71. A research organization tested microwave ovens. At $\alpha = 0.10$, is there a significant difference in the average prices of the three types of ovens?

Watts		
1000	900	800
270	240	180
245	135	155
190	160	200
215	230	120
250	250	140
230	200	180
	200	140
	210	130

12–72. The time it takes (in minutes) to treat randomly selected patients in an emergency room for three shifts is recorded here. At $\alpha = 0.05$, can one conclude that there is a significant difference in the mean time it takes to treat patients for the three shifts?

Morning (7–3)	Afternoon (3–11)	Night (11–7)
12	9	6
18	8	15
18	16	8
21	20	9
19	15	5

12–73. Four types of pain relief medications are given to randomly selected patients. The time it takes to relieve the pain (in minutes) is shown here for each medication. At $\alpha = 0.01$, can one conclude that there is a difference in the pain relief ability of the medications?

A	B	C	D
3	8	7	14
2	12	4	16
5	15	9	8
4	9	2	15
3	6	1	12

12–5

Summary

The uses of the chi-square distribution were explained in this chapter. It can be used as goodness-of-fit test, in order to determine whether the frequencies of a distribution are the same as the hypothesized frequencies. For example, is the number of defective parts produced by a factory the same each day? This test is always a right-tailed test.

The test of independence is used to determine whether two variables are related or are independent. This test uses a contingency table and is always a right-tailed test. An example of its use is a test to determine whether the attitudes of urban residents about the recycling of trash differ from the attitudes of rural residents.

The homogeneity of proportions test is used to determine if several proportions are all equal when samples are selected from different populations.

The F test, as shown in Chapter 10, can be used to compare two sample variances to determine whether they are equal. It can also be used to compare three or more means. When three or more means are compared, the technique is called analysis of variance (ANOVA). The ANOVA technique uses two estimates of the population variance. The between-group variance is the variance of the sample means; the within-group variance is the overall variance of all the values. When there is no significant difference among the means, the two estimates will be approximately equal, and the F test value will be close to 1. If there is a significant difference among the means, the between-group variance estimate will be larger than the within-group variance estimate, and a significant test value will result.

Important Terms

analysis of variance (ANOVA) 498

ANOVA summary table 501

between-group variance 499

contingency table 486

expected frequency 475

goodness-of-fit test 475

homogeneity of proportions test 491

independence test 485

mean square 501

observed frequency 475

sum of squares between groups 501

sum of squares within groups 501

test of normality 481

within-group variance 499

Important Formulas

Formula for the chi-square test for goodness of fit:

$$\chi^2 = \sum \frac{(O - E)^2}{E}$$

with degrees of freedom equal to the number of categories minus 1 and where

O = **observed frequency**
E = **expected frequency**

Formula for the chi-square independence and homogeneity of proportions tests:

$$\chi^2 = \sum \frac{(O - E)^2}{E}$$

with degrees of freedom equal to (rows − 1)(columns − 1). Formula for the expected value for each cell:

$$E = \frac{\textbf{(row sum)(column sum)}}{\textbf{grand total}}$$

Formulas for the ANOVA test:

$$\overline{X}_{\text{GM}} = \frac{\sum X}{N}$$

$$F = \frac{s_B^2}{s_W^2}$$

where

$$s_B^2 = \frac{\sum n_i (\overline{X}_i - \overline{X}_{\text{GM}})^2}{k - 1} \qquad s_W^2 = \frac{\sum (n_i - 1)s_i^2}{\sum (n_i - 1)}$$

$$\text{d.f.N.} = k - 1 \qquad\qquad N = n_1 + n_2 + \cdots + n_k$$

$$\text{d.f.D.} = N - k \qquad\qquad k = \textbf{number of groups}$$

Review Exercises

For Exercises 12–74 through 12–91, follow these steps.

a. State the hypotheses and identify the claim.
b. Find the critical value(s).
c. Compute the test value.
d. Make the decision.
e. Summarize the results.

12–74. A company owner wishes to determine whether the number of sales of a product is equally distributed over five regions. A month is selected at random, and the number of sales is recorded. The data are as shown here. At $\alpha = 0.05$, can the owner conclude that the number of items sold in each region is the same?

Region	NE	SE	MW	NW	SW
Sales	236	324	182	221	365

12–75. An ad is placed in newspapers in four counties asking for volunteers to test a new medication for reducing blood pressure. The number of inquiries received in each area is as shown here. At $\alpha = 0.01$, can one conclude that all the ads produced the same number of responses?

County	1	2	3	4
No. of inquiries	87	62	56	93

12–76. The federal government has proposed labeling tires by fuel efficiency to save fuel and cut emissions. A survey was taken to see who would use these labels. At $\alpha = 0.10$, is the gender of the individual related to whether or not a person would use these labels? The data from a sample are shown here.

Gender	Yes	No	Undecided
Men	114	30	6
Women	136	16	8

Source: *USA Today* Snapshot, September 11, 1995.

12–77. A survey at a county fair shows the following selection of condiments for the hamburgers that are purchased. At $\alpha = 0.10$, is the condiment chosen independent of the gender of the individual?

	Condiment		
Gender	Relish	Catsup	Mustard
Men	15	18	10
Women	25	14	8

12–78. A survey was taken on how a lump-sum pension would be invested by 45-year-olds and 65-year-olds. The data are shown here. At $\alpha = 0.05$, is there a relationship between the age of the investor and the way the money would be invested?

	Large company stock funds	Small company stock funds	Inter-national stock funds	CDs or money market funds	Bonds
Age 45	20	10	10	15	45
Age 65	42	24	24	6	24

Source: *USA Today*, September 11, 1995.

12–79. A pet-store owner wishes to determine whether the type of pet a person selects is related to the individual's gender. The data obtained from a sample are shown here. At $\alpha = 0.10$, is the gender of the purchaser related to the type of pet purchased?

Gender of purchaser	Pet purchased		
	Dog	Cat	Bird
Male	32	27	16
Female	23	4	8

12–80. A guidance counselor wishes to determine if the proportions of high school girls in his school district who have jobs are equal to the national average of 36%. He surveys 80 female students, ages 16 through 18, to determine if they work or not. The results are shown below. At $\alpha = 0.01$, can it be concluded that the proportions of girls who work are equal?

	16-year-olds	17-year-olds	18-year-olds
Work	45	31	38
Don't work	35	49	42
Total	80	80	80

Source: Michael D. Shook and Robert L. Shook, *The Book of Odds* (New York: Plume Books, 1991), p. 196.

12–81. The risk of injury is higher for males as compared to females (57% versus 43%). A hospital emergency room supervisor wishes to determine if the proportions of injuries to males in his hospital are the same for each of four months. He surveys 100 injuries treated in his ER for each month. The results are shown here. At $\alpha = 0.05$, can he reject the claim that the proportions of injuries for males are equal for each of the four months?

	May	June	July	August
Male	51	47	58	63
Female	49	53	42	37
Total	100	100	100	100

Source: Michael D. Shook and Robert L. Shook, *The Book of Odds* (New York: Plume Books, 1991), p. 98.

12–82. A researcher surveyed 50 randomly selected subjects in four cities and asked if they felt that their anger was the most difficult *thing* to control. The results are shown below. At $\alpha = 0.10$, is there enough evidence to reject the claim that the proportion of those who felt this way in each city is the same?

	City A	City B	City C	City D
Yes	12	15	10	21
No	38	35	40	29
Total	50	50	50	50

12–83. In order to see if the proportion of transactions that result in a glitch is the same for three stores, a researcher sampled 200 transactions at each store and asked if the purchaser had any problems. The results are shown here. At $\alpha = 0.01$, can it be concluded that the proportions are equal?

	Store 1	Store 2	Store 3
Yes	87	56	43
No	113	144	157
Total	200	200	200

12–84. The number of cars that park in three city-owned lots is being compared. A week is selected at random, and the number of cars parked each day is shown here. Test the claim, at $\alpha = 0.01$, that there is no difference in the number of cars parked in each lot.

Lot A	Lot B	Lot C
203	319	89
162	321	126
190	271	115
219	194	92
188	342	106
209	423	100
212	199	94

12–85. The prices of three types of men's shoes are shown here. At $\alpha = 0.05$, can one conclude that there is a significant difference among the mean prices of the three groups?

Dress	Casual	Moccasins
$110	$ 80	$64
95	100	66
95	135	70
265	90	92
59	80	
70		
50		

Source: *Consumer Reports,* January 1995, p. 62.

12–86. The weights in ounces of four types of women's shoes are shown here. At $\alpha = 0.05$, can one conclude that there is no difference in the mean weights of the groups?

Dress heels	Dress flats	Casual heels	Casual flats
8	6	11	6
7	6	12	9
7	7	12	7
6			

Source: *Consumer Reports,* January 1995, p. 62.

12–87. A plant owner wants to see whether the average time (in minutes) it takes his employees to commute to work is different for three groups. The data are shown here. At $\alpha = 0.05$, can the owner conclude that there is a significant difference among the means?

Managers	Salespeople	Stock clerks
35	9	15
18	3	6
27	12	27
24	6	22
	14	
	8	
	21	

12–88. The coliform levels (in parts per million) of three lakes were checked for a period of five days. The data are shown here. At $\alpha = 0.05$, is there a difference in the means of the coliform levels of the lakes?

Sunset Lake	South Lake	Indian Lake
62	97	33
53	82	35
41	99	31
38	84	28
55	79	26

12–89. Students are randomly assigned to three reading classes. Each class is taught by a different method. At the end of the course, a comprehensive reading examination is given, and the results are shown here. At $\alpha = 0.05$, is there a significant difference in the means of the examination results?

Class A	Class B	Class C
87	82	97
92	78	90
61	41	83
83	65	92
47	63	91

12–90. Four hospitals are being compared to see whether there is any significant difference in the mean number of operations performed in each. A sample of six days provided the following number of operations performed each day. Can one conclude, at $\alpha = 0.05$, that there is no difference in the means?

Hospital A	Hospital B	Hospital C	Hospital D
8	4	5	10
5	9	6	12
6	3	3	13
3	1	7	9
2	0	7	0
7	1	3	1

12–91. Three composition instructors recorded the number of grammatical errors their students made on a term paper. The data are shown here. At $\alpha = 0.01$, is there a significant difference in the average number of errors in the three instructors' classes?

Instructor A	Instructor B	Instructor C
2	6	1
3	7	4
5	12	0
4	4	1
8	9	2
	1	2
	0	

Statistics Today

Statistics and Heredity   Revisited

Using probability, Mendel predicted the following:

	Smooth		Wrinkled	
	Yellow	Green	Yellow	Green
Expected	0.5625	0.1875	0.1875	0.0625

The observed results were

	Smooth		Wrinkled	
	Yellow	Green	Yellow	Green
Observed	0.5666	0.1942	0.1816	0.0556

Using chi-square tests on the data, Mendel found that his predictions were accurate in most cases (i.e., a good fit), thus supporting his theory. He reported many highly successful experiments. Mendel's genetic theory is simple but useful in predicting the results of hybridization.

A Fly in the Ointment

Although Mendel's theory is basically correct, an English statistician, R. A. Fisher, examined Mendel's data some 50 years later. He found that the observed (actual) results agreed too closely with the expected (theoretical) results and concluded that the data had in some way been falsified. The results are too good to be true. Several explanations have been proposed, ranging from deliberate misinterpretation to an assistant's error, but no one can be sure why this happened.

Data Analysis

The Data Bank is located in Appendix D.

1. From the Data Bank, test the hypothesis that the marital status of individuals is equally distributed among four groups. Use the chi-square goodness-of-fit-test. Choose a sample of at least 50 individuals.

2. From the Data Bank, use the chi-square test of independence to test the hypothesis that smoking is independent of the gender of the individual. Use a sample size of at least 50 individuals.

3. From the Data Bank, select a random sample of subjects, and test the hypothesis that the mean cholesterol levels of the nonsmokers, less-than-one-pack-a-day smokers, and one-pack-plus smokers are equal. Use an ANOVA test. If the null hypothesis is rejected, conduct the Scheffé test to find where the difference is. Summarize the results.

4. Repeat Exercise 2 for the mean IQs of the various educational levels of the subjects.

Quiz

Determine whether each statement is true or false. If the statement is false, explain why.

1. The chi-square test of independence is always two-tailed.

2. The test values for the chi-square goodness-of-fit test and the independence test are computed using the same formula.

3. When the null hypothesis is rejected in the goodness-of-fit test, it means there is a close agreement between the observed and expected frequencies.

4. In analysis of variance, the null hypothesis should be rejected only when there is a significant difference among all pairs of means.

5. The F test does not use the concept of degrees of freedom.

6. When the F test value is close to 1, the null hypothesis should be rejected.

Select the best answer.

7. The values of the chi-square variable cannot be
a. Positive
b. 0
c. Negative
d. None of the above

8. The null hypothesis for the chi-square test of independence is that the variables are
a. Dependent
b. Independent
c. Related
d. Always 0

9. The degrees of freedom for the goodness-of-fit test are
a. 0
b. 1
c. Sample size - 1
d. Number of categories - 1

10. The analysis of variance uses the _____ test.
a. z
b. t
c. χ^2
d. F

11. The null hypothesis in the ANOVA is that all of the means are _____.
a. Equal
b. Unequal
c. Variable
d. None of the above

12. When one conducts an F test, _____ estimates of the population variance are compared.
a. Two
b. Three
c. Any number of
d. No

Complete the following statements with the best answer.

13. The degrees of freedom for a 4×3 contingency table are _____.

14. An important assumption for the chi-square test is that the observations must be _____.

15. The chi-square goodness-of-fit test is always _____ tailed.

16. In the chi-square independence test, the expected frequency for each class must always be _____.

17. When three or more means are compared, one uses the _____ technique.

For Problems 18 through 29, follow these steps.
a. State the hypotheses.
b. Find the critical value.
c. Compute the test value.
d. Make the decision.
e. Summarize the results.

18. A company owner wishes to determine if the number of advertisements of a product are equally distributed over five geographic locations. A month is selected at random, and the number of ads is recorded. The data are shown here. At $\alpha = 0.05$, can it be concluded that the number of ads is the same in each region?

Region	NE	SE	NW	MW	SW
Sales	215	287	201	193	306

19. An ad is placed in five campus buildings asking for volunteers to test a new weight-loss medication. The number of inquiries received in each building is shown here. At $\alpha = 0.01$, can it be concluded that all the ads produced the same number of responses?

Building	A	B	C	D	E
No. of inquiries	73	82	49	51	68

20. A fast-food restaurant manager decides to test his theory that 48% of the customers order hamburgers, 33% order chicken sandwiches, and 19% order a salad. An hour is selected at random, and the number of orders is recorded here. At $\alpha = 0.05$, is the manager's theory valid?

Order	Hamburger	Chicken	Salad
Number	106	82	30

21. A recent survey shows the following number of each type of gifts purchased for Mother's Day for a randomly selected week. The data are shown here.

At $\alpha = 0.01$, can it be concluded that each gift was purchased with equal frequency?

Gift	Clothing	Flowers	Candy
No. purchased	208	318	423

22. A bookstore manager wishes to determine if there is a difference in the type of novels selected by male and female customers. A random sample of males and females provides the following data. At $\alpha = 0.05$, can it be concluded that the type of novel selected is independent of the sex of the individual?

	Type of novel		
	Science fiction	Mystery	Romance
Males	482	303	185
Females	291	257	405

23. A pizza-shop owner wishes to determine if the type of pizza a person selects is related to the age of the individual. The data obtained from a sample are shown here. At $\alpha = 0.10$, is the age of the purchaser related to the type of pizza ordered?

	Type of pizza			
Age	Plain	Pepperoni	Mushroom	Double cheese
10–15	12	21	39	71
15–25	18	76	52	87
25–40	24	50	40	47
40–65	52	30	12	28

24. A survey at a ballpark shows the following selection of pennants sold to fans. The data are presented here. At $\alpha = 0.10$, is the color of the pennant purchased independent of the sex of the individual?

	Blue	Yellow	Red
Men	519	659	876
Women	487	702	787

25. The average number of orders per week from three takeout restaurants is being compared. A week is selected at random, and the number of orders is shown here. Test the claim at $\alpha = 0.05$ that there is no difference in the average number of orders in each restaurant.

Restaurant A	Restaurant B	Restaurant C
467	502	419
318	293	392
384	392	298
427	387	501
504	419	438
309	512	488
356	398	251

26. Four hospitals are being compared to see if there is any significant difference in the mean number of physical therapy sessions held in each. A sample of six days provided the following number of therapy sessions each day. Can it be concluded at $\alpha = 0.05$ that there is no difference in the means?

Hospital A	Hospital B	Hospital C	Hospital D
7	5	3	11
4	10	7	13
5	4	7	14
2	2	3	9
1	0	6	2
6	1	5	8

27. Three composition instructors recorded the number of spelling errors their students made on a research paper. The data are shown here. At $\alpha = 0.01$, is there a significant difference in the average number of errors in the three classes?

Instructor 1	Instructor 2	Instructor 3
2	4	5
3	6	2
5	8	3
0	4	2
8	9	3
	0	3
	2	

28. A college official wishes to see if the average time (in minutes) it takes people to commute to the college is different for three groups. The data are shown here. At $\alpha = 0.05$, can it be concluded that there is a significant difference among the means?

Students	Faculty	Staff
12	57	15
28	43	12
47	12	28
15	10	35
35	25	49
	19	55
	38	19

29. The bacteria levels (in parts per million) of three pools were tested for a period of five days. The data are shown here. At $\alpha = 0.05$, is there a difference in the means of the bacteria levels of the pools?

Sunset Pool	Blue Spruce Pool	Rainbow Pool
58	98	34
61	87	38
42	98	39
37	83	29
58	76	28

Critical Thinking Challenges

1. At Random Use your calculator or the MINITAB random number generator and generate 100 two-digit random numbers. Make a grouped frequency distribution using the chi-square goodness-of-fit test to see if the distribution is random. In order to do this, use an expected frequency of 10 for each class. Can it be concluded that the distribution is random? Explain.

2. Want to Bet? Simulate the state lottery by using your calculator or MINITAB to generate 100 three-digit random numbers. Group these numbers 100–199, 200–299, etc. Use the chi-square goodness-of-fit test to see if the numbers are random. The expected frequency for each class should be 10. Explain why.

Data Projects

Where appropriate, use MINITAB, the TI-83, or a computer program of your choice to complete the following exercises.

1. Select a variable and collect some data over a period of a week or several months. For example, you may want to record the number of phone calls you received over seven days, or the number of times you used your credit card each month for the past several months. Using the chi-square goodness-of-fit test, see if the occurrences are equally distributed over the period.
a. State the purpose of the study.
b. Define the population.
c. State how the sample was selected.
d. State the hypotheses.
e. Select an α value.
f. Compute the chi-square test value.
g. Make the decision.
h. Write a paragraph summarizing the results.

2. Collect some data on a variable and construct a frequency distribution.
a. Using the method shown in Section 12–2, decide if the variable you have chosen is approximately normally distributed.
b. Write a short paper describing your findings and cite some reasons why the variable you selected is or is not approximately normally distributed.

3. Collect some data on a variable that can be divided into groups. For example, you may want to see if there is a difference in the color of cars men own versus the color of cars women own. Using the chi-square independence test, determine if the one variable is independent of the other.
a. State the purpose of the study.
b. Define the population.
c. State how the sample was selected.

d. State the hypotheses for the study.
e. Select an α value.
f. Compute the chi-square test value.
g. Make the decision.
h. Summarize the results.

4. Select a variable and record data for two different groups. For example, you may want to use the pulse rates or blood pressures of males and females. Compute and compare the variances of the data for each group, and then complete the following:
a. What is the purpose of the study?
b. Define the population.
c. State how the samples were selected.
d. State the hypotheses for comparing the two variances.
e. Select an α value.
f. Compute the value for the F test.
g. Decide whether or not the null hypothesis should be rejected.
h. Write a brief summary of the findings.

5. Select a variable and collect data for at least three different groups (samples). For example, you could ask students, faculty, and clerical staff how many cups of coffee they drink per day or how many hours they watch television per day. Compare the means using the one-way ANOVA technique, and then complete the following:
a. What is the purpose of the study?
b. Define the population.
c. How were the samples selected?
d. What α value was used?
e. State the hypotheses.
f. What was the F test value?
g. What was the decision?
h. Summarize the results.

TI-83 Calculator

Chi-Square Test for Contingency Tables

1. Press **MATRX** and move the cursor to Edit, then press **ENTER**.

2. Enter the number of rows and columns. Then press **ENTER**.

3. Enter the values in the matrix as they appear in the contingency table.

4. Press **STAT** and move the cursor to TESTS. Press **C** (**ALPHA PRGM**). Make sure the observed matrix is [A] and the expected measure is [B].

5. Move the cursor to Calculate and press **ENTER**.

Example 34: Using the data shown below, test the claim of independence at $\alpha = 0.05$.

75	25
54	46
60	40

Input

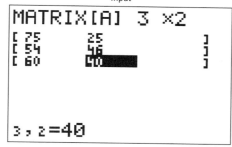

Output

The test value is 10.03861004.
The *P*-value is 0.0066091183. The decision is to reject the null hypothesis, since this value is less than 0.05.
Note: Find the expected values by pressing **MATRX**, moving the cursor to [B], and pressing **ENTER** twice.

One-Way ANOVA Test

1. Enter the data into L_1, L_2, L_3, etc.

2. Press **STAT** and move the cursor to TESTS.

3. Press **F (ALPHA COS)**.

4. Enter each list followed by a comma. End with **)** and press **ENTER**.

Example 35: Test the claim H_0: $\mu_1 = \mu_2 = \mu_3$ at $\alpha = 0.05$ for the following data:

A	B	C
9	8	12
6	7	15
15	12	18
4	3	9
3	5	10

Input

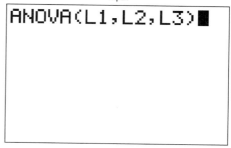

Input

Output

```
One-way ANOVA
 F=3.245360825
 p=.0747075184
 Factor
  df=2
  SS=104.933333
↓ MS=52.4666667
■
```

```
 Error
  df=12
  SS=194
  MS=16.1666667
 Sxp=4.02077936
```

The F test value is 3.245360825. The P-value is 0.0747075184, which is not significant at $\alpha = 0.05$. The factor variable has
d.f. = 2
SS = 104.933333
MS = 52.4666667
The error has
d.f. = 12
SS = 194
MS = 16.1666667

Hypothesis-Testing Summary 2*

7. Test of the significance of the correlation coefficient.

Example: $H_0: \rho = 0$

Use a t test:

$$t = r\sqrt{\frac{n-2}{1-r^2}} \quad \text{with} \quad \text{d.f.} = n - 2$$

8. Comparison of a sample distribution with a specific population.

Example: H_0: There is no difference between the two distributions.

Use the chi-square goodness-of-fit test:

$$\chi^2 = \Sigma \frac{(O-E)^2}{E} \quad \text{with d.f.}$$
$$= \text{no. of categories} - 1$$

9. Comparison of the independence of two variables.

Example: H_0: Variable A is independent of variable B.

Use the chi-square independence test:

$$\chi^2 = \Sigma \frac{(O-E)^2}{E} \quad \text{with d.f.}$$
$$= (R-1)(C-1)$$

10. Test for homogeneity of proportions.

Example: $H_0: p_1 = p_2 = p_3$

Use the chi-square test:

$$\chi^2 = \Sigma \frac{(O-E)^2}{E} \quad \text{with d.f.}$$
$$= (R-1)(C-1)$$

11. Comparison of three or more sample means.

Example: $H_0: \mu_1 = \mu_2 = \mu_3$

Use the analysis of variance test:

$$F = \frac{s_B^2}{s_W^2}$$

where

$$s_B^2 = \frac{\Sigma n_i(\overline{X}_i - \overline{X}_{\text{GM}})^2}{k-1}$$

$$s_W^2 = \frac{\Sigma(n_i - 1)s_i^2}{\Sigma(n_i - 1)}$$

d.f.N. $= k - 1$ $N = n_1 + n_2 + \cdots + n_k$
d.f.D. $= N - k$ $k = $ number of groups

Algebra Review

A–1 Factorials
A–2 Summation Notation
A–3 The Line

A–1 Factorials

Definition and Properties of Factorials

The notation called factorial notation is used in probability. *Factorial notation* uses the exclamation point and involves multiplication. For example,

$$5! = 5 \cdot 4 \cdot 3 \cdot 2 \cdot 1 = 120$$
$$4! = 4 \cdot 3 \cdot 2 \cdot 1 = 24$$
$$3! = 3 \cdot 2 \cdot 1 = 6$$
$$2! = 2 \cdot 1 = 2$$
$$1! = 1$$

In general, a factorial is evaluated as follows:

$$n! = n(n - 1)(n - 2) \cdots 3 \cdot 2 \cdot 1$$

Note that the factorial is the product of n factors, with the number decreased by one for each factor.

One property of factorial notation is that it can be stopped at any point by using the exclamation point. For example,

$5! = 5 \cdot 4!$	since	$4! = 4 \cdot 3 \cdot 2 \cdot 1$
$\quad = 5 \cdot 4 \cdot 3!$	since	$3! = 3 \cdot 2 \cdot 1$
$\quad = 5 \cdot 4 \cdot 3 \cdot 2!$	since	$2! = 2 \cdot 1$
$\quad = 5 \cdot 4 \cdot 3 \cdot 2 \cdot 1$		

Thus,
$$\begin{aligned} n! &= n(n - 1)! \\ &= n(n - 1)(n - 2)! \\ &= n(n - 1)(n - 2)(n - 3)! \quad \text{etc.} \end{aligned}$$

Another property of factorials is

$$0! = 1$$

This fact is needed for formulas.

Operations with Factorials

Factorials cannot be added or subtracted directly. They must be multiplied out. Then the products can be added or subtracted.

A

Example A–1

Evaluate $3! + 4!$.

Solution

$$\begin{aligned} 3! + 4! &= (3 \cdot 2 \cdot 1) + (4 \cdot 3 \cdot 2 \cdot 1) \\ &= 6 + 24 = 30 \end{aligned}$$

Note: $3! + 4! \neq 7!$, since $7! = 5040$.

Example A–2

Evaluate $5! - 3!$.

Solution

$$\begin{aligned} 5! - 3! &= (5 \cdot 4 \cdot 3 \cdot 2 \cdot 1) - (3 \cdot 2 \cdot 1) \\ &= 120 - 6 = 114 \end{aligned}$$

Note: $5! - 3! \neq 2!$, since $2! = 2$.

Factorials cannot be multiplied directly. Again, one must multiply them out and then multiply the products.

Example A–3

Evaluate $3! \cdot 2!$.

Solution

$$\begin{aligned} 3! \cdot 2! &= (3 \cdot 2 \cdot 1) \cdot (2 \cdot 1) \\ &= 6 \cdot 2 = 12 \end{aligned}$$

Note: $3! \cdot 2! \neq 6!$, since $6! = 720$.

Finally, factorials cannot be divided directly unless they are equal.

Example A–4

Evaluate $6! \div 3!$.

Solution

$$\frac{6!}{3!} = \frac{6 \cdot 5 \cdot 4 \cdot 3 \cdot 2 \cdot 1}{3 \cdot 2 \cdot 1} = \frac{720}{6} = 120$$

Note: $\quad \dfrac{6!}{3!} \neq 2! \quad$ since $\quad 2! = 2$

But $\quad \dfrac{3!}{3!} = \dfrac{3 \cdot 2 \cdot 1}{3 \cdot 2 \cdot 1} = \dfrac{6}{6} = 1$

A

In division, one can take some shortcuts, as shown:

$$\frac{6!}{3!} = \frac{6 \cdot 5 \cdot 4 \cdot 3!}{3!} \quad \text{and} \quad \frac{3!}{3!} = 1$$
$$= 6 \cdot 5 \cdot 4 = 120$$

$$\frac{8!}{6!} = \frac{8 \cdot 7 \cdot 6!}{6!} \quad \text{and} \quad \frac{6!}{6!} = 1$$
$$= 8 \cdot 7 = 56$$

Another shortcut that can be used with factorials is cancellation, after factors have been expanded. For example,

$$\frac{7!}{(4!)(3!)} = \frac{7 \cdot 6 \cdot 5 \cdot 4!}{3 \cdot 2 \cdot 1 \cdot 4!}$$

Now cancel both 4!s. Then cancel the $3 \cdot 2$ in the denominator with the 6 in the numerator.

$$\frac{7 \cdot \overset{1}{\cancel{6}} \cdot 5 \cdot \cancel{4!}}{\underset{1}{\cancel{3}} \cdot \underset{1}{\cancel{2}} \cdot 1 \cdot \underset{1}{\cancel{4!}}} = 7 \cdot 5 = 35$$

Example A–5

Evaluate $10! \div (6!)(4!)$.

Solution

$$\frac{10!}{(6!)(4!)} = \frac{10 \cdot \overset{3}{\cancel{9}} \cdot \overset{1}{\cancel{8}} \cdot 7 \cdot \cancel{6!}}{\underset{1}{\cancel{4}} \cdot \underset{1}{\cancel{3}} \cdot \underset{1}{\cancel{2}} \cdot 1 \cdot \underset{1}{\cancel{6!}}} = 10 \cdot 3 \cdot 7 = 210$$

Exercises

Evaluate each expression.

A–1. $9!$

A–2. $7!$

A–3. $5!$

A–4. $0!$

A–5. $1!$

A–6. $3!$

A–7. $\dfrac{12!}{9!}$

A–8. $\dfrac{10!}{2!}$

A–9. $\dfrac{5!}{3!}$

A–10. $\dfrac{11!}{7!}$

A–11. $\dfrac{9!}{(4!)(5!)}$

A–12. $\dfrac{10!}{(7!)(3!)}$

A–13. $\dfrac{8!}{(4!)(4!)}$

A–14. $\dfrac{15!}{(12!)(3!)}$

A–15. $\dfrac{10!}{(10!)(0!)}$

A–16. $\dfrac{5!}{(3!)(2!)(1!)}$

A–17. $\dfrac{8!}{(3!)(3!)(2!)}$

A–18. $\dfrac{11!}{(7!)(2!)(2!)}$

A–19. $\dfrac{10!}{(3!)(2!)(5!)}$

A–20. $\dfrac{6!}{(2!)(2!)(2!)}$

A–2 Summation Notation

In mathematics, the symbol Σ (Greek letter sigma) means to add or find the sum. For example, ΣX means to add the numbers represented by the variable X. Thus, when X represents 5, 8, 2, 4, and 6, then ΣX means $5 + 8 + 2 + 4 + 6 = 25$.

Sometimes, a subscript notation is used, such as

$$\sum_{i=1}^{5} X_i$$

This notation means to find the sum of five numbers represented by X, as shown:

$$\sum_{i=1}^{5} X_i = X_1 + X_2 + X_3 + X_4 + X_5$$

When the number of values is not known, the unknown number can be represented by n, such as

$$\sum_{i=1}^{n} X_i = X_1 + X_2 + X_3 + \cdots + X_n$$

There are several important types of summation used in statistics. The notation ΣX^2 means to square each value before summing. For example, if the values of X are 2, 8, 6, 1, and 4, then

$$\Sigma X^2 = 2^2 + 8^2 + 6^2 + 1^2 + 4^2$$
$$= 4 + 64 + 36 + 1 + 16 = 121$$

The notation $(\Sigma X)^2$ means to find the sum of X's and then square the answer. For instance, if the values for X are 2, 8, 6, 1, and 4, then

$$(\Sigma X)^2 = (2 + 8 + 6 + 1 + 4)^2$$
$$= (21)^2 = 441$$

Another important use of summation notation is in finding the mean (shown in Section 3–2). The mean, $\bar{X}$, is defined as

$$\bar{X} = \frac{\Sigma X}{n}$$

For example, to find the mean of 12, 8, 7, 3, and 10, use the formula and substitute the values, as shown:

$$\bar{X} = \frac{\Sigma X}{n} = \frac{12 + 8 + 7 + 3 + 10}{5} = \frac{40}{5} = 8$$

A

The notation $\sum (X - \overline{X})^2$ means to perform the following steps.

STEP 1   Find the mean.

STEP 2   Subtract the mean from each value.

STEP 3   Square the answers.

STEP 4   Find the sum.

Example A–6

Find the value of $\sum (X - \overline{X})^2$ for the values 12, 8, 7, 3, and 10 of X.

Solution

STEP 1   Find the mean.

$$\overline{X} = \frac{12 + 8 + 7 + 3 + 10}{5} = \frac{40}{5} = 8$$

STEP 2   Subtract the mean from each value.

$$12 - 8 = 4 \qquad 7 - 8 = -1 \qquad 10 - 8 = 2$$
$$8 - 8 = 0 \qquad 3 - 8 = -5$$

STEP 3   Square the answers.

$$4^2 = 16 \qquad 0^2 = 0 \qquad (-1)^2 = 1$$
$$(-5)^2 = 25 \qquad 2^2 = 4$$

STEP 4   Find the sum.

$$16 + 0 + 1 + 25 + 4 = 46$$

Example A–7

Find $\sum (X - \overline{X})^2$ for the following values of X: 5, 7, 2, 1, 3, 6.

Solution

Find the mean:

$$\overline{X} = \frac{5 + 7 + 2 + 1 + 3 + 6}{6} = \frac{24}{6} = 4$$

Then the steps in Example A–6 can be shortened as follows:

$$\begin{aligned}
\sum (X - \overline{X})^2 &= (5 - 4)^2 + (7 - 4)^2 + (2 - 4)^2 \\
&\quad + (1 - 4)^2 + (3 - 4)^2 + (6 - 4)^2 \\
&= 1^2 + 3^2 + (-2)^2 + (-3)^2 \\
&\quad + (-1)^2 + 2^2 \\
&= 1 + 9 + 4 + 9 + 1 + 4 = 28
\end{aligned}$$

Exercises

For each set of values, find $\sum X$, $\sum X^2$, $(\sum X)^2$, and $\sum (X - \overline{X})^2$.

A–21. 9, 17, 32, 16, 8, 2, 9, 7, 3, 18

A–22. 4, 12, 9, 13, 0, 6, 2, 10

A–23. 5, 12, 8, 3, 4

A–24. 6, 2, 18, 30, 31, 42, 16, 5

A–25. 80, 76, 42, 53, 77

A–26. 123, 132, 216, 98, 146, 114

A–27. 53, 72, 81, 42, 63, 71, 73, 85, 98, 55

A–28. 43, 32, 116, 98, 120

A–29. 12, 52, 36, 81, 63, 74

A–30. −9, −12, 18, 0, −2, −15

A–3 The Line

The following figure shows the *rectangular coordinate system* or *Cartesian plane.* This figure consists of two axes: the horizontal axis, called the x axis, and the vertical axis, called the y axis. Each axis has numerical scales. The point of intersection of the axes is called the *origin.*

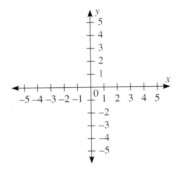

Points can be graphed by using coordinates. For example, the notation for point P(3, 2) means that the x coordinate is 3 and the y coordinate is 2. Hence, P is located at the intersection of x = 3 and y = 2, as shown below.

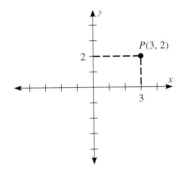

Other points, such as Q(−5, 2), R(4, 1), S(−3, −4) can be plotted as shown in the next figure.

When a point lies on the y axis, the x coordinate is 0, as in (0, 6) (0, − 3), etc. When a point lies on the x axis, the y coordinate is 0, as in (6, 0) (−8, 0), etc., as shown at top of next page.

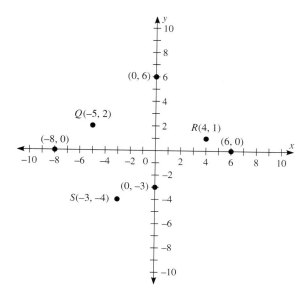

The slopes of lines can be positive, negative, or zero. A line going uphill from left to right has a positive slope. A line going downhill from left to right has a negative slope. And a line that is horizontal has a slope of zero.

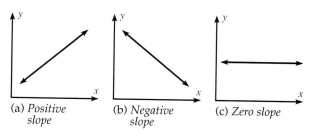

(a) *Positive slope*   (b) *Negative slope*   (c) *Zero slope*

A point *a* where the line crosses the *x* axis is called the *x intercept* and has the coordinates (*a*, 0). A point *b* where the line crosses the *y* axis is called the *y intercept* and has the coordinates (0, *b*).

Two points determine a line. There are two properties of a line: its slope and its equation. The *slope m* of a line is determined by the ratio of the rise (called Δy) and the run (Δx).

$$m = \frac{\text{rise}}{\text{run}} = \frac{\Delta y}{\Delta x}$$

For example, the slope of the line shown below is $\frac{3}{2}$, or 1.5, since the height, Δy, is 3 units and the run Δx is 2 units.

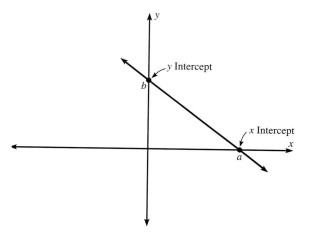

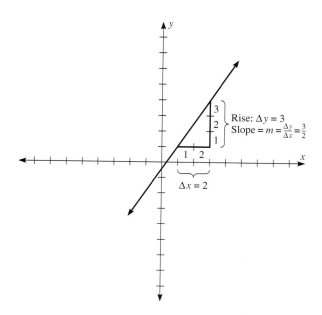

Rise: $\Delta y = 3$
Slope $= m = \frac{\Delta y}{\Delta x} = \frac{3}{2}$

$\Delta x = 2$

Every line has a unique equation of the form $y = a + bx$. For example, the equations

$$y = 5 + 3x$$
$$y = 8.6 + 3.2x$$
$$y = 5.2 - 6.1x$$

all represent different, unique lines. The number represented by *a* is the *y* intercept point; the number represented by *b* is the slope. The line whose equation is $y = 3 + 2x$ has a *y* intercept at 3 and a slope of 2, or 2/1. This line can be shown as in the graph on the next page.

A

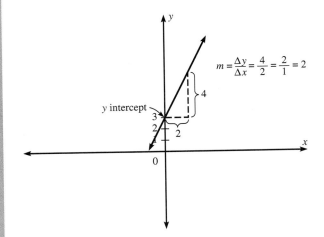

Then,

$$y = 3 + 2x = 3 + 2(0) = 3$$

Hence, when $x = 0$, then $y = 3$, and the line passes through the point $(0, 3)$.

Now select any other value of x, say $x = 2$.

$$y = 3 + 2x = 3 + 2(2) = 7$$

Hence, a second point is $(2, 7)$. Then plot the points and graph the line.

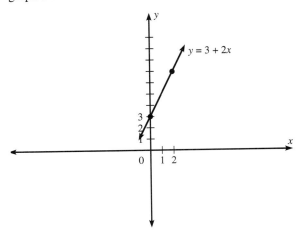

If two points are known, then the graph of the line can be plotted. For example, to find the graph of a line passing through the points $P(2, 1)$ and $Q(3, 5)$, plot the points and connect them as shown below.

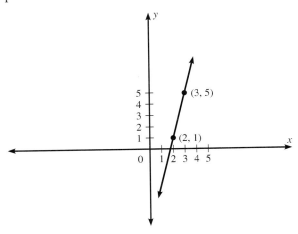

Given the equation of a line, one can graph the line by finding two points and then plotting them.

Example A–8

Plot the graph of the line whose equation is $y = 3 + 2x$.

Solution

Select any number as an x value and substitute it in the equation to get the corresponding y value. Let $x = 0$.

Exercises

Plot the line passing through each set of points.

A–31. $P(3, 2), Q(1, 6)$ **A–34.** $P(-1, -2), Q(-7, 8)$
A–32. $P(0, 5), Q(8, 0)$ **A–35.** $P(6, 3), Q(10, 3)$
A–33. $P(-2, 4), Q(3, 6)$

Find at least two points on each line, and then graph the line containing these points.

A–36. $y = 5 + 2x$ **A–39.** $y = -2 - 2x$
A–37. $y = -1 + x$ **A–40.** $y = 4 - 3x$
A–38. $y = 3 + 4x$

Appendix B

Writing the Research Report

After conducting a statistical study, a researcher must write a final report explaining how the study was conducted and giving the results. The formats of research reports, theses, and dissertations vary from school to school; however, they tend to follow the general format explained here.

Front materials

The front materials typically include the following items:

Title page
Copyright page
Acknowledgments
Table of contents
Table of appendixes
List of tables
List of figures

Chapter 1: Nature and background of the study

This chapter should introduce the reader to the nature of the study and present some discussion on the background. It should contain the following information:

Introduction
Statement of the problem
Background of the problem
Rationale for the study
Research questions and/or hypotheses
Assumptions, limitations, and deliminations
Definitions of terms

Chapter 2: Review of literature

This chapter should explain what has been done in previous research related to the study. It should contain the following information:

Prior research
Related literature

Chapter 3: Methodology

This chapter should explain how the study was conducted. It should contain the following information:

Development of questionnaires, tests, survey instruments, etc.
Definition of the population
Sampling methods used
How the data were collected
Research design used
Statistical tests that will be used to analyze the data

Chapter 4: Analysis of data

This chapter should explain the results of the statistical analysis of the data. It should state whether or not the null hypothesis should be rejected. Any statistical tables used to analyze the data should be included here.

Chapter 5: Summary, conclusions, and recommendations

This chapter summarizes the results of the study and explains any conclusions that have resulted from the statistical analysis of the data. The researchers should cite and explain any shortcomings of the study. Recommendations obtained from the study should be included here, and further studies should be suggested.

Appendix C

Tables

Table A Factorials	
n	*n*!
0	1
1	1
2	2
3	6
4	24
5	120
6	720
7	5,040
8	40,320
9	362,880
10	3,628,800
11	39,916,800
12	479,001,600
13	6,227,020,800
14	87,178,291,200
15	1,307,674,368,000
16	20,922,789,888,000
17	355,687,428,096,000
18	6,402,373,705,728,000
19	121,645,100,408,832,000
20	2,432,902,008,176,640,000

Table B — The Binomial Distribution

n	x	\(p\) 0.05	0.1	0.2	0.3	0.4	0.5	0.6	0.7	0.8	0.9	0.95
2	0	0.902	0.810	0.640	0.490	0.360	0.250	0.160	0.090	0.040	0.010	0.002
	1	0.095	0.180	0.320	0.420	0.480	0.500	0.480	0.420	0.320	0.180	0.095
	2	0.002	0.010	0.040	0.090	0.160	0.250	0.360	0.490	0.640	0.810	0.902
3	0	0.857	0.729	0.512	0.343	0.216	0.125	0.064	0.027	0.008	0.001	
	1	0.135	0.243	0.384	0.441	0.432	0.375	0.288	0.189	0.096	0.027	0.007
	2	0.007	0.027	0.096	0.189	0.288	0.375	0.432	0.441	0.384	0.243	0.135
	3		0.001	0.008	0.027	0.064	0.125	0.216	0.343	0.512	0.729	0.857
4	0	0.815	0.656	0.410	0.240	0.130	0.062	0.026	0.008	0.002		
	1	0.171	0.292	0.410	0.412	0.346	0.250	0.154	0.076	0.026	0.004	
	2	0.014	0.049	0.154	0.265	0.346	0.375	0.346	0.265	0.154	0.049	0.014
	3		0.004	0.026	0.076	0.154	0.250	0.346	0.412	0.410	0.292	0.171
	4			0.002	0.008	0.026	0.062	0.130	0.240	0.410	0.656	0.815
5	0	0.774	0.590	0.328	0.168	0.078	0.031	0.010	0.002			
	1	0.204	0.328	0.410	0.360	0.259	0.156	0.077	0.028	0.006		
	2	0.021	0.073	0.205	0.309	0.346	0.312	0.230	0.132	0.051	0.008	0.001
	3	0.001	0.008	0.051	0.132	0.230	0.312	0.346	0.309	0.205	0.073	0.021
	4			0.006	0.028	0.077	0.156	0.259	0.360	0.410	0.328	0.204
	5				0.002	0.010	0.031	0.078	0.168	0.328	0.590	0.774
6	0	0.735	0.531	0.262	0.118	0.047	0.016	0.004	0.001			
	1	0.232	0.354	0.393	0.303	0.187	0.094	0.037	0.010	0.002		
	2	0.031	0.098	0.246	0.324	0.311	0.234	0.138	0.060	0.015	0.001	
	3	0.002	0.015	0.082	0.185	0.276	0.312	0.276	0.185	0.082	0.015	0.002
	4		0.001	0.015	0.060	0.138	0.234	0.311	0.324	0.246	0.098	0.031
	5			0.002	0.010	0.037	0.094	0.187	0.303	0.393	0.354	0.232
	6				0.001	0.004	0.016	0.047	0.118	0.262	0.531	0.735
7	0	0.698	0.478	0.210	0.082	0.028	0.008	0.002				
	1	0.257	0.372	0.367	0.247	0.131	0.055	0.017	0.004			
	2	0.041	0.124	0.275	0.318	0.261	0.164	0.077	0.025	0.004		
	3	0.004	0.023	0.115	0.227	0.290	0.273	0.194	0.097	0.029	0.003	
	4		0.003	0.029	0.097	0.194	0.273	0.290	0.227	0.115	0.023	0.004
	5			0.004	0.025	0.077	0.164	0.261	0.318	0.275	0.124	0.041
	6				0.004	0.017	0.055	0.131	0.247	0.367	0.372	0.257
	7					0.002	0.008	0.028	0.082	0.210	0.478	0.698
8	0	0.663	0.430	0.168	0.058	0.017	0.004	0.001				
	1	0.279	0.383	0.336	0.198	0.090	0.031	0.008	0.001			
	2	0.051	0.149	0.294	0.296	0.209	0.109	0.041	0.010	0.001		
	3	0.005	0.033	0.147	0.254	0.279	0.219	0.124	0.047	0.009		
	4		0.005	0.046	0.136	0.232	0.273	0.232	0.136	0.046	0.005	
	5			0.009	0.047	0.124	0.219	0.279	0.254	0.147	0.033	0.005
	6			0.001	0.010	0.041	0.109	0.209	0.296	0.294	0.149	0.051
	7				0.001	0.008	0.031	0.090	0.198	0.336	0.383	0.279
	8					0.001	0.004	0.017	0.058	0.168	0.430	0.663

C

Table B   (continued)

n	x	p 0.05	0.1	0.2	0.3	0.4	0.5	0.6	0.7	0.8	0.9	0.95
9	0	0.630	0.387	0.134	0.040	0.010	0.002					
	1	0.299	0.387	0.302	0.156	0.060	0.018	0.004				
	2	0.063	0.172	0.302	0.267	0.161	0.070	0.021	0.004			
	3	0.008	0.045	0.176	0.267	0.251	0.164	0.074	0.021	0.003		
	4	0.001	0.007	0.066	0.172	0.251	0.246	0.167	0.074	0.017	0.001	
	5		0.001	0.017	0.074	0.167	0.246	0.251	0.172	0.066	0.007	0.001
	6			0.003	0.021	0.074	0.164	0.251	0.267	0.176	0.045	0.008
	7				0.004	0.021	0.070	0.161	0.267	0.302	0.172	0.063
	8					0.004	0.018	0.060	0.156	0.302	0.387	0.299
	9						0.002	0.010	0.040	0.134	0.387	0.630
10	0	0.599	0.349	0.107	0.028	0.006	0.001					
	1	0.315	0.387	0.268	0.121	0.040	0.010	0.002				
	2	0.075	0.194	0.302	0.233	0.121	0.044	0.011	0.001			
	3	0.010	0.057	0.201	0.267	0.215	0.117	0.042	0.009	0.001		
	4	0.001	0.011	0.088	0.200	0.251	0.205	0.111	0.037	0.006		
	5		0.001	0.026	0.103	0.201	0.246	0.201	0.103	0.026	0.001	
	6			0.006	0.037	0.111	0.205	0.251	0.200	0.088	0.011	0.001
	7			0.001	0.009	0.042	0.117	0.215	0.267	0.201	0.057	0.010
	8				0.001	0.011	0.044	0.121	0.233	0.302	0.194	0.075
	9					0.002	0.010	0.040	0.121	0.268	0.387	0.315
	10						0.001	0.006	0.028	0.107	0.349	0.599
11	0	0.569	0.314	0.086	0.020	0.004						
	1	0.329	0.384	0.236	0.093	0.027	0.005	0.001				
	2	0.087	0.213	0.295	0.200	0.089	0.027	0.005	0.001			
	3	0.014	0.071	0.221	0.257	0.177	0.081	0.023	0.004			
	4	0.001	0.016	0.111	0.220	0.236	0.161	0.070	0.017	0.002		
	5		0.002	0.039	0.132	0.221	0.226	0.147	0.057	0.010		
	6			0.010	0.057	0.147	0.226	0.221	0.132	0.039	0.002	
	7			0.002	0.017	0.070	0.161	0.236	0.220	0.111	0.016	0.001
	8				0.004	0.023	0.081	0.177	0.257	0.221	0.071	0.014
	9				0.001	0.005	0.027	0.089	0.200	0.295	0.213	0.087
	10					0.001	0.005	0.027	0.093	0.236	0.384	0.329
	11						0.004	0.020	0.086	0.314	0.569	
12	0	0.540	0.282	0.069	0.014	0.002						
	1	0.341	0.377	0.206	0.071	0.017	0.003					
	2	0.099	0.230	0.283	0.168	0.064	0.016	0.002				
	3	0.017	0.085	0.236	0.240	0.142	0.054	0.012	0.001			
	4	0.002	0.021	0.133	0.231	0.213	0.121	0.042	0.008	0.001		
	5		0.004	0.053	0.158	0.227	0.193	0.101	0.029	0.003		
	6			0.016	0.079	0.177	0.226	0.177	0.079	0.016		
	7			0.003	0.029	0.101	0.193	0.227	0.158	0.053	0.004	
	8			0.001	0.008	0.042	0.121	0.213	0.231	0.133	0.021	0.002
	9				0.001	0.012	0.054	0.142	0.240	0.236	0.085	0.017
	10					0.002	0.016	0.064	0.168	0.283	0.230	0.099
	11						0.003	0.017	0.071	0.206	0.377	0.341
	12							0.002	0.014	0.069	0.282	0.540

Table B *(continued)*

n	x	0.05	0.1	0.2	0.3	0.4	0.5	0.6	0.7	0.8	0.9	0.95
13	0	0.513	0.254	0.055	0.010	0.001						
	1	0.351	0.367	0.179	0.054	0.011	0.002					
	2	0.111	0.245	0.268	0.139	0.045	0.010	0.001				
	3	0.021	0.100	0.246	0.218	0.111	0.035	0.006	0.001			
	4	0.003	0.028	0.154	0.234	0.184	0.087	0.024	0.003			
	5		0.006	0.069	0.180	0.221	0.157	0.066	0.014	0.001		
	6		0.001	0.023	0.103	0.197	0.209	0.131	0.044	0.006		
	7			0.006	0.044	0.131	0.209	0.197	0.103	0.023	0.001	
	8			0.001	0.014	0.066	0.157	0.221	0.180	0.069	0.006	
	9				0.003	0.024	0.087	0.184	0.234	0.154	0.028	0.003
	10				0.001	0.006	0.035	0.111	0.218	0.246	0.100	0.021
	11					0.001	0.010	0.045	0.139	0.268	0.245	0.111
	12						0.002	0.011	0.054	0.179	0.367	0.351
	13							0.001	0.010	0.055	0.254	0.513
14	0	0.488	0.229	0.044	0.007	0.001						
	1	0.359	0.356	0.154	0.041	0.007	0.001					
	2	0.123	0.257	0.250	0.113	0.032	0.006	0.001				
	3	0.026	0.114	0.250	0.194	0.085	0.022	0.003				
	4	0.004	0.035	0.172	0.229	0.155	0.061	0.014	0.001			
	5		0.008	0.086	0.196	0.207	0.122	0.041	0.007			
	6		0.001	0.032	0.126	0.207	0.183	0.092	0.023	0.002		
	7			0.009	0.062	0.157	0.209	0.157	0.062	0.009		
	8			0.002	0.023	0.092	0.183	0.207	0.126	0.032	0.001	
	9				0.007	0.041	0.122	0.207	0.196	0.086	0.008	
	10				0.001	0.014	0.061	0.155	0.229	0.172	0.035	0.004
	11					0.003	0.022	0.085	0.194	0.250	0.114	0.026
	12					0.001	0.006	0.032	0.113	0.250	0.257	0.123
	13						0.001	0.007	0.041	0.154	0.356	0.359
	14							0.001	0.007	0.044	0.229	0.488
15	0	0.463	0.206	0.035	0.005							
	1	0.366	0.343	0.132	0.031	0.005						
	2	0.135	0.267	0.231	0.092	0.022	0.003					
	3	0.031	0.129	0.250	0.170	0.063	0.014	0.002				
	4	0.005	0.043	0.188	0.219	0.127	0.042	0.007	0.001			
	5	0.001	0.010	0.103	0.206	0.186	0.092	0.024	0.003			
	6		0.002	0.043	0.147	0.207	0.153	0.061	0.012	0.001		
	7			0.014	0.081	0.177	0.196	0.118	0.035	0.003		
	8			0.003	0.035	0.118	0.196	0.177	0.081	0.014		
	9			0.001	0.012	0.061	0.153	0.207	0.147	0.043	0.002	
	10				0.003	0.024	0.092	0.186	0.206	0.103	0.010	0.001
	11				0.001	0.007	0.042	0.127	0.219	0.188	0.043	0.005
	12					0.002	0.014	0.063	0.170	0.250	0.129	0.031
	13						0.003	0.022	0.092	0.231	0.267	0.135
	14							0.005	0.031	0.132	0.343	0.366
	15								0.005	0.035	0.206	0.463

Table B (continued)

n	x	0.05	0.1	0.2	0.3	0.4	0.5	0.6	0.7	0.8	0.9	0.95
16	0	0.440	0.185	0.028	0.003							
	1	0.371	0.329	0.113	0.023	0.003						
	2	0.146	0.275	0.211	0.073	0.015	0.002					
	3	0.036	0.142	0.246	0.146	0.047	0.009	0.001				
	4	0.006	0.051	0.200	0.204	0.101	0.028	0.004				
	5	0.001	0.014	0.120	0.210	0.162	0.067	0.014	0.001			
	6		0.003	0.055	0.165	0.198	0.122	0.039	0.006			
	7			0.020	0.101	0.189	0.175	0.084	0.019	0.001		
	8			0.006	0.049	0.142	0.196	0.142	0.049	0.006		
	9			0.001	0.019	0.084	0.175	0.189	0.101	0.020		
	10				0.006	0.039	0.122	0.198	0.165	0.055	0.003	
	11				0.001	0.014	0.067	0.162	0.210	0.120	0.014	0.001
	12					0.004	0.028	0.101	0.204	0.200	0.051	0.006
	13					0.001	0.009	0.047	0.146	0.246	0.142	0.036
	14						0.002	0.015	0.073	0.211	0.275	0.146
	15							0.003	0.023	0.113	0.329	0.371
	16								0.003	0.028	0.185	0.440
17	0	0.418	0.167	0.023	0.002							
	1	0.374	0.315	0.096	0.017	0.002						
	2	0.158	0.280	0.191	0.058	0.010	0.001					
	3	0.041	0.156	0.239	0.125	0.034	0.005					
	4	0.008	0.060	0.209	0.187	0.080	0.018	0.002				
	5	0.001	0.017	0.136	0.208	0.138	0.047	0.008	0.001			
	6		0.004	0.068	0.178	0.184	0.094	0.024	0.003			
	7		0.001	0.027	0.120	0.193	0.148	0.057	0.009			
	8			0.008	0.064	0.161	0.185	0.107	0.028	0.002		
	9			0.002	0.028	0.107	0.185	0.161	0.064	0.008		
	10				0.009	0.057	0.148	0.193	0.120	0.027	0.001	
	11				0.003	0.024	0.094	0.184	0.178	0.068	0.004	
	12				0.001	0.008	0.047	0.138	0.208	0.136	0.017	0.001
	13					0.002	0.018	0.080	0.187	0.209	0.060	0.008
	14						0.005	0.034	0.125	0.239	0.156	0.041
	15						0.001	0.010	0.058	0.191	0.280	0.158
	16							0.002	0.017	0.096	0.315	0.374
	17								0.002	0.023	0.167	0.418

Table B   (continued)

n	x	0.05	0.1	0.2	0.3	0.4	0.5	0.6	0.7	0.8	0.9	0.95
18	0	0.397	0.150	0.018	0.002							
	1	0.376	0.300	0.081	0.013	0.001						
	2	0.168	0.284	0.172	0.046	0.007	0.001					
	3	0.047	0.168	0.230	0.105	0.025	0.003					
	4	0.009	0.070	0.215	0.168	0.061	0.012	0.001				
	5	0.001	0.022	0.151	0.202	0.115	0.033	0.004				
	6		0.005	0.082	0.187	0.166	0.071	0.015	0.001			
	7		0.001	0.035	0.138	0.189	0.121	0.037	0.005			
	8			0.012	0.081	0.173	0.167	0.077	0.015	0.001		
	9			0.003	0.039	0.128	0.185	0.128	0.039	0.003		
	10			0.001	0.015	0.077	0.167	0.173	0.081	0.012		
	11				0.005	0.037	0.121	0.189	0.138	0.035	0.001	
	12				0.001	0.015	0.071	0.166	0.187	0.082	0.005	
	13					0.004	0.033	0.115	0.202	0.151	0.022	0.001
	14					0.001	0.012	0.061	0.168	0.215	0.070	0.009
	15						0.003	0.025	0.105	0.230	0.168	0.047
	16						0.001	0.007	0.046	0.172	0.284	0.168
	17							0.001	0.013	0.081	0.300	0.376
	18								0.002	0.018	0.150	0.397
19	0	0.377	0.135	0.014	0.001							
	1	0.377	0.285	0.068	0.009	0.001						
	2	0.179	0.285	0.154	0.036	0.005						
	3	0.053	0.180	0.218	0.087	0.017	0.002					
	4	0.011	0.080	0.218	0.149	0.047	0.007	0.001				
	5	0.002	0.027	0.164	0.192	0.093	0.022	0.002				
	6		0.007	0.095	0.192	0.145	0.052	0.008	0.001			
	7		0.001	0.044	0.153	0.180	0.096	0.024	0.002			
	8			0.017	0.098	0.180	0.144	0.053	0.008			
	9			0.005	0.051	0.146	0.176	0.098	0.022	0.001		
	10			0.001	0.022	0.098	0.176	0.146	0.051	0.005		
	11				0.008	0.053	0.144	0.180	0.098	0.071		
	12				0.002	0.024	0.096	0.180	0.153	0.044	0.001	
	13				0.001	0.008	0.052	0.145	0.192	0.095	0.007	
	14					0.002	0.022	0.093	0.192	0.164	0.027	0.002
	15					0.001	0.007	0.047	0.149	0.218	0.080	0.011
	16						0.002	0.017	0.087	0.218	0.180	0.053
	17							0.005	0.036	0.154	0.285	0.179
	18							0.001	0.009	0.068	0.285	0.377
	19								0.001	0.014	0.135	0.377

Table B   *(concluded)*

n	x	0.05	0.1	0.2	0.3	0.4	0.5	0.6	0.7	0.8	0.9	0.95
							p					
20	0	0.358	0.122	0.012	0.001							
	1	0.377	0.270	0.058	0.007							
	2	0.189	0.285	0.137	0.028	0.003						
	3	0.060	0.190	0.205	0.072	0.012	0.001					
	4	0.013	0.090	0.218	0.130	0.035	0.005					
	5	0.002	0.032	0.175	0.179	0.075	0.015	0.001				
	6		0.009	0.109	0.192	0.124	0.037	0.005				
	7		0.002	0.055	0.164	0.166	0.074	0.015	0.001			
	8			0.022	0.114	0.180	0.120	0.035	0.004			
	9			0.007	0.065	0.160	0.160	0.071	0.012			
	10			0.002	0.031	0.117	0.176	0.117	0.031	0.002		
	11				0.012	0.071	0.160	0.160	0.065	0.007		
	12				0.004	0.035	0.120	0.180	0.114	0.022		
	13				0.001	0.015	0.074	0.166	0.164	0.055	0.002	
	14					0.005	0.037	0.124	0.192	0.109	0.009	
	15					0.001	0.015	0.075	0.179	0.175	0.032	0.002
	16						0.005	0.035	0.130	0.218	0.090	0.013
	17						0.001	0.012	0.072	0.205	0.190	0.060
	18							0.003	0.028	0.137	0.285	0.189
	19								0.007	0.058	0.270	0.377
	20								0.001	0.012	0.122	0.358

Note: All values of 0.0005 or less are omitted.

Source: John E. Freund, *Modern Elementary Statistics,* 8th ed., © 1992. Reprinted by permission of Prentice Hall, Inc., Upper Saddle River, New Jersey.

Table C   Permutations and Combinations

This table contains the number of permutations of n objects taking r at a time or $_nP_r$

n \ r	0	1	2	3	4	5	6	7	8	9	10
0	1										
1	1	1									
2	1	2	2								
3	1	3	6	6							
4	1	4	12	24	24						
5	1	5	20	60	120	120					
6	1	6	30	120	360	720	720				
7	1	7	42	210	840	2520	5040	5040			
8	1	8	56	336	1680	6720	20160	40320	40320		
9	1	9	72	504	3024	15120	60480	181440	362880	362880	
10	1	10	90	720	5040	30240	151200	604800	1814400	3628800	3628800

This table contains the number of combinations of n objects taking r at a time or $_nC_r$

n \ r	0	1	2	3	4	5	6	7	8	9	10
1	1	1									
2	1	2	1								
3	1	3	3	1							
4	1	4	6	4	1						
5	1	5	10	10	5	1					
6	1	6	15	20	15	6	1				
7	1	7	21	35	35	21	7	1			
8	1	8	28	56	70	56	28	8	1		
9	1	9	36	84	126	126	84	36	9	1	
10	1	10	45	120	210	252	210	120	45	10	1

To find the value of $_nP_r$ or $_nC_r$, locate n in the left column and r in the top row. The answer is found where the row and column intersect in the table.

Examples: $_4P_3 = 24$ $_5C_2 = 10$

C

Table D   Random Numbers

10480	15011	01536	02011	81647	91646	67179	14194	62590	36207	20969	99570	91291	90700
22368	46573	25595	85393	30995	89198	27982	53402	93965	34095	52666	19174	39615	99505
24130	48360	22527	97265	76393	64809	15179	24830	49340	32081	30680	19655	63348	58629
42167	93093	06243	61680	07856	16376	39440	53537	71341	57004	00849	74917	97758	16379
37570	39975	81837	16656	06121	91782	60468	81305	49684	60672	14110	06927	01263	54613
77921	06907	11008	42751	27756	53498	18602	70659	90655	15053	21916	81825	44394	42880
99562	72905	56420	69994	98872	31016	71194	18738	44013	48840	63213	21069	10634	12952
96301	91977	05463	07972	18876	20922	94595	56869	69014	60045	18425	84903	42508	32307
89579	14342	63661	10281	17453	18103	57740	84378	25331	12566	58678	44947	05585	56941
85475	36857	43342	53988	53060	59533	38867	62300	08158	17983	16439	11458	18593	64952
28918	69578	88231	33276	70997	79936	56865	05859	90106	31595	01547	85590	91610	78188
63553	40961	48235	03427	49626	69445	18663	72695	52180	20847	12234	90511	33703	90322
09429	93969	52636	92737	88974	33488	36320	17617	30015	08272	84115	27156	30613	74952
10365	61129	87529	85689	48237	52267	67689	93394	01511	26358	85104	20285	29975	89868
07119	97336	71048	08178	77233	13916	47564	81056	97735	85977	29372	74461	28551	90707
51085	12765	51821	51259	77452	16308	60756	92144	49442	53900	70960	63990	75601	40719
02368	21382	52404	60268	89368	19885	55322	44819	01188	65255	64835	44919	05944	55157
01011	54092	33362	94904	31273	04146	18594	29852	71585	85030	51132	01915	92747	64951
52162	53916	46369	58586	23216	14513	83149	98736	23495	64350	94738	17752	35156	35749
07056	97628	33787	09998	42698	06691	76988	13602	51851	46104	88916	19509	25625	58104
48663	91245	85828	14346	09172	30168	90229	04734	59193	22178	30421	61666	99904	32812
54164	58492	22421	74103	47070	25306	76468	26384	58151	06646	21524	15227	96909	44592
32639	32363	05597	24200	13363	38005	94342	28728	35806	06912	17012	64161	18296	22851
29334	27001	87637	87308	58731	00256	45834	15398	46557	41135	10367	07684	36188	18510
02488	33062	28834	07351	19731	92420	60952	61280	50001	67658	32586	86679	50720	94953
81525	72295	04839	96423	24878	82651	66566	14778	76797	14780	13300	87074	79666	95725
29676	20591	68086	26432	46901	20849	89768	81536	86645	12659	92259	57102	80428	25280
00742	57392	39064	66432	84673	40027	32832	61362	98947	96067	64760	64584	96096	98253
05366	04213	25669	26422	44407	44048	37937	63904	45766	66134	75470	66520	34693	90449
91921	26418	64117	94305	26766	25940	39972	22209	71500	64568	91402	42416	07844	69618
00582	04711	87917	77341	42206	35126	74087	99547	81817	42607	43808	76655	62028	76630
00725	69884	62797	56170	86324	88072	76222	36086	84637	93161	76038	65855	77919	88006
69011	65797	95876	55293	18988	27354	26575	08625	40801	59920	29841	80150	12777	48501
25976	57948	29888	88604	67917	48708	18912	82271	65424	69774	33611	54262	85963	03547
09763	83473	73577	12908	30883	18317	28290	35797	05998	41688	34952	37888	38917	88050
91567	42595	27958	30134	04024	86385	29880	99730	55536	84855	29080	09250	79656	73211
17955	56349	90999	49127	20044	59931	06115	20542	18059	02008	73708	83517	36103	42791
46503	18584	18845	49618	02304	51038	20655	58727	28168	15475	56942	53389	20562	87338
92157	89634	94824	78171	84610	82834	09922	25417	44137	48413	25555	21246	35509	20468
14577	62765	35605	81263	39667	47358	56873	56307	61607	49518	89656	20103	77490	18062
98427	07523	33362	64270	01638	92477	66969	98420	04880	45585	46565	04102	46880	45709
34914	63976	88720	82765	34476	17032	87589	40836	32427	70002	70663	88863	77775	69348
70060	28277	39475	46473	23219	53416	94970	25832	69975	94884	19661	72828	00102	66794
53976	54914	06990	67245	68350	82948	11398	42878	80287	88267	47363	46634	06541	97809
76072	29515	40980	07391	58745	25774	22987	80059	39911	96189	41151	14222	60697	59583
90725	52210	83974	29992	65831	38857	50490	83765	55657	14361	31720	57375	56228	41546
64364	67412	33339	31926	14883	24413	59744	92351	97473	89286	35931	04110	23726	51900
08962	00358	31662	25388	61642	34072	81249	35648	56891	69352	48373	45578	78547	81788
95012	68379	93526	70765	10593	04542	76463	54328	02349	17247	28865	14777	62730	92277
15664	10493	20492	38391	91132	21999	59516	81652	27195	48223	46751	22923	32261	85653

Table E   The Standard Normal Distribution

z	.00	.01	.02	.03	.04	.05	.06	.07	.08	.09
0.0	.0000	.0040	.0080	.0120	.0160	.0199	.0239	.0279	.0319	.0359
0.1	.0398	.0438	.0478	.0517	.0557	.0596	.0636	.0675	.0714	.0753
0.2	.0793	.0832	.0871	.0910	.0948	.0987	.1026	.1064	.1103	.1141
0.3	.1179	.1217	.1255	.1293	.1331	.1368	.1406	.1443	.1480	.1517
0.4	.1554	.1591	.1628	.1664	.1700	.1736	.1772	.1808	.1844	.1879
0.5	.1915	.1950	.1985	.2019	.2054	.2088	.2123	.2157	.2190	.2224
0.6	.2257	.2291	.2324	.2357	.2389	.2422	.2454	.2486	.2517	.2549
0.7	.2580	.2611	.2642	.2673	.2704	.2734	.2764	.2794	.2823	.2852
0.8	.2881	.2910	.2939	.2967	.2995	.3023	.3051	.3078	.3106	.3133
0.9	.3159	.3186	.3212	.3238	.3264	.3289	.3315	.3340	.3365	.3389
1.0	.3413	.3438	.3461	.3485	.3508	.3531	.3554	.3577	.3599	.3621
1.1	.3643	.3665	.3686	.3708	.3729	.3749	.3770	.3790	.3810	.3830
1.2	.3849	.3869	.3888	.3907	.3925	.3944	.3962	.3980	.3997	.4015
1.3	.4032	.4049	.4066	.4082	.4099	.4115	.4131	.4147	.4162	.4177
1.4	.4192	.4207	.4222	.4236	.4251	.4265	.4279	.4292	.4306	.4319
1.5	.4332	.4345	.4357	.4370	.4382	.4394	.4406	.4418	.4429	.4441
1.6	.4452	.4463	.4474	.4484	.4495	.4505	.4515	.4525	.4535	.4545
1.7	.4554	.4564	.4573	.4582	.4591	.4599	.4608	.4616	.4625	.4633
1.8	.4641	.4649	.4656	.4664	.4671	.4678	.4686	.4693	.4699	.4706
1.9	.4713	.4719	.4726	.4732	.4738	.4744	.4750	.4756	.4761	.4767
2.0	.4772	.4778	.4783	.4788	.4793	.4798	.4803	.4808	.4812	.4817
2.1	.4821	.4826	.4830	.4834	.4838	.4842	.4846	.4850	.4854	.4857
2.2	.4861	.4864	.4868	.4871	.4875	.4878	.4881	.4884	.4887	.4890
2.3	.4893	.4896	.4898	.4901	.4904	.4906	.4909	.4911	.4913	.4916
2.4	.4918	.4920	.4922	.4925	.4927	.4929	.4931	.4932	.4934	.4936
2.5	.4938	.4940	.4941	.4943	.4945	.4946	.4948	.4949	.4951	.4952
2.6	.4953	.4955	.4956	.4957	.4959	.4960	.4961	.4962	.4963	.4964
2.7	.4965	.4966	.4967	.4968	.4969	.4970	.4971	.4972	.4973	.4974
2.8	.4974	.4975	.4976	.4977	.4977	.4978	.4979	.4979	.4980	.4981
2.9	.4981	.4982	.4982	.4983	.4984	.4984	.4985	.4985	.4986	.4986
3.0	.4987	.4987	.4987	.4988	.4988	.4989	.4989	.4989	.4990	.4990

Note: Use 0.4999 for z values above 3.09.

Source: Frederick Mosteller and Robert E. K. Rourke, *Sturdy Statistics,* Table A–1 (Reading, Mass.: Addison-Wesley, 1973). Reprinted with permission of the copyright owners.

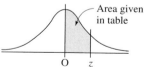
Area given in table

Table F   The *t* Distribution

Confidence intervals	50%	80%	90%	95%	98%	99%
One tail, α	0.25	0.10	0.05	0.025	0.01	0.005
d.f. Two tails, α	0.50	0.20	0.10	0.05	0.02	0.01
1	1.000	3.078	6.314	12.706	31.821	63.657
2	.816	1.886	2.920	4.303	6.965	9.925
3	.765	1.638	2.353	3.182	4.541	5.841
4	.741	1.533	2.132	2.776	3.747	4.604
5	.727	1.476	2.015	2.571	3.365	4.032
6	.718	1.440	1.943	2.447	3.143	3.707
7	.711	1.415	1.895	2.365	2.998	3.499
8	.706	1.397	1.860	2.306	2.896	3.355
9	.703	1.383	1.833	2.262	2.821	3.250
10	.700	1.372	1.812	2.228	2.764	3.169
11	.697	1.363	1.796	2.201	2.718	3.106
12	.695	1.356	1.782	2.179	2.681	3.055
13	.694	1.350	1.771	2.160	2.650	3.012
14	.692	1.345	1.761	2.145	2.624	2.977
15	.691	1.341	1.753	2.131	2.602	2.947
16	.690	1.337	1.746	2.120	2.583	2.921
17	.689	1.333	1.740	2.110	2.567	2.898
18	.688	1.330	1.734	2.101	2.552	2.878
19	.688	1.328	1.729	2.093	2.539	2.861
20	.687	1.325	1.725	2.086	2.528	2.845
21	.686	1.323	1.721	2.080	2.518	2.831
22	.686	1.321	1.717	2.074	2.508	2.819
23	.685	1.319	1.714	2.069	2.500	2.807
24	.685	1.318	1.711	2.064	2.492	2.797
25	.684	1.316	1.708	2.060	2.485	2.787
26	.684	1.315	1.706	2.056	2.479	2.779
27	.684	1.314	1.703	2.052	2.473	2.771
28	.683	1.313	1.701	2.048	2.467	2.763
(z) ∞	.674	1.282[a]	1.645[b]	1.960	2.326[c]	2.576[d]

[a]This value has been rounded to 1.28 in the textbook.
[b]This value has been rounded to 1.65 in the textbook.
[c]This value has been rounded to 2.33 in the textbook.
[d]This value has been rounded to 2.58 in the textbook.

Source: Adapted from W. H. Beyer, *Handbook of Tables for Probability and Statistics*, 2nd ed., CRC Press, Boca Raton, Florida, 1986. Reprinted with permission.

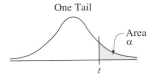

One Tail

Area α

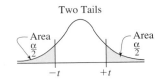

Two Tails

Area $\frac{\alpha}{2}$ Area $\frac{\alpha}{2}$

$-t$ $+t$

Table G The Chi-Square Distribution

Degrees of freedom	α									
	0.995	0.99	0.975	0.95	0.90	0.10	0.05	0.025	0.01	0.005
1	—	—	0.001	0.004	0.016	2.706	3.841	5.024	6.635	7.879
2	0.010	0.020	0.051	0.103	0.211	4.605	5.991	7.378	9.210	10.597
3	0.072	0.115	.0216	0.352	0.584	6.251	7.815	9.348	11.345	12.838
4	0.207	0.297	0.484	0.711	1.064	7.779	9.488	11.143	13.277	14.860
5	0.412	0.554	0.831	1.145	1.610	9.236	11.071	12.833	15.086	16.750
6	0.676	0.872	1.237	1.635	2.204	10.645	12.592	14.449	16.812	18.548
7	0.989	1.239	1.690	2.167	2.833	12.017	14.067	16.013	18.475	20.278
8	1.344	1.646	2.180	2.733	3.490	13.362	15.507	17.535	20.090	21.955
9	1.735	2.088	2.700	3.325	4.168	14.684	16.919	19.023	21.666	23.589
10	2.156	2.558	3.247	3.940	4.865	15.987	18.307	20.483	23.209	25.188
11	2.603	3.053	3.816	4.575	5.578	17.275	19.675	21.920	24.725	26.757
12	3.074	3.571	4.404	5.226	6.304	18.549	21.026	23.337	26.217	28.299
13	3.565	4.107	5.009	5.892	7.042	19.812	22.362	24.736	27.688	29.819
14	4.075	4.660	5.629	6.571	7.790	21.064	23.685	26.119	29.141	31.319
15	4.601	5.229	6.262	7.261	8.547	22.307	24.996	27.488	30.578	32.801
16	5.142	5.812	6.908	7.962	9.312	23.542	26.296	28.845	32.000	34.267
17	5.697	6.408	7.564	8.672	10.085	24.769	27.587	30.191	33.409	35.718
18	6.265	7.015	8.231	9.390	10.865	25.989	28.869	31.526	34.805	37.156
19	6.844	7.633	8.907	10.117	11.651	27.204	30.144	32.852	36.191	38.582
20	7.434	8.260	9.591	10.851	12.443	28.412	31.410	34.170	37.566	39.997
21	8.034	8.897	10.283	11.591	13.240	29.615	32.671	35.479	38.932	41.401
22	8.643	9.542	10.982	12.338	14.042	30.813	33.924	36.781	40.289	42.796
23	9.262	10.196	11.689	13.091	14.848	32.007	35.172	38.076	41.638	44.181
24	9.886	10.856	12.401	13.848	15.659	33.196	36.415	39.364	42.980	45.559
25	10.520	11.524	13.120	14.611	16.473	34.382	37.652	40.646	44.314	46.928
26	11.160	12.198	13.844	15.379	17.292	35.563	38.885	41.923	45.642	48.290
27	11.808	12.879	14.573	16.151	18.114	36.741	40.113	43.194	46.963	49.645
28	12.461	13.565	15.308	16.928	18.939	37.916	41.337	44.461	48.278	50.993
29	13.121	14.257	16.047	17.708	19.768	39.087	42.557	45.722	49.588	52.336
30	13.787	14.954	16.791	18.493	20.599	40.256	43.773	46.979	50.892	53.672
40	20.707	22.164	24.433	26.509	29.051	51.805	55.758	59.342	63.691	66.766
50	27.991	29.707	32.357	34.764	37.689	63.167	67.505	71.420	76.154	79.490
60	35.534	37.485	40.482	43.188	46.459	74.397	79.082	83.298	88.379	91.952
70	43.275	45.442	48.758	51.739	55.329	85.527	90.531	95.023	100.425	104.215
80	51.172	53.540	57.153	60.391	64.278	96.578	101.879	106.629	112.329	116.321
90	59.196	61.754	65.647	69.126	73.291	107.565	113.145	118.136	124.116	128.299
100	67.328	70.065	74.222	77.929	82.358	118.498	124.342	129.561	135.807	140.169

Source: Donald B. Owen, *Handbook of Statistics Tables,* © 1962, by Addison-Wesley Publishing Co., Inc., Reading, Massachusetts. Table A–5. Reprinted by permission of Addison-Wesley Longman.

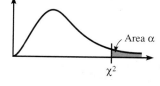

Table H The F Distribution

$\alpha = 0.005$

d.f.D.: degrees of freedom, denominator	d.f.N.: Degrees of freedom, numerator																		
	1	2	3	4	5	6	7	8	9	10	12	15	20	24	30	40	60	120	∞
1	16211	20000	21615	22500	23056	23437	23715	23925	24091	24224	24426	24630	24836	24940	25044	25148	25253	25359	25465
2	198.5	199.0	199.2	199.2	199.3	199.3	199.4	199.4	199.4	199.4	199.4	199.4	199.4	199.5	199.5	199.5	199.5	199.5	199.5
3	55.55	49.80	47.47	46.19	45.39	44.84	44.43	44.13	43.88	43.69	43.39	43.08	42.78	42.62	42.47	42.31	42.15	41.99	41.83
4	31.33	26.28	24.26	23.15	22.46	21.97	21.62	21.35	21.14	20.97	20.70	20.44	20.17	20.03	19.89	19.75	19.61	19.47	19.32
5	22.78	18.31	16.53	15.56	14.94	14.51	14.20	13.96	13.77	13.62	13.38	13.15	12.90	12.78	12.66	12.53	12.40	12.27	12.14
6	18.63	14.54	12.92	12.03	11.46	11.07	10.79	10.57	10.39	10.25	10.03	9.81	9.59	9.47	9.36	9.24	9.12	9.00	8.88
7	16.24	12.40	10.88	10.05	9.52	9.16	8.89	8.68	8.51	8.38	8.18	7.97	7.75	7.65	7.53	7.42	7.31	7.19	7.08
8	14.69	11.04	9.60	8.81	8.30	7.95	7.69	7.50	7.34	7.21	7.01	6.81	6.61	6.50	6.40	6.29	6.18	6.06	5.95
9	13.61	10.11	8.72	7.96	7.47	7.13	6.88	6.69	6.54	6.42	6.23	6.03	5.83	5.73	5.62	5.52	5.41	5.30	5.19
10	12.83	9.43	8.08	7.34	6.87	6.54	6.30	6.12	5.97	5.85	5.66	5.47	5.27	5.17	5.07	4.97	4.86	4.75	4.64
11	12.23	8.91	7.60	6.88	6.42	6.10	5.86	5.68	5.54	5.42	5.24	5.05	4.86	4.76	4.65	4.55	4.44	4.34	4.23
12	11.75	8.51	7.23	6.52	6.07	5.76	5.52	5.35	5.20	5.09	4.91	4.72	4.53	4.43	4.33	4.23	4.12	4.01	3.90
13	11.37	8.19	6.93	6.23	5.79	5.48	5.25	5.08	4.94	4.82	4.64	4.46	4.27	4.17	4.07	3.97	3.87	3.76	3.65
14	11.06	7.92	6.68	6.00	5.56	5.26	5.03	4.86	4.72	4.60	4.43	4.25	4.06	3.96	3.86	3.76	3.66	3.55	3.44
15	10.80	7.70	6.48	5.80	5.37	5.07	4.85	4.67	4.54	4.42	4.25	4.07	3.88	3.79	3.69	3.58	3.48	3.37	3.26
16	10.58	7.51	6.30	5.64	5.21	4.91	4.69	4.52	4.38	4.27	4.10	3.92	3.73	3.64	3.54	3.44	3.33	3.22	3.11
17	10.38	7.35	6.16	5.50	5.07	4.78	4.56	4.39	4.25	4.14	3.97	3.79	3.61	3.51	3.41	3.31	3.21	3.10	2.98
18	10.22	7.21	6.03	5.37	4.96	4.66	4.44	4.28	4.14	4.03	3.86	3.68	3.50	3.40	3.30	3.20	3.10	2.99	2.87
19	10.07	7.09	5.92	5.27	4.85	4.56	4.34	4.18	4.04	3.93	3.76	3.59	3.40	3.31	3.21	3.11	3.00	2.89	2.78
20	9.94	6.99	5.82	5.17	4.76	4.47	4.26	4.09	3.96	3.85	3.68	3.50	3.32	3.22	3.12	3.02	2.92	2.81	2.69
21	9.83	6.89	5.73	5.09	4.68	4.39	4.18	4.01	3.88	3.77	3.60	3.43	3.24	3.15	3.05	2.95	2.84	2.73	2.61
22	9.73	6.81	5.65	5.02	4.61	4.32	4.11	3.94	3.81	3.70	3.54	3.36	3.18	3.08	2.98	2.88	2.77	2.66	2.55
23	9.63	6.73	5.58	4.95	4.54	4.26	4.05	3.88	3.75	3.64	3.47	3.30	3.12	3.02	2.92	2.82	2.71	2.60	2.48
24	9.55	6.66	5.52	4.89	4.49	4.20	3.99	3.83	3.69	3.59	3.42	3.25	3.06	2.97	2.87	2.77	2.66	2.55	2.43
25	9.48	6.60	5.46	4.84	4.43	4.15	3.94	3.78	3.64	3.54	3.37	3.20	3.01	2.92	2.82	2.72	2.61	2.50	2.38
26	9.41	6.54	5.41	4.79	4.38	4.10	3.89	3.73	3.60	3.49	3.33	3.15	2.97	2.87	2.77	2.67	2.56	2.45	2.33
27	9.34	6.49	5.36	4.74	4.34	4.06	3.85	3.69	3.56	3.45	3.28	3.11	2.93	2.83	2.73	2.63	2.52	2.41	2.29
28	9.28	6.44	5.32	4.70	4.30	4.02	3.81	3.65	3.52	3.41	3.25	3.07	2.89	2.79	2.69	2.59	2.48	2.37	2.25
29	9.23	6.40	5.28	4.66	4.26	3.98	3.77	3.61	3.48	3.38	3.21	3.04	2.86	2.76	2.66	2.56	2.45	2.33	2.24
30	9.18	6.35	5.24	4.62	4.23	3.95	3.74	3.58	3.45	3.34	3.18	3.01	2.82	2.73	2.63	2.52	2.42	2.30	2.18
40	8.83	6.07	4.98	4.37	3.99	3.71	3.51	3.35	3.22	3.12	2.95	2.78	2.60	2.50	2.40	2.30	2.18	2.06	1.93
60	8.49	5.79	4.73	4.14	3.76	3.49	3.29	3.13	3.01	2.90	2.74	2.57	2.39	2.29	2.19	2.08	1.96	1.83	1.69
120	8.18	5.54	4.50	3.92	3.55	3.28	3.09	2.93	2.81	2.71	2.54	2.37	2.19	2.09	1.98	1.87	1.75	1.61	1.43
∞	7.88	5.30	4.28	3.72	3.35	3.09	2.90	2.74	2.62	2.52	2.36	2.19	2.00	1.90	1.79	1.67	1.53	1.36	1.00

Table H (continued)

$\alpha = 0.01$

d.f.N.: Degrees of freedom, numerator

d.f.D.: degrees of freedom, denominator	1	2	3	4	5	6	7	8	9	10	12	15	20	24	30	40	60	120	∞
1	4052	4999.5	5403	5625	5764	5859	5928	5982	6022	6056	6106	6157	6209	6235	6261	6287	6313	6339	6366
2	98.50	99.00	99.17	99.25	99.30	99.33	99.36	99.37	99.39	99.40	99.42	99.43	99.45	99.46	99.47	99.47	99.48	99.49	99.50
3	34.12	30.82	29.46	28.71	28.24	27.91	27.67	27.49	27.35	27.23	27.05	26.87	26.69	26.60	26.50	26.41	26.32	26.22	26.13
4	21.20	18.00	16.69	15.98	15.52	15.21	14.98	14.80	14.66	14.55	14.37	14.20	14.02	13.93	13.84	13.75	13.65	13.56	13.46
5	16.26	13.27	12.06	11.39	10.97	10.67	10.46	10.29	10.16	10.05	9.89	9.72	9.55	9.47	9.38	9.29	9.20	9.11	9.02
6	13.75	10.92	9.78	9.15	8.75	8.47	8.26	8.10	7.98	7.87	7.72	7.56	7.40	7.31	7.23	7.14	7.06	6.97	6.88
7	12.25	9.55	8.45	7.85	7.46	7.19	6.99	6.84	6.72	6.62	6.47	6.31	6.16	6.07	5.99	5.91	5.82	5.74	5.65
8	11.26	8.65	7.59	7.01	6.63	6.37	6.18	6.03	5.91	5.81	5.67	5.52	5.36	5.28	5.20	5.12	5.03	4.95	4.86
9	10.56	8.02	6.99	6.42	6.06	5.80	5.61	5.47	5.35	5.26	5.11	4.96	4.81	4.73	4.65	4.57	4.48	4.40	4.31
10	10.04	7.56	6.55	5.99	5.64	5.39	5.20	5.06	4.94	4.85	4.71	4.56	4.41	4.33	4.25	4.17	4.08	4.00	3.91
11	9.65	7.21	6.22	5.67	5.32	5.07	4.89	4.74	4.63	4.54	4.40	4.25	4.10	4.02	3.94	3.86	3.78	3.69	3.60
12	9.33	6.93	5.95	5.41	5.06	4.82	4.64	4.50	4.39	4.30	4.16	4.01	3.86	3.78	3.70	3.62	3.54	3.45	3.36
13	9.07	6.70	5.74	5.21	4.86	4.62	4.44	4.30	4.19	4.10	3.96	3.82	3.66	3.59	3.51	3.43	3.34	3.25	3.17
14	8.86	6.51	5.56	5.04	4.69	4.46	4.28	4.14	4.03	3.94	3.80	3.66	3.51	3.43	3.35	3.27	3.18	3.09	3.00
15	8.68	6.36	5.42	4.89	4.56	4.32	4.14	4.00	3.89	3.80	3.67	3.52	3.37	3.29	3.21	3.13	3.05	2.96	2.87
16	8.53	6.23	5.29	4.77	4.44	4.20	4.03	3.89	3.78	3.69	3.55	3.41	3.26	3.18	3.10	3.02	2.93	2.84	2.75
17	8.40	6.11	5.18	4.67	4.34	4.10	3.93	3.79	3.68	3.59	3.46	3.31	3.16	3.08	3.00	2.92	2.83	2.75	2.65
18	8.29	6.01	5.09	4.58	4.25	4.01	3.84	3.71	3.60	3.51	3.37	3.23	3.08	3.00	2.92	2.84	2.75	2.66	2.57
19	8.18	5.93	5.01	4.50	4.17	3.94	3.77	3.63	3.52	3.43	3.30	3.15	3.00	2.92	2.84	2.76	2.67	2.58	2.49
20	8.10	5.85	4.94	4.43	4.10	3.87	3.70	3.56	3.46	3.37	3.23	3.09	2.94	2.86	2.78	2.69	2.61	2.52	2.42
21	8.02	5.78	4.87	4.37	4.04	3.81	3.64	3.51	3.40	3.31	3.17	3.03	2.88	2.80	2.72	2.64	2.55	2.46	2.36
22	7.95	5.72	4.82	4.31	3.99	3.76	3.59	3.45	3.35	3.26	3.12	2.98	2.83	2.75	2.67	2.58	2.50	2.40	2.31
23	7.88	5.66	4.76	4.26	3.94	3.71	3.54	3.41	3.30	3.21	3.07	2.93	2.78	2.70	2.62	2.54	2.45	2.35	2.26
24	7.82	5.61	4.72	4.22	3.90	3.67	3.50	3.36	3.26	3.17	3.03	2.89	2.74	2.66	2.58	2.49	2.40	2.31	2.21
25	7.77	5.57	4.68	4.18	3.85	3.63	3.46	3.32	3.22	3.13	2.99	2.85	2.70	2.62	2.54	2.45	2.36	2.27	2.17
26	7.72	5.53	4.64	4.14	3.82	3.59	3.42	3.29	3.18	3.09	2.96	2.81	2.66	2.58	2.50	2.42	2.33	2.23	2.13
27	7.68	5.49	4.60	4.11	3.78	3.56	3.39	3.26	3.15	3.06	2.93	2.78	2.63	2.55	2.47	2.38	2.29	2.20	2.10
28	7.64	5.45	4.57	4.07	3.75	3.53	3.36	3.23	3.12	3.03	2.90	2.75	2.60	2.52	2.44	2.35	2.26	2.17	2.06
29	7.60	5.42	4.54	4.04	3.73	3.50	3.33	3.20	3.09	3.00	2.87	2.73	2.57	2.49	2.41	2.33	2.23	2.14	2.03
30	7.56	5.39	4.51	4.02	3.70	3.47	3.30	3.17	3.07	2.98	2.84	2.70	2.55	2.47	2.39	2.30	2.21	2.11	2.01
40	7.31	5.18	4.31	3.83	3.51	3.29	3.12	2.99	2.89	2.80	2.66	2.52	2.37	2.29	2.20	2.11	2.02	1.92	1.80
60	7.08	4.98	4.13	3.65	3.34	3.12	2.95	2.82	2.72	2.63	2.50	2.35	2.20	2.12	2.03	1.94	1.84	1.73	1.60
120	6.85	4.79	3.95	3.48	3.17	2.96	2.79	2.66	2.56	2.47	2.34	2.19	2.03	1.95	1.86	1.76	1.66	1.53	1.38
∞	6.63	4.61	3.78	3.32	3.02	2.80	2.64	2.51	2.41	2.32	2.18	2.04	1.88	1.79	1.70	1.59	1.47	1.32	1.00

Table H (continued)

$\alpha = 0.025$

d.f.N.: Degrees of freedom, numerator

d.f.D.: degrees of freedom, denominator	1	2	3	4	5	6	7	8	9	10	12	15	20	24	30	40	60	120	∞
1	647.8	799.5	864.2	899.6	921.8	937.1	948.2	956.7	963.3	968.6	976.7	984.9	993.1	997.2	1001	1006	1010	1014	1018
2	38.51	39.00	39.17	39.25	39.30	39.33	39.36	39.37	39.39	39.40	39.41	39.43	39.45	39.46	39.46	39.47	39.48	39.49	39.50
3	17.44	16.04	15.44	15.10	14.88	14.73	14.62	14.54	14.47	14.42	14.34	14.25	14.17	14.12	14.08	14.04	13.99	13.95	13.90
4	12.22	10.65	9.98	9.60	9.36	9.20	9.07	8.98	8.90	8.84	8.75	8.66	8.56	8.51	8.46	8.41	8.36	8.31	8.26
5	10.01	8.43	7.76	7.39	7.15	6.98	6.85	6.76	6.68	6.62	6.52	6.43	6.33	6.28	6.23	6.18	6.12	6.07	6.02
6	8.81	7.26	6.60	6.23	5.99	5.82	5.70	5.60	5.52	5.46	5.37	5.27	5.17	5.12	5.07	5.01	4.96	4.90	4.85
7	8.07	6.54	5.89	5.52	5.29	5.12	4.99	4.90	4.82	4.76	4.67	4.57	4.47	4.42	4.36	4.31	4.25	4.20	4.14
8	7.57	6.06	5.42	5.05	4.82	4.65	4.53	4.43	4.36	4.30	4.20	4.10	4.00	3.95	3.89	3.84	3.78	3.73	3.67
9	7.21	5.71	5.08	4.72	4.48	4.32	4.20	4.10	4.03	3.96	3.87	3.77	3.67	3.61	3.56	3.51	3.45	3.39	3.33
10	6.94	5.46	4.83	4.47	4.24	4.07	3.95	3.85	3.78	3.72	3.62	3.52	3.42	3.37	3.31	3.26	3.20	3.14	3.08
11	6.72	5.26	4.63	4.28	4.04	3.88	3.76	3.66	3.59	3.53	3.43	3.33	3.23	3.17	3.12	3.06	3.00	2.94	2.88
12	6.55	5.10	4.47	4.12	3.89	3.73	3.61	3.51	3.44	3.37	3.28	3.18	3.07	3.02	2.96	2.91	2.85	2.79	2.72
13	6.41	4.97	4.35	4.00	3.77	3.60	3.48	3.39	3.31	3.25	3.15	3.05	2.95	2.89	2.84	2.78	2.72	2.66	2.60
14	6.30	4.86	4.24	3.89	3.66	3.50	3.38	3.29	3.21	3.15	3.05	2.95	2.84	2.79	2.73	2.67	2.61	2.55	2.49
15	6.20	4.77	4.15	3.80	3.58	3.41	3.29	3.20	3.12	3.06	2.96	2.86	2.76	2.70	2.64	2.59	2.52	2.46	2.40
16	6.12	4.69	4.08	3.73	3.50	3.34	3.22	3.12	3.05	2.99	2.89	2.79	2.68	2.63	2.57	2.51	2.45	2.38	2.32
17	6.04	4.62	4.01	3.66	3.44	3.28	3.16	3.06	2.98	2.92	2.82	2.72	2.62	2.56	2.50	2.44	2.38	2.32	2.25
18	5.98	4.56	3.95	3.61	3.38	3.22	3.10	3.01	2.93	2.87	2.77	2.67	2.56	2.50	2.44	2.38	2.32	2.26	2.19
19	5.92	4.51	3.90	3.56	3.33	3.17	3.05	2.96	2.88	2.82	2.72	2.62	2.51	2.45	2.39	2.33	2.27	2.20	2.13
20	5.87	4.46	3.86	3.51	3.29	3.13	3.01	2.91	2.84	2.77	2.68	2.57	2.46	2.41	2.35	2.29	2.22	2.16	2.09
21	5.83	4.42	3.82	3.48	3.25	3.09	2.97	2.87	2.80	2.73	2.64	2.53	2.42	2.37	2.31	2.25	2.18	2.11	2.04
22	5.79	4.38	3.78	3.44	3.22	3.05	2.93	2.84	2.76	2.70	2.60	2.50	2.39	2.33	2.27	2.21	2.14	2.08	2.00
23	5.75	4.35	3.75	3.41	3.18	3.02	2.90	2.81	2.73	2.67	2.57	2.47	2.36	2.30	2.24	2.18	2.11	2.04	1.97
24	5.72	4.32	3.72	3.38	3.15	2.99	2.87	2.78	2.70	2.64	2.54	2.44	2.33	2.27	2.21	2.15	2.08	2.01	1.94
25	5.69	4.29	3.69	3.35	3.13	2.97	2.85	2.75	2.68	2.61	2.51	2.41	2.30	2.24	2.18	2.12	2.05	1.98	1.91
26	5.66	4.27	3.67	3.33	3.10	2.94	2.82	2.73	2.65	2.59	2.49	2.39	2.28	2.22	2.16	2.09	2.03	1.95	1.88
27	5.63	4.24	3.65	3.31	3.08	2.92	2.80	2.71	2.63	2.57	2.47	2.36	2.25	2.19	2.13	2.07	2.00	1.93	1.85
28	5.61	4.22	3.63	3.29	3.06	2.90	2.78	2.69	2.61	2.55	2.45	2.34	2.23	2.17	2.11	2.05	1.98	1.91	1.83
29	5.59	4.20	3.61	3.27	3.04	2.88	2.76	2.67	2.59	2.53	2.43	2.32	2.21	2.15	2.09	2.03	1.96	1.89	1.81
30	5.57	4.18	3.59	3.25	3.03	2.87	2.75	2.65	2.57	2.51	2.41	2.31	2.20	2.14	2.07	2.01	1.94	1.87	1.79
40	5.42	4.05	3.46	3.13	2.90	2.74	2.62	2.53	2.45	2.39	2.29	2.18	2.07	2.01	1.94	1.88	1.80	1.72	1.64
60	5.29	3.93	3.34	3.01	2.79	2.63	2.51	2.41	2.33	2.27	2.17	2.06	1.94	1.88	1.82	1.74	1.67	1.58	1.48
120	5.15	3.80	3.23	2.89	2.67	2.52	2.39	2.30	2.22	2.16	2.05	1.94	1.82	1.76	1.69	1.61	1.53	1.43	1.31
∞	5.02	3.69	3.12	2.79	2.57	2.41	2.29	2.19	2.11	2.05	1.94	1.83	1.71	1.64	1.57	1.48	1.39	1.27	1.00

Table H *(continued)*

$\alpha = 0.05$

d.f.N.: Degrees of freedom, numerator

d.f.D.: degrees of freedom, denominator	1	2	3	4	5	6	7	8	9	10	12	15	20	24	30	40	60	120	∞
1	161.4	199.5	215.7	224.6	230.2	234.0	236.8	238.9	240.5	241.9	243.9	245.9	248.0	249.1	250.1	251.1	252.2	253.3	254.3
2	18.51	19.00	19.16	19.25	19.30	19.33	19.35	19.37	19.38	19.40	19.41	19.43	19.45	19.45	19.46	19.47	19.48	19.49	19.50
3	10.13	9.55	9.28	9.12	9.01	8.94	8.89	8.85	8.81	8.79	8.74	8.70	8.66	8.64	8.62	8.59	8.57	8.55	8.53
4	7.71	6.94	6.59	6.39	6.26	6.16	6.09	6.04	6.00	5.96	5.91	5.86	5.80	5.77	5.75	5.72	5.69	5.66	5.63
5	6.61	5.79	5.41	5.19	5.05	4.95	4.88	4.82	4.77	4.74	4.68	4.62	4.56	4.53	4.50	4.46	4.43	4.40	4.36
6	5.99	5.14	4.76	4.53	4.39	4.28	4.21	4.15	4.10	4.06	4.00	3.94	3.87	3.84	3.81	3.77	3.74	3.70	3.67
7	5.59	4.74	4.35	4.12	3.97	3.87	3.79	3.73	3.68	3.64	3.57	3.51	3.44	3.41	3.38	3.34	3.30	3.27	3.23
8	5.32	4.46	4.07	3.84	3.69	3.58	3.50	3.44	3.39	3.35	3.28	3.22	3.15	3.12	3.08	3.04	3.01	2.97	2.93
9	5.12	4.26	3.86	3.63	3.48	3.37	3.29	3.23	3.18	3.14	3.07	3.01	2.94	2.90	2.86	2.83	2.79	2.75	2.71
10	4.96	4.10	3.71	3.48	3.33	3.22	3.14	3.07	3.02	2.98	2.91	2.85	2.77	2.74	2.70	2.66	2.62	2.58	2.54
11	4.84	3.98	3.59	3.36	3.20	3.09	3.01	2.95	2.90	2.85	2.79	2.72	2.65	2.61	2.57	2.53	2.49	2.45	2.40
12	4.75	3.89	3.49	3.26	3.11	3.00	2.91	2.85	2.80	2.75	2.69	2.62	2.54	2.51	2.47	2.43	2.38	2.34	2.30
13	4.67	3.81	3.41	3.18	3.03	2.92	2.83	2.77	2.71	2.67	2.60	2.53	2.46	2.42	2.38	2.34	2.30	2.25	2.21
14	4.60	3.74	3.34	3.11	2.96	2.85	2.76	2.70	2.65	2.60	2.53	2.46	2.39	2.35	2.31	2.27	2.22	2.18	2.13
15	4.54	3.68	3.29	3.06	2.90	2.79	2.71	2.64	2.59	2.54	2.48	2.40	2.33	2.29	2.25	2.20	2.16	2.11	2.07
16	4.49	3.63	3.24	3.01	2.85	2.74	2.66	2.59	2.54	2.49	2.42	2.35	2.28	2.24	2.19	2.15	2.11	2.06	2.01
17	4.45	3.59	3.20	2.96	2.81	2.70	2.61	2.55	2.49	2.45	2.38	2.31	2.23	2.19	2.15	2.10	2.06	2.01	1.96
18	4.41	3.55	3.16	2.93	2.77	2.66	2.58	2.51	2.46	2.41	2.34	2.27	2.19	2.15	2.11	2.06	2.02	1.97	1.92
19	4.38	3.52	3.13	2.90	2.74	2.63	2.54	2.48	2.42	2.38	2.31	2.23	2.16	2.11	2.07	2.03	1.98	1.93	1.88
20	4.35	3.49	3.10	2.87	2.71	2.60	2.51	2.45	2.39	2.35	2.28	2.20	2.12	2.08	2.04	1.99	1.95	1.90	1.84
21	4.32	3.47	3.07	2.84	2.68	2.57	2.49	2.42	2.37	2.32	2.25	2.18	2.10	2.05	2.01	1.96	1.92	1.87	1.81
22	4.30	3.44	3.05	2.82	2.66	2.55	2.46	2.40	2.34	2.30	2.23	2.15	2.07	2.03	1.98	1.94	1.89	1.84	1.78
23	4.28	3.42	3.03	2.80	2.64	2.53	2.44	2.37	2.32	2.27	2.20	2.13	2.05	2.01	1.96	1.91	1.86	1.81	1.76
24	4.26	3.40	3.01	2.78	2.62	2.51	2.42	2.36	2.30	2.25	2.18	2.11	2.03	1.98	1.94	1.89	1.84	1.79	1.73
25	4.24	3.39	2.99	2.76	2.60	2.49	2.40	2.34	2.28	2.24	2.16	2.09	2.01	1.96	1.92	1.87	1.82	1.77	1.71
26	4.23	3.37	2.98	2.74	2.59	2.47	2.39	2.32	2.27	2.22	2.15	2.07	1.99	1.95	1.90	1.85	1.80	1.75	1.69
27	4.21	3.35	2.96	2.73	2.57	2.46	2.37	2.31	2.25	2.20	2.13	2.06	1.97	1.93	1.88	1.84	1.79	1.73	1.67
28	4.20	3.34	2.95	2.71	2.56	2.45	2.36	2.29	2.24	2.19	2.12	2.04	1.96	1.91	1.87	1.82	1.77	1.71	1.65
29	4.18	3.33	2.93	2.70	2.55	2.43	2.35	2.28	2.22	2.18	2.10	2.03	1.94	1.90	1.85	1.81	1.75	1.70	1.64
30	4.17	3.32	2.92	2.69	2.53	2.42	2.33	2.27	2.21	2.16	2.09	2.01	1.93	1.89	1.84	1.79	1.74	1.68	1.62
40	4.08	3.23	2.84	2.61	2.45	2.34	2.25	2.18	2.12	2.08	2.00	1.92	1.84	1.79	1.74	1.69	1.64	1.58	1.51
60	4.00	3.15	2.76	2.53	2.37	2.25	2.17	2.10	2.04	1.99	1.92	1.84	1.75	1.70	1.65	1.59	1.53	1.47	1.39
120	3.92	3.07	2.68	2.45	2.29	2.17	2.09	2.02	1.96	1.91	1.83	1.75	1.66	1.61	1.55	1.50	1.43	1.35	1.25
∞	3.84	3.00	2.60	2.37	2.21	2.10	2.01	1.94	1.88	1.83	1.75	1.67	1.57	1.52	1.46	1.39	1.32	1.22	1.00

C

Table H (concluded)

α = 0.10

d.f.N.: Degrees of freedom, numerator

d.f.D.: degrees of freedom, denominator	1	2	3	4	5	6	7	8	9	10	12	15	20	24	30	40	60	120	∞
1	39.86	49.50	53.59	55.83	57.24	58.20	58.91	59.44	59.86	60.19	60.71	61.22	61.74	62.00	62.26	62.53	62.79	63.06	63.33
2	8.53	9.00	9.16	9.24	9.29	9.33	9.35	9.37	9.38	9.39	9.41	9.42	9.44	9.45	9.46	9.47	9.47	9.48	9.49
3	5.54	5.46	5.39	5.34	5.31	5.28	5.27	5.25	5.24	5.23	5.22	5.20	5.18	5.18	5.17	5.16	5.15	5.14	5.13
4	4.54	4.32	4.19	4.11	4.05	4.01	3.98	3.95	3.94	3.92	3.90	3.87	3.84	3.83	3.82	3.80	3.79	3.78	3.76
5	4.06	3.78	3.62	3.52	3.45	3.40	3.37	3.34	3.32	3.30	3.27	3.24	3.21	3.19	3.17	3.16	3.14	3.12	3.10
6	3.78	3.46	3.29	3.18	3.11	3.05	3.01	2.98	2.96	2.94	2.90	2.87	2.84	2.82	2.80	2.78	2.76	2.74	2.72
7	3.59	3.26	3.07	2.96	2.88	2.83	2.78	2.75	2.72	2.70	2.67	2.63	2.59	2.58	2.56	2.54	2.51	2.49	2.47
8	3.46	3.11	2.92	2.81	2.73	2.67	2.62	2.59	2.56	2.54	2.50	2.46	2.42	2.40	2.38	2.36	2.34	2.32	2.29
9	3.36	3.01	2.81	2.69	2.61	2.55	2.51	2.47	2.44	2.42	2.38	2.34	2.30	2.28	2.25	2.23	2.21	2.18	2.16
10	3.29	2.92	2.73	2.61	2.52	2.46	2.41	2.38	2.35	2.32	2.28	2.24	2.20	2.18	2.16	2.13	2.11	2.08	2.06
11	3.23	2.86	2.66	2.54	2.45	2.39	2.34	2.30	2.27	2.25	2.21	2.17	2.12	2.10	2.08	2.05	2.03	2.00	1.97
12	3.18	2.81	2.61	2.48	2.39	2.33	2.28	2.24	2.21	2.19	2.15	2.10	2.06	2.04	2.01	1.99	1.96	1.93	1.90
13	3.14	2.76	2.56	2.43	2.35	2.28	2.23	2.20	2.16	2.14	2.10	2.05	2.01	1.98	1.96	1.93	1.90	1.88	1.85
14	3.10	2.73	2.52	2.39	2.31	2.24	2.19	2.15	2.12	2.10	2.05	2.01	1.96	1.94	1.91	1.89	1.86	1.83	1.80
15	3.07	2.70	2.49	2.36	2.27	2.21	2.16	2.12	2.09	2.06	2.02	1.97	1.92	1.90	1.87	1.85	1.82	1.79	1.76
16	3.05	2.67	2.46	2.33	2.24	2.18	2.13	2.09	2.06	2.03	1.99	1.94	1.89	1.87	1.84	1.81	1.78	1.75	1.72
17	3.03	2.64	2.44	2.31	2.22	2.15	2.10	2.06	2.03	2.00	1.96	1.91	1.86	1.84	1.81	1.78	1.75	1.72	1.69
18	3.01	2.62	2.42	2.29	2.20	2.13	2.08	2.04	2.00	1.98	1.93	1.89	1.84	1.81	1.78	1.75	1.72	1.69	1.66
19	2.99	2.61	2.40	2.27	2.18	2.11	2.06	2.02	1.98	1.96	1.91	1.86	1.81	1.79	1.76	1.73	1.70	1.67	1.63
20	2.97	2.59	2.38	2.25	2.16	2.09	2.04	2.00	1.96	1.94	1.89	1.84	1.79	1.77	1.74	1.71	1.68	1.64	1.61
21	2.96	2.57	2.36	2.23	2.14	2.08	2.02	1.98	1.95	1.92	1.87	1.83	1.78	1.75	1.72	1.69	1.66	1.62	1.59
22	2.95	2.56	2.35	2.22	2.13	2.06	2.01	1.97	1.93	1.90	1.86	1.81	1.76	1.73	1.70	1.67	1.64	1.60	1.57
23	2.94	2.55	2.34	2.21	2.11	2.05	1.99	1.95	1.92	1.89	1.84	1.80	1.74	1.72	1.69	1.66	1.62	1.59	1.55
24	2.93	2.54	2.33	2.19	2.10	2.04	1.98	1.94	1.91	1.88	1.83	1.78	1.73	1.70	1.67	1.64	1.61	1.57	1.53
25	2.92	2.53	2.32	2.18	2.09	2.02	1.97	1.93	1.89	1.87	1.82	1.77	1.72	1.69	1.66	1.63	1.59	1.56	1.52
26	2.91	2.52	2.31	2.17	2.08	2.01	1.96	1.92	1.88	1.86	1.81	1.76	1.71	1.68	1.65	1.61	1.58	1.54	1.50
27	2.90	2.51	2.30	2.17	2.07	2.00	1.95	1.91	1.87	1.85	1.80	1.75	1.70	1.67	1.64	1.60	1.57	1.53	1.49
28	2.89	2.50	2.29	2.16	2.06	2.00	1.94	1.90	1.87	1.84	1.79	1.74	1.69	1.66	1.63	1.59	1.56	1.52	1.48
29	2.89	2.50	2.28	2.15	2.06	1.99	1.93	1.89	1.86	1.83	1.78	1.73	1.68	1.65	1.62	1.58	1.55	1.51	1.47
30	2.88	2.49	2.28	2.14	2.05	1.98	1.93	1.88	1.85	1.82	1.77	1.72	1.67	1.64	1.61	1.57	1.54	1.50	1.46
40	2.84	2.44	2.23	2.09	2.00	1.93	1.87	1.83	1.79	1.76	1.71	1.66	1.61	1.57	1.54	1.51	1.47	1.42	1.38
60	2.79	2.39	2.18	2.04	1.95	1.87	1.82	1.77	1.74	1.71	1.66	1.60	1.54	1.51	1.48	1.44	1.40	1.35	1.29
120	2.75	2.35	2.13	1.99	1.90	1.82	1.77	1.72	1.68	1.65	1.60	1.55	1.48	1.45	1.41	1.37	1.32	1.26	1.19
∞	2.71	2.30	2.08	1.94	1.85	1.77	1.72	1.67	1.63	1.60	1.55	1.49	1.42	1.38	1.34	1.30	1.24	1.17	1.00

From M. Merrington and C. M. Thompson (1943). Table of Percentage Points of the Inverted Beta (*F*) Distribution. *Biometrika* 33, pp. 74–87. Reprinted with permission from Biometrika.

Table I	Critical Values for the PPMC

Reject H_0: $\rho = 0$ if the absolute value of r is greater than the value given in the table. The values are for a two-tailed test; d.f. $= n - 2$.

d.f.	$\alpha = 0.05$	$\alpha = 0.01$
1	0.999	0.999
2	0.950	0.999
3	0.878	0.959
4	0.811	0.917
5	0.754	0.875
6	0.707	0.834
7	0.666	0.798
8	0.632	0.765
9	0.602	0.735
10	0.576	0.708
11	0.553	0.684
12	0.532	0.661
13	0.514	0.641
14	0.497	0.623
15	0.482	0.606
16	0.468	0.590
17	0.456	0.575
18	0.444	0.561
19	0.433	0.549
20	0.423	0.537
25	0.381	0.487
30	0.349	0.449
35	0.325	0.418
40	0.304	0.393
45	0.288	0.372
50	0.273	0.354
60	0.250	0.325
70	0.232	0.302
80	0.217	0.283
90	0.205	0.267
100	0.195	0.254

Source: From *Biometrika Tables for Statisticians* Vol. 1 (1962) p. 138. Reprinted with permission.

Appendix D

Data Bank

Data Bank Values

The following list explains the values given for the categories in the Data Bank.

1. "Age" is given in years.

2. "Educational level" values are defined as follows:

   0 = no high school degree   2 = college graduate
   1 = high school graduate   3 = graduate degree

3. "Smoking status" values are defined as follows:

   0 = does not smoke
   1 = smokes less than one pack per day
   2 = smokes one or more than one pack per day

4. "Exercise" values are defined as follows:

   0 = none   2 = moderate
   1 = light   3 = heavy

5. "Weight" is given in pounds.

6. "Serum cholesterol" is given in milligram percent (mg%).

7. "Systolic pressure" is given in millimeters of mercury (mmHg).

8. "IQ" is given in standard IQ test score values.

9. "Sodium" is given in milliequivalents per liter (mEq/1).

10. "Gender" is listed as male (M) or female (F).

11. "Marital status" values are defined as follows:

    M = married   S = single
    W = widowed   D = divorced

ID number	Age	Educational level	Smoking status	Exercise	Weight	Serum cholesterol	Systolic pressure	IQ	Sodium	Gender	Marital status
01	27	2	1	1	120	193	126	118	136	F	M
02	18	1	0	1	145	210	120	105	137	M	S
03	32	2	0	0	118	196	128	115	135	F	M
04	24	2	0	1	162	208	129	108	142	M	M
05	19	1	2	0	106	188	119	106	133	F	S
06	56	1	0	0	143	206	136	111	138	F	W
07	65	1	2	0	160	240	131	99	140	M	W
08	36	2	1	0	215	215	163	106	151	M	D
09	43	1	0	1	127	201	132	111	134	F	M
10	47	1	1	1	132	215	138	109	135	F	D

Data Bank *(continued)*

ID number	Age	Educational level	Smoking status	Exercise	Weight	Serum cholesterol	Systolic pressure	IQ	Sodium	Gender	Marital status
11	48	3	1	2	196	199	148	115	146	M	D
12	25	2	2	3	109	210	115	114	141	F	S
13	63	0	1	0	170	242	149	101	152	F	D
14	37	2	0	3	187	193	142	109	144	M	M
15	40	0	1	1	234	208	156	98	147	M	M
16	25	1	2	1	199	253	135	103	148	M	S
17	72	0	0	0	143	288	156	103	145	F	M
18	56	1	1	0	156	164	153	99	144	F	D
19	37	2	0	2	142	214	122	110	135	M	M
20	41	1	1	1	123	220	142	108	134	F	M
21	33	2	1	1	165	194	122	112	137	M	S
22	52	1	0	1	157	205	119	106	134	M	D
23	44	2	0	1	121	223	135	116	133	F	M
24	53	1	0	0	131	199	133	121	136	F	M
25	19	1	0	3	128	206	118	122	132	M	S
26	25	1	0	0	143	200	118	103	135	M	M
27	31	2	1	1	152	204	120	119	136	M	M
28	28	2	0	0	119	203	118	116	138	F	M
29	23	1	0	0	111	240	120	105	135	F	S
30	47	2	1	0	149	199	132	123	136	F	M
31	47	2	1	0	179	235	131	113	139	M	M
32	59	1	2	0	206	260	151	99	143	M	W
33	36	2	1	0	191	201	148	118	145	M	D
34	59	0	1	1	156	235	142	100	132	F	W
35	35	1	0	0	122	232	131	106	135	F	M
36	29	2	0	2	175	195	129	121	148	M	M
37	43	3	0	3	194	211	138	129	146	M	M
38	44	1	2	0	132	240	130	109	132	F	S
39	63	2	2	1	188	255	156	121	145	M	M
40	36	2	1	1	125	220	126	117	140	F	S
41	21	1	0	1	109	206	114	102	136	F	M
42	31	2	0	2	112	201	116	123	133	F	M
43	57	1	1	1	167	213	141	103	143	M	W
44	20	1	2	3	101	194	110	111	125	F	S
45	24	2	1	3	106	188	113	114	127	F	D
46	42	1	0	1	148	206	136	107	140	M	S
47	55	1	0	0	170	257	152	106	130	F	M
48	23	0	0	1	152	204	116	95	142	M	M
49	32	2	0	0	191	210	132	115	147	M	M
50	28	1	0	1	148	222	135	100	135	M	M
51	67	0	0	0	160	250	141	116	146	F	W
52	22	1	1	1	109	220	121	103	144	F	M
53	19	1	1	1	131	231	117	112	133	M	S
54	25	2	0	2	153	212	121	119	149	M	D
55	41	3	2	2	165	236	130	131	152	M	M

Data Bank *(concluded)*

ID number	Age	Educational level	Smoking status	Exercise	Weight	Serum cholesterol	Systolic pressure	IQ	Sodium	Gender	Marital status
56	24	2	0	3	112	205	118	100	132	F	S
57	32	2	0	1	115	187	115	109	136	F	S
58	50	3	0	1	173	203	136	126	146	M	M
59	32	2	1	0	186	248	119	122	149	M	M
60	26	2	0	1	181	207	123	121	142	M	S
61	36	1	1	0	112	188	117	98	135	F	D
62	40	1	1	0	130	201	121	105	136	F	D
63	19	1	1	1	132	237	115	111	137	M	S
64	37	2	0	2	179	228	141	127	141	F	M
65	65	3	2	1	212	220	158	129	148	M	M
66	21	1	2	2	99	191	117	103	131	F	S
67	25	2	2	1	128	195	120	121	131	F	S
68	68	0	0	0	167	210	142	98	140	M	W
69	18	1	1	2	121	198	123	113	136	F	S
70	26	0	1	1	163	235	128	99	140	M	M
71	45	1	1	1	185	229	125	101	143	M	M
72	44	3	0	0	130	215	128	128	137	F	M
73	50	1	0	0	142	232	135	104	138	F	M
74	63	0	0	0	166	271	143	103	147	F	W
75	48	1	0	3	163	203	131	103	144	M	M
76	27	2	0	3	147	186	118	114	134	M	M
77	31	3	1	1	152	228	116	126	138	M	D
78	28	2	0	2	112	197	120	123	133	F	M
79	36	2	1	2	190	226	123	121	147	M	M
80	43	3	2	0	179	252	127	131	145	M	D
81	21	1	0	1	117	185	116	105	137	F	S
82	32	2	1	0	125	193	123	119	135	F	M
83	29	2	1	0	123	192	131	116	131	F	D
84	49	2	2	1	185	190	129	127	144	M	M
85	24	1	1	1	133	237	121	114	129	M	M
86	36	2	0	2	163	195	115	119	139	M	M
87	34	1	2	0	135	199	133	117	135	F	M
88	36	0	0	1	142	216	138	88	137	F	M
89	29	1	1	1	155	214	120	98	135	M	S
90	42	0	0	2	169	201	123	96	137	M	D
91	41	1	1	1	136	214	133	102	141	F	D
92	29	1	1	0	112	205	120	102	130	F	M
93	43	1	1	0	185	208	127	100	143	M	M
94	61	1	2	0	173	248	142	101	141	M	M
95	21	1	1	3	106	210	111	105	131	F	S
96	56	0	0	0	149	232	142	103	141	F	M
97	63	0	1	0	192	193	163	95	147	M	M
98	74	1	0	0	162	247	151	99	151	F	W
99	35	2	0	1	151	251	147	113	145	F	M
100	28	2	0	3	161	199	129	116	138	M	M

Appendix E

Glossary

alpha   the probability of a type I error, represented by the Greek letter α

alternative hypothesis   a statistical hypothesis that states a specific difference between a parameter and a specific value or states that there is a difference between two parameters

analysis of variance (ANOVA)   a statistical technique used to test a hypothesis concerning the means of three or more populations

ANOVA summary table   the table used to summarize the results of an ANOVA test

bar graph   a graph that uses vertical or horizontal bars to represent the frequencies of a distribution

beta   the probability of a type II error, represented by the Greek letter β

between-group variance   a variance estimate using the means of the groups or between the groups in an F test

biased sample   a sample for which some type of systematic error has been made in the selection of subjects for the sample

binomial distribution   the outcomes of a binomial experiment and the corresponding probabilities of these outcomes

binomial experiment   a probability experiment in which each trial has only two outcomes, there is a fixed number of trials, the outcomes of the trials are independent, and the probability of success remains the same for each trial

box plot   a graph used to represent a data set when the data set contains a small number of values

categorical frequency distribution   a frequency distribution used when the data are categorical (nominal)

central limit theorem   a theorem that states that as the sample size increases, the shape of the distribution of the sample means taken from the population with mean μ and standard deviation σ will approach a normal distribution; the distribution will have a mean μ and a standard deviation σ

Chebyshev's theorem   a theorem that states that the proportion of values from a data set that fall within k standard deviations of the mean will be at least $1 - 1/k^2$, where k is a number greater than 1

chi-square distribution   a probability distribution obtained from the values of $(n - 1)s^2/\sigma^2$ when random samples are selected from a normally distributed population whose variance is σ^2

class boundaries   the upper and lower values of a class for a grouped frequency distribution whose values have one additional decimal place more than the data and end in a 5-digit

class midpoint   a value for a class in a frequency distribution obtained by adding the lower and upper class boundaries (or the lower and upper class limits) and dividing by 2

class width   the difference between the upper class boundary and the lower class boundary for a class in a frequency distribution

classical probability   the type of probability that uses sample spaces to determine the numerical probability that an event will happen

cluster sample   a sample obtained by selecting a preexisting or natural group, called a cluster, and using the members in the cluster for the sample

coefficient of determination   a measure of the variation of the dependent variable that is explained by the regression line and the independent variable; the ratio of the explained variation to the total variation

coefficient of variation   the standard deviation divided by the mean; the result is expressed as a percentage

combination   a selection of objects without regard to order

complement of an event   the set of outcomes in the sample space that are not in the outcomes of the event itself

E

compound event an event that consists of two or more outcomes or simple events

conditional probability the probability that an event *B* occurs after an event *A* has already occurred

confidence interval a specific interval estimate of a parameter determined by using data obtained from a sample and the specific confidence level of the estimate

confidence level the probability that a parameter will fall within the specified interval estimate of the parameter

consistent estimator an estimator whose value approaches the value of the parameter estimated as the sample size increases

contingency table data arranged in table form for the chi-square independence test, with *R* rows and *C* columns

continuous variable a variable that can assume all values between any two specific values; a variable obtained by measuring

convenience sample sample of subjects used because they are convenient and available

correction for continuity a correction employed when a continuous distribution is used to approximate a discrete distribution

correlation a statistical method used to determine whether a relationship exists between variables

correlation coefficient a statistic or parameter that measures the strength and direction of a relationship between two variables

critical or **rejection region** the range of values of the test value that indicates that there is a significant difference and the null hypothesis should be rejected in a hypothesis test

critical value (C.V.) a value that separates the critical region from the noncritical region in a hypothesis test

cumulative frequency the sum of the frequencies accumulated up to the upper boundary of a class in a frequency distribution

data measurements or observations for a variable

data array a data set that has been ordered

data set a collection of data values

data value or **datum** a value in a data set

decile a location measure of a data value; it divides the distribution into 10 groups

degrees of freedom the number of values that are free to vary after a sample statistic has been computed; used when a distribution (such as the *t* distribution) consists of a family of curves

dependent events events for which the outcome or occurrence of the first event affects the outcome or occurrence of the second event in such a way that the probability is changed

dependent samples samples in which the subjects are paired or matched in some way; i.e., the samples are related

dependent variable a variable in correlation and regression analysis that cannot be controlled or manipulated

descriptive statistics a branch of statistics that consists of the collection, organization, summarization, and presentation of data

discrete variable a variable that assumes values that can be counted

double sampling a sampling method in which a very large population is given a questionnaire to determine those who meet the qualifications for a study; the questionnaire is reviewed, a second smaller population is defined, and a sample is selected from this group

empirical probability the type of probability that uses frequency distributions based on observations to determine numerical probabilities of events

empirical rule a rule that states that when a distribution is bell-shaped (normal), approximately 68% of the data values will fall within one standard deviation of the mean; approximately 95% of the data values will fall within two standard deviations of the mean; and approximately 99.7% of the data values will fall within three standard deviations of the mean

equally likely events the events that have the same probability of occurring

estimation the process of estimating the value of a parameter from information obtained from a sample

estimator a statistic used to estimate a parameter

event one or more outcomes of a probability experiment

expected frequency the frequency obtained by calculation (as if there were no preference) and used in the chi-square test

expected value the theoretical average of a variable that has a probability distribution

exploratory data analysis the act of analyzing data to determine what information can be obtained using stem and leaf plots, medians, interquartile ranges, and box plots

F distribution the sampling distribution of the variances when two independent samples are selected from two normally distributed populations in which the variances are equal and the variances s_1^2 and s_2^2 are compared as $s_1^2 \div s_2^2$

F test   a statistical test used to compare two variances or three or more means

five-number summary   five specific values for a data set that consist of the lowest and highest values, the lower and upper hinges, and the median

frequency   the number of values in a specific class of a frequency distribution

frequency distribution   an organization of raw data in table form, using classes and frequencies

frequency polygon   a graph that displays the data by using lines that connect points plotted for the frequencies at the midpoints of the classes

goodness-of-fit test   a chi-square test used to see whether a frequency distribution fits a specific pattern

grouped frequency distribution   a distribution used when the range is large and classes of several units in width are needed

histogram   a graph that displays the data by using vertical bars of various heights to represent the frequencies of a distribution

homogeneity of proportions test   a test used to determine the equality of three or more proportions

hypothesis testing   a decision-making process for evaluating claims about a population

independence test   a chi-square test used to test the independence of two variables when data are tabulated in table form in terms of frequencies

independent events   events for which the probability of the first occurring does not affect the probability of the second occurring

independent samples   samples that are not related

independent variable   a variable in correlation and regression analysis that can be controlled or manipulated

inferential statistics   a branch of statistics that consists of generalizing from samples to populations, performing hypothesis testing, determining relationships among variables, and making predictions

interquartile range $Q_3 - Q_1$

interval estimate   a range of values used to estimate a parameter

interval level of measurement   a measurement level that ranks data and in which precise differences between units of measure exist. *See also* nominal, ordinal, and ratio levels of measurement

law of large numbers   when a probability experiment is repeated a large number of times, the relative frequency probability of an outcome will approach its theoretical probability

left-tailed test   a test used on a hypothesis when the critical region is on the left side of the distribution

level of significance   the maximum probability of commiting a type I error in hypothesis testing

lower class boundary   the lower value of a class in a frequency distribution that has one more decimal place value than the data and ends in a 5 digit

lower class limit   the lower value of a class in a frequency distribution that has the same decimal place value as the data

lower hinge   the median of all values less than or equal to the median when the data set has an odd number of values, or the median of all values less than the median when the data set has an even number of values

maximum error of estimate   the maximum difference between the point estimate of a parameter and the actual value of the parameter

mean   the sum of the values divided by the total number of values

mean square   the variance found by dividing the sum of the squares of a variable by the corresponding degrees of freedom; used in ANOVA

measurement scales   a type of classification that tells how variables are categorized, counted, or measured; the four types of scales are nominal, ordinal, interval, and ratio

median   the midpoint of a data array

median class   the class of a frequency distribution that contains the median

midrange   the sum of the lowest and highest data values divided by 2

modal class   the class with the largest frequency

mode   the value that occurs most often in a data set

multiple relationship   a relationship in which many variables are under study

multistage sampling   a sampling technique that uses a combination of sampling methods

mutually exclusive events   probability events that cannot occur at the same time

negative relationship   a relationship between variables such that as one variable increases, the other variable decreases, and vice versa

negatively skewed distribution   a distribution in which the majority of the data values fall to the right of the mean

E

nominal level of measurement a measurement level that classifies data into mutually exclusive (nonoverlapping) exhaustive categories in which no order or ranking can be imposed on them. *See also* interval, ordinal, and ratio levels of measurement

noncritical or **nonrejection region** the range of values of the test value that indicates that the difference was probably due to chance and the null hypothesis should not be rejected

nonrejection region *see* noncritical region

normal distribution a continuous, symmetric, bell-shaped distribution of a variable

null hypothesis a statistical hypothesis that states that there is no difference between a parameter and a specific value or that there is no difference between two parameters

observed frequency the actual frequency value obtained from a sample and used in the chi-square test

ogive a graph that represents the cumulative frequencies for the classes in a frequency distribution

one-tailed test a test that indicates that the null hypothesis should be rejected when the test statistic value is in the critical region on one side of the mean

open-ended distribution a frequency distribution that has no specific beginning value or no specific ending value

ordinal level of measurement a measurement level that classifies data into categories that can be ranked; however, precise differences between the ranks do not exist. *See also* interval, nominal, and ratio levels of measurement

outcome the result of a single trial of a probability experiment

outlier an extreme value in a data set; it is omitted from a box plot

parameter a characteristic or measure obtained by using all the data values for a specific population

Pareto chart chart that uses vertical bars to represent frequencies for a categorical variable

Pearson product moment correlation coefficient (PPMCC) a statistic used to determine the strength of a relationship when the variables are normally distributed

percentile a location measure of a data value; it divides the distribution into 100 groups

permutation an arrangement of n objects in a specific order

phi correlation coefficient a statistic used to determine the strength of a relationship when both variables are truly dichotomous

pictograph a graph that uses symbols or pictures to represent data

pie graph a circle that is divided into sections or wedges according to the percentage of frequencies in each category of the distribution

point estimate a specific numerical value estimate of a parameter

pooled estimate of the variance a weighted average of the variance using the two sample variances and their respective degrees of freedom as the weights

population the totality of all subjects possessing certain common characteristics that are being studied

population correlation coefficient the value of the correlation coefficient computed by using all possible pairs of data values (x, y) taken from a population

positive relationship a relationship between two variables such that as one variable increases, the other variable increases or as one variable decreases, the other decreases

positively skewed distribution a distribution in which the majority of the data values fall to the left of the mean

prediction interval a confidence interval for a predicted value y

probability the chance of an event occurring

probability distribution the values a random variable can assume and the corresponding probabilities of the values

probability experiment a process that leads to well-defined results called outcomes

proportion a part of a whole, represented by a fraction, a decimal, or a percentage

***P*-value** the actual probability of getting the sample mean value if the null hypothesis is true

qualitative variable a variable that can be placed into distinct categories, according to some characteristic or attribute

quantitative variable a variable that is numerical in nature and that can be ordered or ranked

quartile a location measure of a data value; it divides the distribution into four groups

random sample a sample obtained by using random or chance methods; a sample for which every member of the population has an equal chance of being selected

random variable a variable whose values are determined by chance

range the highest data value minus the lowest data value

ranking   the positioning of a data value in a data array according to some rating scale

ratio level of measurement   a measurement level that possesses all the characteristics of interval measurement and a true zero; it also has true ratios between different units of measure. *See also* interval, nominal, and ordinal levels of measurement

raw data   data collected in original form

regression   a statistical method used to describe the nature of the relationship between variables—i.e., a positive or negative, linear or nonlinear relationship

regression line   the line of best fit of the data

rejection region   *see* critical region

relative frequency graph   a graph using proportions instead of raw data as frequencies

relatively efficient estimator   an estimator that has the smallest variance from among all the statistics that can be used to estimate a parameter

right-tailed test   a test used on a hypothesis when the critical region is on the right side of the distribution

sample   a subgroup or subset of the population

sample space   the set of all possible outcomes of a probability experiment

sampling distribution of sample means   a distribution obtained by using the means computed from random samples taken from a population

sampling error   the difference between the sample measure and the corresponding population measure due to the fact that the sample is not a perfect representation of the population

scatter plot   a graph of the independent and dependent variables in regression and correlation analysis

sequence sampling   a sampling technique used in quality control in which successive units are taken from production lines and tested to see whether they meet the standards set by the manufacturing company

simple event   an outcome that results from a single trial of a probability experiment

simple relationship   a relationship in which only two variables are under study

standard deviation   the square root of the variance

standard error of estimate   the standard deviation of the observed y values about the predicted y' values in regression and correlation analysis

standard error of the mean   the standard deviation of the sample means for samples taken from the same population

standard normal distribution   a normal distribution for which the mean is equal to 0 and the standard deviation is equal to 1

standard score   the difference between a data value and the mean divided by the standard deviation

statistic   a characteristic or measure obtained by using the data values from a sample

statistical hypothesis   a conjecture about a population parameter, which may or may not be true

statistical test   a test that uses data obtained from a sample to make a decision about whether or not the null hypothesis should be rejected

stem and leaf plot   a data plot that uses part of a data value as the stem and part of the data value as the leaf to form groups or classes

stratified sample   a sample obtained by dividing the population into subgroups, called strata, according to various homogeneous characteristics and then selecting members from each stratum

subjective probability   the type of probability that uses a probability value based on an educated guess or estimate, employing opinions and inexact information

sum of squares between groups   a statistic computed in the numerator of the fraction used to find the between-group variance in ANOVA

sum of squares within groups   a statistic computed in the numerator of the fraction used to find the within-group variance in ANOVA

symmetrical distribution   a distribution in which the data values are uniformly distributed about the mean

systematic sample   a sample obtained by numbering each element in the population and then selecting every kth number from the population to be included in the sample

t **distribution**   a family of bell-shaped curves based on degrees of freedom, similar to the standard normal distribution with the exception that the variance is greater than 1; used when one is testing small samples and when the population standard deviation is unknown

t **test**   a statistical test for the mean of a population, used when the population is normally distributed, the population standard deviation is unknown, and the sample size is less than 30

test value   the numerical value obtained from a statistical test, computed from (observed value − expected value) ÷ standard error

time series graph   a graph that represents data that occur over a specific period of time

treatment groups   the groups used in an ANOVA study

tree diagram   a device used to list all possibilities of a sequence of events in a systematic way

two-tailed test   a test that indicates that the null hypothesis should be rejected when the test value is in either of the two critical regions

type I error   the error that occurs if one rejects the null hypothesis when it is true

type II error   the error that occurs if one does not reject the null hypothesis when it is false

unbiased estimator   an estimator whose value approximates the expected value of a population parameter, used for the variance or standard deviation when the sample size is less than 30; an estimator whose expected value or mean must be equal to the mean of the parameter being estimated

unbiased sample   a sample chosen at random from the population that is, for the most part, representative of the population

ungrouped frequency distribution   a distribution that uses individual data and has a small range of data

upper class boundary   the upper value of a class in a frequency distribution that has one more decimal place value than the data and ends in a 5 digit

upper class limit   the upper value of a class in a frequency distribution that has the same decimal place value as the data

upper hinge   the median of all values greater than or equal to the median when the data set has an odd number of values, or the median of all values greater

than the median when the data set has an even number of values

variable   a characteristic or attribute that can assume different values

variance   the average of the squares of the distance each value is from the mean

Venn diagram   a diagram used as a pictorial representative for a probability concept or rule

weighted mean   the mean found by multiplying each value by its corresponding weight and dividing by the sum of the weights

within-group variance   a variance estimate using all the sample data for an F test; it is not affected by differences in the means

z **distribution**   *see* standard normal distribution

z **score**   *see* standard score

z **test**   a statistical test for means and proportions of a population, used when the population is normally distributed and the population standard deviation is known or the sample size is 30 or more

z **value**   same as *z* score

Appendix F

Bibliography

Aczel, Amir D. *Complete Business Statistics,* 3rd ed. Chicago: Irwin, 1996.

Beyer, William H. *CRC Handbook of Tables for Probability and Statistics,* 2nd ed. Boca Raton, Fla.: CRC Press, 1986.

Chao, Lincoln L. *Introduction to Statistics.* Monterey, Calif.: Brooks/Cole, 1980.

Daniel, Wayne W., and James C. Terrell. *Business Statistics,* 4th ed. Boston: Houghton Mifflin, 1986.

Edwards, Allan L. *An Introduction to Linear Regression and Correlation,* 2nd ed. New York: Freeman, 1984.

Eves, Howard. *An Introduction to the History of Mathematics,* 3rd ed. New York: Holt, Rinehart and Winston, 1969.

Famighetti, Robert, ed. *The World Almanac and Book of Facts 1996.* New York: Pharos Books, 1995.

Freund, John E., and Gary Simon. *Statistics—A First Course,* 6th ed. Englewood Cliffs, N.J.: Prentice-Hall, 1995.

Gibson, Henry R. *Elementary Statistics.* Dubuque, Ia.: Wm. C. Brown Publishers, 1994.

Glass, Gene V., and Kenneth D. Hopkins. *Statistical Methods in Education and Psychology,* 2nd ed. Englewood Cliffs, N.J.: Prentice-Hall, 1984.

Guilford, J. P. *Fundamental Statistics in Psychology and Education,* 4th ed. New York: McGraw-Hill, 1965.

Haack, Dennis G. *Statistical Literacy: A Guide to Interpretation.* Boston: Duxbury Press, 1979.

Hartwig, Frederick, with Brian Dearing. *Exploratory Data Analysis.* Newbury Park, Calif.: Sage Publications, 1979.

Henry, Gary T. *Graphing Data: Techniques for Display and Analysis.* Thousand Oaks, Calif.: Sage Publications, 1995.

Isaac, Stephen, and William B. Michael. *Handbook in Research and Evaluation,* 2nd ed. San Diego: EdITS, 1990.

Johnson, Robert. *Elementary Statistics,* 6th ed. Boston: PWS–Kent, 1992.

Kachigan, Sam Kash. *Statistical Analysis.* New York: Radius Press, 1986.

Khazanie, Ramakant. *Elementary Statistics in a World of Applications,* 3rd ed. Glenview, Ill.: Scott, Foresman, 1990.

Kuzma, Jan W. *Basic Statistics for the Health Sciences.* Mountain View, Calif.: Mayfield, 1984.

Lapham, Lewis H., Michael Pollan, and Eric Ethridge. *The Harper's Index Book.* New York: Henry Holt, 1987.

Lipschultz, Seymour. *Schaum's Outline of Theory and Problems of Probability.* New York: McGraw-Hill, 1968.

Marascuilo, Leonard A., and Maryellen McSweeney. *Nonparametric and Distribution-Free Methods for the Social Sciences.* Monterey, Calif.: Brooks/Cole, 1977.

Marzillier, Leon F. *Elementary Statistics.* Dubuque, Ia.: Wm. C. Brown Publishers, 1990.

Mason, Robert D., Douglas A. Lind, and William G. Marchal. *Statistics: An Introduction.* New York: Harcourt Brace Jovanovich, 1988.

MINITAB. *MINITAB Reference Manual.* State College, Pa.: MINITAB, Inc., 1994.

Minium, Edward W. *Statistical Reasoning in Psychology and Education.* New York: Wiley, 1970.

Moore, David S. *The Basic Practice of Statistics.* New York: W. H. Freeman and Co., 1995.

Newmark, Joseph. *Statistics and Probability in Modern Life.* New York: Saunders, 1988.

Pagano, Robert R. *Understanding Statistics,* 3rd ed. New York: West, 1990.

Phillips, John L., Jr. *How to Think about Statistics.* New York: Freeman, 1988.

Reinhardt, Howard E., and Don O. Loftsgaarden. *Elementary Probability and Statistical Reasoning.* Lexington, Mass.: Heath, 1977.

Roscoe, John T. *Fundamental Research Statistics for the Behavioral Sciences,* 2nd ed. New York: Holt, Rinehart and Winston, 1975.

Rossman, Allan J. *Workshop Statistics, Discovery with Data.* New York: Springer, 1996.

Runyon, Richard P., and Audrey Haber. *Fundamentals of Behavioral Statistics,* 6th ed. New York: Random House, 1988.

Shulte, Albert P., 1981 yearbook editor, and James R. Smart, general yearbook editor. *Teaching Statistics and Probability, 1981 Yearbook.* Reston, Va.: National Council of Teachers of Mathematics, 1981.

Smith, Gary. *Statistical Reasoning.* Boston: Allyn and Bacon, 1985.

Spiegel, Murray R. *Schaum's Outline of Theory and Problems of Statistics.* New York: McGraw-Hill, 1961.

Texas Instruments. *TI-83 Graphing Calculator Guidebook.* Temple, Tx.: Texas Instruments, 1996.

Triola, Mario G. *Elementary Statistics,* 6th ed. Reading, Mass.: Addison-Wesley, 1995.

Wardrop, Robert L. *Statistics: Learning in the Presence of Variation.* Dubuque, Ia.: Wm. C. Brown Publishers, 1995.

Warwick, Donald P., and Charles A. Lininger. *The Sample Survey: Theory and Practice.* New York: McGraw-Hill, 1975.

Weiss, Daniel Evan. *100% American.* New York: Poseidon Press, 1988.

Williams, Jack. *The USA Today Weather Almanac 1995.* New York: Vintage Books, 1994.

Wright, John W., ed. *The Universal Almanac 1995.* Kansas City, Mo.: Andrews & McMeel, 1994.

Appendix G

Photo Credits

Appendix H

Selected Answers*

Chapter 1

1–1. Descriptive statistics describe the data set. Inferential statistics use the data to draw conclusions about the population.

1–3. Answers will vary.

1–5. Samples are used to save time and money when the population is large and when the units must be destroyed to gain information.

1–6. *a.* Descriptive *d.* Inferential *g.* Descriptive*
    b. Descriptive* *e.* Inferential *h.* Descriptive
    c. Inferential *f.* Inferential
    *The entire population was not used.

1–7. *a.* Ratio *d.* Ratio *g.* Ratio *i.* Ordinal
    b. Ordinal *e.* Ratio *h.* Ratio *j.* Ratio
    c. Interval *f.* Nominal

1–8. *a.* Qualitative *c.* Qualitative *e.* Quantitative
    b. Quantitative *d.* Quantitative

1–9. *a.* Discrete *d.* Continuous *f.* Continuous
    b. Continuous *e.* Continuous *g.* Discrete
    c. Discrete

1–11. Random, systematic, stratified, cluster

1–12. *a.* Cluster *c.* Random *e.* Stratified
    b. Systematic *d.* Systematic

1–13. Answers will vary.

1–15. Answers will vary.

1–17. Answers will vary.

Quiz—Chapter 1

1. True
2. False
3. False
4. False
5. False
6. True
7. False
8. *c.*
9. *b.*
10. *d.*
11. *a.*
12. Descriptive, inferential
13. Gambling, insurance
14. Population
15. Sample

Note: These answers to odd-numbered and selected even-numbered exercises include *all* quiz answers.
*Answers may vary due to rounding.

16. *a.* Saves time
    b. Saves money
    c. Use when population is infinite

17. *a.* Random
    b. Systematic
    c. Cluster
    d. Stratified

18. *a.* Inferential
    b. Descriptive
    c. Inferential
    d. Descriptive
    e. Inferential

19. *a.* Ratio
    b. Ordinal
    c. Interval
    d. Ratio
    e. Nominal

20. *a.* Continuous
    b. Discrete
    c. Discrete
    d. Continuous
    e. Continuous
    f. Discrete

21. *a.* 3.15–3.25
    b. 17.5–18.5
    c. 8.5–9.5
    d. 0.265–0.275
    e. 35.5–36.5

Chapter 2

2–1. To organize data in a meaningful way, to determine the shape of the distribution, to facilitate computational procedures for statistics, to make it easier to draw charts and graphs, to make comparisons among different sets of data.

2–3. *a.* 10.5–15.5; 13; 5
    b. 16.5–39.5; 28; 23
    c. 292.5–353.5; 323; 61
    d. 11.75–14.75; 13.25; 3
    e. 3.125–3.935; 3.53; 0.81

2–5. *a.* Class width is not uniform.
    b. Class limits overlap, and class width is not uniform.
    c. A class has been omitted.
    d. Class width is not uniform.

2–7.

Class	*f*
15130	5
15131	3
15132	3
15133	7
15134	2
	20

2–9.

Class		f
0	−0.5–0.5	5
1	0.5–1.5	8
2	1.5–2.5	10
3	2.5–3.5	2
4	3.5–4.5	3
5	4.5–5.5	2
		30

2–11.

Limits	Boundaries	f	cf
21–53	20.5–53.5	2	2
54–86	53.5–86.5	0	2
87–119	86.5–119.5	1	3
120–152	119.5–152.5	1	4
153–185	152.5–185.5	2	6
186–218	185.5–218.5	10	16
219–251	218.5–251.5	9	25
252–284	251.5–284.5	9	34
285–317	284.5–317.5	5	39
318–350	317.5–350.5	4	43
351–383	350.5–383.5	5	48
384–416	383.5–416.5	1	49
		49	

2–13.

Limits	Boundaries	f	cf
27–33	26.5–33.5	7	7
34–40	33.5–40.5	14	21
41–47	40.5–47.5	15	36
48–54	47.5–54.5	11	47
55–61	54.5–61.5	3	50
62–68	61.5–68.5	3	53
69–75	68.5–75.5	2	55
		55	

2–15.

Limits	Boundaries	f	cf
2.7–3.1	2.65–3.15	1	1
3.2–3.6	3.15–3.65	4	5
3.7–4.1	3.65–4.15	8	13
4.2–4.6	4.15–4.65	13	26
4.7–5.1	4.65–5.15	8	34
5.2–5.6	5.15–5.65	3	37
5.7–6.1	5.65–6.15	3	40
		40	

2–17.

Limits	Boundaries	f	cf
150–1276	149.5–1276.5	2	2
1277–2403	1276.5–2403.5	2	4
2404–3530	2403.5–3530.5	5	9
3531–4657	3530.5–4657.5	8	17
4658–5784	4657.5–5784.5	7	24
5785–6911	5784.5–6911.5	3	27
6912–8038	6911.5–8038.5	7	34
8039–9165	8038.5–9165.5	3	37
9166–10,292	9165.5–10,292.5	3	40
10,293–11,419	10,292.5–11,419.5	2	42
		42	

2–19.

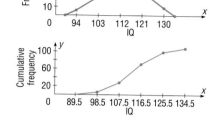

2–21.

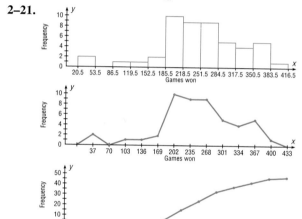

2–23.

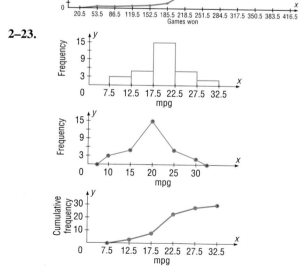

2–25.

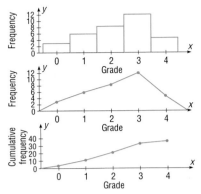

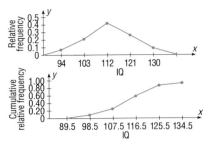

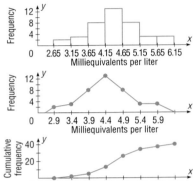

2–33.

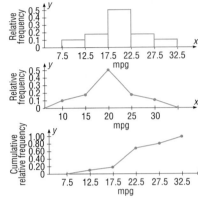

2–27.

2–35.

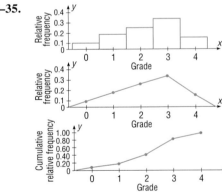

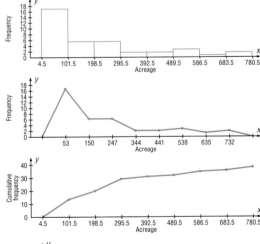

2–37. *a.* 0   *b.* 14   *c.* 10   *d.* 16

2–39.

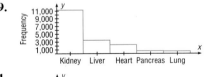

2–41.

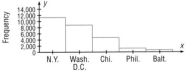

2–31.

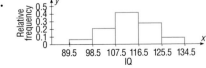

2–43.

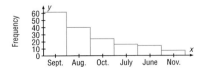

2–45.

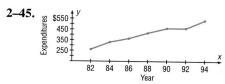

2–47.

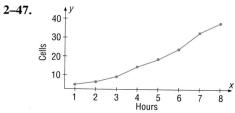

2–49.

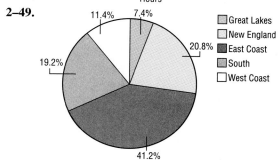

2–51.

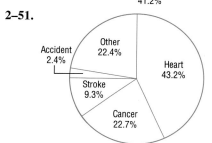

2–53.

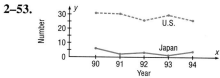

2–55.

Class	f
Newspaper	7
Television	5
Radio	7
Magazine	6
	25

2–57.

Class	f
Baseballs	4
Golf balls	5
Tennis balls	6
Soccer balls	5
Footballs	5
	25

2–59.

Class	f	cf
11	1	1
12	2	3
13	2	5
14	2	7
15	1	8
16	2	10
17	4	14
18	2	16
19	2	18
20	1	19
21	0	19
22	1	20
	20	

2–61.

Limits	Boundaries	f	cf
40–42	39.5–42.5	2	2
43–45	42.5–45.5	7	9
46–48	45.5–48.5	10	19
49–51	48.5–51.5	8	27
52–54	51.5–54.5	3	30
		30	

2–63.

Limits	Boundaries	f	cf
120–128	119.5–128.5	2	2
129–137	128.5–137.5	1	3
138–146	137.5–146.5	4	7
147–155	146.5–155.5	0	7
156–164	155.5–164.5	13	20
165–173	164.5–173.5	10	30

2–65.

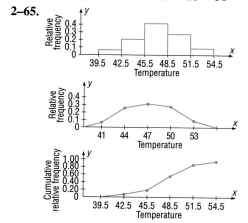

2–67.

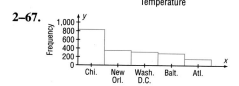

2–69.

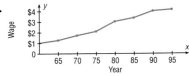

2–71.

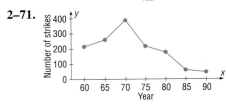

2–73.

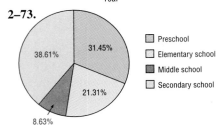

Preschool
Elementary school
Middle school
Secondary school

Quiz—Chapter 2

1. False
2. False
3. False
4. True
5. True
6. False
7. False
8. *c.*
9. *c.*
10. *b.*
11. *b.*
12. Categorical, ungrouped, grouped
13. 5, 20
14. Categorical
15. Time series
16. Raw
17. Vertical or *y*

18.

	f
H	6
A	5
M	6
C	8
	25

19.

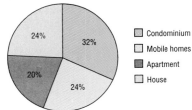

Condominium
Mobile homes
Apartment
House

20.

Class	*f*	cf
0.5–1.5	1	1
1.5–2.5	5	6
2.5–3.5	3	9
3.5–4.5	4	13
4.5–5.5	2	15
5.5–6.5	6	21
6.5–7.5	2	23
7.5–8.5	3	26
8.5–9.5	4	30
	30	

21.

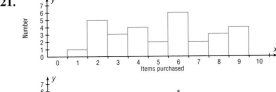

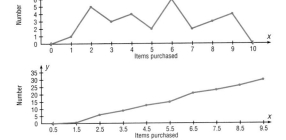

22.

Class		mp	*f*	cf
101.5–116.5	102–116	109	4	4
116.5–131.5	117–131	124	3	7
131.5–146.5	132–146	139	1	8
146.5–161.5	147–161	154	4	12
161.5–176.5	162–176	169	11	23
176.5–191.5	177–191	184	7	30
			30	

23.

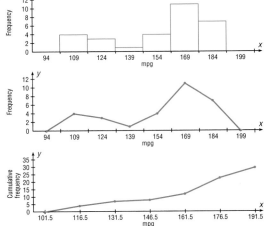

24.

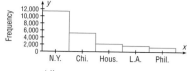

25.

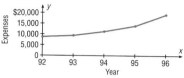

Chapter 3

3–1. *a.* 8.8 *b.* 8 *c.* 8 *d.* 10.5

3–3. *a.* 43.9 *b.* 45.5 *c.* 51 *d.* 44.5

3–5. *a.* 124.3 *b.* 121 *c.* 121 *d.* 124

3–7. *a.* 2.780 *b.* 2.715 *c.* none *d.* 2.91

3–9. *a.* 105 *b.* 110 *c.* 110 *d.* 97.5

3–11. *a.* 6.63 *b.* 6.45 *c.* none *d.* 6.7

3–13. *a.* 19.8 *b.* 19.3 *c.* 15.5–18.5

3–15. *a.* 2.0 *b.* 2 *c.* 2

3–17. *a.* 85.1 *b.* 84.2 *c.* 74.5–85.5

3–19. *a.* 33.8 *b.* 31.5 *c.* 26.5–33.5

3–21. *a.* 23.7 *b.* 23.4 *c.* 21.5–24.5

3–23. 44.8; 43.5; 40.5–47.5

3–25. 4.45; 4.42; 4.15–4.65

3–27. 2.896

3–29. $545,666.67

3–31. 82.7

3–33. *a.* Median *c.* Mode *e.* Mode
  b. Mean *d.* Mode *f.* Mean

3–35. Greek letters, μ

3–37. 6

3–39. *a.* 36 mph *b.* 30.77 mph *c.* $16.67

3–41. 5.48

3–43. The square root of the variance is the standard deviation.

3–45. σ^2; σ

3–47. When the sample size is less than 30, the formula for the variance of the sample will underestimate the population variance.

3–49. 9; 6.2; 2.5

3–51. 10; 11.3; 3.4

3–53. 30; 77.1; 8.8

3–55. 5; 3.19; 1.79

3–57. 620; 40,135.9; 200.3

3–59. 305; 10,325.9; 101.6

3–61. 133.6; 11.6

3–63. 1.1; 1.0

3–65. 211.2; 14.5

3–67. 80.3; 9.0

3–69. 11.7; 3.4

3–71. 10%; 10%; they are equal

3–73. 23.1%; 12.9%; age is more variable

3–75. *a.* 96% *b.* 93.75%

3–77. $4.84–$5.20

3–79. 89–101

3–81. 86%

3–83. All the data values fall within two standard deviations of the means.

3–85. 56%; 75%; 84%; 88.89%; 92%

3–87. 4.36

3–89. A *z* score tells how many standard deviations the data value is above or below the mean.

3–91. A percentile is a relative measurement of position; a percentage is an absolute measure of the part to the total.

3–93. $Q_1 = P_{25}$; $Q_2 = P_{50}$; $Q_3 = P_{75}$

3–95. $D_1 = P_{10}$; $D_2 = P_{20}$; $D_3 = P_{30}$; etc.

3–97. *a.* 1.00 *b.* 1.47 *c.* −0.47 *d.* 0 *e.* −1.00

3–99. *a.* 0.75 *b.* −1.25 *c.* 2.25 *d.* −2 *e.* −0.5

3–101. *a.* 1 *b.* 0.6; grade in part *a* is higher

3–103. *a.* −0.93 *c.* −1.4; score in part *b* is highest
  b. −0.85

3–105. *a.* 21st *b.* 58th *c.* 77th *d.* 29th

3–106. *a.* 7 *b.* 25 *c.* 64 *d.* 76 *e.* 93

3–107. *a.* 235 *b.* 255 *c.* 261 *d.* 275 *e.* 283

3–108. *a.* 376 *b.* 389 *c.* 432 *d.* 473 *e.* 498

3–109. *a.* 17th *b.* 39th *c.* 53rd *d.* 79th *e.* 91st

3–111. 82

3–113. 47

3–115. 12

3–117. *a.* 12; 20.5; 32; 22; 20 *b.* 62; 94; 99; 80.5; 37

3–119. Stem and leaf plot, median, interquartile range and box plot

3–121. Resistant statistics are relatively more unaffected by outliers than are nonresistant statistics.

3–123.

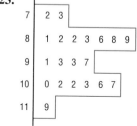

3–125.

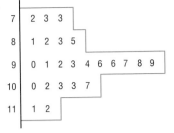

```
 7 | 2  3  3
 8 | 1  2  3  5
 9 | 0  1  2  3  4  6  6  7  8  9
10 | 0  2  3  3  7
11 | 1  2
```

3–127.

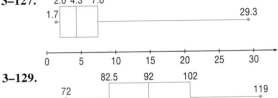

```
     2.0 4.3 7.6
   1.7            29.3

   0   5   10  15  20  25  30
```

3–129.

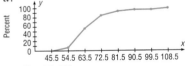

```
        82.5   92   102
   72                      119

   70  80  90  100  110  120
```

3–131. *a.* 1203.9   *d.* 1400.5   *g.* 689.6
          b. 1164   *e.* 1885
          c. No mode   *f.* 475,610.1

3–133. *a.* 3.8 *b.* 4 *c.* 4 *d.* 4 *e.* 1.1 *f.* 1.1

3–135. *a.* 55.5   *c.* 57.5–72.5   *e.* 23.8
          b. 59.4   *d.* 566.1

3–137. 1.1

3–139. 6

3–141. Magazine variance, 0.214; year variance, 0.417; years are more variable

3–143. *a.*

```
Percent
100
 80
 60
 40
 20
  0
    45.5 54.5 63.5 72.5 81.5 90.5 99.5 108.5
```

b. 60; 67; 75
c. 4th; 9th; 42nd

3–145. 0.26–0.38

3–147. 56%

3–149. 88.89%

3–151.

```
2 | 9  9
3 | 2  4  5  6  8  8
4 | 1  2  3  7  7
5 | 1  3  5  8
6 | 2  2  2  3  7
7 | 2  3
```

3–153.

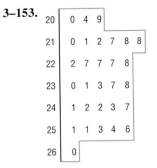

```
20 | 0  4  9
21 | 0  1  2  7  8  8
22 | 2  7  7  7  8
23 | 0  1  3  7  8
24 | 1  2  2  3  7
25 | 1  1  3  4  6
26 | 0
```

3–155.

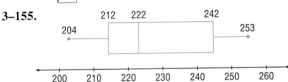

```
            212   222          242
      204                            253

   200  210  220  230  240  250  260
```

Quiz—Chapter 3

1. True
2. False
3. False
4. False
5. False
6. False
7. False
8. False
9. False
10. *c.*
11. *c.*
12. *a* and *b.*
13. *b.*
14. *d.*
15. *b.*
16. Statistic
17. Parameters, statistics
18. Standard deviation
19. σ
20. Midrange
21. Positively
22. Outlier

23. *a.* 84.1   *c.* none   *e.* 12   *g.* 4.1
    b. 85   *d.* 84   *f.* 17.1

24. *a.* 4.8 *b.* 5 *c.* 5 *d.* 4 *e.* 1.3 *f.* 1.1

25. *a.* 51.4 *b.* 49.6 *c.* 35.5–50.5 *d.* 451.5
    e. 21.2

26. *a.* 8.2 *b.* 7.7 *c.* 7–9 *d.* 21.6 *e.* 4.6

27. 1.6

28. 4.5

29. 0.33; 0.162; newspapers

30. 0.3125; 0.229; brands

31. −0.75; −1.67; science

32. *a.* 0.5 *b.* 1.6 *c.* 15, *c* is higher

33. *a.* 6; 19; 31; 44; 56; 69; 81; 94
    b. 27
    c.

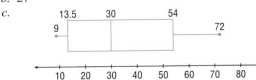

```
          13.5    30       54
      9                        72

   10  20  30  40  50  60  70  80
```

34. *a.*

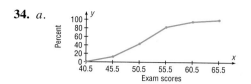

b. 47; 53; 65
c. 60th, 6th, 98th percentile

35.

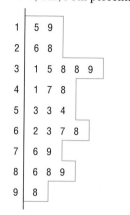

Chapter 4

4–1.

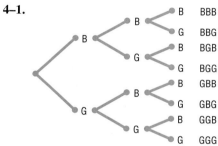

4–3.

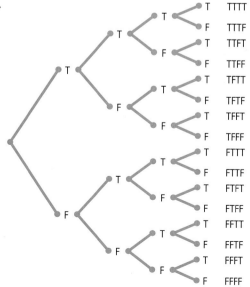

4–5.

4–7.

4–9.

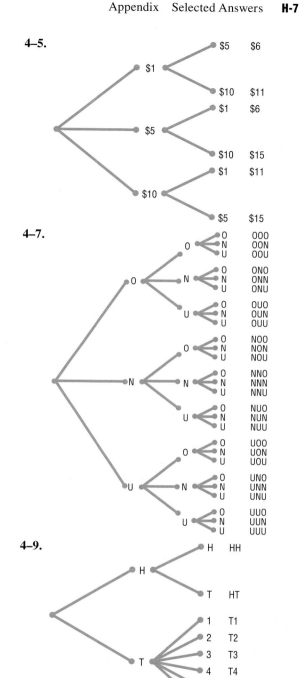

4–11.

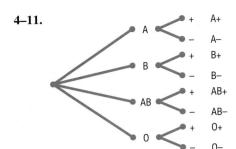

4–13. 100,000; 30,240

4–15. 5,040

4–17. 40,320

4–19. 120

4–21. 1000; 72

4–23. 10

4–25. 64

4–27. 15

4–29. 7 chickens, 8 cows

4–31. *a.* 40,320   *c.* 24   *e.* 1
      b. 3,628,800   *d.* 1

4–32. *a.* 56   *c.* 11,880   *e.* 1   *g.* 1   *i.* 990
      b. 2520   *d.* 60   *f.* 720   *h.* 40,320   *j.* 30

4–33. 840

4–35. 151,200

4–37. 24

4–39. 120

4–41. 120

4–43. 1,860,480

4–45. 2,520

4–46. *a.* 10   *e.* 15   *i.* 66
      b. 56   *f.* 1   *j.* 4
      c. 35   *g.* 1
      d. 15   *h.* 36

4–47. 22,100

4–49. 41,580

4–51. 120

4–53. 462

4–55. 166,320

4–57. 14,400

4–59. 194,040

4–61. 53,130

4–63. 126

4–65. 45

4–67. 24,310

4–69. *a.* 4   *b.* 40   *c.* 624   *d.* 3,744

4–71. 120

4–73. 78 (only two cards are drawn)

4–75. 792

4–77.

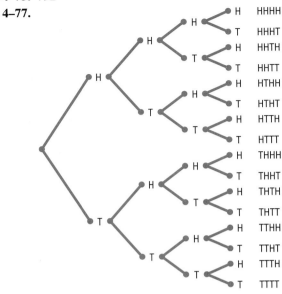

4–79. 40,320

4–81. 800

4–83.

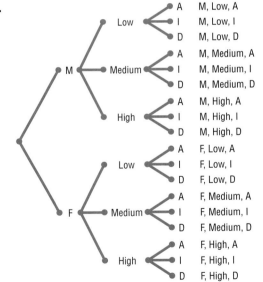

4–85. 720

4–87. 24

4–89. 25; 20

4–91. 4

4–93. 30

4–95. 495

4–97. 15,504

Quiz—Chapter 4

1. False
2. False
3. True
4. True
5. False
6. *b.*
7. *d.*
8. *d.*
9. *b.*
10. *b.*
11. Tree diagram
12. 12
13. n^k
14. $n!$
15. Permutations
16. 1,188,137,600; 710,424,000
17. 720
18. 33,554,432
19. 35
20. 8

```
      P   < H    PH
          < GS   P, GS
      W   < H    WH
          < GS   W, GS
      Chi < H    Chi, H
          < GS   Chi, GS
      Cha < H    Cha, H
          < GS   Cha, GS
```

21. 2,646
22. 40,320
23.

```
          L < SN   ML, SN
            < HN   ML, HN
            < CN   ML, CN
      M   M < SN   MM, SN
            < HN   MM, HN
            < CN   MM, CN
          H < SN   MH, SN
            < HN   MH, HN
            < CN   MH, CN
          L < SN   FL, SN
            < HN   FL, HN
            < CN   FL, CN
      F   M < SN   FM, SN
            < HN   FM, HN
            < CN   FM, CN
          H < SN   FH, SN
            < HN   FH, HN
            < CN   FH, CN
```

24. 1,365
25. 64; 24
26. 4, or 6 if he wears a sweater over the shirt
27. 120
28. 5,005
29. 60
30. 15

Chapter 5

5–1. A probability experiment is a process that leads to well-defined outcomes.

5–3. An outcome is the result of a single trial of a probability experiment, but an event consists of one or more outcomes.

5–5. The range of values is 0 to 1, inclusive.

5–7. 0

5–9. 0.55

5–11. *a.* Empirical   *d.* Classical   *f.* Empirical
 b. Classical   *e.* Empirical   *g.* Subjective
 c. Empirical

5–12. *a.* $\frac{1}{6}$ *c.* $\frac{1}{3}$ *e.* 1   *f.* $\frac{5}{6}$ *g.* $\frac{1}{6}$
 b. $\frac{1}{2}$ *d.* 1

5–13. *a.* $\frac{5}{36}$ *b.* $\frac{1}{6}$ *c.* $\frac{2}{9}$ *d.* $\frac{1}{6}$ *e.* $\frac{1}{6}$

5–14. *a.* $\frac{1}{13}$ *c.* $\frac{1}{52}$ *e.* $\frac{4}{13}$ *g.* $\frac{1}{2}$ *i.* $\frac{7}{13}$
 b. $\frac{1}{4}$ *d.* $\frac{2}{13}$ *f.* $\frac{4}{13}$ *h.* $\frac{1}{26}$ *j.* $\frac{1}{26}$

5–15. *a.* $\frac{1}{2}$ *b.* $\frac{3}{10}$ *c.* $\frac{7}{10}$ *d.* $\frac{7}{10}$ *e.* $\frac{1}{2}$

5–17. $\frac{7}{50}$

5–19. 47%

5–21. *a.* $\frac{1}{8}$ *b.* $\frac{1}{4}$ *c.* $\frac{3}{4}$ *d.* $\frac{3}{4}$

5–23. $\frac{1}{9}$

5–25. *a.* $\frac{9}{19}$ *b.* $\frac{11}{38}$ *c.* $\frac{7}{19}$

5–27. 18

5–29. $\frac{1}{36}$

5–31. *a.* 20%   *b.* 43%   *c.* 47%   *d.* 40%

5–33. The statement is probably not based on empirical probability, and is probably not true.

5–35. Actual outcomes will differ; however, each number should occur approximately $\frac{1}{6}$ of the time.

5–37. *a.* 1:5, 5:1   *d.* 1:1, 1:1   *g.* 1:1, 1:1
 b. 1:1, 1:1   *e.* 1:12, 12:1
 c. 1:3, 3:1   *f.* 1:3, 3:1

5–39. Answers will vary.

5–41. $\frac{1}{6}$

5–43. $\frac{11}{19}$

5–45. *a.* $\frac{8}{17}$ *b.* $\frac{6}{17}$ *c.* $\frac{9}{17}$ *d.* $\frac{12}{17}$

5–47. 0.93

5–49. *a.* $\frac{6}{7}$ *b.* $\frac{4}{7}$ *c.* 1

5–51. *a.* $\frac{67}{118}$ *b.* $\frac{81}{118}$ *c.* $\frac{44}{59}$

5–53. *a.* $\frac{38}{45}$ *b.* $\frac{22}{45}$ *c.* $\frac{2}{3}$

5–55. *a.* $\frac{14}{31}$ *b.* $\frac{23}{31}$ *c.* $\frac{19}{31}$

5–57. *a.* $\frac{1}{15}$ *b.* $\frac{1}{3}$ *c.* $\frac{5}{6}$ *d.* $\frac{5}{6}$ *e.* $\frac{1}{3}$

5–59. *a.* $\frac{5}{12}$ *b.* $\frac{1}{8}$ *c.* $\frac{2}{3}$ *d.* $\frac{23}{24}$

5–61. *a.* $\frac{3}{13}$ *b.* $\frac{3}{4}$ *c.* $\frac{19}{52}$ *d.* $\frac{7}{13}$ *e.* $\frac{15}{26}$

5–63. $\frac{7}{10}$

5–65. 0.06

5–67. 0.30

5–69. *a.* Independent   *e.* Independent
 b. Dependent   *f.* Dependent
 c. Dependent   *g.* Dependent
 d. Dependent   *h.* Independent

5–71. 31.4%

5–73. 0.003

H

5–75. $\frac{1}{12}$

5–77. $\frac{1}{1728}$

5–79. $\frac{1}{133,225}$

5–81. $\frac{4}{15}$

5–83. $\frac{243}{1024}$

5–85. $\frac{5}{28}$

5–87. $\frac{22}{117}$

5–89. 0.116

5–91. 0.03

5–93. $\frac{49}{72}$

5–95. 0.6

5–97. 89%

5–99. 70%

5–101. 82%

5–103. *a.* $\frac{18}{47}$ *b.* $\frac{14}{23}$

5–105. $\frac{55}{56}$

5–107. $\frac{31}{32}$

5–109. $\frac{15}{16}$

5–111. $\frac{14,498}{20,825}$

5–113. 26.6%

5–115. $\frac{63}{64}$

5–117. $\frac{91}{216}$

5–119. $\frac{7}{8}$

5–121. *a.* $\frac{1}{6}$ *b.* $\frac{1}{6}$ *c.* $\frac{2}{3}$

5–123. $\frac{16}{45}$

5–125. $\frac{17}{30}$

5–127. *a.* $\frac{1}{10}$ *b.* $\frac{11}{30}$ *c.* $\frac{13}{15}$ *d.* $\frac{13}{15}$

5–129. 0.98

5–131. 28.9%

5–133. *a.* $\frac{2}{17}$ *b.* $\frac{11}{850}$ *c.* $\frac{1}{5525}$

5–135. $\frac{5}{13}$

5–137. 0.4

5–139. 0.51

5–141. 57.3%

5–143. *a.* $\frac{6}{67}$ *b.* $\frac{7}{50}$

5–145. 99.7%

5–147. 55.6%

Quiz—Chapter 5

1. False
2. False
3. True
4. False
5. False
6. False
7. *b.*
8. *b* and *d*
9. *d.*
10. *b.*

11. *c.*
12. Sample space
13. Zero, one
14. Zero
15. One
16. Mutually exclusive

17. *a.* $\frac{1}{13}$ *b.* $\frac{1}{13}$ *c.* $\frac{4}{13}$

18. *a.* $\frac{1}{4}$ *b.* $\frac{4}{13}$ *c.* $\frac{1}{52}$ *d.* $\frac{1}{13}$ *e.* $\frac{1}{2}$

19. *a.* $\frac{12}{31}$ *b.* $\frac{12}{31}$ *c.* $\frac{27}{31}$ *d.* $\frac{24}{31}$

20. *a.* $\frac{11}{36}$ *b.* $\frac{5}{18}$ *c.* $\frac{11}{36}$ *d.* $\frac{1}{3}$ *e.* 0   *f.* $\frac{11}{12}$

21. 0.84
22. 0.002

23. *a.* $\frac{253}{9996}$ *b.* $\frac{33}{66,640}$ *c.* 0

24. 0.54
25. 0.53
26. 0.81
27. 0.056

28. *a.* $\frac{1}{2}$ *b.* $\frac{3}{7}$
29. 0.99
30. 0.518
31. 0.999886

Chapter 6

6–1. A random variable is a variable whose values are determined by chance. Examples will vary.

6–3. The number of commercials a radio station plays during each hour. The number of times a student uses his or her calculator during a mathematics exam. The number of leaves on a specific type of tree.

6–5. A probability distribution is a distribution that consists of the values a random variable can assume along with the corresponding probabilities of these values.

6–7. Yes

6–9. Yes

6–11. No, the probability values cannot be greater than one.

6–13. Discrete

6–15. Continuous

6–17. Discrete

6–19.

X	0	1	2	3
P(X)	$\frac{6}{15}$	$\frac{5}{15}$	$\frac{3}{15}$	$\frac{1}{15}$

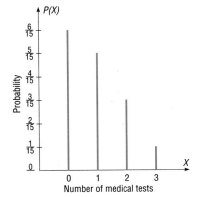

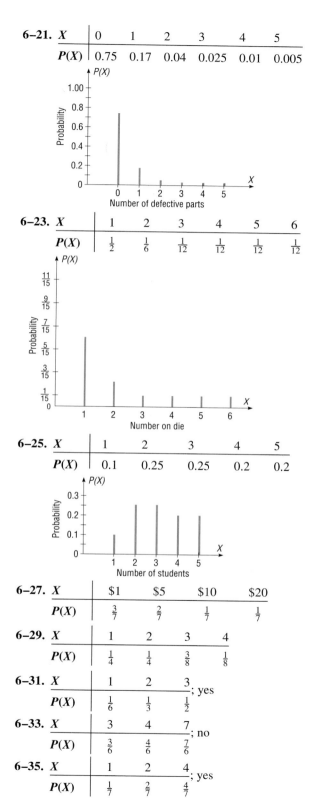

6–21.

X	0	1	2	3	4	5
P(X)	0.75	0.17	0.04	0.025	0.01	0.005

6–23.

X	1	2	3	4	5	6
P(X)	$\frac{1}{2}$	$\frac{1}{6}$	$\frac{1}{12}$	$\frac{1}{12}$	$\frac{1}{12}$	$\frac{1}{12}$

6–25.

X	1	2	3	4	5
P(X)	0.1	0.25	0.25	0.2	0.2

6–27.

X	$1	$5	$10	$20
P(X)	$\frac{3}{7}$	$\frac{2}{7}$	$\frac{1}{7}$	$\frac{1}{7}$

6–29.

X	1	2	3	4
P(X)	$\frac{1}{4}$	$\frac{1}{4}$	$\frac{3}{8}$	$\frac{1}{8}$

6–31.

X	1	2	3
P(X)	$\frac{1}{6}$	$\frac{1}{3}$	$\frac{1}{2}$

; yes

6–33.

X	3	4	7
P(X)	$\frac{3}{6}$	$\frac{4}{6}$	$\frac{7}{6}$

; no

6–35.

X	1	2	4
P(X)	$\frac{1}{7}$	$\frac{2}{7}$	$\frac{4}{7}$

; yes

6–37. 0.2; 0.3; 0.6
6–39. 9.7; 1.6; 1.3
6–41. 2; 1.6; 1.3
6–43. 6.6; 1.3; 1.1
6–45. 13.9; 1.3; 1.1
6–47. −$3.00
6–49. $0.83
6–51. −$1.00
6–53. −$0.50; −$0.52
6–55. All answers are −$0.05
6–57. 10.5
6–59. Answers will vary.
6–61. Answers will vary.
6–63. *a.* Yes *c.* Yes *e.* No *g.* Yes *i.* No
        b. Yes *d.* No *f.* Yes *h.* Yes *j.* Yes
6–64. *a.* 0.420 *c.* 0.590 *e.* 0.000 *g.* 0.418 *i.* 0.246
        b. 0.346 *d.* 0.251 *f.* 0.250 *h.* 0.176
6–65. *a.* 0.0005 *c.* 0.342 *e.* 0.173
        b. 0.131 *d.* 0.007
6–67. 0.377
6–69. 0.267
6–71. 0.071
6–73. *a.* 0.346 *b.* 0.913 *c.* 0.663 *d.* 0.683
6–75. *a.* 0.878 *b.* 0.201 *c.* 0.033
6–76. *a.* 75; 18.8; 4.3   *e.* 100; 90; 9.5
        b. 90; 63; 7.9   *f.* 125; 93.75; 9.7
        c. 10; 5; 2.2   *g.* 20; 12; 3.5
        d. 8; 1.6; 1.3   *h.* 6; 5; 2.2
6–77. 8; 7.9; 2.8
6–79. 10; 9.8; 3.1
6–81. 210; 165.9; 12.9
6–83. 0.199
6–85. 0.559
6–87. 0.018
6–89. 0.770
6–91.

X	0	1	2	3	4	5
P(X)	0.328	0.410	0.205	0.051	0.006	0.00

6–93. No; the sum of the probabilities is greater than one.

6–95. No; the sum of the probabilities is greater than one.

6–97.

X	0	1	2	3	4
$P(X)$	0.05	0.30	0.45	0.12	0.08

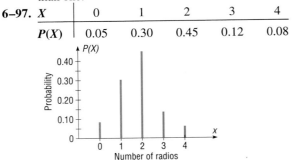

6–99.

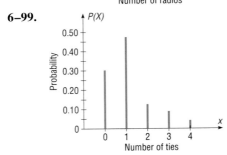

6–101. 9.9; 1.5; 1.2

6–103. 24.2; 1.5; 1.2

6–105. \$7.23

6–107. *a.* 0.122   *b.* 0.989   *c.* 0.043

6–109. 135; 33.8; 5.8

6–111. 0.886

6–113. 0.190

Quiz—Chapter 6

1. True
2. False
3. False
4. True
5. Chance
6. $n \cdot p$
7. One
8. *c.*
9. *c.*
10. No
11. Yes
12. Yes
13. Yes

14.

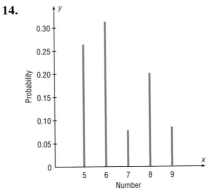

15.

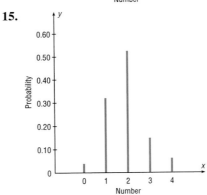

16. 2.0; 1.3; 1.1
17. 32.2; 1.1; 1.0
18. 5.2
19. \$9.65
20. 0.124
21. *a.* 0.075
    b. 0.872
    c. 0.126
22. 240; 48; 6.9
23. 9; 7.9; 2.8

Chapter 7

7–1. The characteristics of the normal distribution are as follows:
   a. It is bell-shaped.
   b. It is symmetric about the mean.
   c. Its mean, median, and mode are equal.
   d. It is continuous.
   e. It never touches the *x* axis.
   f. The area under the curve is equal to 1.
   g. It is unimodal.

7–3. 1, or 100%

7–5. 68%; 95%; 99.7%

7–7. 0.2123

7–9. 0.4808

7–11. 0.4090

7–13. 0.0764

7–15. 0.1145

7–17. 0.0258

7–19. 0.8417

7–21. 0.9846

7–23. 0.5714

7–25. 0.3015

7–27. 0.2486

7–29. 0.4418

7–31. 0.0023

7–33. 0.0655

7–35. 0.9522

7–37. 0.0706

7–39. 0.9222

7–41. −1.94

7–43. −2.13

7–45. −1.26

7–47. −1.86

7–49. *a.* $z = +1.96$ and $z = -1.96$
b. $z = +1.65$ and $z = -1.65$, approximately
c. $z = +2.58$ and $z = -2.58$, approximately

7–51.

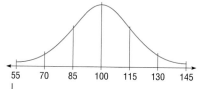

55   70   85   100   115   130   145

7–53.

x	−2	−1.5	−1	−0.5	0	0.5	1	1.5	2
y	0.05	0.13	0.24	0.35	0.4	0.35	0.24	0.13	0.05

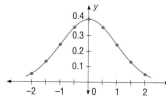

7–55. *a.* 0.3859   *b.* 0.0838

7–57. *a.* 0.0427   *b.* 0.0537

7–59. *a.* 0.0049   *b.* 0.1624   *c.* 0.7486

7–61. *a.* 0.1525   *b.* 0.7745   *c.* 0.1865

7–63. *a.* 0.9772   *b.* 0.6915

7–65. *a.* 0.5691   *b.* 0.6412

7–67. *a.* 0.9938   *b.* 0.1894   *c.* 0.9637

7–69. *a.* 0.5987   *b.* 0.8413   *c.* 0.2432

7–71. 89.95

7–73. The maximum size is 1927.76 square feet; the minimum size is 1692.24 square feet.

7–75. 62.21 minutes

7–77. The maximum price is $9,222 and the minimum price is $7,290.

7–79. 76.18

7–81. $18,840.48

7–83. 18.6 months

7–85. *a.* $\mu = 120$, $\sigma = 20$ *b.* $\mu = 15$, $\sigma = 2.5$
c. $\mu = 30$, $\sigma = 5$

7–87. There are several mathematics tests that can be used.

7–89. 2.59

7–91. $\mu = 45$, $\sigma = 1.34$

7–93. The distribution is called the sampling distribution of sample means.

7–95. The mean of the sample means is equal to the population mean.

7–97. The distribution will be approximately normal when the sample size is large.

7–99. $z = \dfrac{\overline{X} - \mu}{\sigma / n}$

7–101. 0.2486

7–103. 0.2327

7–105. 0.0571

7–107. 0.8239

7–109. 0.0951

7–111. 0.1254

7–113. *a.* 0.3446   *c.* Yes
b. 0.0023   *d.* Very unlikely

7–115. *a.* 0.3707   *b.* 0.0475

7–117. *a.* 0.1815   *c.* Means are less variable than
b. 0.3854   individual data.

7–119. $\sigma_{\overline{X}} = 1.5$, $n = 25$

7–121. When p is approximately 0.5, as n increases, the shape of the binomial distribution becomes similar to the normal distribution. The conditions are $n \cdot p$ and $n \cdot q$ both ≥ 5. The correction is necessary because the normal distribution is continuous and the binomial distribution is discrete.

7–122. *a.* 0.0811   *c.* 0.1052   *e.* 0.2327
b. 0.0516   *d.* 0.1711   *f.* 0.9988

7–123. *a.* Yes   *c.* No   *e.* Yes
b. No   *d.* Yes   *f.* No

7–125. 0.166

7–127. *a.* 0.9793   *b.* 0.0094

7–129. 0.1034

7–131. 0.6844

7–133. 0.9875

7–135. *a.* 0.4744   *e.* 0.2139   *i.* 0.0183
b. 0.1443   *f.* 0.8284   *j.* 0.9535
c. 0.0590   *g.* 0.0233
d. 0.8329   *h.* 0.9131

7–137. *a.* 0.2039   *b.* 0.0099   *c.* 0.0918

7–139. *a.* 0.4649 *b.* 0.0228 *c.* 0.1894
7–141. *a.* 0.5745 *b.* 0.2090 *c.* 0.4483
7–143. 92.2–107.8
7–145. 0.0351
7–147. 0.2033
7–149. 0.0465

Quiz—Chapter 7

1. False
2. True
3. True
4. True
5. False
6. False
7. *a.*
8. *a.*
9. *b.*
10. *b.*
11. *c.*
12. 0.5
13. Sampling error
14. The population mean
15. Standard error of the mean
16. 5
17. left

18. *a.* 0.4332 *d.* 0.1029 *g.* 0.0401 *j.* 0.9131
    b. 0.3944 *e.* 0.2912 *h.* 0.8997
    c. 0.0344 *f.* 0.8284 *i.* 0.017
19. *a.* 0.4846 *d.* 0.0188 *g.* 0.0089 *j.* 0.8461
    b. 0.4693 *e.* 0.7461 *h.* 0.9582
    c. 0.9334 *f.* 0.0384 *i.* 0.9788
20. *a.* 0.0531 *b.* 0.1056 *c.* 0.1056 *d.* 0.0994
21. *a.* 0.0668 *b.* 0.0228 *c.* 0.4649 *d.* 0.0934
22. *a.* 0.4525 *b.* 0.3707 *c.* 0.3707 *d.* 0.019
23. *a.* 0.0013 *b.* 0.5 *c.* 0.0081 *d.* 0.5511
24. *a.* 0.0037 *b.* 0.0228 *c.* 0.5 *d.* 0.3232
25. 8.804 centimeters
26. 121.24 is the lowest acceptable score.
27. 0.015   30. 0.0630
28. 0.9738   31. 0.0336
29. 0.2296   32. 0.0838

Chapter 8

8–1. A point estimate of a parameter specifies a specific value, such as $\mu = 87$; an interval estimate specifies a range of values for the parameter, such as $84 < \mu < 90$. The advantage of an interval estimate is that a specific confidence level (say 95%) can be selected, and one can be 95% confident that the interval contains the parameter that is being estimated.
8–3. The maximum error of estimate is the range of values to the right or left of the statistic which may contain the parameter.
8–5. A good estimator should be unbiased, consistent, and relatively efficient.

8–7. For one to be able to determine sample size, the maximum error of estimate and the degree of confidence must be specified and the population standard deviation must be known.
8–9. *a.* 2.58 *b.* 2.33 *c.* 1.96 *d.* 1.65 *e.* 1.88
8–11. *a.* $77 < \mu < 87$ *b.* $75 < \mu < 89$
    c. The 99% confidence interval is larger because the confidence level is larger.
8–13. *a.* $11.9 < \mu < 13.3$
    b. It would be highly unlikely, since this is far larger than 13.3.
8–15. $18.13 < \mu < 18.87
8–17. *a.* $37 < \mu < 39$ *b.* $35 < \mu < 41$
    c. The confidence interval for part *b* is larger because the standard deviation is larger.
8–19. $59.5 < \mu < 62.9$
8–21. 45
8–23. 25
8–25. 5
8–27. W. S. Gossett
8–29. The *t* distribution should be used when σ is unknown and $n < 30$.
8–30. *a.* 2.898 *c.* 2.624 *e.* 2.093
    b. 2.074 *d.* 1.833
8–31. $15 < \mu < 17$
8–33. $20 < \mu < 22$
8–35. $11{,}990 < \mu < 12{,}410$
8–37. $8.7 < \mu < 9.9$
8–39. $17.29 < \mu < 19.77
8–41. $109 < \mu < 121$
8–43. $312.07 < \mu < 328.33
8–45. $58{,}197.00 < \mu < $58{,}241.00$
8–47. *a.* 0.5, 0.5 *c.* 0.46, 0.54 *e.* 0.45, 0.55
    b. 0.45, 0.55 *d.* 0.58, 0.42
8–48. *a.* 0.12, 0.88 *c.* 0.65, 0.35 *e.* 0.67, 0.33
    b. 0.29, 0.71 *d.* 0.53, 0.47
8–49. $0.365 < p < 0.415$
8–51. $0.557 < p < 0.743$
8–53. $0.144 < p < 0.296$
8–55. $0.153 < p < 0.307$
8–57. $0.337 < p < 0.543$
8–59. $0.149 < p < 0.351$
8–61. 3121
8–63. 99
8–65. 95%
8–67. chi-square

8–69. *a.* 6.262; 27.488   *d.* 15.308; 44.461
  b. 0.711; 9.488   *e.* 5.892; 22.362
  c. 8.643; 42.796

8–71. $30.9 < \sigma^2 < 78.2$
  $5.6 < \sigma < 8.8$

8–73. $13 < \sigma^2 < 214$
  $4 < \sigma < 15$

8–75. $1.2 < \sigma^2 < 7.7$
  $1.1 < \sigma < 2.8$

8–77. $4.1 < \sigma < 7.1$

8–79. $3.2 < \sigma^2 < 33.1$
  $1.79 < \sigma < 5.75$

8–81. $\$1,240.00 < \mu < \$1,260.00$

8–83. $7.46 < \mu < 7.54$

8–85. $25 < \mu < 31$

8–87. 28

8–89. $0.343 < p < 0.457$

8–91. 1418

8–93. $0.25 < \sigma < 0.51$

8–95. $0.5 < \sigma^2 < 21.5$

Quiz—Chapter 8

 1. True
 2. True
 3. False
 4. True
 5. *b.*
 6. *a.*
 7. *b.*
 8. Unbiased, consistent, relatively efficient
 9. Maximum error of estimate
10. Point
11. 90; 95; 99
12. $114 < \mu < 118$
13. $\$43.15 < \mu < \$46.45$
14. $3954 < \mu < 4346$
15. $45.7 < \mu < 51.5$
16. $418 < \mu < 458$
17. $26 < \mu < 36$
18. 155
19. 25
20. $0.604 < p < 0.810$
21. $0.295 < p < 0.425$
22. $0.441 < p < 0.559$
23. 545
24. $7 < \sigma < 13$
25. $3 < \sigma^2 < 47$
26. $1.8 < \sigma < 3.2$

Chapter 9

9–1. The null hypothesis states that there is no difference between a parameter and a specific value or that there is no difference between two parameters. The alternative hypothesis states that there is a specific difference between a parameter and a specific value or that there is a difference between two parameters. Examples will vary.

9–3. A statistical test uses the data obtained from a sample to make a decision about whether or not the null hypothesis should be rejected.

9–5. The critical region is the range of values of the test statistic that indicates that there is a significant difference and the null hypothesis should be rejected. The noncritical region is the range of values of the test statistic that indicates that the difference was probably due to chance and the null hypothesis should not be rejected.

9–7. α, β

9–9. A one-tailed test should be used when a specific direction, such as greater than or less than, is being hypothesized; when no direction is specified, a two-tailed test should be used.

9–11. Hypotheses can be proved true only when the entire population is used to compute the test statistic. In most cases, this is impossible.

9–12. *a.* ± 2.58 *d.* -1.28 *g.* -2.33 *i.* $+2.05$
  b. $+1.65$ *e.* ± 1.96 *h.* ± 1.65 *j.* ± 2.33
  c. -2.58 *f.* $+1.75$

9–13. *a.* H_0: $\mu = 36.3$ (claim), H_1: $\mu \neq 36.3$
  b. H_0: $\mu = \$36,250$ (claim), H_1: $\mu \neq \$36,250$
  c. H_0: $\mu \leq 27.6$, H_1: $\mu > 27.6$ (claim)
  d. H_0: $\mu \geq 72$, H_1: $\mu < 72$ (claim)
  e. H_0: $\mu \geq 100$, H_1: $\mu < 100$ (claim)
  f. H_0: $\mu = \$297.75$ (claim), H_1: $\mu \neq \$297.75$
  g. H_0: $\mu \leq \$52.98$, H_1: $\mu > \$52.98$ (claim)
  h. H_0: $\mu \leq 300$ (claim), H_1: $\mu > 300$
  i. H_0: $\mu \geq 3.6$ (claim), H_1: $\mu < 3.6$

9–15. H_0: $\mu = \$69.21$ (claim) and H_1: $\mu \neq \$69.21$; C.V. $= \pm 1.96$; $z = -1.15$; do not reject. There is not enough evidence to reject the claim that the average cost of a hotel stay in Atlanta is $69.21.

9–17. H_0: $\mu \leq 483$ and H_1: $\mu > 483$ (claim); C.V. $= +1.65$; $z = 0.62$; do not reject. No; there is not enough evidence to support the claim that the course increases SAT scores.

9–19. H_0: $\mu \geq 14$ and H_1: $\mu < 14$ (claim); C.V. $= -2.33$; $z = -4.89$; reject. There is enough evidence to support the claim that the average age of the planes in the executive's airline is less than the national average.

9–21. H_0: $\mu = \$915$ (claim) and H_1: $\mu \neq \$915$; C.V. $= \pm 1.96$; $z = 2.66$; reject. There is enough evidence to reject the agent's claim that the average cost of the trip is $915.

9–23. H_0: $\mu = 36$ (claim) and H_1: $\mu \neq 36$; C.V. $= \pm 2.58$; $z = -3.54$; reject. Yes; there is enough evidence to reject the claim that the average lifetime of the lightbulbs is 36 months.

9–25. H_0: $\mu \geq 240$ and H_1: $\mu < 240$ (claim); C.V. $= -2.33$; $z = -3.87$; reject. Yes; there is enough evidence to support the claim that the medication lowers the cholesterol level.

9–27. H_0: $\mu = \$24.44$ and H_1: $\mu \neq \$24.44$ (claim); C.V. $= \pm 2.33$; $z = -2.28$; do not reject. There is not enough evidence to support the claim that the amount spent at a local mall is not equal to the national average of $24.44.

9–29. *a.* Do not reject.   *d.* Reject.
b. Do not reject.   *e.* Reject.
c. Do not reject.

9–31. H_0: $\mu \geq 264$ and H_1: $\mu < 264$ (claim); $z = -2.53$; P-value $= 0.0057$; reject. There is enough evidence to support the claim that the average stopping distance is less than 264 ft.

9–33. H_0: $\mu \leq 84$ and H_1: $\mu > 84$ (claim); $z = 1.1$; P-value $= 0.1357$; do not reject. There is not enough evidence to support the claim that the average lifetime is more than 84 months.

9–35. H_0: $\mu = 6.32$ (claim) and H_1: $\mu \neq 6.32$; $z = 2.49$; P-value $= 0.0128$; reject. There is enough evidence to reject the claim that the average wage is $6.32.

9–37. H_0: $\mu = 30,000$ (claim) and H_1: $\mu \neq 30,000$; $z = 1.71$; P-value $= 0.0872$; reject. There is enough evidence to reject the claim that the customers are adhering to the recommendation. Yes, the 0.10 level is appropriate.

9–39. $\overline{X} = 5.025$, $s = 3.63$; H_0: $\mu \geq 10$ and H_1: $\mu < 10$ (claim); C.V. $= -1.65$; $z = -8.67$; reject. Yes; there is enough evidence to support the claim that the average number of days missed per year is less than 10.

9–41. H_0: $\mu = 8.65$ (claim) and H_1: $\mu \neq 8.65$; C.V. $= \pm 1.96$; $z = -1.35$; do not reject. Yes; there is not enough evidence to reject the claim that the average hourly wage of the employees is $8.65.

9–43. The t distribution differs from the standard normal distribution in that it is a family of curves and the variance is greater than 1; and as the degrees of freedom increase, the t distribution approaches the standard normal distribution.

9–45. The t test statistic uses s instead of σ when $n < 30$.

9–46. *a.* $+1.833$ *c.* -3.365 *e.* ± 2.145 *g.* ± 2.771
b. ± 1.740 *d.* $+2.306$ *f.* -2.819 *h.* ± 2.583

9–47. H_0: $\mu \geq 11.52$ and H_1: $\mu < 11.52$ (claim); C.V. $= -1.833$; d.f. $= 9$; $t = -9.97$; reject. There is enough evidence to support the claim that the amount of rainfall is below average.

9–49. H_0: $\mu = 800$ (claim) and H_1: $\mu \neq 800$; C.V. $= \pm 2.262$; d.f. $= 9$; $t = 9.96$; reject. Yes; there is enough evidence to reject the claim that

the average rent that small-business establishments pay in Eagle City is $800.

9–51. H_0: $\mu \geq 12$ and H_1: $\mu < 12$ (claim); C.V. $= -2.571$; d.f. $= 5$; $t = -0.47$; do not reject. No; there is not enough evidence to support the claim that the muffler can be changed in less than 12 minutes.

9–53. H_0: $\mu = \$750$ (claim) and H_1: $\mu \neq \$750$; C.V. $= \pm 3.106$; $t = -3.67$; reject. Yes; there is enough evidence to reject the claim that the average rent is $750.

9–55. H_0: $\mu \leq 350$ and H_1: $\mu > 350$ (claim); C.V. $= 1.796$; d.f. $= 11$; $t = 1.732$; do not reject. No; there is not enough evidence to support the claim that the average fine is higher than $350.

9–57. H_0: $\mu \geq 37$ (claim) and H_1: $\mu < 37$; C.V. $= -1.701$; d.f. $= 28$; $t = -1.88$; reject. There is enough evidence to reject the claim that the average household receives at least 37 phone calls per month.

9–59. $\overline{x} = 70.85$; $s = 6.56$; H_0: $\mu = 75$ (claim) and H_1: $\mu \neq 75$; C.V. $= \pm 2.861$; d.f. $= 19$; $t = -2.83$; do not reject. There is not enough evidence to reject the claim that the average score on the real estate exam is 75.

9–61. Answers will vary.

9–63. $np \geq 5$ and $nq \geq 5$

9–65. H_0: $p \geq 0.23$ (claim) and H_1: $p < 0.23$; $\mu = 9.2$; $\sigma = 2.66$; C.V. $= -1.65$; $z = -0.83$; do not reject. There is not enough evidence to reject the claim that at least 23% of the 14-year-old residents own a skateboard.

9–67. H_0: $p \geq 0.40$ (claim) and H_1: $p < 0.40$; $\mu = 32$; $\sigma = 4.38$; C.V. $= -1.28$; $z = -0.457$; do not reject. There is not enough evidence to reject the claim that at least 40% of the arsonists are under 21 years old.

9–69. H_0: $p = 0.63$ (claim) and H_1: $p \neq 0.63$; $\mu = 90.1$; $\sigma = 5.773$; C.V. $= \pm 1.96$; $z = -0.88$; do not reject. There is not enough evidence to reject the claim that the percentage is the same.

9–71. H_0: $p \geq 0.15$ (claim) and H_1: $p < 0.15$; $\mu = 12$; $\sigma = 3.19$; C.V. $= -1.65$; $z = -0.94$; do not reject. There is not enough evidence to reject the claim that at least 15% of all eighth-grade students are overweight.

9–73. H_0: $p \leq 0.30$ and H_1: $p > 0.30$ (claim); $\mu = 60$; $\sigma = 6.48$; C.V. $= +1.65$; $z = +1.85$; reject. Yes; there is enough evidence to support the claim that more than 30% of the customers have at least two telephones.

9–75. H_0: $p = 0.18$ (claim) and H_1: $p \neq 0.18$; $\mu = 54$; $\sigma = 6.654$; C.V. $= \pm 1.96$; $z = -0.60$; do not reject. There is not enough evidence to reject the claim that 18% of all high school students smoke at least a pack of cigarettes a day.

9–77. Yes

9–79. a. H_0: $\sigma^2 \leq 225$ and H_1: $\sigma^2 > 225$; C.V. $= 27.587$; d.f. $= 17$
b. H_0: $\sigma^2 \geq 225$ and H_1: $\sigma^2 < 225$; C.V. $= 14.042$; d.f. $= 22$
c. H_0: $\sigma^2 = 225$ and H_1: $\sigma^2 \neq 225$; C.V. $= 5.629$; 26.119; d.f. $= 14$
d. H_0: $\sigma^2 = 225$ and H_1: $\sigma^2 \neq 225$; C.V. $= 2.167$; 14.067; d.f. $= 7$
e. H_0: $\sigma^2 \leq 225$ and H_1: $\sigma^2 > 225$; C.V. $= 32.000$; d.f. $= 16$
f. H_0: $\sigma^2 \geq 225$ and H_1: $\sigma^2 < 225$; C.V. $= 8.907$; d.f. $= 19$
g. H_0: $\sigma^2 = 225$ and H_1: $\sigma^2 \neq 225$; C.V. $= 3.074$; 28.299; d.f. $= 12$
h. H_0: $\sigma^2 \geq 225$ and H_1: $\sigma^2 < 225$; C.V. $= 15.308$; d.f. $= 28$

9–81. H_0: $\sigma = 16.8$ and H_1: $\sigma \neq 16.8$ (claim); C.V. $= 13.091$, 35.172; d.f. $= 23$; $\chi^2 = 12.733$; reject. There is enough evidence to support the claim that the standard deviation has changed.

9–83. H_0: $\sigma^2 \leq 25$ (claim) and H_1: $\sigma^2 > 25$; C.V. $= 27.204$; $\alpha = 0.10$; d.f. $= 19$; $\chi^2 = 27.36$; reject. There is enough evidence to reject the claim that the variance is less than or equal to 25.

9–85. H_0: $\sigma \leq 1.2$ (claim) and H_1: $\sigma > 1.2$; C.V. $= 29.141$; $\alpha = 0.01$; d.f. $= 14$; $\chi^2 = 31.5$; reject. There is enough evidence to reject the claim that the standard deviation is less than or equal to 1.2 minutes.

9–87. H_0: $\sigma \leq 2$ and H_1: $\sigma > 2$ (claim); $s = 2.83$; C.V. $= 24.725$; $\alpha = 0.01$; d.f. $= 11$; $\chi^2 = 22.02$; do not reject. The lot is acceptable, since there is not enough evidence to support the claim that the standard deviation is greater than 2 pounds.

9–89. H_0: $\mu = 1800$ (claim) and H_1: $\mu \neq 1800$; C.V. $= \pm 1.96$; $z = 0.47$; $1706.04 < \mu < 1953.96$; do not reject. There is not enough evidence to reject the claim that the average of the sales is $1800.

9–91. H_0: $\mu = 86$ (claim) and H_1: $\mu \neq 86$; C.V. $= \pm 2.58$; $z = -1.29$; $80.00 < \mu < 88.00$; do not reject. There is not enough evidence to reject the claim that the average monthly maintenance fee is $86.

9–93. H_0: $\mu = 22$ and H_1: $\mu \neq 22$ (claim); C.V. $= \pm 2.58$; $z = -2.32$; $19.47 < \mu < 22.13$; do not reject. There is not enough evidence to support the claim that the average has changed.

9–95. H_0: $\mu = \$16,411.00$ and H_1: $\mu \neq \$16,411.00$ (claim); C.V. $= \pm 1.96$; $z = -\$5.35$; reject. There is enough evidence to support the claim that the average cost is different.

9–97. H_0: $\mu = 14$ (claim) and H_1: $\mu \neq 14$; C.V. $= \pm 1.96$; $z = 4.22$; reject. No, there is enough evidence to reject the claim that the average number of cigarettes a person smokes a day is 14.

9–99. H_0: $\mu = 61.2$ (claim) and H_1: $\mu \neq 61.2$; C.V. $= \pm 2.831$; $t = -4.378$; reject. No, there is enough evidence to reject the claim that the average age is 61.2.

9–101. H_0: $\mu \leq 23.2$ (claim) and H_1: $\mu > 23.2$; C.V. $= 1.74$; d.f. $= 17$; $t = -1.27$; do not reject. No, there is not enough evidence to reject the claim that the average age is less than or equal to 23.2 years.

9–103. H_0: $p \geq 0.3$ (claim) and H_1: $p < 0.3$; $\mu = 18$; $\sigma = 3.55$; C.V. $= -1.65$; $z = -0.85$; do not reject. No, there is not enough evidence to reject the claim that at least 30% of the students are receiving financial aid.

9–105. H_0: $p = 0.80$ (claim) and H_1: $p \neq 0.80$; $\mu = 24$; $\sigma = 2.19$; C.V. $= \pm 2.33$; $z = -1.83$; do not reject. Yes, there is not enough evidence to reject the claim that 80% of the new-home buyers wanted a fireplace.

9–107. H_0: $\mu = 225$ (claim) and H_1: $\mu \neq 225$; $z = 2.36$; P-value $= 0.0182$; do not reject. There is not enough evidence to reject the claim that the mean is 225 pounds.

9–109. H_0: $\sigma = 3.4$ (claim) and H_1: $\sigma \neq 3.4$; C.V. $= 11.689$ and 38.076; d.f. $= 23$; $\chi^2 = 35.1$; do not reject. No, there is not enough evidence to reject the claim that the standard deviation is 3.4 minutes.

9–111. H_0: $\sigma = 18$ (claim) and H_1: $\sigma \neq 18$; C.V. $= 11.143$ and 0.484; d.f. $= 4$; $\chi^2 = 5.44$; do not reject. There is not enough evidence to reject the claim that the standard deviation is 21 minutes.

9–113. H_0: $\mu = 4$ and H_1: $\mu \neq 4$ (claim); C.V. $= \pm 2.58$; $z = 1.49$; $3.85 < \mu < 4.55$; do not reject. There is not enough evidence to support the claim that the growth has changed.

Quiz—Chapter 9

1. True
2. True
3. False
4. True
5. False
6. b.
7. d.

8. c.
9. b.
10. Type I
11. β
12. Statistical hypothesis
13. Right
14. $n - 1$

15. H_0: $\mu = 28.6$ (claim) and H_1: $\mu \neq 28.6$; $z = 1.16$; C.V. $= \pm 1.65$; do not reject. There is not enough evidence to reject the claim that the average is 28.6.

16. H_0: $\mu = \$6500$ (claim) and H_1: $\mu \neq \$6500$; $z = 5.27$; C.V. $= \pm 1.96$; reject. There is enough evidence to reject the agent's claim.

17. H_0: $\mu \leq 8$ and H_1: $\mu > 8$ (claim); $z = 6$; C.V. $= 1.65$; reject. There is enough evidence to support the claim that the average is greater than 8.

18. H_0: $\mu = 21$ (claim) and H_1: $\mu \neq 21$; $t = -6.32$; C.V. $= \pm 2.921$; reject. There is enough evidence to reject the claim that the average is 21.

19. H_0: $\mu \geq 67$ and H_1: $\mu < 67$ (claim); $t = -3.1568$; C.V. $= -1.729$; reject. There is enough evidence to support the claim that the average height is less than 67 inches.

20. H_0: $\mu \geq 12.4$ and H_1: $\mu < 12.4$ (claim); $t = -2.324$; C.V. $= -1.345$; reject. There is enough evidence to support the claim that the average is less than the company claimed.

21. H_0: $\mu \leq 63.5$ and H_1: $\mu > 63.5$ (claim); $t = 0.47075$; C.V. $= 1.729$; do not reject. There is not enough evidence to support the claim that the average is greater than 63.5.

22. H_0: $\mu = 26$ (claim) and H_1: $\mu \neq 26$; $t = -1.5$; C.V. $= \pm 2.492$; do not reject. There is not enough evidence to reject the claim that the average is 26.

23. H_0: $p \geq 0.25$ (claim) and H_1: $p < 0.25$; $z = -0.6928$; C.V. $= -1.65$; do not reject. There is not enough evidence to reject the claim that the proportion is at least 0.25.

24. H_0: $p \geq 0.55$ (claim) and H_1: $p < 0.55$; $z = -0.8989$; C.V. $= -1.28$; do not reject. There is not enough evidence to reject the survey's claim.

25. H_0: $p = 0.7$ (claim) and H_1: $p \neq 0.7$; $z = 0.7968$; C.V. $= \pm 2.33$; do not reject. There is not enough evidence to reject the claim that the proportion is 0.7.

26. H_0: $p = 0.75$ (claim) and H_1: $p \neq 0.75$; $z = 2.6833$; C.V. $= \pm 2.58$; reject. There is enough evidence to reject the claim.

27. $P = 0.246$

28. $P = 0.0002$

29. H_0: $\sigma \leq 6$ and H_1: $\sigma > 6$ (claim); $\chi^2 = 54$; C.V. $= 36.415$; reject. There is enough evidence to support the claim.

30. H_0: $\sigma = 8$ (claim) and H_1: $\sigma \neq 8$; $\chi^2 = 33.2$; C.V. $= 27.991, 79.490$; do not reject. There is not enough evidence to reject the claim that $\sigma = 8$.

31. H_0: $\sigma \geq 2.3$ and H_1: $\sigma < 2.3$ (claim); $\chi^2 = 13$; C.V. $= 10.117$; Do not reject. There is not enough evidence to support the claim that the standard deviation is less than 2.3.

32. H_0: $\sigma = 9$ (claim) and H_1: $\sigma \neq 9$; $\chi^2 = 13.4$; C.V. $= 2.7, 19.023$; do not reject. There is not enough evidence to reject the claim that $\sigma = 9$.

33. $28.3 < \mu < 30.1$

34. $\$6562.81 < \mu < \$6637.19$

Chapter 10

10–1. Testing a single mean involves comparing a sample mean to a specific value such as $\mu = 100$; testing the difference between two means involves comparing the difference of the means of two samples, to $\mu_1 - \mu_2$.

10–3. The samples must be independent of each other, and the populations must be normally distributed; s_1 and s_2 can be used in place of σ_1 and σ_2 when σ_1 and σ_2 are unknown and both samples are each greater than or equal to 30.

10–5. H_0: $\mu_1 = \mu_2$ and H_1: $\mu_1 \neq \mu_2$ (claim); C.V. $= \pm 2.58$; $z = -2.5$; do not reject. There is not enough evidence to support the claim that there is a difference in the speeds of the two companies.

10–7. H_0: $\mu_1 \leq \mu_2$ and H_1: $\mu_1 > \mu_2$ (claim); C.V. $= +1.65$; $z = 2.56$; reject. Yes, there is enough evidence to support the claim that pulse rates of smokers are higher than pulse rates of nonsmokers.

10–9. H_0: $\mu_1 \leq \mu_2$ and H_1: $\mu_1 > \mu_2$ (claim); C.V. $= +2.05$; $z = 1.12$; do not reject. There is not enough evidence to support the claim that the noise levels in the corridors are higher than the noise levels in the clinics.

10–11. H_0: $\mu_1 \geq \mu_2$ and H_1: $\mu_1 < \mu_2$ (claim); C.V. $= -1.65$; $z = -2.01$; reject. There is enough evidence to support the claim that the stayers had a higher grade point average.

10–13. H_0: $\mu_1 = \mu_2$ and H_1: $\mu_1 \neq \mu_2$ (claim); C.V. $= \pm 1.65$; $z = -20.97$; reject. There is enough evidence to support the claim that there is a difference in the cranking power of the automobile batteries.

10–15. H_0: $\mu_1 = \mu_2$ (claim) and H_1: $\mu_1 \neq \mu_2$;
C.V. = ± 1.96; $z = 1.01$; do not reject. There is
not enough evidence to reject the claim of no
difference.

10–17. H_0: $\mu_1 - \mu_2 \leq 8$ (claim) and H_1: $\mu_1 - \mu_2 > 8$;
C.V. = $+1.65$; $z = -0.73$; do not reject. There
is not enough evidence to reject the claim that
private-school students have an IQ that is at
most 8 points higher than that of students in
public schools.

10–19. $-7.3 < \mu_1 - \mu_2 < -1.3$

10–21. $2.8 < \mu_1 - \mu_2 < 6.0$

10–23. It should be the larger of the two variances.

10–25. The F test is used to test the equality of two
variances.

10–27. The characteristics of the F distribution are:
  a. The values of the F cannot be negative.
  b. The distribution is positively skewed.
  c. The mean value of the F is approximately
     equal to one.
  d. The F is a family of curves based on the
     degrees of freedom.

10–29. H_0: $\sigma_1^2 \leq \sigma_2^2$; H_1: $\sigma_1^2 > \sigma_2^2$ (claim); $\alpha = 0.05$;
C.V. = 2.23; d.f.N. = 19; d.f.D. = 19;
$F = 1.41$; do not reject. There is not enough
evidence to support the claim that variance of
the exam scores of the students who had the
word processing will be larger than the
variance of the exam scores of the students who
did not have word processing in conjunction
with a composition course.

10–31. H_0: $\sigma_1^2 \leq \sigma_2^2$; H_1: $\sigma_1^2 > \sigma_2^2$ (claim);
C.V. = 2.23; $\alpha = 0.05$; d.f.N. = 21;
d.f.D. = 17; $F = 2.29$; reject. There is enough
evidence to support the claim that variance
of the IQ scores of women who major in
psychology is larger than the variance of the
IQ scores of men who major in psychology.

10–33. H_0: $\sigma_1^2 = \sigma_2^2$; H_1: $\sigma_1^2 \neq \sigma_2^2$ (claim); $\alpha = 0.10$;
C.V. = 2.53; d.f.N. = 14; d.f.D. = 14;
$F = 2.09$; do not reject. There is not enough
evidence to support the claim that variations of
the lengths of newborn males differ from the
variations of the lengths of newborn females.

10–35. H_0: $\sigma_1^2 \leq \sigma_2^2$; H_1: $\sigma_1^2 > \sigma_2^2$ (claim);
C.V. = 2.66; $\alpha = 0.01$; d.f.N. = 27;
d.f.D. = 24; $F = 5.27$; reject. There is enough
evidence to support the claim that the variation
of blood pressure of overweight individuals is
greater than the variation of blood pressure of
normal-weight individuals.

10–37. H_0: $\sigma_1^2 \leq \sigma_2^2$; H_1: $\sigma_1^2 > \sigma_2^2$ (claim); $\alpha = 0.05$;
C.V. = 2.90; d.f.N. = 9; d.f.D. = 11; $F = 1.72$;
do not reject. There is not enough evidence to
support the claim that the variation in pounds
lost following Diet A is greater than the
variation in pounds lost from Diet B.

10–39. H_0: $\sigma_1^2 = \sigma_2^2$; H_1: $\sigma_1^2 \neq \sigma_2^2$ (claim);
C.V. = 6.42; $\alpha = 0.01$; d.f.N. = 11; d.f.D. = 9;
$F = 1.88$; do not reject. There is not enough
evidence to support the claim that the variances
are different.

10–41. H_0: $\sigma_1^2 = \sigma_2^2$ (claim); H_1: $\sigma_1^2 \neq \sigma_2^2$;
C.V. = 3.05; $\alpha = 0.05$; d.f.N. = 14;
d.f.D. = 14; $F = 5.321$; reject. There is enough
evidence to reject the claim that the variances
are equal.

10–43. H_0: $\sigma_1^2 = \sigma_2^2$; C.V. = 4.03; $F = 1.93$; do not
reject. H_0: $\mu_1 = \mu_2$ and H_1: $\mu_1 \neq \mu_2$ (claim);
C.V. = ± 2.101; d.f. = 18; $t = -4.02$; reject.
There is enough evidence to support the claim
that there is a significant difference in the
values of the homes based on their values.
$-\$7762 < \mu_1 - \mu_2 < -\$2434$

10–45. H_0: $\sigma_1^2 = \sigma_2^2$; C.V. = 2.72; $F = 5.06$; reject.
H_0: $\mu_1 \geq \mu_2$ and H_1: $\mu_1 < \mu_2$ (claim);
C.V. = -1.363; d.f. = 11; $t = -1.24$; do not
reject. There is not enough evidence to support
the claim that math majors can write programs
faster than business majors.

10–47. H_0: $\sigma_1^2 = \sigma_2^2$; C.V. = 6.54; $F = 1.31$; do not
reject. H_0: $\mu_1 \geq \mu_2$ and H_1: $\mu_1 < \mu_2$ (claim);
C.V. = -2.552; d.f. = 18; $t = -0.89$; do not
reject. There is not enough evidence to support
the claim that Brand A costs less than Brand B
to repair.

10–49. H_0: $\sigma_1^2 = \sigma_2^2$; C.V. = 4.19; $F = 1.70$; do not
reject. H_0: $\mu_1 = \mu_2$ and H_1: $\mu_1 \neq \mu_2$ (claim);
C.V. = ± 2.508; d.f. = 22; $t = -2.97$; reject.
There is enough evidence to support the claim
that there is a difference in the average times of
the two groups. $-11.1 < \mu_1 - \mu_2 < -0.94$

10–51. H_0: $\sigma_1^2 = \sigma_2^2$; C.V. = 3.17; $F = 3.67$; reject.
H_0: $\mu_1 \geq \mu_2$ and H_1: $\mu_1 < \mu_2$ (claim);
C.V. = -1.796; d.f. = 11; $t = -4.98$; reject.
There is enough evidence to support the claim
that the nurses pay more for insurance than the
administrators.

10–53. H_0: $\sigma_1^2 = \sigma_2^2$; C.V. = 7.15; $F = 1.22$; do not
reject. White mice: $\overline{X}_1 = 17$, $s_1 = 4.56$; brown
mice: $\overline{X}_2 = 16.67$, $s_2 = 5.05$. H_0: $\mu_1 = \mu_2$ and
H_1: $\mu_1 \neq \mu_2$ (claim); C.V. = ± 2.228; d.f. = 10;
$t = 0.119$; do not reject. There is not enough
evidence to support the claim that the

color of the mice made a difference.
$-5.9 < \mu_1 - \mu_2 < 6.5$

10–55. a. Dependent   d. Dependent
b. Dependent   e. Independent
c. Independent

10–57. H_0: $\mu_D \geq 0$ and H_1: $\mu_D < 0$ (claim);
C.V. = -1.397; d.f. = 8; $t = -2.8$; reject.
There is enough evidence to support the claim
that the seminar increased the number of hours
students studied.

10–59. H_0: $\mu_D \leq 0$ and H_1: $\mu_D > 0$ (claim);
C.V. = $+1.833$; d.f. = 9; $t = 5.435$; reject.
There is enough evidence to support the claim
that the diet reduced the sodium level of the
patients.

10–61. H_0: $\mu_D \leq 0$ and H_1: $\mu_D > 0$ (claim);
C.V. = 2.571; d.f. = 5; $t = 2.24$; do not reject.
There is not enough evidence to support the
claim that the errors have been reduced.

10–63. H_0: $\mu_D = 0$ and H_1: $\mu_D \neq 0$ (claim);
C.V. = ±3.499; d.f. = 7; $t = 0.978$; do not
reject. There is not enough evidence to support
the claim that there is a difference in the pulse
rates. $-3.22 < \mu_D < 5.72$

10–65. Using the previous problem $\overline{D} = 8.5$, whereas
the mean of the before values is 89.8 and the
mean of the after values is 81.3; hence,
$\overline{D} = 89.8 - 81.3 = 8.5$.

10–67. a. 16 b. 4 c. 4.8 d. 104 e. 30

10–69. $\hat{p}_1 = 0.5333$; $\hat{p}_2 = 0.3$; $\overline{p} = 0.44$; $\overline{q} = 0.56$; H_0:
$p_1 = p_2$ and H_1: $p_1 \neq p_2$ (claim); C.V. = ±1.96;
$z = 3.64$; reject. There is enough evidence to
support the claim that there is a significant
difference in the proportions.

10–71. $\hat{p}_1 = 0.43$; $\hat{p}_2 = 0.58$; $\overline{p} = 0.505$; $\overline{q} = 0.495$;
H_0: $p_1 = p_2$ and H_1: $p_1 \neq p_2$ (claim);
C.V. = ±1.96; $z = -2.12$; reject. There is
enough evidence to support the claim that the
proportions are different.

10–73. $\hat{p}_1 = 0.83$; $\hat{p}_2 = 0.75$; $\overline{p} = 0.79$; $\overline{q} = 0.21$; H_0:
$p_1 = p_2$ (claim) and H_1: $p_1 \neq p_2$; C.V. = ±1.96;
$z = 1.39$; do not reject. There is not enough
evidence to reject the claim that the proportions
are equal. $-0.032 < p_1 - p_2 < 0.192$

10–75. $\hat{p}_1 = 0.55$; $\hat{p}_2 = 0.45$; $\overline{p} = 0.497$; $\overline{q} = 0.503$;
H_0: $p_1 = p_2$ and H_1: $p_1 \neq p_2$ (claim);
C.V. = ±2.58; $z = 1.302$; do not reject.
There is not enough evidence to support the
claim that the proportions are different.
$-0.097 < p_1 - p_2 < 0.297$

10–77. $\hat{p}_1 = 0.5625$; $\hat{p}_2 = 0.525$; $\overline{p} = 0.54$; $\overline{q} = 0.46$;
H_0: $p_1 = p_2$ and H_1: $p_1 \neq p_2$ (claim);

C.V. = ±1.96; $z = 0.521$; do not reject. There
is not enough evidence to support the claim that
there is a difference in the proportions.
$-0.103 < p_1 - p_2 < 0.178$

10–79. $\hat{p}_1 = 0.25$; $\hat{p}_2 = 0.31$; $\overline{p} = 0.286$; $\overline{q} = 0.714$;
H_0: $p_1 = p_2$ and H_1: $p_1 \neq p_2$ (claim);
C.V. = ±2.58; $z = -1.45$; do not reject.
There is not enough evidence to support the
claim that the proportions are different.
$-0.165 < p_1 - p_2 < 0.045$

10–81. No, p_1 could equal p_3.

10–83. $0.077 < p_1 - p_2 < 0.323$

10–85. H_0: $\mu_1 \leq \mu_2$ and H_1: $\mu_1 > \mu_2$ (claim);
C.V. = $+2.33$; $z = +15.85$; reject. There is
enough evidence to support the claim that
single people do more pleasure driving than do
married people.

10–87. H_0: $\sigma_1 = \sigma_2$ and H_1: $\sigma_1 \neq \sigma_2$ (claim);
C.V. = 2.77; $\alpha = 0.10$; d.f.N. = 23;
d.f.D. = 10; $F = 10.365$; reject. There is
enough evidence to support the claim that there
is a difference in the standard deviations.

10–89. H_0: $\sigma_1^2 \leq \sigma_2^2$ and H_1: $\sigma_1^2 > \sigma_2^2$ (claim);
C.V. = 3.18; $\alpha = 0.05$; d.f.N. = 9; d.f.D. = 9;
$F = 5.06$; reject. There is enough evidence
to support the claim that the variance of the
number of speeding tickets issued on Route 19
is greater than the variance of the number of
speeding tickets issued on Route 22.

10–91. H_0: $\sigma_1^2 \leq \sigma_2^2$ and H_1: $\sigma_1^2 > \sigma_2^2$ (claim);
C.V. = 1.47; $\alpha = 0.10$; d.f.N. = 64;
d.f.D. = 41; $F = 2.32$; reject. There is enough
evidence to support the claim that the variation
in the number of days factory workers miss per
year due to illness is greater than the variation
in the number of days hospital workers miss
per year.

10–93. H_0: $\sigma_1^2 = \sigma_2^2$ (claim); H_1: $\sigma_1^2 \neq \sigma_2^2$;
C.V. = 2.97; $F = 2.78$; do not reject. There is
not enough evidence to reject the claim that the
variances are equal. H_0: $\mu_1 \leq \mu_2$ and H_1: $\mu_1 >$
μ_2 (claim); C.V. = $+2.33$; d.f. = 48; $t = 2.572$;
reject. There is enough evidence to support the
claim that it is warmer in Birmingham.

10–95. H_0: $\sigma_1^2 = \sigma_2^2$ (claim); H_1: $\sigma_1^2 \neq \sigma_2^2$;
C.V. = 3.33; $F = 39.66$; reject. There is
enough evidence to reject the claim that the
variances are equal. H_0: $\mu_1 \leq \mu_2$ and H_1: $\mu_1 >$
μ_2 (claim); C.V. = $+1.796$; d.f. = 11; $t = 3.74$;
reject. There is enough evidence to support the
claim that incomes of city residents are greater
than incomes of suburban residents.

10–97. $H_0: \mu_D \geq 0$ and $H_1: \mu_D < 0$ (claim);
C.V. $= -1.895$; d.f. $= 7$; $\overline{D} = -2$; $s_D = 2.07$;
$t = -2.73$; reject. There is enough evidence to
support the claim that the music has increased
production.

10–99. $\hat{p}_1 = 0.64$; $\hat{p}_2 = 0.40$; $\overline{p} = 0.509$; $\overline{q} = 0.491$;
$H_0: p_1 = p_2$ (claim) and $H_1: p_1 \neq p_2$;
C.V. $= \pm 1.96$; $z = +2.51$; reject. There
is enough evidence to reject the claim
that the proportions are equal.
$0.058 < p_1 - p_2 < 0.422$

Quiz—Chapter 10

1. False
2. False
3. True
4. False
5. *d.*
6. *a.*
7. *c.*
8. *b.*
9. $\mu_1 = \mu_2$
10. Pooled
11. Normal
12. Negative
13. $\dfrac{s_1^2}{s_2^2}, s_1^2 > s_2^2$

14. $H_0: \mu_1 = \mu_2$ and $H_1: \mu \neq \mu_2$ (claim);
$z = -3.69274$; C.V. $= \pm 2.58$; reject. There is
enough evidence to support the claim that there
is a difference in the cholesterol levels of the two
groups. $-10.2 < \mu_1 - \mu_2 < -1.8$.

15. $H_0: \mu_1 \leq \mu_2$ and $H_1: \mu > \mu_2$ (claim); $z = 2.23$;
C.V. $= 2.33$; do not reject. There is not enough
evidence to support the claim that single men do
more work-related driving than married men.

16. $H_0: \sigma_1^2 = \sigma_2^2$ and $H_1: \sigma_1^2 \neq \sigma_2^2$ (claim); $F = 1.637$;
C.V. $= 2.95$; do not reject. There is not enough
evidence to support the claim that the variances
are different.

17. $H_0: \sigma_1^2 = \sigma_2^2$ and $H_1: \sigma_1^2 \neq \sigma_2^2$ (claim); $F = 1.296$;
C.V. $= 1.90$; do not reject. There is not enough
evidence to support the claim that the variances are
different.

18. $H_0: \sigma_1^2 = \sigma_2^2$ (claim) and $H_1: \sigma_1^2 \neq \sigma_2^2$; $F = 2.2117$;
C.V. $= 7.15$; do not reject. There is not enough
evidence to reject the claim that the variances
are equal.

19. $H_0: \sigma_1^2 = \sigma_2^2$ and $H_1: \sigma_1^2 \neq \sigma_2^2$ (claim); $F = 1.94$;
C.V. $= 3.01$; do not reject. There is not enough
evidence to support the claim that the variances
are not equal.

20. $H_0: \sigma_1^2 \leq \sigma_2^2$ and $H_1: \sigma_1^2 > \sigma_2^2$ (claim); $F = 1.474$;
C.V. $= 1.44$; reject. There is enough evidence to
support the claim that the variance of the days
missed by teachers is greater than the variance
of the days missed by nurses.

21. $H_0: \sigma_1 = \sigma_2$ and $H_1: \sigma_1 \neq \sigma_2$ (claim); $F = 1.65$;
C.V. $= 2.46$; do not reject. There is not enough

evidence to support the claim that the standard
deviations are different.

22. $H_0: \sigma_1^2 = \sigma_2^2$ and $H_1: \sigma_1^2 \neq \sigma_2^2$; C.V. $= 5.05$;
$F = 1.23$; do not reject. $H_0: \mu_1 = \mu_2$ and H_1:
$\mu_1 \neq \mu_2$ (claim); $t = 10.922$; C.V. $= \pm 2.779$;
reject. There is enough evidence to support the
claim that the average prices are different.
$0.298 < \mu_1 - \mu_2 < 0.502$

23. $H_0: \sigma_1^2 = \sigma_2^2$ and $H_1: \sigma_1^2 \neq \sigma_2^2$; C.V. $= 3.59$; $F =$
1.56; do not reject. $H_0: \mu_1 \leq \mu_2$ and $H_1: \mu_1 > \mu_2$
(claim); $t = 2.794$; C.V. $= 2.33$; reject. There is
enough evidence to support the claim that it is
warmer in Pasadena.

24. $H_0: \sigma_1^2 = \sigma_2^2$ and $H_1: \sigma_1^2 \neq \sigma_2^2$; C.V. $= 4.02$;
$F = 6.155$; reject. $H_0: \mu_1 = \mu_2$ and $H_1: \mu_1 \neq \mu_2$
(claim); $t = 9.807$; C.V. $= \pm 2.718$; reject. There is
enough evidence to support the claim that the
salaries are different. $\$6,653 < \mu_1 - \mu_2 < \$11,757$

25. $H_0: \sigma_1^2 = \sigma_2^2$ and $H_1: \sigma_1^2 \neq \sigma_2^2$; C.V. $= 3.62$;
$F = 23.08$; reject. $H_0: \mu_1 \leq \mu_2$ and $H_1: \mu_1 > \mu_2$
(claim); $t = 0.874$; C.V. $= 1.812$; do not reject.
There is not enough evidence to support the claim
that the incomes of city residents are greater than
the incomes of rural residents.

26. $H_0: \mu_1 \geq \mu_2$ and $H_1: \mu_1 < \mu_2$ (claim); $t = -4.172$;
C.V. $= -2.821$; reject. There is enough evidence to
support the claim that the sessions improved math
skills.

27. $H_0: \mu_1 \geq \mu_2$ and $H_1: \mu_1 < \mu_2$ (claim); $t = -1.714$;
C.V. $= -1.833$; do not reject. There is not enough
evidence to support the claim that egg production
was increased.

28. $H_0: p_1 = p_2$ and $H_1: p_1 \neq p_2$ (claim); $z = -0.69$;
C.V. $= \pm 1.65$; do not reject. There is not enough
evidence to support the claim that the proportions
are different. $-0.105 < p_1 - p_2 < 0.045$

29. $H_0: p_1 = p_2$ (claim) and $H_1: p_1 \neq p_2$; $z = 2.58$;
C.V. $= \pm 1.96$; reject. There is enough evidence to
reject the claim that the proportions are equal.
$0.069 < p_1 - p_2 < 0.449$

Chapter 11

11–1. Two variables are related when a discernible
pattern exists between them.

11–3. r, ρ (rho)

11–5. A positive relationship means that as x
increases, y increases. A negative relationship
means that as x increases, y decreases.

11–7. Answers will vary.

11–9. Pearson product moment correlation
coefficient

11–11. There are many other possibilities, such as chance, or relationship to a third variable.

11–13. H_0: $\rho = 0$ and H_1: $\rho \neq 0$; $r = -0.832$; C.V. $= \pm 0.811$; reject. There is a significant relationship between a person's age and the number of hours he or she exercises.

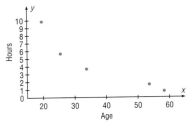

11–15. H_0: $\rho = 0$ and H_1: $\rho \neq 0$; $r = -0.883$; C.V. $= \pm 0.811$; d.f. $= 4$; reject. There is a significant relationship between a person's age and his or her contribution.

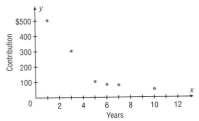

11–17. H_0: $\rho = 0$ and H_1: $\rho \neq 0$; $r = 0.716$; C.V. $= \pm 0.707$; d.f. $= 6$; reject. There is a significant relationship between test scores and GPA.

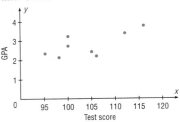

11–19. H_0: $\rho = 0$ and H_1: $\rho \neq 0$; $r = 0.814$; C.V. $= \pm 0.878$; d.f. $= 3$; do not reject. There is not a significant relationship between the variables.

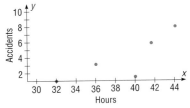

11–21. H_0: $\rho = 0$ and H_1: $\rho \neq 0$; $r = -0.143$; C.V. $= \pm 0.754$; do not reject. There is no

significant relationship between the calories of a sandwich and its sodium content.

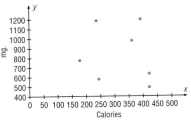

11–23. H_0: $\rho = 0$ and H_1: $\rho \neq 0$; $r = 0.913$; C.V. $= \pm 0.754$; reject. There is a significant relationship between the EER and the cost of the air conditioner.

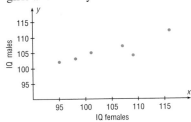

11–25. H_0: $\rho = 0$ and H_1: $\rho \neq 0$; $r = +0.873$; C.V. $= \pm 0.811$; d.f. $= 4$; reject. There is a significant relationship between the IQs of the girls and the boys.

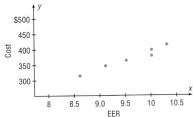

11–27. H_0: $\rho = 0$ and H_1: $\rho \neq 0$; $r = -0.909$; C.V. $= \pm 0.811$; d.f. $= 4$; reject. There is a significant relationship between the years of service and the number of resignations.

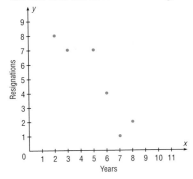

11–29. $r = 1.00$: All values fall in a straight line. $r = 1.00$: The relationship between x and y is the same when the values are interchanged.

11–31. A scatter plot should be drawn and the value of the correlation coefficient should be tested to see whether it is significant.

11–33. $y' = a + bx$

11–35. It is the line that is drawn through the points on the scatter plot such that the sum of the squares of the vertical distances from each point to the line is at a minimum.

11–37. When r is positive, b will be positive. When r is negative, b will be negative.

11–39. The closer r is to $+1$ or -1, the more accurate the predicted value will be.

11–41. $y' = 10.499 - 0.18x$; 4.2

11–43. $y' = 453.176 - 50.439x$; 251.42

11–45. $y' = -3.759 + 0.063x$; 2.793

11–47. Since r is not significant, no regression analysis should be done.

11–49. No regression should be done.

11–51. $y' = -167.07 + 55.99x$; 336.84

11–53. $y' = 63.193 + 0.405x$; 105.313

11–55. $y' = 10.770 - 1.149x$; 6.174

11–57. $r = +0.956$; $H_0: \rho = 0$ and $H_1: \rho \neq 0$; C.V. $= \pm 0.754$; d.f. $= 5$; $y' = -10.944 + 1.969x$; when $x = 30$, $y' = 48.126$; reject. There is a significant relationship between the amount of lung damage and the number of years a person has been smoking.

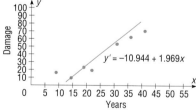

11–59. $H_0: \rho = 0$ and $H_1: \rho \neq 0$; $r = 0.896$; C.V. $= \pm 0.754$; d.f. $= 5$; reject. There is a significant relationship between income and the amount a person spends on recreation; $y' = -222.493 + 0.355x$.

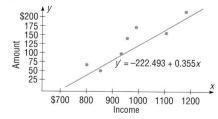

11–61. $H_0: \rho = 0$ and $H_1: \rho \neq 0$; $r = -0.981$; C.V. $= \pm 0.811$; d.f. $= 4$; reject. There is a significant relationship between the number of absences and the final grade; $y' = 96.784 - 2.668x$.

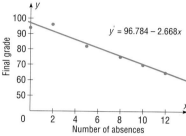

11–63. $H_0: \rho = 0$ and $H_1: \rho \neq 0$; $r = 0.821$; C.V. $= \pm 0.878$; d.f. $= 3$; do not reject. There is no significant relationship between the number of years of experience and monthly sales. Since r is not significant, no regression analysis should be done.

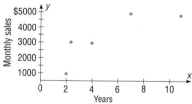

11–65. 453.173, 21.1, -3.7895

11–67. Explained variation is the variation due to the relationship. It is computed by $\Sigma(y' - \bar{y})^2$.

11–69. Total variation is the sum of the squares of the vertical distances of the points from the mean. It is computed by $\Sigma(y - \bar{y})^2$.

11–71. The coefficient of determination is found by squaring the value of the correlation coefficient.

11–73. The coefficient of nondetermination is found by subtracting r^2 from 1.

11–75. $r^2 = 0.5625$; 56.25% of the variation of y is due to the variation of x; 43.75% of the variation of y is due to chance.

11–77. $r^2 = 0.1369$; 13.69% of the variation of y is due to the variation of x; 86.31% of the variation of y is due to chance.

11–79. $r^2 = 0.0025$; 0.25% of the variation of y is due to the variation of x; 99.75% of the variation of y is due to the variation of x.

11–81. 2.092

11–83. 94.22

11–85. $1.59 < y < 12.21$

11–87. $30.46 < y < 472.38$

H

11–89. H_0: $\rho = 0$ and H_1: $\rho \neq 0$; $r = +0.895$;
C.V. $= \pm 0.875$; d.f. $= 5$; reject. There is a
significant relationship between the number of
hours a student studies and the student's GPA;
$y' = 1.572 + 0.152x$; when $x = 10$, $y' = 3.1$.

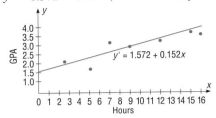

11–91. $r = 0.873$; H_0: $\rho = 0$ and H_1: $\rho \neq 0$;
C.V. $= \pm 0.834$; d.f. $= 6$; reject. There is a
significant relationship between the mother's
age and the number of children she has;
$y' = -2.457 + 0.187x$; $y' = 3.9$.

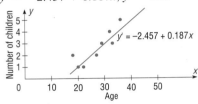

11–93. $r = -0.974$; H_0: $\rho = 0$ and H_1: $\rho \neq 0$;
C.V. $= \pm 0.708$; d.f. $= 10$; reject. There is a
significant relationship between speed and
time; $y' = 14.086 - 0.137x$; $y' = 4.2222$.

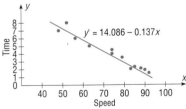

11–95. $r = -0.397$; H_0: $\rho = 0$ and H_1: $\rho \neq 0$;
C.V. $= \pm 0.798$; d.f. $= 7$; do not reject. Since r
is not significant, no regression analysis should
be done.

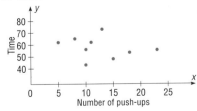

11–97. 0.513
11–99. $3.25 < y < 5.19$

Quiz—Chapter 11

1. False
2. True
3. True
4. False
5. False
6. True
7. *a.*
8. *a.*
9. *d.*
10. *c.*
11. *b.*
12. Scattered diagram
13. Independent
14. $-1, +1$
15. *b.*
16. Line of best fit
17. $+1, -1$
18. $r = 0.857$; H_0: $\rho = 0$ and H_1: $\rho \neq 0$;
C.V. $= \pm 0.707$; reject. $y' = -2.818 + 0.19x$;
3.84 or 4.

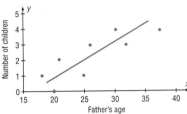

19. $r = -0.078$; H_0: $\rho = 0$ and H_1: $\rho \neq 0$;
C.V. $= \pm 0.754$; do not reject. No regression should
be done.

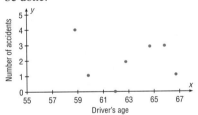

20. $r = 0.842$; H_0: $\rho = 0$ and H_1: $\rho \neq 0$;
C.V. $= \pm 0.811$; reject. $y' = -1.918 + 0.551x$;
4.14 or 4.

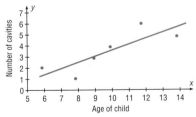

21. $r = 0.602$; H_0: $\rho = 0$ and H_1: $\rho \neq 0$;
C.V. $= \pm 0.707$; do not reject. No regression should
be done.

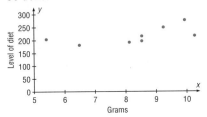

22. 1.129

23. 29.5

24. $0 < y < 5$

25. 217.5 (Average of y' values)

Chapter 12

12–1. The variance test compares a sample variance with a hypothesized population variance; the goodness-of-fit test compares a distribution obtained from a sample with a hypothesized distribution.

12–3. The expected values are computed on the basis of what the null hypothesis states about the distribution.

12–5. H_0: The number of accidents is equally distributed throughout the week (claim). H_1: The number of accidents is not equally distributed throughout the week. C.V. = 12.592; d.f. = 6; χ^2 = 28.887; reject. There is enough evidence to reject the claim that the number of accidents is equally distributed during the week.

12–7. H_0: The flavors are selected with equal frequency (claim). H_1: The flavors are not selected with equal frequency. C.V. = 11.345; d.f. = 3; χ^2 = 13.266; reject. There is enough evidence to reject the claim that the flavors are selected with equal frequency.

12–9. H_0: The blood types of the individuals are 42% type O, 44% type A, 10% type B, and 4% type AB (claim). H_1: The null hypothesis is not true. C.V. = 6.251; d.f. = 3; χ^2 = 100.275; reject. There is enough evidence to reject the claim that the distribution of blood types is the same as the national distribution.

12–11. H_0: The types of loans are distributed as follows: 21% for home mortgages, 39% for automobile purchases, 20% for credit card, 12% for real estate, and 8% for miscellaneous (claim). H_1: The distribution is different from that stated in the null hypothesis. C.V. = 9.488; d.f. = 4; χ^2 = 4.7862; do not reject. There is not enough evidence to reject the claim that the distribution is the same as reported in the newspaper.

12–13. H_0: The types of payments of adult shoppers for purchases are distributed as follows: 53% pay cash, 30% use checks, 16% use credit cards, and 1% have no preference (claim). H_1: The distribution is not the same as stated in the null hypothesis. C.V. = 11.345; d.f. = 3; χ^2 = 36.8897; reject. There is enough evidence

to reject the claim that the distribution at the large store is the same as in the survey.

12–15. H_0: The distribution of college majors is as follows: 40% business, 25% computer science, 15% science, 10% social science, 5% liberal arts, and 5% general studies (claim). H_1: The null hypothesis is not true. C.V. = 9.236; d.f. = 5; χ^2 = 5.613; do not reject. There is not enough evidence to reject the dean's hypothesis.

12–17. H_0: 50% of customers purchase word-processing programs, 25% purchase spreadsheet programs, and 25% purchase data base programs (claim). H_1: The null hypothesis is not true. C.V. = 5.991; d.f. = 2; χ^2 = 0.6; do not reject. There is not enough evidence to reject the department manager's assumption.

12–19. Answers will vary.

12–21. d.f. = (rows − 1) (columns − 1)

12–23. H_0: The variables are independent (or not related). H_1: The variables are dependent (or related).

12–25. The expected values are computed as (row total × column total) ÷ grand total.

12–27. H_0: $p_1 = p_2 = p_3 = p_4 = \ldots = p_n$. H_1: At least one proportion is different from the others.

12–29. H_0: Milk consumption of individuals is independent of their age. H_1: Milk consumption is dependent on age (claim). C.V. = 10.645; d.f. = 6; χ^2 = 13.365; reject. Yes, there is enough evidence to support the claim that milk consumption is dependent on the age of the individual.

12–31. H_0: The amount of liability a person carries is independent of that person's age (claim). H_1: The amount of liability is dependent on the age of the person. C.V. = 12.592; d.f. = 6; χ^2 = 11.878; do not reject. There is not enough evidence to reject the claim that the amount of coverage is independent of the age of the driver.

12–33. H_0: The number of plants sold is independent of the location of the stand. H_1: The number of plants sold is dependent on the location of the stand (claim). C.V. = 13.277; d.f. = 4; χ^2 = 28.588; reject. Yes, there is enough evidence to support the claim that the number of plants sold is dependent on the location of the stand.

12–35. H_0: The student's rating of the instructor is independent of the type of degree the instructor has. H_1: The rating is dependent on the type of degree the instructor has (claim). C.V. = 7.779; d.f. = 4; χ^2 = 19.507; reject. Yes, there is enough evidence to support the claim that the instructors' degrees are related to students' opinions about their effectiveness.

12–37. H_0: The type of video rented is independent of the person's age. H_1: The type of video rented is dependent on the person's age (claim). C.V. = 13.362; d.f. = 8; χ^2 = 46.733; reject. Yes, there is enough evidence to support the claim that the type of movie selected is related to the age of the customer.

12–39. H_0: The condiment preference is independent of the gender of the purchaser (claim). H_1: The condiment preference is dependent on the gender of the individual. C.V. = 4.605; d.f. = 2; χ^2 = 3.05; do not reject. There is not enough evidence to reject the claim that the condiment choice is independent of the gender of the individual.

12–41. H_0: The subject selected is independent of the student's gender (claim). H_1: The subject selected is dependent on the gender of the student. C.V. = 5.991; d.f. = 2; χ^2 = 114.373; reject. There is enough evidence to reject the claim that the subject selected is independent of gender.

12–43. H_0: $p_1 = p_2 = p_3 = p_4$ (claim). H_1: At least one proportion is different. C.V. = 7.851; d.f. = 3; χ^2 = 5.317; do not reject. There is not enough evidence to reject the claim that the proportions are equal.

12–45. H_0: $p_1 = p_2 = p_3 = p_4$ (claim). H_1: At least one proportion is different. C.V. = 7.815; d.f. = 3; χ^2 = 3.403; do not reject. There is not enough evidence to reject the claim that the proportions are equal.

12–47. H_0: $p_1 = p_2 = p_3 = p_4$ (claim). H_1: At least one proportion is different. C.V. = 6.251; d.f. = 3; χ^2 = 12.755; reject. There is enough evidence to reject the claim that the proportions are equal.

12–49. H_0: $p_1 = p_2 = p_3 = p_4$ (claim). H_1: At least one proportion is different. C.V. = 7.851; d.f. = 3; χ^2 = 1.734; do not reject. There is not enough evidence to reject the claim that the proportions are equal.

12–51. H_0: $p_1 = p_2 = p_3$ (claim). H_1: At least one proportion is different. C.V. = 4.605; d.f. = 2; χ^2 = 2.401; do not reject. There is not enough evidence to reject the claim that the proportions are equal.

12–53. χ^2 = 1.075

12–55. The analysis of variance using the F test can be used to compare three or more means.

12–57. The populations from which the samples were obtained must be normally distributed. The samples must be independent of each other. The variances of the populations must be equal.

12–59. $F = \dfrac{s_B^2}{s_W^2}$

12–61. One.

12–63. H_0: $\mu_1 = \mu_2 = \mu_3$. H_1: At least one mean is different from the others. C.V. = 3.35; α = 0.05; d.f. = 2, 27; F = 7.456; reject.

12–65. H_0: $\mu_1 = \mu_2 = \mu_3$. H_1: At least one mean is different from the others. C.V. = 3.55; α = 0.05; d.f.N. = 2; d.f.D. = 18; F = 19.05; reject.

12–67. H_0: $\mu_1 = \mu_2 = \mu_3$. H_1: At least one mean is different from the others. C.V. = 2.64; α = 0.10; d.f.N. = 2; d.f.D. = 17; F = 1.28; do not reject. There is not enough evidence to support the claim that at least one mean is different from the rest.

12–69. H_0: $\mu_1 = \mu_2 = \mu_3 = \mu_4$. H_1: At least one mean is different from the others. C.V. = 2.28; α = 0.10; d.f.N. = 3; d.f.D. = 31; F = 234.5; reject.

12–71. H_0: $\mu_1 = \mu_2 = \mu_3$. H_1: At least one mean is different from the others (claim). C.V. = 2.61; α = 0.10; d.f.N. = 2; d.f.D. = 19; F = 10.12; reject.

12–73. H_0: $\mu_1 = \mu_2 = \mu_3 = \mu_4$. H_1: At least one mean is different from the others. C.V. = 5.29; α = 0.01; d.f.N. = 3; d.f.D. = 16; F = 11.675; reject.

12–75. H_0: The ad produced the same number of responses in each county (claim). H_1: The null hypothesis is not true. C.V. = 11.345; d.f. = 3; χ^2 = 13.38; reject the null hypothesis. There is enough evidence to reject the claim that the ad produced the same number of responses in each county.

12–77. H_0: The condiment preference is independent of the purchaser's gender (claim). H_1: The condiment preference is dependent on the purchaser's gender. C.V. = 4.605; d.f. = 2; χ^2 = 3.050; do not reject. There is not enough evidence to reject the claim that the condiment chosen is independent of the purchaser's gender.

12–79. H_0: The type of pet purchased is independent of the purchaser's gender. H_1: The type of pet purchased is dependent on the purchaser's gender (claim). C.V. = 4.605; d.f. = 2; χ^2 = 7.674; reject. There is enough evidence to support the claim that the type of pet purchased is related to the purchaser's gender.

12–81. H_0: $p_1 = p_2 = p_3 = p_4$ (claim). H_1: At least one proportion is different. C.V. = 7.851; d.f. = 3; χ^2 = 6.166; do not reject. There is not enough evidence to reject the claim that the proportions are equal.

12–83. H_0: $p_1 = p_2 = p_3$ (claim). H_1: At least one proportion is different. C.V. = 9.210; d.f. = 2; χ^2 = 23.89; reject. There is enough evidence to reject the claim that the proportions are equal.

12–85. H_0: $\mu_1 = \mu_2 = \mu_3$. H_1: At least one mean is different from the others (claim). C.V. = 3.81; α = 0.05; d.f.N. = 2; d.f.D. = 13; F = 0.533; do not reject. There is not enough evidence to support the claim that at least one mean is different from the others.

12–87. H_0: $\mu_1 = \mu_2 = \mu_3$. H_1: At least one mean is different from the others. C.V. = 3.89; α = 0.05; d.f.N. = 2; d.f.D. = 12; F = 6.141; reject.

12–89. H_0: $\mu_1 = \mu_2 = \mu_3$. H_1: At least one mean is different. C.V. = 3.89; α = 0.05; d.f.N. = 2; d.f.D. = 12; F = 3.673; do not reject. There is not enough evidence to conclude that there is a difference in the means.

12–91. H_0: $\mu_1 = \mu_2 = \mu_3$. H_1: At least one mean is different from the others. C.V. = 6.36; α = 0.01; d.f.N. = 2; d.f.D. = 15; F = 2.704; do not reject.

Quiz—Chapter 12

1. False
2. True
3. False
4. False
5. False
6. False
7. *c.*
8. *b.*
9. *d.*
10. *d.*
11. *a.*
12. *a.*
13. 6
14. Independent
15. Right
16. At least five
17. ANOVA

18. H_0: The number of ads is equally distributed over five geographic regions (claim). H_1: The number of ads is not equally distributed over five geographic regions. χ^2 = 45.4; C.V. = 9.488; reject. There is enough evidence to reject the claim that the number of ads is equally distributed over five geographic regions.

19. H_0: The ads produced the same number of responses (claim). H_1: The ads produced different numbers of responses. χ^2 = 12.6; C.V. = 13.277;

do not reject. There is not enough evidence to reject the claim that the ads produced the same number of responses.

20. H_0: 48% of the customers order hamburgers, 33% order chicken, and 19% order salad (claim). H_1: The distribution is not the same as stated in the null hypothesis. χ^2 = 4.6; C.V. = 5.991; do not reject. There is not enough evidence to refute the manager's theory.

21. H_0: All gifts were purchased with the same frequency (claim). H_1: The gifts were not purchased with the same frequency. χ^2 = 73.1; C.V. = 9.21; reject. There is enough evidence to reject the claim that all gifts were purchased with the same frequency.

22. H_0: The type of novel purchased is independent of the gender of the purchaser (claim); H_1: The type of novel purchased is dependent on the gender of the purchaser. χ^2 = 132.9; C.V. = 5.991; reject. There is enough evidence to reject the claim that the type of novel purchased is independent of the gender of the purchaser.

23. H_0: The type of pizza ordered is independent of the age of the individual who purchases it. H_1: The type of pizza ordered is dependent on the age of the individual who purchases it (claim). χ^2 = 107.3; C.V. = 14.684; reject. There is enough evidence to support the claim that the pizza purchased is related to the age of the purchaser.

24. H_0: The color of the pennant purchased is independent of the gender of the purchaser (claim). H_1: The color of the pennant purchased is dependent on the gender of the purchaser. χ^2 = 5.6; C.V. = 4.605; reject. There is enough evidence to reject the claim that the color of the pennant purchased is independent of the gender of the purchaser.

25. H_0: $\mu_1 = \mu_2 = \mu_3$. H_1: At least one mean is different from the others. F = 0.119; C.V. = 3.55; do not reject.

26. H_0: $\mu_1 = \mu_2 = \mu_3 = \mu_4$. H_1: At least one mean is different from the others. F = 4.19; C.V. = 3.10; reject.

27. H_0: $\mu_1 = \mu_2 = \mu_3$. H_1: At least one mean is different from the others. F = 0.708; C.V. = 6.36; do not reject.

28. H_0: $\mu_1 = \mu_2 = \mu_3$. H_1: At least one mean is different from the others. F = 0.049; C.V. = 3.63; do not reject.

29. H_0: $\mu_1 = \mu_2 = \mu_3$. H_1: At least one mean is different from the others. F = 49.689; C.V. = 3.89; reject.

Index